Applied Probability and Statistics (*Continued*)

BHAT · Elements of Applied Stochastic Processes
BOX and DRAPER · Evolutionary Operation: A Statistical Method for Process Improvement
BROWNLEE · Statistical Theory and Methodology in Science and Engineering, *Second Edition*
CHAKRAVARTI, LAHA and ROY · Handbook of Methods of Applied Statistics, Vol. II
CHERNOFF and MOSES · Elementary Decision Theory
CHIANG · Introduction to Stochastic Processes in Biostatistics
CLELLAND, deCANI, BROWN, BURSK, and MURRAY · Basic Statistics with Business Applications, *Second Edition*
COCHRAN · Sampling Techniques, *Second Edition*
COCHRAN and COX · Experimental Designs, *Second Edition*
COX · Planning of Experiments
COX and MILLER · The Theory of Stochastic Processes
DANIEL and WOOD · Fitting Equations to Data
DAVID · Order Statistics
DEMING · Sample Design in Business Research
DODGE and ROMIG · Sampling Inspection Tables, *Second Edition*
DRAPER and SMITH · Applied Regression Analysis
DUNN and CLARK · Applied Statistics: Analysis of Variance and Regression
ELANDT-JOHNSON · Probability Models and Statistical Methods in Genetics
FLEISS · Statistical Methods for Rates and Proportions
GOLDBERGER · Econometric Theory
GUTTMAN, WILKS and HUNTER · Introductory Engineering Statistics, *Second Edition*
HAHN and SHAPIRO · Statistical Models in Engineering
HALD · Statistical Tables and Formulas
HALD · Statistical Theory with Engineering Applications
HOEL · Elementary Statistics, *Third Edition*
HOLLANDER and WOLFE · Nonparametric Statistical Methods
HUANG · Regression and Econometric Methods
JOHNSON and KOTZ · Distributions in Statistics
Discrete Distributions
Continuous Univariate Distributions-1
Continuous Univariate Distributions-2
Continuous Multivariate Distributions
JOHNSON and LEONE · Statistics and Experimental Design: In Engineering and the Physical Sciences, Volumes I and II
LANCASTER · The Chi Squared Distribution
LANCASTER · An Introduction to Medical Statistics
LEWIS · Stochastic Point Processes
MILTON · Rank Order Probabilities: Two-Sample Normal Shift Alternatives
OTNES and ENOCHSON · Digital Time Series Analysis
RAO and MITRA · Generalized Inverse of Matrices and Its Applications

continued on back

Characterization Problems in Mathematical Statistics

A. M. KAGAN & YU. V. LINNIK
Steklov Institute of Mathematics, Leningrad

C. RADHAKRISHNA RAO
Indian Statistical Institute, Calcutta

Translated from Russian text by
B. RAMACHANDRAN
Indian Statistical Institute, Calcutta

A Wiley-Interscience Publication

JOHN WILEY & SONS
New York London Sydney Toronto

Library of Congress Cataloging in Publication Data

Kagan, Abram Meerovich.
Characterization problems in mathematical statistics.

(Probability and Mathematical statistics series.
Probablity & statistics section)

Translation of Kharakterizatsionnye zadachi matematicheskoi statistiki.

"A Wiley-Interscience publication."
Bibliography: p.

1. Mathematical statistics. I. Linnik, IUrii Vladimirovich, 1915–1972, joint author. II. Rao, Calyampudi Radhakrishna, joint author. III. Title.

QA276.K213 519.5 73–9643

ISBN 0-471-45421-4

Printed in the United States of America

10 9 8 7 6 5 4 3 2 1

We owe a deep debt of gratitude to the late
Academician Yuri Vladimiovich Linnik
who conceived of this book and provided
the main inspiration in writing it.
The world has lost a remarkably great
mathematician with his premature death on June 30th, 1972.

A. M. Kagan
C. R. Rao

Preface

In many problems of mathematical statistics, conclusions are based on the circumstance that certain special distributions possess important properties which permit the reduction of the original problem to a substantially simpler one.

The natural question as to how fully such a reduction utilizes the special nature of the parent distribution leads to the study of the characteristic properties of the principal distributions of mathematical statistics.

The present work is concerned with the analytical theory of characterization problems and their connection with various areas of mathematical statistics such as the theory of estimation, testing of hypotheses, and sequential analysis. Here are collected together many results (mostly of recent origin) and also formulated problems which appear to the authors to be of interest and importance.

The book is primarily intended for specialists in the areas of probability theory and mathematical statistics. A large part of it should also be accessible to students of advanced courses. It may also prove useful to workers in the area of applications of mathematical statistics.

A. M. Kagan
Yu. V. Linnik (deceased)
C. R. Rao

Contents

CHAPTER 12

Characterizations of Sequential Estimation Plans 367

CHAPTER 13

Other Characteristic Properties of Distributions................ 405

CHAPTER 14

Unsolved Problems 460

Addendum A... 467

Introduction

The standard procedure for making statistical inferences about a population under study is as follows: first of all, from certain *a priori* considerations, the class of possible parent distributions is fixed. This class determines the class of distributions P_s of the sample s and the class of distributions P_T of any chosen statistic T. The observed value of the statistic T and prior information consisting in the knowledge of the class P_s serve as the basis for statistical inferences concerning the population. It is clear that the choice of T must be coordinated in a definite manner with the class P_s fixed beforehand.

If, for instance, the statistic T is used for testing the hypothesis that the sample s comes from one of the distributions of the class P_s, then it is highly desirable that P_s can be reconstructed from P_T.

Or, if the statistic T possesses some desirable property when the sample s has one of the distributions P_s, then it is natural to enquire: what is the widest class P_s in which this property of the statistic T is preserved?

Let us turn to some examples.

For constructing goodness of fit tests for normal, Poisson, and other distributions, conditional distributions free of nuisance parameters are used. Clearly, tests based on conditional distributions are unsatisfactory, at least from a theoretical point of view, if only because of the lack of one-one correspondence between conditional and parent distributions. Thus it becomes necessary to study distributions which correspond to given conditional distributions for fixed values of some statistics.

It is well-known that if $X_1, X_2, \ldots, X_n$ are independent and identically distributed random variables (r.v.'s) following the normal law $N(0, \sigma^2)$, then the statistic $t = \sqrt{n}\,\bar{X}/s$ (in standard notation) follows a Student's t-distribution with $(n-1)$ degrees of freedom. It can be shown that for $n = 2$, the same is true if the density of the r.v.'s has one of the forms

$$\frac{\sqrt{2}}{\pi\sigma\left[1 + \left(\frac{x}{\sigma}\right)^4\right]} \quad \text{or} \quad \frac{\sqrt{2}\,x^2}{\sigma^3\left[1 + \left(\frac{x}{\sigma}\right)^4\right]}.$$

In fact, the statistic t will have the t-distribution with one degree of freedom if the distribution of the r.v.'s X_i belongs to a certain infinite class which includes the normal laws. Therefore, inferences based on the t-distribution are valid not only for normal variables. A complete description of the class of all distributions for which, for $n > 2$, the statistic t has the Student's t-distribution, remains an unsolved problem.

If $X_1, X_2, \ldots, X_n$ are independent and identically distributed according to the law $N(\mu, \sigma^2)$, then the vector

$$\mathbf{Y} = \left(\frac{X_1 - \bar{X}}{s}, \ldots, \frac{X_n - \bar{X}}{s}\right)$$

is uniformly distributed on the unit sphere. Suppose now that there are independent sets of observations $(x_1^{(i)}, \ldots, x_n^{(i)})$ where, within the i-th set, the observations follow a law with d.f. $F[(x - \mu_i)/\sigma_i]$, with μ_i, σ_i unknown. Then it is possible to construct a sequence $\{y^{(i)}\}$ and test the hypothesis that the sequence arises from a population with uniform distribution on the unit sphere. If this hypothesis is accepted, then does it follow that the hypothesis that the d.f. F is a normal law is also to be accepted? Clearly it would, if, from the uniformity of distribution of $\mathbf{Y}$ on the unit sphere, the normality of the r.v.'s X_i follows (concerning this characterization of the normal law, *vide* Section 13.5).

It is well-known that the sample mean $\bar{X}$ is an unbiased estimator with minimum variance for the population mean in samples from a normal population. Is this a property solely of the normal law? It is shown in Chapter 7 that, for any $n \geqq 3$, this is so.

We have so far cited typical examples of problems considered in this book. We shall now briefly describe the contents of the individual chapters.

In Chapter 1, we present general information from the theory of characteristic functions and also some apparatus for solving certain nonlinear differential and functional equations which will be encountered in the book. These equations appear to us to be of interest also from a purely analytical point of view, and some of the results of this chapter appear to be potentially useful for solving other problems.

Chapters 2, 3, and 5 are mostly concerned with characterizations of distributions through stochastic properties of pairs of linear statistics of independent observations. The property of identical distribution is considered in Chapter 2, that of independence in Chapter 3, and that of constancy of regression of one linear statistic on another in Chapter 5. The normal law plays a central role in all these chapters, but, at the same time, it will appear that certain other laws behave, in respect of the properties studied, similarly to the normal law.

The problem of conditions for the identical distribution of two linear statistics was studied by Yu. V. Linnik [81] several years ago. A full description of laws admitting the property of equidistribution was given there, and it was further shown that under certain supplementary conditions this property distinguishes the normal law. The proof of Yu. V. Linnik's was simplified by A. A. Zinger [37], and the latter's results are essentially used in Chapter 2.

The characterization of the normal law through the independence of two linear statistics was originally considered by Bernstein, Darmois, and Skitovich. The original proofs were simplified by Zinger and Linnik [38]. The corresponding theorems for linear statistics in a finite or denumerable number of scalar or vector r.v.'s are given in Chapter 3. Clearly, if in place of $X_1, \ldots, X_n$, we consider $\phi(X_1), \ldots, \phi(X_n)$, where ϕ is a measurable function, and require the independence of two linear functions in the arguments $\phi(X_1), \ldots, \phi(X_n)$, then we obtain a characterization of such distributions of the r.v.'s X_i for which $\phi(X_i)$ is normally distributed. Thus, for any ϕ, we obtain a "two-parameter" family of distributions.

In Chapter 4, we study the independence of linear statistics and also the related problem of independence of linear statistics whose coefficients are themselves r.v.'s. The analysis of this phenomenon requires the solving of some nonlinear integrofunctional equations. For the problem of the independence of a tube statistic with finite basis and the sample mean, the corresponding equation was solved by Anosov [1]. In some cases, we show that the parent distribution is normal; in others, we only establish some analytical properties of the distribution. We may remark that, in this circle of problems, only the simplest have been solved so far.

The problem of constancy of regression of one linear statistic on another arises as a natural generalization of the Kagan-Linnik-Rao theorem [56], which characterizes the normal law by the condition

$$E(\overline{X} \mid X_1 - \overline{X}, \ldots, X_n - \overline{X}) = \text{constant}, n \geqq 3.$$

In a certain sense, constancy of regression is a weaker requirement than independence. But independence is not linked to any condition on the moments, whereas in the regression problem it is necessary to assume beforehand the existence of the first moment. The results of Chapter 5 are based on the works of Rao [130], Ramachandran and Rao [122, 123], and Shimizu [164]. It is interesting to note that the functional equation arising in the regression problem is analogous to the one which arises in the study of the problem of identical distribution of two linear statistics. In special cases, the solutions constitute a subclass of infinitely divisible distributions, which we call "generalized stable laws."

In Chapter 6 we consider the problem of constancy of regression for some nonlinear statistics. Various characterizations of the gamma distribution are obtained here.

In Chapter 7, we study conditions of admissibility and optimality of some estimators commonly used in statistics. The original result (cf. [56]), which stimulated subsequent research in the area, is as follows: let the observations have the form $x_i = \theta + \varepsilon_i$, where θ is an unknown parameter and ε_i are errors of observation. Then, for $n \geqq 3$, the sample mean is an admissible estimator for θ, under quadratic loss, if and only if the ε_i are normal; otherwise it is possible to suggest an estimator better than $\bar{X}$ for θ. This chapter contains many results of this type, relating to various schemes of observation and to various loss-functions. From an analytical standpoint, Chapter 7 borders closely on Chapters 5 and 6. An important result in estimation theory is established that *linear estimators of location parameters are admissible iff the r.v.'s are normally distributed.*

Chapter 8 concerns the exponential family of distributions, which are the only solutions of the functional equation

$$\prod_{i=1}^{n} f(x_i, \theta) = R[T(x_1, \ldots, x_n); \theta] \cdot r(x_1, \ldots, x_n)$$

arising in the theory of sufficient statistics. A general method of solving this equation is described, which is essentially due to Dynkin [26]; other approaches are also presented, applicable in the case of special parameters (shift and scale). Some characteristic properties of sufficient statistics are mentioned, and some extensions of the concept of sufficiency are also studied.

In Chapter 9, the "stability" of certain characteristic properties discussed in the preceding chapters are analysed. For example, if, instead of the independence of $X_1 - X_2$ and $X_1 + X_2$, is assumed only their "ε-independence" (in a suitable sense), then what can be said about the distribution of the random variables under study? It will be shown that X_1 and X_2 will be "$h(\varepsilon)$-normal" with a certain $h(\varepsilon)$. Analogous stability holds for characterizations of the normal law through the admissibility of $\bar{X}$ as an estimator for the shift parameter.

The problem of characterizing the distribution of the structure r.v.'s in linear models studied extensively by Rao [128, 175, 176] is considered in Chapter 10. It is shown that the normal random vector is characterized by the nonuniqueness of linear structure, both for a given number of structure r.v.'s and with respect to their number.

Conditions for the uniqueness of linear structure are examined. It is shown that every random vector possessing a linear structure can be expressed as the sum of two random vectors, of which one has unique structure and the other is normal. These results have important significance in some problems of biological and psychological measurement.

In Chapter 11, we study polynomial and rational statistics of a normal sample, and algebraic transformations preserving normality. It is shown that "almost all" (in a certain sense) pairs of independent polynomials can be "unlinked" provided only that the degree of one is significantly larger than that of the other.

Some isolated results constitute Chapter 12. Characterization problems from the theory of sequential estimation are studied there. We consider Markovian stopping times of the type of "first passage" for Binomial, Multinomial, Poisson, and Wienerian parameters; the problem consists in characterizing sequential estimation plans through some properties of the Markovian stopping times corresponding to them.

In Chapter 13, some results of a miscellaneous character are collected together. The normal, gamma, and other distributions are characterized here through the properties of "minimum Fisherian information," "maximum entropy," etc. Various properties of order statistics are used for characterizing the exponential and geometric distributions. Some characteristic properties of the important discrete distributions are also cited. The set of results given here are important in problems of testing of hypotheses.

Finally, some unsolved problems, which appear to the authors to be of interest and importance, are given in Chapter 14.

The chapters of the book bear the Indo-Arabic numbers: 1, 2, . . . ; sections within chapters bear double numbers (for example, 7.1 denotes the first section of Chapter 7): articles within sections (if there be such) bear triple numbers (1.1.1 denotes the first article of Section 1.1). Formulas, theorems and lemmas in each section have been numbered sequentially, independently of the presence or otherwise of articles or subsections: for example, formula (7.7.1.), Theorem 7.5.1., etc. The same practice is followed in making references.

We express our thanks to A. A. Zinger, A. L. Rukhin, and V. N. Sudakov for help in the compilation of the various sections of the book and also to A. P. Khusu, S. I. Chirkunova and T. M. Gryaznova for help in preparing the manuscript.

The English edition closely follows the Russian text except for some remarks, notes and stylistic changes, and the new material given at the end of the book following Chapter 14, as addenda A and B.

CHAPTER 1

Preliminary Information and Auxiliary Tools

1.1 SOME LEMMAS ON CHARACTERISTIC FUNCTIONS

1.1.1 Characteristic Functions

For any given one-dimensional *distribution function* (d.f.), F, consider the function f of the real variable t given by

$$f(t) = \int e^{itx}\, dF(x). \tag{1.1.1}$$

Since $e^{itx} = \cos tx + i \sin tx$ has bounded real and imaginary parts, the integral (1.1.1) exists. The complex-valued function f of the real variable t is called the *characteristic function* (c.f.) of the d.f. F. The function f is uniformly continuous on the entire real line. It is also referred to as the Fourier-Stieltjes transform of the d.f. F. Some important results concerning c.f.'s are stated below without proofs—which are available in standard textbooks.

(i) $f(0) = 1, f(-t) = \overline{f(t)},\ |f(t)| \leqq f(0) = 1$.

(ii) If $f_1, \ldots, f_n$ are c.f.'s, the a_j are nonnegative real numbers with their sum equal to unity, then $f = \Sigma\, a_j f_j$ is also a c.f.

(iii) (Uniqueness theorem) Two d.f.'s coincide if and only if their c.f.'s coincide.

(iv) (Inversion theorem) Let the d.f. F have f as its c.f. Then, if a and b are continuity points of F, we have

$$F(b) - F(a) = \lim_{T \to \infty} \frac{1}{2\pi} \int_{-T}^{T} \frac{e^{-iat} - e^{-ibt}}{it} f(t)\, dt. \tag{1.1.2}$$

It is possible to check whether the point x is a continuity point of F or not by means of the formula

$$F(x) - F(x-0) = \lim_{T \to \infty} (1/2T) \int_{-T}^{T} e^{-itx} f(t)\, dt$$

(v) (A sufficient condition for the existence of a density function) If f is the c.f. of the d.f. F, and f is absolutely integrable over $(-\infty, \infty)$, then F is absolutely continuous and a version of the density function is given by

$$F'(x) = (1/2\pi) \int e^{-itx} f(t)\, dt. \tag{1.1.3}$$

(vi) (The continuity theorem) A sequence $\{F_n\}$ of d.f.'s converges weakly to a d.f. F if and only if the corresponding sequence of c.f.'s $\{f_n\}$ converges pointwise to a function f which is continuous at the origin. f is then the c.f. of F.

(vii) (A variant of the continuity theorem) A sequence $\{F_n\}$ of d.f.'s converges weakly to a d.f. F if and only if the corresponding sequence of c.f.'s $\{f_n\}$ converges to a function f uniformly on every finite real interval $[-T, T]$. f is then the c.f. of the limiting d.f. F.

(viii) (Existence of moments) If f, the c.f. of the d.f. F, has the k-th order derivative at the origin, then F has moments of all orders up to k if k is even, and up to $(k-1)$ if k is odd.

Conversely, if the s-th moment of the d.f. F exists, then f can be differentiated s times for all real t, and

$$f^{(s)}(t) = i^s \int x^s e^{itx}\, dF(x) \tag{1.1.4}$$

so that, by the bounded convergence theorem, $f^{(s)}$ is a continuous function. In particular, the s-th moment of F is given by: $\mu_s = i^{-s} f^{(s)}(0)$. The existence of the derivatives of all orders for f at the origin is equivalent to the existence of the moments of all orders for F.

(ix) (Symmetric d.f.'s) A d.f. F is called *symmetric* if

$$\tilde{F}(x) \equiv 1 - F(-x-0) = F(x) \qquad \text{for all real } x. \tag{1.1.5}$$

If f is the c.f. of such a d.f., then $f(-t) = f(t)$ and hence f is real-valued; the converse is also true.

1.1.2 Infinitely Divisible Distributions

The c.f. f will be called *infinitely divisible* (i.d.) if, for every natural number n, there exists a c.f. f_n such that

$$f(t) = [f_n(t)]^n \qquad \text{for all real } t.$$

We state below some properties of i.d.c.f.'s.

(i) An i.d.c.f. never vanishes.

(ii) The product of two, and hence of any finite number of, i.d.c.f.'s is again an i.d.c.f.

(iii) A c.f. which is the pointwise limit of a sequence of i.d.c.f.'s is i.d.

(iv) If f is i.d., then $|f|$ and $f^\alpha = \exp(\alpha \log f)$ for any positive α are i.d.c.f.'s.

(v) The Kolmogorov representation: if f is i.d. and the second moment of the corresponding d.f. F exists, then $\log f$ admits the representation

$$\log f(t) = i\beta t + \int \frac{e^{itu} - 1 - itu}{u^2} \, dK(u) \tag{1.1.6}$$

where β is a real constant, and the function K coincides with a d.f. to within a nonnegative multiplicative constant. At $u = 0$, the integrand is defined by continuity to be $-t^2/2$. The representation (1.1.6) is unique, i.e., f uniquely defines β and K.

(vi) The Levy representation $L(\beta, \sigma^2, M, N)$: for an arbitrary i.d.c.f. f, we have

$$\log f(t) = i\beta t - (1/2)\sigma^2 t^2 + \int_{(0,\infty)} h(t, u) \, dN(u) + \int_{(-\infty,0)} h(t, u) \, dM(u) \tag{1.1.7}$$

where β is real, $\sigma \geqq 0$, $h(t, u) = e^{itu} - 1 - itu/(1 + u^2)$, and the functions M and N satisfy the following conditions:

(a) M and N are nondecreasing and right-continuous on the intervals $(-\infty, 0)$ and $(0, \infty)$, respectively;

(b) $M(-\infty) = N(\infty) = 0$; and

(c) the integrals $\int_{(-a,0)} u^2 \, dM(u)$ and $\int_{(0,a)} u^2 \, dN(u)$ are finite for any $a > 0$.

The representation (1.1.7) is unique, i.e., f uniquely defines β, σ, M, and N.

(vii) The Levy-Khinchin representation: for an arbitrary i.d.c.f., f,

$$\log f(t) = i\beta t + \int \left(e^{itu} - 1 - \frac{itu}{1 + u^2} \right) \frac{1 + u^2}{u^2} \, dG(u) \tag{1.1.8}$$

where β is real and G coincides with a d.f. to within a nonnegative multiplicative constant. The integrand is defined at $u = 0$ by continuity to be $-t^2/2$. The representation (1.1.8) is also unique.

(viii) Stable laws: a d.f. F is called *stable* if, to any positive a_1 and a_2 and any real b_1 and b_2, there correspond constants $a > 0$ and b such that the relation

$$F(a_1 x + b_1) * F(a_2 x + b_2) = F(ax + b) \tag{1.1.9}$$

is satisfied; or, in terms of c.f.'s,

$$e^{-i(b_1 + b_2)t} f(t/a_1) f(t/a_2) = e^{-ibt} f(t/a) \qquad \text{for all } t. \tag{1.1.10}$$

(a) In order for a d.f. F to be stable, it is necessary and sufficient that its c.f. f be representable in the form

$$\log f(t) = i\alpha t - c|t|^{\lambda}\,[1 + i\gamma(t/|t|)w(t, \lambda)] \tag{1.1.11}$$

where λ, γ, α, and c are constants: α is some real number, $c \geqq 0$, $-1 \leqq \gamma \leqq 1$, and $0 < \lambda \leqq 2$; and

$$w(t, \lambda) = \begin{cases} \tan(\pi\lambda/2) & \text{if } \lambda \neq 1, \\ (2/\pi)\log|t| & \text{if } \lambda = 1. \end{cases}$$

(b) A (nondegenerate) stable law is i.d., and the functions M and N and the constant σ in the Levy representation have the following forms:

$$\text{for}\quad 0 < \lambda < 2,\ M(u) = c_1/|u|^{\lambda},\ N(u) = -c_2/u^{\lambda},\ \sigma = 0, \quad (c_1 \geqq 0,\ c_2 \geqq 0, \text{ and } c_1 + c_2 > 0); \tag{1.1.12}$$

$$\text{for}\quad \lambda = 2,\quad M(u) = N(u) = 0,\quad \sigma \geqq 0. \tag{1.1.13}$$

(The case $\lambda = 2$ corresponds to the normal law—possibly degenerate.)
(c) A stable law will be called a *symmetric* stable law if

$$\log f(t) = -c|t|^{\lambda} \tag{1.1.14}$$

which corresponds to $\alpha = \gamma = 0$ in the representation (1.1.11) and $\alpha = 0$, $c_1 = c_2$ in the representation (1.1.12). For the Cauchy law, $\lambda = 1$.
(d) Semistable laws of P. Levy: f is said to be the c.f. of a *semistable* law if it satisfies the equation

$$f(t) = [f(\beta t)]^{\gamma}, \qquad \gamma > 0, \qquad 0 < |\beta| < 1 \tag{1.1.15}$$

for all real t. Let λ be the unique real root of the equation $\gamma|\beta|^{\lambda} = 1$. Then, as will be shown in Chapter 5 (also, cf. [122]), a semistable law is infinitely divisible, and in the Levy representation $L(\mu, \sigma^2, M, N)$ for $\log f$ given by (1.1.7),

$$\sigma = 0, \qquad M = N = 0 \quad \text{if}\quad \lambda < 0 \quad \text{or} \quad \lambda > 2;$$

$$\sigma \geqq 0, \qquad M = N = 0 \quad \text{if}\quad \lambda = 2;$$

$$\sigma = 0, \qquad M(u) = \xi(\log|u|)/|u|^{\lambda}, \qquad N(u) = -\eta(\log u)/u^{\lambda}$$

if $0 < \lambda < 2$, $\beta > 0$, where ξ and η are nonnegative right-continuous functions with period $-\log\beta$; and

$$\sigma = 0, \qquad M(u) = \xi(\log|u|)/|u|^{\lambda}, \qquad N(u) = -\xi(\log u - \log|\beta|)/u^{\lambda}$$

where ξ is a nonnegative right-continuous function with period $-2\log|\beta|$, if $0 < \lambda < 2$, $\beta < 0$.
(e) Generalized stable laws: f will be called the c.f. of a *generalized stable law* if it is nonvanishing and satisfies for all real t an equation of the form

$$[f(\beta_1 t)]^{\alpha_1} \cdots [f(\beta_k t)]^{\alpha_k} = [f(\beta_{k+1} t)]^{\alpha_{k+1}} \cdots [f(\beta_n t)]^{\alpha_n} \tag{1.1.16}$$

where $\alpha_i > 0$ for $i = 1, 2, \ldots, n$. Of special interest is the case where there occurs only one factor on (say) the left-hand side. Representations for generalized stable laws will be considered for the general case in Chapter 2, and for the above-mentioned special case in Chapter 5.

1.1.3 The Conditions of Linearity of Regression and Homoscedasticity

LEMMA 1.1.1. Let X and Y be random variables and EY exist. Y has constant regression on X if and only if the relation

$$E(Ye^{itX}) = EY \cdot Ee^{itX} \tag{1.1.17}$$

holds for all real t.

***Proof*:** The necessity is established immediately by multiplying both sides of the equality $E(Y \mid X) = EY$ by e^{itX} and then taking the expectation on both sides of the relation obtained.

For the proof of the sufficiency, first consider the case $EY \neq 0$. Let P denote the marginal distribution of X. Then (1.1.17) can be rewritten as

$$\int_{-\infty}^{\infty} e^{itx} \frac{E(Y \mid x)}{EY}\, dP = \int_{-\infty}^{\infty} e^{itx}\, dP. \tag{1.1.18}$$

Introducing the set function

$$Q(A) = \int_A \frac{E(Y \mid x)}{EY}\, dP \tag{1.1.19}$$

which is of bounded variation, we find from (1.1.18) and (1.1.19) that

$$\int e^{itx}\, dQ = \int e^{itx}\, dP. \tag{1.1.20}$$

By the uniqueness theorem for the Fourier-Stieltjes transforms of functions of bounded variation, we have from (1.1.20)

$$\int_A \frac{E(Y \mid x)}{EY}\, dP = \int_A dP. \tag{1.1.21}$$

whence $E(Y \mid x)/EY = 1$ a.e. (almost everywhere), which means constancy of regression.

We pass on to the case $EY = 0$. Then (1.1.17) can be written in the form

$$\int_{-\infty}^{\infty} e^{itx} E(Y \mid x)\, dP = 0. \tag{1.1.22}$$

Let $Q(A) = \int_A E(Y \mid x)\, dP$; then, from (1.1.22),

$$\int_{-\infty}^{\infty} e^{itx}\, dQ = 0 \tag{1.1.23}$$

which means constancy of the function $Q(A)$, so that $Q(A) = Q(R_1) = EY = 0$, i.e., $E(Y|X) = 0$, as desired to prove. ∎

LEMMA 1.1.2 (cf. [125]). Let (X, Y) be a two-dimensional random vector with $EX = EY = 0$. A necessary and sufficient condition for the linearity of the regression of Y on X is the existence of a constant β such that, for all real t_1,

$$\left.\frac{\partial\phi(t_1, t_2)}{\partial t_2}\right|_{t_2=0} = \beta\frac{d\phi(t_1, 0)}{dt_1} \tag{1.1.24}$$

where $\phi(t_1, t_2)$ is the c.f. of (X, Y).

***Proof*:** In view of Lemma 1.1.1, a necessary and sufficient condition for $E(Y - \beta X|X) = 0$ for some constant β is that

$$E(Y - \beta X)e^{itX} = 0 \qquad \text{for all real } t$$

which is easily seen to reduce to (1.1.24). ∎

LEMMA 1.1.3 (cf. [125]). In order for the two-dimensional random vector (X, Y) to satisfy the conditions

$$\begin{aligned} E(Y|X) &= \alpha + \beta X \\ \text{Var}\,(Y|X) &= \sigma^2 = \text{constant} \end{aligned} \tag{1.1.25}$$

it is necessary and sufficient that

$$\left.\begin{aligned} \left.\frac{\partial\phi(t_1, t_2)}{\partial t_2}\right|_{t_2=0} &= i\alpha\phi(t_1, 0) + \beta\frac{d\phi(t_1, 0)}{dt_1} \\ \left.\frac{\partial^2\phi(t_1, t_2)}{\partial t_2^2}\right|_{t_2=0} &= -(\sigma^2 + \alpha^2)\phi(t_1, 0) + 2i\alpha\beta\frac{d\phi(t_1, 0)}{dt_1} + \beta^2\frac{d^2\phi(t_1, 0)}{dt_1^2} \end{aligned}\right\}. \tag{1.1.26}$$

***Proof*:** The conditions (1.1.25) are equivalent to

$$\begin{aligned} E(Y - \beta X|X) &= \alpha \\ E[Y^2 - (\alpha + \beta X)^2|X] &= \sigma^2 \end{aligned}$$

which, in view of Lemma 1.1.1, are easily seen to be equivalent to the conditions (1.1.26). ∎

Lemmas 1.1.2 and 1.1.3 can be generalized to the case of several variables. Let $X_0, X_1, \ldots, X_p$ be random variables with zero expectation. A necessary and sufficient condition for the relation

$$E(X_0|X_1, \ldots, X_p) = \beta_1 X_1 + \cdots + \beta_p X_p$$

to be satisfied is the following

$$\left.\frac{\partial\phi(t_0, t_1, \ldots, t_p)}{\partial t_0}\right|_{t_0=0} = \sum_1^p \beta_j \frac{\partial\phi(0, t_1, \ldots, t_p)}{\partial t_j}$$

for all real $t_1, \ldots, t_p$. Similarly, for the relations

$$E(X_0 | X_1, \ldots, X_p) = \beta_1 X_1 + \cdots + \beta_p X_p,$$
$$\text{Var}\,(X_0 | X_1, \ldots, X_p) = \sigma_0^2 = \text{constant},$$

to hold, it is necessary and sufficient that

$$\left.\frac{\partial \phi(t_0, t_1, \ldots, t_p)}{\partial t_0}\right|_{t_0=0} = \sum_1^p \beta_j \frac{\partial \phi(0, t_1, \ldots, t_p)}{\partial t_j}$$

$$\left.\frac{\partial^2 \phi(t_0, t_1, \ldots, t_p)}{\partial t_0^2}\right|_{t_0=0} = -\sigma_0^2\, \phi(0, t_1, \ldots, t_p) + \sum\sum \beta_j \beta_k \frac{\partial^2 \phi(0, t_1, \ldots, t_p)}{\partial t_j\, \partial t_k}.$$

In the above, $\phi(t_0, t_1, \ldots, t_p)$ is the c.f. of the vector $(X_0, X_1, \ldots, X_p)$.

1.1.4 Some Lemmas Concerning the Existence of Moments

LEMMA 1.1.4 (Ramachandran and Rao [122]). Let f be a c.f. such that $f^{(2n-1)}$, the $(2n-1)$-th derivative of f, is defined for all real t. If

$$[f^{(2n-1)}(t) - f^{(2n-1)}(0)]/t$$

is bounded in some deleted neighborhood of the origin, then F has the moment of order $2n$, and conversely.

Proof. We shall prove the lemma for the case $n = 1$. The general case then follows from the fact that if $f^{(2n-1)}$ exists, then $f^{(2n-2)}/f^{(2n-2)}(0)$ is a c.f. with its first derivative defined for all t.

Let $\phi(t) = [f'(t) - f'(0)]/t$ so that $|\phi(t)| \leqq c$ for $0 < |t| < \delta$, where c and δ are positive constants. By the mean value theorem,

$$f(t) + f(-t) - 2f(0) = tf'[t\theta(t)] - tf'[-t\theta(-t)]$$

where $0 < \theta(\pm t) < 1$, and hence

$$= t^2\{\theta(t)\phi[t\theta(t)] + \theta(-t)\phi[-t\theta(-t)]\}$$

so that

$$\left|\frac{f(t) + f(-t) - 2f(0)}{t^2}\right| \leqq 2c \qquad \text{for} \quad 0 < |t| < \delta.$$

It follows from Fatou's lemma that $\int x^2\, dF(x) < \infty$.

The converse is trivial since $[f'(t) - f(0)]/t \to f''(0)$ as $t \to 0$ if $\int x^2 fF(x) < \infty$. ∎

LEMMA 1.1.5. Let $\{t_n\}$, $t_n \to 0$ as $n \to \infty$, be a sequence of values of t.

(a) If $|t_n|^{-\lambda} \cdot \log |f(t_n)| \leqq -c' < 0$ for some $\lambda < 2$, then $\int |x|^\delta\, dF(x)$ does not exist for any $\delta > \lambda$.

(b) If $|t_n|^{-\lambda} \cdot \log|f(t_n)| \geqq -c' > -\infty$ for a sequence $\{t_n\}$ which is such that

(i) $\sum |t_n|^{\varepsilon} < \infty$ for every $\varepsilon > 0$, and

(ii) $\{t_{n-1}/t_n\}$ is bounded,

then F has absolute moments of all orders $< \lambda$. In particular, if $\log|f(t)|/|t|^{\lambda}$ is bounded in a deleted neighborhood of the origin, then F has moments of all orders $<\lambda$.

(c) If $\log|f(t_n)|/t_n^2$ is bounded (for some sequence $\{t_n\}$ tending to zero), then the second moment of F exists; the converse is true for all such sequences.

(d) If $\log|f(t_n)|/t_n^2 \to 0$ as $n \to \infty$, then F is degenerate; the converse is trivially true for all such sequences.

Proof. (a) For some $c > 0$, $|f(t_n)|^2 \leqq \exp(-c|t_n|^{\lambda})$. Let $\lambda < \delta \leqq 2$. Denote by F^* the convolution of F with its conjugate d.f., so that the c.f. of F^* is $|f|^2$. Suppose the moment of order δ exists. Set $u_n = (1/2)t_n$. Then, by Fatou's lemma,

$$\begin{aligned}
\infty > 2\int |x|^{\delta}\, dF^*(x) &= 2\int \limsup_{n\to\infty} \left|\frac{\sin u_n x}{u_n}\right|^{\delta} dF^*(x) \\
&\geqq 2 \limsup_{n\to\infty} \int \left|\frac{\sin u_n x}{u_n}\right|^{\delta} dF^*(x) \\
&\geqslant 2 \limsup_{n\to\infty} \int \frac{\sin^2 u_n x}{|u_n|^{\delta}}\, dF^*(x) \\
&= \limsup_{n\to\infty} 2^{\delta}|t_n|^{-\delta}[1 - |f(t_n)|^2] \\
&\geqq 2^{\delta} \lim_{n\to\infty} |t_n|^{-\delta}[1 - \exp(-c|t_n|^{\lambda})] = \infty
\end{aligned}$$

where we have used the assumption of the integrability of $|(\sin u_n x)/u_n|^{\delta} \leqq |x|^{\delta}$ and the condition $\delta > \lambda$. This means that F^* and hence F does not have moments of order $> \lambda$.

(b) Suppose, without loss of generality, that $1 \geqq t_{n-1} > t_n > 0$ for all $n \geqq 2$. Then, for some $c > 0$,

$$1 - |f(t_n)|^2 \leqq 1 - \exp(-ct_n^{\lambda}) \leqq ct_n^{\lambda}$$

for all sufficiently large n. Let, as above, $u_n = (1/2)t_n$, and $c_1, c_2, \ldots$ denote positive constants. Then we have

$$\int \sin^2 u_n x\, dF^*(x) \leqq c_1 u_n^{\lambda}.$$

Since $\sin^2\theta \geqq \theta^2 \sin^2 1$ if $0 \leqq \theta \leqq 1$, setting $x_n = 1/u_n$, we have

$$\int_{x_{n-1}}^{x_n} x^2\, dF^*(x) \leqq c_2 \int_{x_{n-1}}^{x_n} \frac{\sin^2(u_n x)}{u_n^2}\, dF^*(x) \leqq c_3 u_n^{\lambda-2}$$

so that

$$\int_{x_{n-1}}^{x_n} x^\delta \, dF^*(x) \leqq c_3 \frac{x_n^\delta}{x_{n-1}^2} u_n^{\lambda-2} = c_3 \frac{u_{n-1}^2}{u_n^2} u_n^{\lambda-\delta}$$

$$\leqq c_4 u_n^{\lambda-\delta}$$

because of the assumption that the sequence $\{u_{n-1}/u_n\}$ is bounded. Further, $\sum u_n^\varepsilon < \infty$ for any $\varepsilon > 0$, so that $\int |x|^\delta \, dF^*(x) < \infty$ for any $\delta < \lambda$, whence F^* and, therefore, F has moments of all orders $< \lambda$. If, in particular, the function $\log |f(t)|/|t|^\lambda$ is bounded for $0 < |t| < t_0$, then the sequence $\{\beta^n\}$ may be chosen as our $\{t_n\}$ for some $0 < \beta < 1$.

(c) For some $c > 0$, $|f(t_n)|^2 \geqq \exp(-ct_n^2)$, so that, by Fatou's lemma,

$$\int x^2 \, dF^*(x) = 2 \int \liminf_{n\to\infty} \frac{1 - \cos t_n x}{t_n^2} \, dF^*(x)$$

$$\leqq 2 \liminf_{n\to\infty} \int \frac{1 - \cos t_n x}{t_n^2} \, dF^*(x)$$

$$= 2 \liminf_{n\to\infty} \frac{1 - |f(t_n)|^2}{t_n^2}$$

$$\leqq 2 \liminf_{n\to\infty} \frac{1 - \exp(-ct_n^2)}{t_n^2} = 2c.$$

Hence F^* and consequently F has finite second moment. The converse is trivial.

(d) In this case, in view of (c) above, $\int x^2 \, dF^*(x) \leqq 2c$ for any $c > 0$, and hence is zero, so that F^* and hence F is degenerate. The converse is trivial since $\log |f(t)|$ vanishes identically if F is degenerate. ∎

LEMMA 1.1.6 (Ramachandran [119]). Let F be a symmetric d.f. having the moment of order $2n$ ($n \geqq 0$ integer). Then, for F to have the (absolute) moment of order $2n + \lambda$, $0 < \lambda < 2$, it is necessary (but not sufficient) that

$$\lim_{t\to 0} |t|^{-\lambda} \log [f^{(2n)}(t)/f^{(2n)}(0)] = 0.$$

Proof. It is sufficient to consider only the case $n = 0$; the general case will follow from the fact that if $\int x^{2n} \, dF(x)$ exists, then $f^{(2n)}/f^{(2n)}(0)$ is the c.f. of a d.f. which will have the moment of order λ if and only if F has the moment of order $2n + \lambda$. We begin by noting that if $\int |x|^\lambda \, dF(x) < \infty$, then

$$x^\lambda[1 - F(x)] \leqq \int_x^\infty u^\lambda \, dF(u) \to 0 \text{ as } x \to \infty;$$

therefore, if for some sequence $\{s_k\} \to \infty$, $s_k^\lambda[1 - F(s_k)]$ does not tend to zero, then $\int |x|^\lambda \, dF(x) = \infty$.

Let us now suppose that $|t|^{-\lambda} \log f(t)$ does not tend to zero as $t \to 0$. We shall denote by $c_1, c_2, \ldots$ suitably chosen positive constants. There exist c_1 and a sequence $\{t_n\} \downarrow 0$ such that $-t_n^\lambda \log f(t_n) \geqq c_1$. Since, for $1/2 \leqq x \leqq 1$, $2(1 - x) \geqq -\log x \geqq 1 - x$, we have

$$[1 - f(t_n)] \cdot t_n^{-\lambda} \geqq c_2 \qquad \text{for all sufficiently large } n,$$

i.e., $\int_0^\infty (1 - \cos t_n x)\, dF(x) \geqq c_3 t_n^\lambda$. [We note that the integrand vanishes at the origin so that

$$\int_{-\infty}^{\infty} (1 - \cos t_n x)\, dF(x) = 2 \int_0^\infty (1 - \cos t_n x)\, dF(x)].$$

Integrating by parts, we obtain

$$\lim_{X \to \infty} \int_0^X \sin t_n x[1 - F(x)]\, dx \geqq c_3 t_n^{\lambda - 1}.$$

Since $1 - F(x)$ is a nonincreasing function, $\sin [t_n(x + \pi/t_n)] = -\sin t_n x$, and $(0, X] = (0, \pi/t_n] \cup (\pi/t_n, 2\pi/t_n] \cup \ldots$, it is easily seen that the left side of the above inequality is $\leqq \int_0^{\pi/t_n} \sin t_n x\, [1 - F(x)]\, dx$. Setting

$$h(x, t) = [1 - F(x/t)](x/t)^\lambda,$$

we have

$$\int_0^\pi h(x, t_n) x^{-\lambda} \sin x\, dx \geqq c_3 .$$

Now there are two possibilities: (A) $h(x, t_n)$ is uniformly bounded for all x in $(0, \pi)$ and all n, or (B) there exist sequences $\{x_k\}$ and $\{t_{n_k}\}$ such that $h(x_k, t_{n_k}) \to \infty$. In case (B), setting $s_k = x_k/t_{n_k}$, we find that $[1 - F(s_k)]s_k^\lambda \to \infty$ (s_k necessarily tends to infinity) and therefore $\int |x|^\lambda\, dF(x) = \infty$. In case (A), from the uniform boundedness of the function $h(x, t_n)$ and the integrability of the function $x^{-\lambda} \sin x$ over $(0, \pi)$ (remembering that $0 < \lambda < 2$) we conclude from Fatou's lemma that

$$\int_0^\pi \left[\limsup_{n \to \infty} h(x, t_n)\right] x^{-\lambda} \sin x\, dx \geqq c_3$$

so that $\limsup_{n \to \infty} h(x, t_n) > 0$ on a set of positive Lebesgue measure. For us to conclude that $\int |x|^\lambda\, dF(x) = \infty$, it is sufficient that

$$\limsup_{n \to \infty} h(x, t_n) > 0$$

for *some* $x > 0$. ▌

Corollary. If, for some sequence $\{t_n\} \to 0$, and $0 < \lambda < 2$,

$$|t_n|^{-\lambda} \log |f^{(2k)}(t_n)/f^{(2k)}(0)|$$

is bounded away from zero, then $\int |x|^{2k+\lambda}\, dF(x) = \infty$. In particular, a nonnormal stable law with exponent α does not have the moment of order α (and consequently of any higher order).

The following result is an easy consequence of the foregoing lemmas. For brevity, we use the phrase "bounded" in place of the more appropriate "bounded in some deleted neighborhood of the origin." Also, F is assumed nondegenerate.

LEMMA 1.1.7. Let $\int x^{2n}\, dF(x) < \infty$ where $n \geqq 0$ is an integer, and let

$$\lambda = \sup \{\delta: |t|^{-\delta} \log |f^{(2n)}(t)/f^{(2n)}(0)| \text{ is bounded}\}.$$

Then

(a) $\lambda \leqq 2$;

(b) F has moments of all orders $< 2n + \lambda$;

(c) if $\lambda < 2$, then F does not have moments of order $> 2n + \lambda$; and

(d) if $\lambda = 2$ and the function $t^{-2} \log |f^{(2n)}(t)/f^{(2n)}(0)|$ is bounded, then $\int x^{2n+2}\, dF(x) < \infty$ and conversely. (The boundedness of this function at some sequence of points $\to 0$ is a sufficient condition for the existence of the moment of order $2n + 2$.)

We cite yet another condition for the existence of the moments of all orders for a d.f.; this is due to Zinger [31] and is formulated, not in terms of c.f.'s, but in terms of the convergence of an integral of a special nature. We first formulate a general property of d.f.s.

LEMMA 1.1.8. For any nonnegative random variable X with d.f. F, the following relation holds:

$$\int_0^\infty \frac{dF(x)}{[1 - F(x - 0)]^\delta} < \infty \qquad \text{for any } \delta, \qquad 0 < \delta < 1. \tag{1.1.27}$$

Proof. We consider separately three cases:

(1) the distribution of X is discrete;

(2) F is a continuous function; and

(3) the general case.

Case 1. Here again we first consider the case where

$$F(x) = \sum p_j \varepsilon(x - x_j)$$

where the x_j can be arranged in an increasing sequence. Then

$$\int_0^\infty \frac{dF(x)}{[1 - F(x - 0)]^\delta} = \sum \frac{p_i}{(p_i + p_{i+1} + \cdots)^\delta}. \tag{1.1.28}$$

We shall show that (1.1.28) is bounded. For simplicity, we introduce the notation $\alpha_i = p_i + p_{i+1} + \cdots$. Evidently,

$$\alpha_i \leqq 1,\ \alpha_i > \alpha_{i+1} > 0,\ \lim_{i\to\infty} \alpha_i = 0,$$

$$\begin{aligned}\frac{p_i}{(p_i + \cdots)^\delta} &= \frac{\alpha_i - \alpha_{i+1}}{\alpha_i^\delta} = (\alpha_i^{1-\delta} - \alpha_{i+1}^{1-\delta})\frac{1-(\alpha_{i+1}/\alpha_i)}{1-(\alpha_{i+1}/\alpha_i)^{1-\delta}} \\ &\leqq \frac{\alpha_i^{1-\delta} - \alpha_{i+1}^{1-\delta}}{1-\delta}.\end{aligned} \tag{1.1.29}$$

From (1.1.29) it easily follows in view of the properties of the α_i that

$$\int_0^\infty \frac{dF(x)}{[1 - F(x-0)]^\delta} \leqq \frac{1}{1-\delta}. \tag{1.1.30}$$

Passing to the case of an arbitrary discrete d.f., let, for every n,

$$x_1^{(n)} < x_2^{(n)} < \cdots < x_n^{(n)}$$

be the natural ordering of the set $\{x_1, \ldots, x_n\}$. Let

$$F_n(x) = \sum_1^n p_i\, \varepsilon\,(x - x_i) = \sum_1^n p_i^{(n)}\, \varepsilon\,(x - x_i^{(n)})$$

where $p_i^{(n)}$ is the jump of F at $x_i^{(n)}$. Though F_n may not be a d.f., our discussion above can be applied to it to yield the relation

$$\int_0^\infty \frac{dF_n(x)}{[F_n(\infty) - F_n(x-0)]^\delta} \leqq \frac{[F_n(\infty)]^{1-\delta}}{1-\delta} \leqq \frac{1}{1-\delta}.$$

But the left-most member of the above relation

$$\sum_1^n p_i^{(n)}[p_i^{(n)} + p_{i+1}^{(n)} + \cdots + p_n^{(n)}]^{-\delta}$$

which is obviously $\geqq \sum_1^n p_i^{(n)}[1 - F(x_i^{(n)} - 0)]^{-\delta} = \sum_1^n p_i[1 - F(x_i - 0)]^{-\delta}$. Letting $n \to \infty$, we have $\sum_1^\infty p_i[1 - F(x_i - 0)]^{-\delta} \leqq 1/(1-\delta)$, thus proving the lemma in this case also.

Case 2. We choose some $x_0 > 0$ and construct F_n in the following manner.

$$\begin{aligned} F_n(x) = F(kx_0/n) \qquad &\text{for} \quad kx_0/n \leqq x < (k+1)x_0/n \\ &\text{for} \quad k = 0, 1, 2, \ldots; \qquad n > 0. \end{aligned} \tag{1.1.31}$$

Evidently $F_n(x) \leqq F(x)$ and $\lim_{n\to\infty} F_n(x) = F(x)$. Choose and fix an $\varepsilon > 0$. Corresponding to ε and x_0, there exists n_0 such that for all $n > n_0$ and for all x with $0 \leqq x \leqq x_0$

$$0 \leqq F(x) - F_n(x) \leqq (1/2)\varepsilon[1 - F(x_0)]^{1+\delta}. \tag{1.1.32}$$

By the well-known Helly's theorem

$$\lim_{n\to\infty} \int_0^{x_0} \frac{dF_n(x)}{[1-F(x)]^\delta} = \int_0^{x_0} \frac{dF(x)}{[1-F(x)]^\delta}. \tag{1.1.33}$$

Therefore, it is possible to find n' such that for all $n > n'$

$$\int_0^{x_0} \frac{dF(x)}{[1-F(x)]^\delta} - \int_0^{x_0} \frac{dF_n(x)}{[1-F(x)]^\delta} < \frac{\varepsilon}{2}. \tag{1.1.34}$$

We set up the difference

$$\int_0^{x_0} \frac{dF(x)}{[1-F(x)]^\delta} - \int_0^{x_0} \frac{dF_n(x)}{[1-F_n(x)]^\delta}$$

$$= \left\{\int_0^{x_0} \frac{dF(x)}{[1-F(x)]^\delta} - \int_0^{x_0} \frac{dF_n(x)}{[1-F(x)]^\delta}\right\} + \left\{\int_0^{x_0} \frac{dF_n(x)}{[1-F(x)]^\delta} - \int_0^{x_0} \frac{dF_n(x)}{[1-F_n(x)]^\delta}\right\}. \tag{1.1.35}$$

The first term on the right-hand side of (1.1.35) can be estimated by means of (1.1.34); for estimating the second term, we may use the inequality

$$\frac{1}{[1-F(x)]^\delta} - \frac{1}{[1-F_n(x)]^\delta} \leqq \frac{F(x)-F_n(x)}{[1-F(x)]^{1+\delta}} \leqq \frac{F(x)-F_n(x)}{[1-F(x_0)]^{1+\delta}} \tag{1.1.36}$$

and also (1.1.32). Then, from (1.1.35), we obtain

$$\int_0^{x_0} \frac{dF(x)}{[1-F(x)]^\delta} \leqq \int_0^{x_0} \frac{dF_n(x)}{[1-F_n(x)]^\delta} + \varepsilon$$

or, using Case 1,

$$\int_0^{x_0} \frac{dF(x)}{[1-F(x)]^\delta} \leqq \frac{1}{1-\delta} + \varepsilon. \tag{1.1.37}$$

Since x_0 and ε are arbitrary, it follows hence that

$$\int_0^{\infty} \frac{dF(x)}{[1-F(x)]^\delta} \leqq \frac{1}{1-\delta} < \infty. \tag{1.1.38}$$

Case 3. According to a well-known theorem, every d.f. F can be decomposed into a discrete and a continuous component,

$$F = gG + hH \tag{1.1.39}$$

where G and H are respectively discrete and continuous d.f.'s and g and h are nonnegative real numbers with $g + h = 1$.

Taking (1.1.39) into account, we find

$$\int_0^\infty \frac{dF(x)}{[1-F(x)]^\delta} = g\int_0^\infty \frac{dG(x)}{[1-F(x)]^\delta} + h\int_0^\infty \frac{dH(x)}{[1-F(x)]^\delta}$$

$$\leqq g^{1-\delta}\int_0^\infty \frac{dG(x)}{[1-G(x)]^\delta} + h^{1-\delta}\int_0^\infty \frac{dH(x)}{[1-H(x)]^\delta}$$

$$\leqq \frac{g^{1-\delta}+h^{1-\delta}}{1-\delta} \leqq \frac{2^\delta}{1-\delta} < \infty \tag{1.1.40}$$

This proves Lemma 1.1.8.▌

LEMMA 1.1.9. Let X be a nonnegative r.v. with F as its d.f. If, for some α, $0 < \alpha < 1$, and $\Delta > 1$, the relation

$$\int_0^\infty \frac{dF(x)}{[1-F(\alpha x)]^\Delta} < \infty \tag{1.1.41}$$

holds, then, for $x > 0$, the relation

$$P(X \geqq x)\, O[\exp = (-x^\beta)] \tag{1.1.42}$$

is satisfied for some $\beta > 0$.

Proof. In view of (1.1.41), there exists x_0 such that for all $x > x_0$

$$\int_x^\infty \frac{dF(u)}{[1-F(\alpha u)]^\Delta} \leqq 1. \tag{1.1.43}$$

Because of the monotonicity of F, we hence obtain

$$\frac{1-F(x)}{[1-F(\alpha x)]^\Delta} \leqq 1. \tag{1.1.44}$$

Let us write $1 - F(\alpha x) = \phi(y)$, $y = \alpha x$ for brevity. Then, in place of (1.1.44), we can write

$$\phi(y/\alpha) \leqq \phi^\Delta(y) \tag{1.1.45}$$

valid for $y \geqq y_0 = x_0/\alpha$. Taking successively, $y = y_0, y_0/\alpha, \ldots$, we obtain from (1.1.45) the sequence of relations

$$\phi(y_0/\alpha^k) \leqq [\phi(y_0)]^{\Delta^k}. \tag{1.1.46}$$

For any $y \geqq y_0$, an integer $k \geqq 0$ exists such that

$$y_0/\alpha^k \leqq y < y_0/\alpha^{k+1} \tag{1.1.47}$$

Using the monotonicity of ϕ and also (1.1.46) and (1.1.47) leads us to the assertion of the lemma in an obvious manner.▌

Remark. As is evident from the proof of the lemma, the conditions $\Delta > 1$, $0 < \alpha < 1$ are essential. For $\alpha = 1$ and $\Delta \geqq 1$, the integral in (1.1.41) always diverges. If we take $\Delta = 1$ but $0 < \alpha < 1$, then an argument, wholly similar to the one above, yields only the estimate, valid for $x > 0$,

$$P[X \geqq x] = O(x^{-A}) \tag{1.1.48}$$

where $A > 0$ is arbitrary.

1.1.5 Hamburger's Theorem on Moments

Let

$$\alpha_{i_1, i_2, \ldots, i_s} = \int \cdots \int x_1^{i_1} \cdots x_s^{i_s} \, dF(x_1, \ldots, x_s)$$

be the compound moment of order $(i_1, \ldots, i_s)$ for some s-dimensional distribution. An answer to the question: "Under what conditions does the set of all moments uniquely define the s-dimensional d.f. F?" is given by the following lemma.

LEMMA 1.1.10 If the moments $\alpha_{i_1, \ldots, i_s}$ of the d.f., F are such that $\sum_1^\infty \lambda_{2n}^{-1/2n}$ diverges, where

$$\lambda_{2n} = \alpha_{2n, 0, \ldots, 0} + \alpha_{0, 2n, 0, \ldots, 0} + \cdots + \alpha_{0, 0, \ldots, 2n}$$

then F is the only d.f. with the given moments.

A proof may be found in Shohat and Tamarkin [167].▮

1.1.6 Esseen's Lemma

The following lemma is due to Esseen; it gives an estimate of the closeness of two d.f.'s F and G in terms of the closeness of the c.f.'s on an arbitrary interval $[-T, T]$.

LEMMA 1.1.11. (Esseen [168]; also cf. [91], p. 299). Let F and G be d.f.'s, and f and g their c.f.'s. If G has the first derivative, then

$$\sup_x |F(x) - G(x)| \leqq \frac{1}{\pi} \int_{-T}^{T} \left| \frac{f(t) - g(t)}{t} \right| dt + \frac{24}{\pi T} \sup_x |G'(x)|.$$

1.2 AN α-DECOMPOSITION THEOREM AND CONSEQUENCES

Lemma 1.2.1 given below is one of the variants of a theorem on α-decomposition presented in Linnik [85].

LEMMA 1.2.1. (α-decomposition theorem). Let the function $\phi(z)$ of the complex variable z be regular and nonvanishing on the disc $|z| < R$ and possess the Hermitian property: $\phi(-z) = \overline{\phi(\bar{z})}$. If $\phi_1, \ldots, \phi_s$ be c.f.s, and

$\alpha_1, \ldots, \alpha_s$ be positive numbers such that for some sequence $\{t_n\}$ of real numbers tending to zero the relation

$$[\phi_1(t)]^{\alpha_1} \cdots [\phi_s(t)]^{\alpha_s} = \phi(t) \tag{1.2.1}$$

is satisfied, then the functions ϕ_j are regular and nonvanishing in $|z| < R$, and relation (1.2.1) is valid throughout the disc.

If in (1.2.1), ϕ is a function of the form $\exp Q(t)$, where $Q(t)$ is a polynomial with the Hermitian property, then every ϕ_j is a normal c.f.

The above assertions continue to be valid if the left member of (1.2.1) consists of a countable number of factors, provided only that α_j are bounded away from zero.

Corollaries 1 and 2 below, though formally speaking not corollaries to lemma 1.2.1, can be easily derived with the help of arguments used for the proof of the lemma (cf. [85], pp. 79–80) as pointed out in Ramachandran and Rao [123].

Corollary 1. Let g be a c.f., corresponding to the d.f. G, which has moments of all orders and is further uniquely defined by them. (In particular, g may be any analytic c.f.) If the c.f. f is such that $f(t) = g(t)$ at some sequence $\{t_n\}$ of values of t tending to zero, then f coincides with g.

Corollary 2. Let $f(t) = \exp(-ct^2)$, $c \geqq 0$ some constant, for $t = t_n$, $n = 1, 2, \ldots$, where $t_n \to 0$ as $n \to \infty$. Then $f(t) \equiv \exp(-ct^2)$.

***Proof*:** Let f be the c.f. of F. Since all the moments of G are finite, g has derivatives of all orders. In particular, by Fatou's lemma,

$$\int x^2\, dF(x) \leqq \liminf_{n\to\infty} \frac{2 - f(t_n) - f(-t_n)}{t_n^2} = -g''(0)$$

so that $\int x^2\, dF(x) < \infty$ and hence f has the first and second derivatives defined for all real t. By Rolle's theorem, we conclude $\operatorname{Re} f' = \operatorname{Re} g'$ at a sequence of points tending to zero, and consequently the same assertion is true for f'' and g''; let $\operatorname{Re} f''(u_n) = \operatorname{Re} g''(u_n)$ at a sequence $\{u_n\}$ of points tending to zero; we again have by Fatou's lemma that

$$\int x^4\, dF(x) \leqq \liminf_{n\to\infty} \frac{f''(u_n) + f''(-u_n) - 2f''(0)}{u_n^2} = g^{(4)}(0)$$

so that $\int x^4\, dF(x) < \infty$, and hence f has derivatives of the third and fourth orders for all real t. Then, it follows from Rolle's theorem that $\operatorname{Re} f^{(3)}$ and $\operatorname{Re} g^{(3)}$ coincide at a sequence of points tending to zero, and again the same assertion is true for $\operatorname{Re} f^{(4)}$ and $\operatorname{Re} g^{(4)}$. Continuing this process, repeatedly using Fatou's lemma and Rolle's theorem, we establish that F has moments of all even orders, and so of all orders, and that f has derivatives of all orders; further, by separately considering real and imaginary parts of successive derivatives of f and g $f^{(k)}(0) = g^{(k)}(0)$ for all positive integers k, so that the corresponding moments of F and G coincide. Since by assumption G is uniquely defined by its moments, $F = G$ thus establishing Corollary 1.▮

Corollary 2 follows from Corollary 1.

1.3 INFORMATION FROM ALGEBRAIC GEOMETRY

Contemporary algebraic geometry considers algebraic sets over an arbitrary field K. In this book, we shall need only either the real field R^1 or the complex field C^1, so that K will be one of these two fields.

The information from algebraic geometry needed by us will be given without proof except for certain special lemmas which find application in Chapter 11 and are not available in the literature in the form in which we require them.

We shall take for granted the concepts of topological dimension and topological components of sets in Euclidean spaces.

Let $Q = Q(x_1, \ldots, x_m) = \sum_{|\alpha|=q} q_\alpha x_1^{\alpha_1} \cdots x_m^{\alpha_m}$ be a form of degree q in m variables with coefficients q_α either real or complex. Here $\alpha = (\alpha_1, \ldots, \alpha_m)$ is a multiple index; $\alpha_1 + \cdots + \alpha_m = q$, $\alpha_i \geqq 0$.

LEMMA 1.3.1. The number of possible coefficients in form Q equals

$$D_1 = m(m+1) \cdots (m+q-1)/q! \tag{1.3.1}$$

***Proof*:** See, for instance, Webir [9], pp. 65–66.▮

Thus the coefficients $\{q_\alpha\}$ form a linear space of dimension D_1.

Let $\mathbf{F} = \mathbf{F}_{mm} = ((f_{ij}))$ be an $m \times m$ orthogonal matrix and let us consider the orthogonal transformation $\mathbf{x} = \mathbf{F}\mathbf{y}$, where $\mathbf{x}^T = (x_1, \ldots, x_m)$ and $\mathbf{y}^T = (y_1, \ldots, y_m)$. Then the tensor of coefficients $Q : \{q_\alpha\}$ undergoes a corresponding transformation: $\{q_\alpha\} \to \{q'_\alpha\}$, q'_α will be a linear function of the q_α and a polynomial of degree q in the f_{ij}. The coefficients $\{q'_\alpha\}$ form a linear space of the same dimension D_1.

Let $m \geqq 3$ henceforward. Consider in the space $\{q'_\alpha\}$ a subspace L given by the relations

$$q'_\alpha = 0, \qquad \alpha = (\alpha_1, \ldots, \alpha_m), \qquad \alpha_m > 0 \tag{1.3.2}$$

i.e., we require the vanishing of the q'_α corresponding to any monomial $y_1^{\alpha_1} \ldots y_m^{\alpha_m}$ with $\alpha_m > 0$.

LEMMA 1.3.2. The dimension of the subspace L equals

$$D'_1 = (m-1)m \ldots (m+q-2)/q! \tag{1.3.3}$$

***Proof*:** Indeed, consider the form $Q = \sum_{|\beta|=q} q'_\beta y_1^{\beta_1} \ldots y_{m-1}^{\beta_{m-1}}$ where $\beta_1 + \cdots + \beta_{m-1} = q$ and q'_β runs through values in R^1 or C^1. The dimension of $\{q'_\beta\}$ is

D_1' and they may be obtained from $\{q_\alpha\}$ by using the inverse transformation: $\mathbf{y} = \mathbf{F}^{-1}\mathbf{x}$.▮

LEMMA 1.3.3. (the Hilbert basis theorem). In the polynomial ring $K[x_1, \ldots, x_m]$ of m variables over a field K, every ideal has a finite basis.
***Proof*:** See for instance Van der Waerden [8], pp. 27–29.▮

The set of common real roots $(\xi_1, \ldots, \xi_m)$ of a system of r polynomial equations

$$P_1(x_1, \ldots, x_m) = 0, \ldots, P_r(x_1, \ldots, x_m) = 0$$

is called a *real algebraic set* over K. It is clear that the set of equations can be given by just one equation

$$\sum_{i=1}^{r} [P_i(x_1, \ldots, x_m)]^2 = 0$$

but this is not always expedient.

The set of common complex zeros of a system of polynomials such as the above is called a *complex algebraic set* over K.

A real or complex algebraic set A is called an *irreducible algebraic set* or an *algebraic variety* if it cannot be represented in the form $A = A_1 \cup A_2$, where A_1 and A_2 are nonempty algebraic sets.

LEMMA 1.3.4. A real or complex algebraic set A can be represented as the union of a finite number of algebraic varieties: $A = A_1 \cup \cdots \cup A_N$.

LEMMA 1.3.5. The intersection of real (respectively, complex) algebraic varieties is a real (respectively complex) algebraic variety.

For a proof for the complex case, see for example Lang [96], p. 25; for the real case, it proceeds analogously.▮

By the *dimension* of a real algebraic set $A \subset R^n$ will be meant its *topological dimension* and denoted by dim A.

LEMMA 1.3.6. A real algebraic set has a finite number of topological components.

For a proof, see Whitney [151].▮

LEMMA 1.3.7. Let $A \subset R^n$ be an algebraic set of dimension d. Then it is possible to indicate n algebraic functions $\xi_i(\tau_1, \ldots, \tau_d)$, $i = 1, 2, \ldots, n$, and a number $\varepsilon > 0$ such that the point $(\xi_1(\tau_1, \ldots, \tau_d), \ldots, \xi_n(\tau_1, \ldots, \tau_d))$ belongs to A if $|\tau_i - \tau_i^0| \leqq \varepsilon$; the functions $\xi_i(\tau_1, \ldots, \tau_d)$ are continuously differentiable for $|\tau_i - \tau_i^0| \leqq \varepsilon$, and the rank of the matrix

$$\left(\left(\frac{\partial \xi_i}{\partial \tau_j}\right)\right); \qquad i = 1, 2, \ldots, n; \qquad j = 1, 2, \ldots, d$$

is d. Here $(\tau_1^0, \ldots, \tau_d^0)$ is a suitably chosen parametric point.

For a proof, *vide* [151].▌

Let A_1 and A be algebraic sets, $A_1 \subset A$ and $\dim A_1 < \dim A$. Consider the set $B = A - A_1$. Generally speaking, it will not be an algebraic set; of course, it is nonempty since $\dim A_1 < \dim A$.

LEMMA 1.3.8. For the set $B = A - A_1$, the assertion of Lemma 1.3.7 is valid, with A replaced by B and d by $\dim A$.

In fact, by Lemma 1.3.6, the algebraic sets A and A_1 have a finite number of topological components each, where the component of A with the highest dimension has dimension greater than $\dim A_1$. Therefore, an application of Lemma 1.3.7 to the algebraic set A gives us the desired assertion; here the corresponding set, $|\tau_i - \tau_i^0| \leqq \varepsilon_1$, will be contained in the set $|\tau_i - \tau_i^0| \leqq \varepsilon$, $i = 1, \ldots, d$.▌

For a complex algebraic variety $V \subset C^n$, the maximal number of algebraically independent parameters defining the variety V will be called its *complex algebraic dimension*. It equals twice the topological dimension of V considered in R^{2n}. We shall denote it by $\dim_C V$. For a complex algebraic set $A \subset C^n$, the complex dimension of its maximal irreducible component will be its dimension $\dim_C A$.

Let a real algebraic set A be given by the system of polynomial relations over R^1

$$P_i(x_1, \ldots, x_m) = 0; \qquad i = 1, \ldots, r.$$

By the *complexification* of A will be meant its imbedding in a complex algebraic set A^*, i.e., the construction of a complex algebraic set A^* such that it reduces to A if only its real points are considered. Such a complexification can be constructed in various ways. Let, for example, A be given by the equation $x_1^2 + x_2^2 = 0$. Then A consists of the single point (0, 0) and $\dim A = 0$. We may complexify A by treating x_1 and x_2 to be complex; then A^* consists of two straight lines $x_1 + ix_2 = 0$ and $x_1 - ix_2 = 0$, and $\dim_C A^* = 1$. If A is given in the form $x_1 = 0$, $x_2 = 0$, then complexification gives A^* as: $x_1 = 0$, $x_2 = 0$, so that $\dim_C A^* = 0$.

Whitney [151] proved the following theorem on "economical complexification."

LEMMA 1.3.9. Let a real algebraic variety A be given, with $\dim A = d$. Then it can be complexified "economically," i.e., a complex algebraic variety $A^* \supset A$ can be constructed such that $\dim_C A^* = \dim A = d$.

For a proof, *vide* [151].∎

We shall also require

LEMMA 1.3.10. Let V_1 and V_2 be real algebraic varieties with $V_2 \subset V_1$ and $\dim V_2 = \dim V_1$. Then $V_2 = V_1$.

This lemma follows at once from Lemma 1.3.9.∎

LEMMA 1.3.11. (the intersection theorem). Let V_1 and V_2 be complex algebraic varieties in the complex space C^n, having points in common. Then

$$\dim_C(V_1 \cap V_2) \geqq \dim_C V_1 + \dim_C V_2 - n. \tag{1.3.5}$$

For a proof, *vide* [96], pp. 38–39.∎

1.4 INFORMATION FROM THE THEORY OF ENTIRE FUNCTIONS AND FROM THE THEORY OF DIFFERENTIAL EQUATIONS

The material given below will be used mainly in Chapters 2, 5, 6, and 12.

LEMMA 1.4.1. (the Paley-Wiener theorem). In order that the entire function f of exponential type belong to $L^2(-\infty, \infty)$ on the real axis, i.e., for

$$\int |f(x)|^2 \, dx < \infty$$

it is necessary and sufficient that it admits the representation

$$f(z) = \int_{-\tau}^{\tau} e^{izt}\phi(t) \, dt$$

for some $\phi \in L^2(-\tau, \tau)$.

For a proof, see, for instance, [85], pp. 35–36.∎

LEMMA 1.4.2. (Marcinkiewicz' theorem). If a c.f. ϕ has the form $\exp P(t)$ where P is a polynomial, then P is a quadratic polynomial, and ϕ is the c.f. of a normal law.

For a proof, *vide* for instance [85], pp. 64–65.∎

In Chapter 6, we shall require the following lemma from the theory of differential equations.

LEMMA 1.4.3. In the differential equation

$$L(y) \equiv z^n y^{(n)} + \sum_{j=1}^{n} a_j z^{n-j} y^{(n-j)} = f(z) \tag{1.4.1}$$

let the a_j be constants and f a function analytic in the angle:

$$\mathscr{A} = \{\text{Re } z \geqq a > 0, \ |\arg z| \leqq \pi - \delta_0, \delta_0 > 0\}$$

and let further

$$|f(z)| \leqq A_0 \frac{[\log |z|]^{q_0}}{|z|^\beta} \tag{1.4.2}$$

where $q_0 \geq 0$ and $\beta > 0$ are constants. Then, there exists a particular solution $y_0(z)$ of the equation (1.4.1) in $\mathscr{A}$, satisfying the inequalities

$$|y_0^{(j)}(z)| \leqq AA_0 \frac{[\log |z|]^{q_0+(j+1)s_0}}{|z|^{\beta+j}}; \qquad j = 1, \ldots, n \tag{1.4.3}$$

where the constant s_0 does not depend on f. If $\beta \geqq \beta_0 > 0$, where β_0 is a sufficiently large number, then A also is independent of f.

Proof. We shall solve the equation (1.4.1) by the method of variation of parameters. For this purpose, we first consider the homogeneous equation—$L(y)$ is defined in (1.4.1)—

$$L(y) = 0. \tag{1.4.4}$$

The characteristic equation for (1.4.4) is given by

$$\lambda(\lambda - 1) \cdots (\lambda - n + 1) + \sum_{j=1}^{n} a_j \lambda(\lambda - 1) \cdots (\lambda - n + j - 1) = 0. \tag{1.4.5}$$

Let λ^* be a root of the equation (1.4.5) of multiplicity k^*. As is well-known, to the root λ^* there correspond k^* linearly independent solutions

$$z^{\lambda^*}, z^{\lambda^*} \ln z, \ldots, z^{\lambda^*}(\ln z)^{k^*-1}. \tag{1.4.6}$$

Let us denote by $y_1(z), \ldots, y_n(z)$ a fundamental system of solutions of the equation (1.4.4), which we construct in the manner described above corresponding to the various distinct roots of the characteristic equation (1.4.5). We seek for the solution of (1.4.1) in the form

$$y(z) = \sum_{1}^{n} c_p y_p(z) \tag{1.4.7}$$

where the $c_p = c_p(z)$ are functions to be determined. Since, for the determination of the functions $c_p(z)$, $p = 1, \ldots, n$, there exists only one condition, viz., (1.4.1), it is natural to set

$$\sum_{p=1}^{n} c_p'(z) y_p^{(j-1)}(z) = 0, \qquad j = 1, \ldots, (n-1). \tag{1.4.8}$$

Then

$$y^{(j)}(z) = \sum_{p=1}^{n} c_p(z) y_p^{(j)}(z); \qquad j = 1, \ldots, (n-1), \tag{1.4.9}$$

$$y^{(n)}(z) = \sum_{p=1}^{n} c_p'(z) y_p^{(n-1)}(z) + \sum_{p=1}^{n} c_p(z) y_p^{(n)}(z). \tag{1.4.10}$$

Substituting all these expressions in (1.4.1) and using the equality $L(y_p) = 0$, we find

$$\sum_{p=1}^{n} c_p'(z) y_p^{(n-1)}(z) = f(z)/z^n. \tag{1.4.11}$$

The preceding relation together with (1.4.8) gives us a system of linear equations for determining the functions $c_p'(z)$, $p = 1, \ldots, n$. We obtain

$$c_p'(z) = D_p^*(z)/D(z)$$

where we have adopted the notation: $D(z)$ is the determinant of the Wronskian of the fundamental system of solutions of (1.4.4).

$$D(z) = \begin{vmatrix} y_1 & \cdots & y_n \\ y_1' & \cdots & y_n' \\ \cdots & & \cdots \\ y_1^{(n-1)} & \cdots & y_n^{(n-1)} \end{vmatrix} \tag{1.4.12}$$

$$D_p^*(z) = D_p(z)(f(z)/z^n)$$

and $D_p(z)$ is the cofactor of the p-th element of the last row of the determinant $D(z)$. Thus

$$c_p'(z) = \frac{D_p(z) f(z)}{D(z) z^n}. \tag{1.4.13}$$

As is well-known, the determinant of the Wronskian is

$$D(z) = c_0 z^{-a_1} \tag{1.4.14}$$

where $c_0 \neq 0$ is a constant, and a_1 is the coefficient appearing in (1.4.1). We recall that every solution $y_p(z)$ is one of the functions given by (1.4.6) with an appropriate λ_p. If we substitute in (1.4.12) the appropriate expression for $y_p^{(j)}$, then we can take out a factor z^{1-j} from the j-th row of the determinant, and a factor z^{λ_p} from the p-th column, so that

$$D(z) = z^{\Sigma_1^n \lambda_p - n(n-1)/2} D^*(z) \tag{1.4.15}$$

with positive upper and lower bounds for the determinant $D^*(z)$. But $\sum_1^n \lambda_p$ is the sum of the roots of the characteristic equation (1.4.5), so that it is equal to $a_1 + (1/2)n(n-1)$, and in view of (1.4.14) and (1.4.15),

$$D(z) = c_0 z^{-(1/2)n(n-1) + \Sigma_1^n \lambda_p}.$$

Let us now consider the determinant $D_p(z)$. We can take out a factor z^{λ_q}, $q \neq p$, from its q-th column, and a factor z^{1-j} from its j-th row. Thus

$$D_p(z) = z^{\Sigma_1^n \lambda_q - \lambda_p - n(n-1)/2 + (n-1)} \tilde{D}_p(z). \tag{1.4.16}$$

In view of (1.4.6), in every row of the determinant $\tilde{D}_p(z)$, there can occur only elements of the form $\psi(z)(\ln z)^s$, where $\psi(z)$ is bounded in $\mathscr{A}$. Let s_0 be the largest of the powers of $(\ln z)$ occurring in the various elements of $\tilde{D}_p(z)$. Then

$$|D_p(z)| < D_0 |z^{\Sigma \lambda_q - \lambda_p - (1/2)n(n-1) + (n-1)} (\log z)^{s_0}|, \; D_0 > 0, \; s_0 > 0.$$

Since the determinant $D_p(z)$ does not depend on $f(z)$, s_0 and D_0 are also independent of $f(z)$. We also note that $|\arg z| < (\pi/2) - \delta_0$, $\delta_0 > 0$, and hence

$$|z^\alpha| \leqq D^{**} |z|^{\operatorname{Re} \alpha}, \qquad |\ln z| < D^{***} \ln |z|$$

for some positive constants D^{**} and D^{***}. Finally, all these arguments yield the conclusion

$$|D_p(z)| < C^* |z|^{\Sigma \lambda_q - \lambda_p - n(n-1)/2 + (n-1)} (\ln |z|)^{s_0} \tag{1.4.17}$$

where C^* and s_0 are constants not depending on p and $f(z)$. From (1.4.16) and (1.4.17) we obtain

$$|c_p'(z)| < C^* \frac{|f(z)| (\ln |z|)^{s_0}}{|z|^{\operatorname{Re} \lambda_p + 1}}$$

and taking (1.4.2) into account, we conclude that

$$|c_p'(z)| < A_0 C^* \frac{(\ln |z|)^{q_0 + s_0}}{|z|^{\operatorname{Re} \lambda_p + \beta + 1}}. \tag{1.4.18}$$

As is seen from (1.4.13) and (1.4.14), the function $c_p'(z)$ is analytic in $\mathscr{A}$. We shall choose the functions $c_p(z)$ as follows:

(a) if $\operatorname{Re} \lambda_p + \beta + 1 > 0$, then we set

$$c_p(z) = -\int_z^\infty \frac{D_p(z)}{D(z)} \frac{f(z)}{z^n} \, dz \tag{1.4.19}$$

where the integral is taken along the ray (z, ∞) lying along the line passing through the origin of coordinates;

(b) if $\operatorname{Re} \lambda_p + \beta + 1 \leqq 0$, then we set

$$c_p(z) = \int_a^z \frac{D_p(z)}{D(z)} \frac{f(z)}{z^n} \, dz \tag{1.4.20}$$

where the integral is taken along the line-segment $[a, z]$.

In case (a),

$$|c_p(z)| < A_0 C^* \int_{|z|}^{\infty} \frac{(\ln|z|)^{q_0+s_0}}{|z|^{\operatorname{Re}\lambda_p+\beta+1}} d|z|$$

$$= A_0 C^* \int_{|z|}^{\infty} \frac{(\ln|z|)^{q_0+s_0}}{|z|^{\delta}} \frac{d|z|}{|z|^{\operatorname{Re}\lambda_p+\beta+1-\delta}}$$

$$< \frac{A_0 C^{**}}{|\operatorname{Re}\lambda_p+\beta-\delta|} \frac{(\ln|z|)^{q_0+s_0}}{|z|^{\operatorname{Re}\lambda_p+\beta}} \tag{1.4.21}$$

where δ is such that $\operatorname{Re}\lambda_p + \beta + 1 - \delta > 0$, and C^{**} is some constant.

In case (b), we choose δ such that $\operatorname{Re}\lambda_p - \beta + 1 < \delta$. Then we note that on the segment $[a, z]$, $|dz| < \tilde{c}\, d|z|$ for a constant $\tilde{c} > 0$ not depending on z, and obtain

$$|c_p(z)| < A_0 C^* \tilde{c} \int_{|a|}^{|z|} |z|^{\delta} \frac{(\ln|z|)^{q_0+s_0}}{|z|^{\operatorname{Re}\lambda_p+\beta+1-\delta}} d|z|$$

$$< \frac{A_0 C^{**}}{|\operatorname{Re}\lambda_p+\beta-\delta|} \frac{(\ln|z|)^{q_0+s_0}}{|z|^{\operatorname{Re}\lambda_p+\beta}}.$$

Now substituting the expressions (1.4.19) and (1.4.20) in (1.4.9), we find

$$|y^{(j)}(z)| < A_0 C^{**} \sum_{p=1}^{n} \frac{|y_p^{(j)}(z)|}{|\operatorname{Re}\lambda_p+\beta-\delta|} \frac{(\ln|z|)^{q_0+s_0}}{|z|^{\operatorname{Re}\lambda_p+\beta}}, \qquad j = 0, 1, \ldots, (n-1),$$

and, since according to (1.4.6), for a constant C^{***},

$$|(z^{\lambda^*}(\ln z)^{q^*})^{(j)}| < C^{***}|z|^{\operatorname{Re}\lambda^*-j}(\ln|z|)^{q^*}$$

consequently, for a constant A,

$$|y^{(j)}(z)| < A_0 C^{**} C^{***} \sum_{p=1}^{n} \frac{1}{|\operatorname{Re}\lambda_p+\beta-\delta|} \frac{(\ln|z|)^{q_0+s_0}}{|z|^{\operatorname{Re}\lambda_p+\beta}} \frac{(\ln|z|)^{s_0}}{|z|^{j}} |z|^{\operatorname{Re}\lambda_p}$$

$$< AA_0 \frac{(\ln|z|)^{q_0+(j+1)s_0}}{|z|^{\beta+j}}, \qquad j = 0, 1, \ldots, (n-1). \tag{1.4.22}$$

For $j = n$, an estimate of the same type is obtained from (1.4.10).

To conclude the proof, we note that if $\beta > \sum_1^n |\operatorname{Re}\lambda_p|$, then in (1.4.22) and hence in (1.4.3), the constant A may be taken to be independent of β, q_0, and A_0, i.e., of $f(z)$.∎

1.5 SOLUTIONS OF CERTAIN FUNCTIONAL EQUATIONS

LEMMA 1.5.1. (Linnik [85]; Rao [129]). Consider the equation, assumed valid for $|u| < \delta_0$, $|v| < \delta_0$,

$$\psi_1(u + b_1 v) + \cdots + \psi_r(u + b_r v) = A(u) + B(v) + P_k(u, v) \tag{1.5.1}$$

where P_k is a polynomial of degree k; ψ_i, A and B are complex valued functions of two real variables u and v. We assume that

(i) the numbers b_j are all distinct without loss of generality,

(ii) the functions A, B, and the ψ_j are continuous. Then, in some neighborhood of the origin, the functions A, B, and the ψ_j are all polynomials of degree $\leqq \max(r, k)$.

Proof: Following the work [85], we multiply both sides of (1.5.1) by $(x-u)$ and integrate with respect to u from 0 to x, $|x| < \delta_0$. We obtain

$$\sum_1^r \int_0^x (x-u)\psi_j(u+b_j v)\,du = \int_0^x (x-u)A(u)\,du + B(v)\int_0^x (x-u)\,du + \int_0^x (x-u)P_k(u,v)\,du$$
$$= C(x) + (x^2/2)B(v) + P_{k+2}(u,v). \qquad (1.5.2)$$

After the transformation $\tau = u + b_j v$, $|\tau| < \delta_1 < \delta_0$, we can rewrite (1.5.2) as

$$\sum_j \int_{b_j v}^{x+b_j v} (x + b_j v - \tau)\psi_j(\tau)\,d\tau = C(x) + (x^2/2)B(v) + P_{k+2}(u,v). \qquad (1.5.3)$$

Since the left member of (1.5.3) is obviously differentiable with respect to v for every x, the same is true of the right member also. Differentiating first with respect to v, then with respect to x, we reduce (1.5.3) to the form

$$\sum_j b_j\psi_j(u+b_j v) = B_1(v) + uB_2(v) + P_k(u,v) \qquad (1.5.4)$$

valid for $|u| < \delta_2$, $|v| < \delta_2$, $0 < \delta_2 < \delta_1$.

Setting $v = 0$ in (1.5.4), we obtain

$$\sum_j b_j\psi_j(u) = P_{s_1}(u), \qquad s_1 \leqq \max(1, k). \qquad (1.5.5)$$

Starting from (1.5.4) and carrying out the same transformation and a similar argument following it, which led from (1.5.1) to (1.5.5), we obtain

$$\sum b_j^2\psi_j(u) = P_{s_2}(u), \qquad s_2 \leqq \max(2, k). \qquad (1.5.6)$$

Repeating this operation, we obtain r equations

$$\sum b_j^m\psi_j(u) = P_{s_m}(u), \qquad s_m \leqq \max(m, k), \quad m = 1, 2, \ldots, r. \qquad (1.5.7)$$

Since the b_j are all distinct, the equations (1.5.7) have a unique solution for $\psi_j(u)$ in some neighborhood of the origin, namely, a polynomial of degree $\leqq \max(r, k)$.

An alternative proof, due to Darmois and Skitovich and based on the method of finite differences, is of some interest. We outline it below.

Let $a_i^{(1)} = (1 - b_i b_r^{-1})a$. Then, for small a, u, v, we have

$$\sum_1^r \psi_i(u + a_i^{(1)} + b_i v) = \sum_1^r \psi_i[u + a + b_i(v - ab_r^{-1})]$$
$$= A(u + a) + B(v - ab_r^{-1}) + P_k(u + a, v - ab_r^{-1}). \tag{1.5.8}$$

Subtracting (1.5.1) from (1.5.8), we obtain

$$\sum_1^{r-1} \psi_i^{(1)}(u + b_i v) = A^{(1)}(u) + B^{(1)}(v) + P_{k-1}(u, v) \tag{1.5.9}$$

where $\psi_i^{(1)}(x) = \psi_i(x + a_i^{(1)}) - \psi_i(x)$, and the rest of the notation is obvious. Equation (1.5.9) is of the same form as (1.5.1) except that r has been replaced by $(r - 1)$. Repeating this process $(r - 1)$ times, we arrive at

$$\psi_1^{(r-1)}(u + b_1 v) = A^{(r-1)}(u) + B^{(r-1)}(v) + P_c(u, v) \tag{1.5.10}$$

where

$$c = \max(0, k - r + 1) \quad \text{and} \quad \psi_1^{(r-1)}(x) = \psi_1^{(r-2)}(x + a_1^{(r-1)}) - \psi_1^{(r-2)}(x). \tag{1.5.11}$$

Here $a_1^{(r-1)}$ is an arbitrary number restricted to a small neighborhood of the origin. Equation (1.5.10) shows that $\psi_1^{(r-1)}$ is a polynomial of degree $\leqq \max(1, k - r + 1)$, and (1.5.11) that $\psi_1^{(r-2)}$ is a polynomial of degree $\leqq \max(2, k - r + 2)$, and so on. Noting that $a_1^{(1)}, a_1^{(2)}, \ldots$ are arbitrary, we finally obtain that ψ_1 is a polynomial of degree $\leqq \max(r, k)$. The same is true of $\psi_2, \ldots, \psi_r$. Then A and B also turn out to be polynomials of the same kind. ∎

Remark. If only $s <$ of the b_i are different, then A and B are polynomials of degree $\leq \max(s, k)$.

Corollary. If the equation (1.5.1) of Lemma 1.5.1 has the form

$$\sum_1^r \psi_i(u + b_i v) = au + cv + d \tag{1.5.12}$$

with $r \leqq 3$, then, under the conditions of Lemma 1.5.1, all the ψ_i, $i = 1, \ldots, r$, are linear functions.

Proof: In fact, Lemma 1.5.1 shows that ψ_i are at most of degree 3. But then, from (1.5.12), we see that the coefficients of the second and third degree terms must be zero. ∎

We now generalize Lemma 1.5.1 to cases of more than two arguments. Let $\mathbf{t}$ and $\boldsymbol{\alpha}_1, \ldots, \boldsymbol{\alpha}_r$ be p-dimensional column vectors. Denoting the components of $\mathbf{t}$ by $t_1, \ldots, t_p$ and the scalar product of $\mathbf{t}$ and $\boldsymbol{\alpha}_i$ by $\boldsymbol{\alpha}_i'\mathbf{t}$, we consider the functional equation

$$\psi_1(\boldsymbol{\alpha}_1'\mathbf{t}) + \cdots + \psi_r(\boldsymbol{\alpha}_r'\mathbf{t}) = \xi_1(t_1) + \cdots + \xi_p(t_p) \tag{1.5.13}$$

valid for $|t_i| < \delta$, $i = 1, \ldots, p$. We denote by $\mathbf{A}$ the $p \times r$ matrix with columns $\boldsymbol{\alpha}_1, \ldots, \boldsymbol{\alpha}_r$.

LEMMA 1.5.2. (Rao [130], Khatri and Rao [74]). Let $\boldsymbol{\alpha}_i$, the i-th column of $\mathbf{A}$, be not proportional to any other column of $\mathbf{A}$ or to any column of the identity matrix $\mathbf{I}_p$ of order p. Then the functions ψ_i satisfying (1.5.13) are necessarily polynomials of degree $\leq r$.

Proof: Without loss of generality, we may take the first component of $\boldsymbol{\alpha}_i$ to be nonzero. Then (cf. [129]) there exists a $2 \times p$ matrix of the form

$$H = \begin{pmatrix} 1 & 0 & \cdots & 0 \\ 0 & h_2 & \ldots & h_p \end{pmatrix} \tag{1.5.14}$$

such that the i-th column of the matrix $\mathbf{B} = \mathbf{HA}$ is not proportional to any other column of $\mathbf{HA}$ or to any column of $\mathbf{H}$. We take $\mathbf{t}' = \mathbf{u}'\mathbf{H}$, where $\mathbf{u}$ is a two-dimensional vector with components u_1 and u_2, and substitute in (1.5.13), obtaining

$$\begin{aligned} \sum \psi_i(b_{i1}u_1 + b_{i2}u_2) &= \xi_1(u_1) + \xi_2(u_2 h_2) + \cdots + \xi_p(u_2 h_p) \\ &= \xi_1(u_1) + \xi_2^*(u_2). \end{aligned} \tag{1.5.15}$$

In (1.5.15), the vector (b_{i1}, b_{i2}) is not proportional either to (b_{j1}, b_{j2}) for any $j \neq i$, or to any of the vectors (1, 0) and (0, 1). The assertion of the present lemma then follows from Lemma 1.5.1. ∎

LEMMA 1.5.3. (cf. [74]). If no column of the matrix $\mathbf{A}$ is proportional to any other column of $\mathbf{A}$ or to any column of the matrix $\mathbf{I}_p$, then the functions $\psi_1, \ldots, \psi_r$ and $\xi_1, \ldots, \xi_p$ are polynomials of degree r at most.

This lemma is a direct consequence of Lemma 1.5.2. ∎

In order to establish, and indeed to state, Lemma 1.5.4 below, we have to define a new matrix product: that of a matrix $\mathbf{C}$ of order $p \times r$ and a matrix $\mathbf{D}$ of order $q \times r$. Let $\boldsymbol{\gamma}_1, \ldots, \boldsymbol{\gamma}_r$ be the columns of $\mathbf{C}$, and $\boldsymbol{\delta}_1, \ldots, \boldsymbol{\delta}_r$ those of $\mathbf{D}$. Then we define the product $\mathbf{C} \odot \mathbf{D}$ as the $(pq \times r)$ matrix whose columns are the Kronecker products $\boldsymbol{\gamma}_i \otimes \boldsymbol{\delta}_i$. Thus, by definition,

$$\mathbf{C} \odot \mathbf{D} = (\boldsymbol{\gamma}_1 \otimes \boldsymbol{\delta}_1; \ldots; \boldsymbol{\gamma}_r \otimes \boldsymbol{\delta}_r). \tag{1.5.16}$$

We also define $\mathbf{C}^{\#}$ as the matrix of order $p(p-1) \times r$, obtained by deleting the p rows involving the square terms, the first, the $(p+2)$-th, ... and the p^2-th rows from $\mathbf{C} \odot \mathbf{C}$.

LEMMA 1.5.4. (cf. [74]). Let the relation (1.5.13) be satisfied for $|t_i| < \delta$, $i = 1, \ldots, p$, where the matrix **A** of the relation is such that rank $\mathbf{A}^{\#} = r$. Then the functions $\psi_1, \ldots, \psi_r$ and $\xi_1, \ldots, \xi_p$ are linear.

Proof: It is easily seen that, if rank $\mathbf{A}^{\#} = r$, then no column of **A** is a multiple of any other, or of any column of the matrix $\mathbf{I}_p$. Then, by Lemma 1.5.3, the ψ_i and ξ_j are polynomials of degree $\leqq r$. Let now

$$\psi_i(u) = \lambda_{ir} u^r + \cdots + \lambda_{io}; \qquad i = 1, 2, \ldots, r \tag{1.5.17}$$

$$\xi_j(u) = \mu_{jr} u^r + \cdots + \mu_{jo}; \qquad j = 1, 2, \ldots, p. \tag{1.5.18}$$

We set $\boldsymbol{\lambda}_i' = (\lambda_{i1}, \ldots, \lambda_{ir})$.

Substituting the expressions (1.5.17) and (1.5.18) into (1.5.15) and collecting the coefficients of $t_i t_j (i \neq j)$, we obtain

$$\mathbf{A}^{\#}\boldsymbol{\lambda}_2 = \mathbf{0} \tag{1.5.19}$$

whence it follows that $\boldsymbol{\lambda}_2 = \mathbf{0}$, since rank $\mathbf{A}^{\#} = r$. Thus there are no terms of degree two in (1.5.17).

Collecting now the coefficients of $t_i^{\pi_1} t_j^{\pi_2} t_k^{\pi_3}$, $i \neq j \neq k$, with $\pi_1 + \pi_2 + \pi_3 = 3$, with at least two of the numbers π_1, π_2, π_3 nonzero, we have

$$(\mathbf{A} \odot \mathbf{A}^{\#})\boldsymbol{\lambda}_3 = \mathbf{0} \qquad \text{or} \qquad (\mathbf{A}^{\#} \odot \mathbf{A})\boldsymbol{\lambda}_3 = \mathbf{0}. \tag{1.5.20}$$

It is easy to see that rank $(\mathbf{A} \odot \mathbf{A}^{\#}) = r$, so that $\boldsymbol{\lambda}_3 = \mathbf{0}$, and so on. Thus all terms of degree higher than one vanish, proving the lemma.∎

LEMMA 1.5.5. (cf. [74]). Let (1.5.13) be satisfied for $|t_i| < \delta$, $i = 1, \ldots, p$, with rank $\mathbf{A} = r \leqq p$, and let every column of **A** have at least two nonzero elements. Then the degree of the polynomials $\psi_1, \ldots, \psi_r$ and $\xi_1, \ldots, \xi_p$ is at most two.

The proof is similar to that of Lemma 1.5.3.∎

We proceed to generalize the results of Lemmas 1.5.2 and 1.5.3, considering q equations with p variables $\mathbf{t}' = (t_1, \ldots, t_p)$.

$$\sum_{u=1}^{p} c_{ju} \xi_u(\mathbf{e}_u' \mathbf{t}) + \sum_{i=1}^{r} b_{ji} \psi_i(\boldsymbol{\alpha}_i' \mathbf{t}) = g_j \,(\text{constant}) \tag{1.5.21}$$

$$j = 1, \ldots, q,$$

where $|t_i| < \delta$, $i = 1, \ldots, p$; $\mathbf{e}_1, \ldots, \mathbf{e}_p$ are the columns of the identity matrix. Let **A** be the matrix made up of the column vectors $\boldsymbol{\alpha}_i$, $i = 1, \ldots, r$; $\mathbf{B} = (b_{ji})$ and $\mathbf{C} = (c_{ju})$.

LEMMA 1.5.6. (cf. [74]) Suppose that every column of the matrices **A** and **B** contains at least one nonzero element and that no column of **A** is proportional either to another column of **A** or to any column of I_p. Then the functions $\psi_1, \ldots, \psi_r$ and $\xi_1, \ldots, \xi_p$ satisfying (1.5.21) are polynomials of degree at most r.

The proof proceeds according to the same scheme as the proofs of Lemmas 1.5.2 through 1.5.5. It should be noted that if some column $\boldsymbol{\alpha}_i$ of the matrix **A** is proportional to some other columns $\boldsymbol{\alpha}_{i_1}, \ldots, \boldsymbol{\alpha}_{i_k}$ of that matrix, then Lemma 1.5.6 is still true if from the corresponding combination of the equations (1.5.21) it is possible to obtain $k+1$ equations, each containing exactly one of the functions $\psi_i, \psi_{i_1}, \ldots, \psi_{i_k}$.

The following lemma can be proved similarly.

LEMMA 1.5.7. (cf. [74]). Let rank $(\mathbf{B} \odot \mathbf{A}^{\#}) = r$. Then the functions $\psi_1, \ldots, \psi_r$ and $\xi_1, \ldots, \xi_p$, satisfying (1.5.21), are linear functions.

We further generalize Lemma 1.5.1 to the case of equations containing a denumerable number of terms.

LEMMA 1.5.8. (Ramachandran [121]). Consider the equation

$$\sum_1^{\infty} \psi_i(u + c_i v) = A(u) + B(v) \tag{1.5.22}$$

where A, B, and the ψ's are continuous complex-valued functions of the real variables u and v, defined in some neighborhood of the origin. If the series $\sum_1^{\infty} \psi_i(c_i t)$ converges uniformly and the c_j are bounded, then $\sum_1^{\infty} c_i^2 \psi_i(x)$ is a polynomial of degree two at most.

Proof: Since the series on the left side of (1.5.22) converges uniformly, we may multiply both sides of (1.5.22) by $(x - u)$ and integrate with respect to u over the interval $(0, x)$, as in the "finite" case covered by Lemma 1.5.1. We have

$$\sum_1^{\infty} \int_{c_j v}^{x + c_j v} \psi_j(t)(x + c_j v - t)\, dt = D(x) + (x^2/2)B(v). \tag{1.5.23}$$

In view of the boundedness of the c_j and the uniformity of convergence of the series $\sum \psi_j(c_j t)$, the validity of the following equality is easy to see:

$$\begin{aligned}
&\sum \int_0^{c_j v} \psi_j(t)(x + c_j v - t)\, dt \\
&\quad = \sum c_j \int_0^{v} (x + c_j v - c_j t)\psi_j(c_j t)\, dt \\
&\quad = \int_0^{v} \sum [c_j(x + c_j v - c_j t)\psi_j(c_j t)]\, dt \\
&\quad = x D(v) + E(v)
\end{aligned} \tag{1.5.24}$$

where $D(v)$ and $E(v)$ are continuous functions of v. From (1.5.23) and (1.5.24) it follows that

$$\sum \int_0^{x + c_j v} (x + c_j v - t)\psi_j(t)\, dt = D(x) + (x^2/2)B_2(v) + xB_1(v) + B_0(v). \tag{1.5.25}$$

On formally differentiating the series on the left termwise with respect to v, we obtain

$$\sum c_j \int_0^{x+c_0 v} \psi_j(t)\,dt = \sum c_j \int_0^x \psi_j(t)\,dt + \sum c_j^2 \int_0^v \psi_j(x + c_j u)\,du. \quad (1.5.26)$$

Formally again, the second derivative of the left member of (1.5.25) with respect to v is

$$\sum c_j^2 \psi_j(x + c_j v). \quad (1.5.27)$$

The series in (1.5.26) and (1.5.27) converge uniformly, so that the series in (1.5.25) can be differentiated termwise twice with respect to v, for every fixed x. Considering three different values of x, we find that the functions B_0, B_1 and B_2 are also twice differentiable. Setting $v = 0$ in the equality obtained by differentiating both sides of (1.5.25) with respect to v, we obtain the relation

$$\sum c_j^2 \psi_j(x) = a + bx + cx^2 \quad (1.5.28)$$

valid for x in some neighborhood of the origin.▌

Also of interest is an extension of Lemma 1.5.1 to the case where the ψ_i are continuous complex-valued functions of real vector variables, defined in a neighborhood of the origin.

LEMMA 1.5.9. (Ghurye and Olkin [19]). Consider the equation:

$$\psi_1(\mathbf{u} + \mathbf{C}_1\mathbf{v}) + \cdots + \psi_r(\mathbf{u} + \mathbf{C}_r\mathbf{v}) = A(\mathbf{u}) + B(\mathbf{v}) \quad (1.5.29)$$

where $\mathbf{u}$, $\mathbf{v}$ are real vector variables. If the $\mathbf{C}_i$ are nonsingular matrices with $|\mathbf{C}_i - \mathbf{C}_j| \neq 0$ for $i \neq j$, then the functions $\psi_1, \ldots, \psi_r$, A and B, satisfying (1.5.29), are polynomials of degree $\leqq r$.

Proof: We shall adopt the method of finite differences used above for proving Lemma 1.5.1. Let $\mathbf{a}_i^{(1)} = (\mathbf{I} - \mathbf{C}_i\mathbf{C}_r^{-1})\mathbf{a}^{(1)}$. Then

$$\begin{aligned}\sum_1^r \psi_i(\mathbf{u} + \mathbf{a}_i^{(1)} + \mathbf{C}_i\mathbf{v}) &= \sum_1^r \psi_i[\mathbf{u} + \mathbf{a}^{(1)} + \mathbf{C}_i(\mathbf{v} - \mathbf{C}_r^{-1}\mathbf{a}^{(1)})] \\ &= A(\mathbf{u} + \mathbf{a}^{(1)}) + B(\mathbf{v} - \mathbf{C}_r^{-1}\mathbf{a}^{(1)}). \quad (1.5.30)\end{aligned}$$

Subtracting (1.5.29) from (1.5.30), we have

$$\sum_1^{r-1} \psi_i^{(1)}(\mathbf{u} + \mathbf{C}_i\mathbf{v}) = A^{(1)}(\mathbf{u}) + B^{(1)}(\mathbf{v}) \quad (1.5.31)$$

where $\psi_i^{(1)}(\mathbf{x}) = \psi_i(\mathbf{x} + \mathbf{a}_i^{(1)}) - \psi_i(\mathbf{x})$. Continuing this process as in the proof of Lemma 1.5.1, we obtain

$$\psi_1^{(r-1)}(\mathbf{u} + \mathbf{C}_1\mathbf{v}) = A^{(r-1)}(\mathbf{u}) + B^{(r-1)}(\mathbf{v}) \quad (1.5.32)$$

where

$$\psi_1^{(r-1)}(\mathbf{x}) = \psi_1^{(r-2)}(\mathbf{x} + \mathbf{a}_1^{(r-1)}) - \psi_1^{(r-2)}(\mathbf{x}). \quad (1.5.33)$$

Equation (1.5.32) shows that $\psi_1^{(r-1)}$ is a linear function, and (1.5.33) that $\psi_1^{(r-2)}$ is a polynomial of at most the second degree, in view of the arbitrariness of $\mathbf{a}_1^{(r-1)}$. Finally we obtain that ψ_1 is a polynomial of degree at most r. Similarly, all the other functions in (1.5.29) are also polynomials of the same type (in some neighborhood of the origin).∎

If we have as the right-hand member of (1.5.29), $A(\mathbf{u}) + B(\mathbf{v}) + P_k(\mathbf{u}, \mathbf{v})$, where $P_k(\mathbf{u}, \mathbf{v})$ is a polynomial of degree k, then the functions ψ_i satisfying (1.5.29) would be polynomials of degree $\leqq \max(r, k)$ in some neighborhood of the origin, as in Lemma 1.5.1.

Remark. (i) If in Lemma 1.5.9, the condition $|\mathbf{C}_i - \mathbf{C}_j| \neq 0$ for $i \neq j$ is not satisfied we can, in general, conclude that $A(\mathbf{u})$ and $B(\mathbf{v})$ are polynomials of degree $\leqq r$. (ii) If $\mathbf{C}_i$ is such that $|\mathbf{C}_i - \mathbf{C}_j| \neq 0$, for $j = 1, \ldots, i-1, i+1, \ldots, r$, then ψ_i is a polynomial of degree $\leqq r$. (iii) If no condition is imposed on $\mathbf{C}_i$, (i.e., if they are not nonsingular), then $B(\mathbf{v})$ is a polynomial of degree $\leq r$, but no definite statement can be made about $A(\mathbf{u})$.

For solutions to more general functional equations, see Addendum A at the end of the book.

LEMMA 1.5.10. (Rao [130]). Suppose $\psi(t)$ satisfies the equation

$$a_1\psi(b_1 t) + \cdots + a_n\psi(b_n t) = 0 \tag{1.5.34}$$

where $\sum a_i b_i = 0$, $|b_1| > \max(|b_2|, \ldots, |b_n|)$ and $a_i b_i$ for $i = 2, \ldots, n$ have one sign while $a_1 b_1$ has the opposite sign. Let further $\psi(t) = c + t\phi(t)$, where $\phi(t)$ is continuous at the origin. Then ψ is a linear function.

Proof: We substitute $c + t\phi(t)$ for $\psi(t)$ in (1.5.34) and divide both sides by t

$$\phi(t) = p_1\phi(\beta_1 t) + \cdots + p_{n-1}\phi(\beta_{n-1}t) \tag{1.5.35}$$

where $p_i > 0$, $\sum p_i = 1$, and $|\beta_i| < 1$, $i = 1, \ldots, (n-1)$. Replacing t by $\beta_i t$ in (1.5.35), we obtain

$$\phi(\beta_i t) = p_1\phi(\beta_1\beta_i t) + \cdots + p_{n-1}\phi(\beta_{n-1}\beta_i t)$$

so that

$$\begin{aligned} \phi(t) = \sum p_i\phi(\beta_i t) &= \sum\sum p_i p_j \phi(\beta_i\beta_j t) \\ &= \sum\sum q_{ij}\phi(\beta_i\beta_j t) \end{aligned} \tag{1.5.36}$$

where $\sum\sum q_{ij} = 1$; hence $\phi(t) - \phi(0) = \sum\sum q_{ij}[\phi(\beta_i\beta_j t) - \phi(0)]$.

Continuing in the same manner, we obtain

$$\phi(t) - \phi(0) = \sum q_{i_1 \cdots i_m}[\phi(\beta_{i_1} \cdots \beta_{i_m} t) - \phi(0)] \tag{1.5.37}$$

where $\sum q_{i_1 \ldots i_m} = 1$ and the summation is taken over all i_k, $1 \leqq i_k \leqq n-1$; $k = 1, \ldots, m$. Now, for fixed $t \neq 0$ and arbitrary $\varepsilon > 0$, we choose m so large that $[\max_i |\beta_i|]^m < \eta/|t|$, where η is such that $|\phi(x) - \phi(0)| < \varepsilon$ for $|x| < \eta$.

Then the absolute value of the right-hand member of (1.5.37) is less than ε, so that $|\phi(t) - \phi(0)| < \varepsilon$. In view of the arbitrariness of ε, we conclude that $\phi(t) = \phi(0) = a$, whence $\psi(t) = c + at$.▌

1.6 A LEMMA FROM THE THEORY OF RANDOM PROCESSES

We present below a lemma due to Sudakov on Markov processes, established by him in [143]; it will be used in Chapter 12. We follow the presentation of [143].

We consider stochastically continuous random processes

$$x = \xi(\omega, t), \qquad t \in [0, \infty), \qquad \omega \in (\Omega, \mu), \qquad x \in X \subset R^m$$

and the family of such processes generated by the family of measures $\{\mu_\theta\}$, $\theta \in \Theta$. We shall take them to be absolutely continuous with respect to some measure μ_0 on any σ-algebra $\sum_T$ generated by the functions $\xi(\omega, t)$ for $t \in [0, T]$, $T < \infty$ (each of the measures μ_θ is defined on the σ-algebra $\sum_\infty$ generated by all the functions $\xi(\omega, t)$, $t \in [0, \infty)$).

As usual, a function $\tau(\omega)$ such that for any t we have $\{\omega\colon \tau(\omega) \leqq t\} \in \sum_t$ is called a Markovian time. For the determination of $\tau(\omega)$ it is necessary to know only $\xi(\omega, t)$, and not μ_θ. An important class of Markovian times is that of first passage (entry) times; we shall consider only the latter in Chapter 12, where Sudakov's lemma will be applied. Such a Markovian time is defined as the time of first entry of a realization of the process into a given, fixed subset S of $[0, \infty) \times X$.

If S is closed and realizations of the process ξ are right-continuous, then the Markovian time

$$\tau(\omega) = \sup \{t'\colon (t, \xi(\omega, t)) \notin S \text{ for } 0 \leqq t < t'\}$$

will be the first passage time into S;

$$(\tau(\omega), \xi(\omega, \tau(\omega))) \in S, \qquad \text{and,} \qquad \text{for } t < \tau(\omega), (t, \xi(\omega, t) \notin S.$$

Without additional restrictions on the process and the set S, it is in general not meaningful to speak of the Markovian time of first passage into S.

Let a process $\xi(\omega, t)$ and a Markovian time $\tau(\omega)$ be given. Let us consider the Borel σ-algebra on $[0, \infty)$ and let the function $\xi(\omega, t)$ be measurable with respect to all its arguments. Then the function $\xi(\omega, \tau(\omega))$ is measurable. Indeed, $\xi(\omega, t)$ as a function of t alone is right-continuous, and for almost all ω, $\xi(\omega, \tau(\omega)) = \lim \xi(\omega, \tau_n(\omega))$, where $\{\tau_n(\omega)\}$ is a sequence of finite-valued functions such that for almost all ω we have $\tau_n(\omega) \downarrow \tau(\omega)$. Further, the functions $\xi(\omega, \tau_n(\omega))$ are measurable, and hence so is the limit function $\xi(\omega, \tau(\omega))$. The Markovian time $\tau(\omega)$ and the stopping point $\xi(\omega, \tau(\omega))$, which are measurable functions, generate a measure m on the space $[0, \infty) \times X$, obtained by the usual principle. For a given mapping:

$$s: \omega \to (\tau(\omega), \xi(\omega, \tau(\omega)))$$

we define for any Borel set $A \subset [0, \infty) \times X$, $m(A) = \mu(s^{-1}A)$. If $\tau(\omega)$ is the Markovian time of first entry into some closed set S, then the support of the measure m is contained in that set.

If we have two random processes, corresponding to two measures μ_0 and μ on Ω, where μ is absolutely continuous with respect to μ_0, then the measure m corresponding to μ is also absolutely continuous with respect to m_0, the measure corresponding to μ_0. In what follows, we shall compute the Radon-Nikodym derivative dm/dm_0 under certain natural assumptions.

We shall say that the last (latest) time defines a sufficient statistic for the processes $(\xi(\omega, t), \Omega, \mu_0)$ and $(\xi(\omega, t), \Omega, \mu_1)$ if for any $t_1, \ldots, t_n$ with $0 \leqq t_1 < \cdots < t_n$ the joint probability distributions $\mu_0(t_1, \ldots, t_n)$ and $\mu_1(t_1, \ldots, t_n)$ in X^n, of the values of the processes at these n points of time are such that the conditional measure with respect to the partition defined by the values of $x(t_n)$ is the same for both the measures and is determined solely by the choice of the elements of the partition, i.e., by the values of $x(t_n)$. In the cases which will be considered in Chapter 12 and also in more general ones, it is possible to speak of the class of conditional processes defined by the value $x(t)$ at a fixed time t; then the requirement that the last time t define a sufficient statistic means that the family of conditional processes on the segment $[0, t]$ defined by the value $x(t)$ is one and the same for each of the processes considered.

LEMMA 1.6.1. Let a family of random processes

$$x = \xi(\omega, t), \qquad t \in [0, \infty), \qquad \omega \in (\Omega, \mu_\theta), \qquad \theta \in \Theta$$

be given, where, for any $T < \infty$, all the measures μ_θ are absolutely continuous with respect to some measure μ_{θ_0} on the σ-algebra $\sum_T$ generated by the functions $\xi(\omega, t)$, $t \in [0, T]$, $\xi \in X \subset R^m$. Let the function $\xi(\omega, t)$ be right-continuous in t for almost all $\omega[\mu_{\theta_0}]$. Let $p_{\theta, \theta_0}(t, x)$ be the density of the distribution of the values of the process at time t corresponding to measure μ_θ with respect to the distribution of the same corresponding to measure μ_{θ_0}. Moreover, let this density be continuous in both its arguments throughout the set $[0, \infty) \times X$. Let this family of processes be such that the last time defines a sufficient statistic. If $\tau(\omega)$ be a Markovian time, then the measures m_θ, $\theta \in \Theta$, generated on the set $[0, \infty) \times X$ by the processes $(\xi(\omega, t), \Omega, \mu_\theta)$ are absolutely continuous with respect to the measure m_{θ_0}, and $dm_\theta/dm_{\theta_0} = p_{\theta, \theta_0}(t, x)$.

Proof: The existence of the densities $p_{\theta, \theta_0}(t, x)$ is guaranteed by the absolute continuity of the processes on any finite interval. We consider a sequence of stopping times $\tau_n(\omega)$ defined as follows: let $t_k^{(n)} = k/2^n$, $k, n = 1, 2, \ldots$. We set $\tau_n = -2^{-n}[-2^n\tau(\omega)]$, $[z]$ being as usual the integral part of z, i.e., $\tau_n(\omega)$ is the smallest of the numbers $t_k^{(n)}$, $k = 1, 2, \ldots$ which is not smaller than $\tau(\omega)$. The function $\xi(\omega, \tau_n(\omega))$ is measurable since its restriction to each of an at most denumerable number of subsets on which $\tau_n(\omega)$ is constant is measurable. $\tau_n(\omega) \downarrow \tau(\omega)$ obviously, and from the assumption that the realizations are right-continuous it follows that for almost all $[\mu_{\theta_0}]$ ω, the relation

$$\lim_{n\to\infty} \xi(\omega, \tau_n(\omega)) = \xi(\omega, \tau(\omega))$$

is satisfied, proving the measurability of the function $\xi(\omega, \tau(\omega))$.

We first show that the theorem holds if $\tau(\omega)$ is replaced by $\tau_n(\omega)$; it is easy to check that τ_n are Markovian times. The measures $m_\theta^{(n)}$, generated by the times $\tau_n(\omega)$, are concentrated on the subsets of the set $[0, \infty) \times X$ given by the conditions $t = t_k^{(n)}$, $k = 1, 2, \ldots$. We fix any one of the points $t_k^{(n)}$ and consider the restriction $m_{k,\theta}^{(n)}$ of the measure $m_\theta^{(n)}$ on the subset $t = t_k^{(n)}$.

Let $Y_k^{(n)} = \{\omega\colon \tau_n(\omega) = t_k^{(n)}\}$. Consider in Ω the σ-algebra $\sum t_k^{(n)}$, generated by the functions $\xi(\omega, t)$, $0 \leqq t \leqq t_k^{(n)}$. Obviously, in view of the Markovian property, the set $Y_k^{(n)}$ is measurable with respect to that σ-algebra. On the same σ-algebra, the measure μ_θ is absolutely continuous with respect to the measure μ_{θ_0}; moreover, from the fact that the last time is a sufficient statistic, it follows that the density of μ_θ relative to μ_{θ_0} depends only on $(t_k^{(n)})$ and hence is equal to $p_{\theta,\theta_0}(t_k^{(n)}, \xi(\omega, t_k^{(n)}))$. The measure $m_{k,\theta_0}^{(n)}$ is the image of the restriction $(\mu_{\theta_0} | Y_k^{(n)})$, of the measure μ_{θ_0} to the set $Y_k^{(n)}$, under the mapping

$$\pi\colon \omega \to (t_k^{(n)}, \xi(\omega, t_k^{(n)})) \in [0, \infty) \times X$$

and similarly for $m_{k,\theta}^{(n)}$. Hence it follows that the relation below is valid for any measurable A.

$$m_{k,\theta_0}^{(n)}(A) = \mu_{\theta_0}(\pi^{-1}A \cap Y_k^{(n)}) = (\mu_{\theta_0} | Y_k^{(n)})(\pi^{-1}A)$$

$$m_{k,\theta}^{(n)}(A) = (\mu_\theta | Y_k^{(n)})(\pi^{-1}A) = \int_{\pi^{-1}A} p_{\theta,\theta_0}(t_k^{(n)}, \xi(\omega, t_k^{(n)}))\, d\mu_{\theta_0}$$

$$= \int_A p_{\theta,\theta_0}(t_k^{(n)}, x)\, dm_{k,\theta_0}^{(n)}.$$

It hence follows that $(dm_{k,\theta}^{(n)})/(dm_{k,\theta_0}^{(n)}) = p_{\theta,\theta_0}(t_k^{(n)}, x)$ and hence $dm_\theta^{(n)}/dm_{\theta_0}^{(n)} = p_{\theta,\theta_0}(t, x)$, which proves the theorem for the case when the function $\tau_n(\omega)$ is considered in place of $\tau(\omega)$.

To conclude the proof, we note that from the almost everywhere convergence of $\xi(\omega, \tau_n(\omega))$ to $\xi(\omega, \tau(\omega))$ it follows that the measures $m_\theta^{(n)}$ generated by the times $\tau_n(\omega)$ converge to a measure m_θ in the sense of weak convergence of measures on $[0, \infty) \times X \subset R^{m+1}$, for every fixed $\theta \in \Theta$. We then verify the following assertion.

Let $\{m_n\}$ be a sequence of measures on R^m, converging weakly to a measure m, and $p(t, x)$ a nonnegative continuous function. Then the sequence $\{m_n'\}$ of measures having the same density $p(t, x)$ with respect to the corresponding members of the sequence $\{m_n\}$ (p is independent of n), converge weakly to a measure m' having the same density $p(t, x)$ with respect to m. This assertion follows immediately from the definition of weak convergence: for any continuous function $f(t, x)$ with compact support, we have

$$\int f(t, x)\, dm_n' = \int p(t, x) f(t, x)\, dm_n \to \int p(t, x) f(t, x)\, dm$$
$$= \int f(t, x)\, dm'.$$

This concludes the proof.∎

Examples of processes of the type indicated in the formulation of Lemma 1.6.1 will be given in Chapter 12.

1.7 SOME MISCELLANEOUS RESULTS

LEMMA 1.7.1. Let $a_1, \ldots, a_k$ be positive numbers, and a their maximum. Let $\gamma_1, \ldots, \gamma_k$ be another sequence of positive numbers. Then

$$\lim_{r\to\infty} \left(\sum_1^k \gamma_j a_j^r\right)^{1/r} = a. \tag{1.7.1}$$

Proof: Let $a = a_m$ (i.e., the maximum corresponds to the suffix m), and let $b_j = a_j/a_m$. Then, since $b_j \leqq 1$ for all j,

$$\sum \gamma_j a_j^r = a_m^r\left(\sum \gamma_j b_j^r\right) \leqq a_m^r \sum \gamma_j.$$

Therefore,

$$\overline{\lim_{r\to\infty}} \left(\sum \gamma_j a_j^r\right)^{1/r} \leqq a_m. \tag{1.7.2}$$

Further,

$$\sum \gamma_j a_j^r \geqq a_m^r \gamma_m.$$

Therefore,

$$\underline{\lim_{r\to\infty}} \left(\sum \gamma_j a_j^r\right)^{1/r} \geqq a_m. \tag{1.7.3}$$

Equation (1.7.1) follows from (1.7.2) and (1.7.3).∎

LEMMA 1.7.2. Let $\{a_n\}$ and $\{\gamma_n\}$ be sequences of positive numbers such that the series $\sum \gamma_n a_n^s$ converges for some $s > 0$, and $a = \sup\{a_n\}$. Then the series $\sum \gamma_n a_n^r$ converges for $r > s$ and

$$\lim_{r\to\infty} \left(\sum \gamma_n a_n^r\right)^{1/r} = a. \tag{1.7.4}$$

Proof: It is easily seen that ($b_n = a_n/a$ as above)

$$\begin{aligned}\sum \gamma_n a_n^r &= a^r\left(\sum \gamma_n b_n^r\right) \leqq a^r\left(\sum \gamma_n b_b^s\right)\\ &= a^{r-s}\left(\sum \gamma_n a_n^s\right) = a^{r-s}b\end{aligned}$$

where $b = \sum \gamma_n a_n^s$. Hence the series $\sum \gamma_n a_n^r$ converges and

$$\overline{\lim_{r\to\infty}} \left(\sum \gamma_n a_n^r\right)^{1/r} \leqq a. \tag{1.7.5}$$

Also, for every n,

$$\left(\sum \gamma_j a_j^r\right)^{1/r} \geqq (a_n^r \gamma_n)^{1/r} = a_n \gamma_n^{1/r}$$

so that

$$\underline{\lim_{r\to\infty}} \left(\sum \gamma_j a_j^r\right)^{1/r} \geqq a_n$$

for every n, and hence

$$\underline{\lim_{r\to\infty}} \left(\sum \gamma_j a_j^r\right)^{1/r} \geqq a. \tag{1.7.6}$$

Then (1.7.4) follows from (1.7.5) and (1.7.6). ∎

LEMMA 1.7.3. Let g be a nonnegative, and h a positive, function defined and continuous on $[a, b]$. Set

$$A_r = \int_a^b h(t)[g(t)]^r \, dt. \tag{1.7.7}$$

Then,

$$\lim_{r\to\infty} A_r^{1/r} \text{ exists and } = g_0 = \max_{a \leqq t \leqq b} g(t). \tag{1.7.8}$$

Proof: It is easily seen that

$$A_r^{1/r} \leqq g_0 \left[\int_a^b h(t)\, dt\right]^{1/r}$$

whence

$$\overline{\lim_{r\to\infty}} A_r^{1/r} \leqq g_0. \tag{1.7.9}$$

Let ε be an arbitrary fixed positive number. Let us denote by I the interval in which $g(t) \geqq g_0 - \varepsilon$. Then

$$A_r^{1/r} \geqq (g_0 - \varepsilon)\left[\int_I h(t)\, dt\right]^{1/r}$$

whence

$$\underline{\lim_{r\to\infty}} A_r^{1/r} \geqq g_0 - \varepsilon. \tag{1.7.10}$$

$\varepsilon > 0$ being arbitrary, (1.7.8) follows from (1.7.9) and (1.7.10).

Finally, we shall also require the following inequality of Jensen's concerning conditional mathematical expectations (cf. for instance [25], p. 37).▮

LEMMA 1.7.4. Let ξ be a random variable given on a probability space $(\Omega, \mathscr{A}, P)$, f a convex function on R^1, $\mathscr{B}$ a subalgebra of $\mathscr{A}$. Then, if $E|f(\xi)| < \infty$, the following relation holds with probability one.

$$f\{E(\xi \mid \mathscr{B})\} \leqq E\{f(\xi) \mid \mathscr{B}\}$$

In particular,

$$Ef\{E(\xi \mid \mathscr{B})\} \leqq Ef(\xi).$$

CHAPTER 2

Identically Distributed Linear Statistics Based on a Random Sample

2.1 STATEMENT OF THE PROBLEM

The simplest types of statistics, admitting a fairly complete study of the phenomenon of identical distribution, are linear statistics based on a random sample.

The earliest work in this direction appears to be the note [114] of G. Polya's in the year 1923, where it was established that only the normal law admits identically distributed linear statistics X_1 and $a_1 X_1 + a_2 X_2$ from a random sample (X_1, X_2).

The note [101] of I. Marcinkiewicz's was published in 1938; it was established there that laws having moments of all orders and admitting the existence of a nontrivial pair of identically distributed linear statistics based on a random sample are normal. In 1953 appeared the work [81] of Yu. V. Linnik, in which a description was given of the class of symmetric laws admitting identically distributed linear statistics, the problem of characterizing the normal law through such statistics was studied in detail, and some applications were given to the theory of testing of hypotheses.

The possibility of a series of applications to other problems of statistics and to statistical physics appeared in [79]. A series of important results connected with linear models, and characterizations of new types of laws were obtained by Rao [129]. Important applications to the theory of limit theorems were given recently by Zinger [34], [35], and [37]. Again, in [37], Zinger developed a method of studying the phenomenon of identical distribution of linear statistics based on a random sample, which is considerably simpler than that of [81]; it is his presentation which we shall follow below.

Let $\mathbf{X}^{(r)} = (X_1, \ldots, X_r)$ be a random vector in r-dimensional Euclidean space, with its components independent and identically distributed according to the distribution law F.

We set up two real linear statistics

$$L_1 = \sum_1^r a_j X_j \quad \text{and} \quad L_2 = \sum_1^r b_j X_j \tag{2.1.1}$$

and consider the family of all laws F for which L_1 and L_2 are identically distributed ($L_1 \cong L_2$). Following [81], we shall denote the family of all such laws by Ω_{L_1, L_2}.

If now we introduce the characteristic function (c.f.) f of F, then the condition $L_1 \cong L_2$ can be rewritten in the equivalent form*

$$\prod_1^r f(a_j t) = \prod_1^r f(b_j t), \qquad t \text{ real.} \tag{2.1.2}$$

Since $f(t) \neq 0$ in some interval $|t| < t'$, we can consider the function $w(t) = \ln f(t)$. For definiteness, we choose that continuous branch of the logarithmic function which takes real values for positive values of the argument.

If t'' be such that $t'' \max \{|a_j|, |b_j|\} \leqq t'$, then we have from (2.1.2) that

$$\sum_1^r w(a_j t) = \sum_1^r w(b_j t), \qquad |t| < t''. \tag{2.1.3}$$

Separating the real and imaginary parts in (2.1.3), we can split it up into two equations:

$$\sum_1^r \phi(|a_j| t) = \sum_1^r \phi(|b_j| t) \tag{2.1.4}$$

$$\sum_1^r \text{sign } a_j \psi(|a_j| t) = \sum_1^r \text{sign } b_j \psi(|b_j| t) \tag{2.1.5}$$

where $\phi(t) = -\ln |f(t)|$, and $\psi(t) = \arg f(t)$.

The problem of constant regression of one linear statistic on another, considered by Ramachandran and Rao ([122], also see Chapter 5), leads to an equation similar to (2.1.2), of the form

$$\prod_1^r f^{\lambda_j}(a_j t) = \prod_1^r f^{\mu_j}(b_j t) \tag{2.1.6}$$

where the λ_j and μ_j are some positive numbers. Then we can obtain from (2.1.6) an equation similar in form to (2.1.3), valid in some neighborhood of the origin.

* We shall freely use the simple basic properties of characteristic functions (cf. for instance [70], p. 33), without explicitly stating them.

Generalizing such equations, we set up an equation of the form

$$\int_{-a'}^{a'} w(at)\, dV_{\mathrm{I}}(a) = \int_{-a'}^{a'} w(at)\, dV_{\mathrm{II}}(a) \tag{2.1.7}$$

where $V_{\mathrm{I}}(a)$ and $V_{\mathrm{II}}(a)$ are bounded nondecreasing functions given on $(-\bar{a}, \bar{a})$ where $0 < a' < \bar{a}$. We shall consider (2.1.7) only for complex-valued functions $w(t)$ which are continuous in some neighborhood of the origin, since we shall be interested (only) in those solutions of (2.1.7) which are representable for $|t| < t'$ in the form

$$w(t) = \ln f(t) \tag{2.1.8}$$

where f is the c.f. of some nondegenerate law F. We shall call such solutions "probabilistic." In this case, as in the case of (2.1.3), (2.1.7) can be split up into a pair of equations

$$\int_{[0, a']} \phi(at)\, dV_1(a) = \int_{[0, a']} \phi(at)\, dV_2(a), \; 0 < t < t' \tag{2.1.9}$$

$$\int_{[0, a']} \psi(at)\, dV_3(a) = \int_{[0, a']} \psi(at)\, dV_4(a) \tag{2.1.10}$$

where $\phi(t) = -\ln |f(t)|$ and $\psi(t) = \arg f(t)$. For negative values of t, these solutions of (2.1.9) and (2.1.10) are extended respectively as even and odd functions.

In (2.1.9) and (2.1.10), $V_j(a)$, $j = 1, 2, 3, 4$, can be taken to be bounded and nondecreasing, and in what follows we shall take them to be such without explicitly stating it.

Let F be a nondegenerate probability law on the real line, and f, its characteristic function, be such that for $0 \leqq t < t'$, $\phi(t) = -\ln f(t)$ is defined, and for $0 \leqq t < t'/a'$ relation (2.1.9) (assumed nondegenerate, i.e., $V_1 - V_2 \not\equiv$ constant) is satisfied, with

$$\int_{[0, a']} a^{-\delta}\, d[V_1(a) + V_2(a)] < \infty \tag{2.1.11}$$

for some $\delta > 0$. We shall denote the family of such laws F by $\mathscr{F}(V_1, V_2, \delta, t')$.

If in addition to the above, a nondegenerate equation of the form (2.1.10) is satisfied for $0 \leqq t < t'/a'$ by $\psi(t) = \arg f(t)$, where V_3 and V_4 satisfy a condition of the form (2.1.11), and further (2.1.10) together with (2.1.9) is derived from (2.1.7), then we shall denote the corresponding family of laws F by $\mathscr{F}_1(V_{\mathrm{I}}, V_{\mathrm{II}}, \delta, t')$.

Equation (2.1.9) can be considered independently of (2.1.7) but whenever they agree $\mathscr{F}_1 \subset \mathscr{F}$.

To study these families, we shall utilize certain expansions for the solutions of the equations considered. These expansions are developed in Section 2.2. In Sections 2.3 and especially 2.4, the properties of such expansions for probabilistic solutions will be studied. The results obtained will be used in Section 2.4 for characterizing the normal law. Some multivariate generalizations are given in Section 2.5, mainly toward the same purpose.

For characterizing the normal law, the case where $f(t) \neq 0$ for any real t is important. In this case, $w(t) = \ln f(t)$ is defined throughout the real line and we can, for instance, discuss (2.1.3) for $0 \leqq t < \infty$. In [81], it was established that such a situation occurs for (2.1.2) under the supplementary condition

$$\max_j |a_j| \neq \max_j |b_j|. \tag{2.1.12}$$

For (2.1.9), a preliminary result of a similar nature can be obtained in the following form.

LEMMA 2.1.1. Let the function g be given and continuous on $[0, \infty)$, take values in $[0, 1]$ and $g(0) = 1$. If

(1) for every $c > 0$ such that $\gamma(t) = \ln g(t)$ exists in $[0, c)$, (2.1.9) and (2.1.11) are satisfied for $0 \leqq t < c/a'$; and

(2) $V_2(a)$ has a positive jump at the point a' and it is possible to indicate an a_1 in $(0, a')$ such that

$$\int_{a_1}^{a'} dV_1(a) = 0. \tag{2.1.13}$$

Then for all $0 \leqq t < \infty$, $g(t) \neq 0$, $\gamma(t)$ exists, is continuous and satisfies (2.1.9); and as $t \to \infty$

$$\int_0^t \gamma(u)\, du = O(t^A) \tag{2.1.14}$$

where $A > 0$ is a suitable constant.

Proof: Suppose g has zeros, and let $c_0 = \inf\{t: g(t) = 0\}$. Then c_0 must obviously be itself a zero and hence $c_0 > 0$. Taking (2.1.9) as $t \to (c_0/a') - 0$ and considering condition (2) into account, we easily arrive at a contradiction. Having established that $\gamma(t)$ exists for all $0 \leqq t < \infty$ and satisfies (2.1.9), we may consider

$$\gamma_1(t) = -\int_0^t t^{\varepsilon - 1}\gamma(t)\, dt \tag{2.1.15}$$

where $0 < \varepsilon < \min(1, \delta)$. In view of condition (2), we immediately obtain from (2.1.9)

$$\gamma_1(a't) \leqq c_1\gamma_1(a_1 t). \tag{2.1.16}$$

Since $a' > a_1$ and $\gamma_1(t)$ is nondecreasing, we may conclude from (2.1.16), using the usual elementary arguments for such situations (c.f. for example [81]), that as $t \to \infty$

$$\gamma_1(t) = O(t^{A_1}) \tag{2.1.17}$$

where $A_1 > 0$ is a suitable constant.

Noting that it follows from (2.1.15) that

$$-\int_0^t \gamma(u)\,du \leqq t^{1-\varepsilon}\gamma_1(t) \tag{2.1.18}$$

and combining (2.1.17) and (2.1.18), we obtain the required (2.1.14).∎

2.2 THE BASIC EXPANSION

Let us turn to the study of continuous solutions of equations of the type (2.1.9). In order to avoid making trivial "exceptions," we shall assume that the equations under consideration are nondegenerate, i.e., $V_1(a) - V_2(a) \not\equiv$ constant, and the solutions under study are not identically zero.

THEOREM 2.2.1. Let ϕ be real-valued and continuous on $[0, t')$ and for $0 \leqq t < t'/a'$, the equation (2.1.9) together with (2.1.11) be satisfied. Then, for any positive Λ, γ, ε, the following expansion holds:

$$\int_0^t u^{\Lambda-1} \ln(t/u)\phi(u)\,du = \pi(t) + \sum_{n=0}^{\infty} S_n(t) + R(t) \tag{2.2.1}$$

for $0 < t < t'$. Here

$$S_n(t) = t^{\Lambda} \sum_{k=0}^{k_n} \{t^{\xi_{nk}} P_{\xi_{nk}}(\ln t) + \overline{t^{\xi_{nk}} P_{\xi_{nk}}(\ln t)}\} \tag{2.2.2}$$

where the complex numbers $\xi_{nk}(k = 0, 1, \ldots, k_n)$ are zeros of the function $\sigma(-z)$,

$$\sigma(z) = \int_0^{\infty} e^{z\alpha}\, d\{V_1(a'e^{-\alpha}) - V_2(a'e^{-\alpha})\} \tag{2.2.3}$$

situated in the rectangle $\{0 \leqq \operatorname{Re} z \leqq \gamma + \varepsilon,\ y_n < \operatorname{Im} z < y_{n+1}\}$ where $y_0 = 0$ and, for $n \geqq 0$, y_{n+1} is chosen suitably from the interval $(n, n+1)$; $P_{\xi_{nk}}$ is a polynomial of degree $m_{\xi_{nk}}$; also, the numbers k_n are bounded by some $\bar{k}$ independent of $n = 0, 1, 2, \ldots$.

$$\pi(t) = t^{\Lambda} \sum_{k=0}^{\bar{k}} t^{\gamma_k} P_{\gamma_k}(\ln t) \tag{2.2.4}$$

$\gamma_k(k = 0, 1, \ldots, \bar{\bar{k}})$ are the real zeros of $\sigma(-z)$ lying in $[0, \gamma]$, P_{γ_k} is a polynomial of degree $m_{\gamma k}(\bar{k}, m_{\gamma_k} \leqq \bar{k})$, and as $t \to 0$, $R(t)$ satisfies the relation

$$R(t) = O(t^{\gamma+\Lambda}). \tag{2.2.5}$$

The series in (2.2.1) converges absolutely and uniformly in every interval $0 < t \leqq t_1 < t'$.

Proof: For proving (2.2.1), we shall use the unilateral Laplace transform, as in [81].*

We start by effecting a change of variables in (2.1.9). Without loss of generality, we may take $a' = t' = 1$. We set

$$\chi(\tau) = \phi(e^{-\tau}); \qquad V_j(\alpha) = -V_j(e^{-\alpha}), \qquad j = 1, 2; \qquad 0 \leqq \tau, \qquad \alpha < \infty. \tag{2.2.6}$$

Then, (2.1.9) can be rewritten in the form

$$\int_0^\infty \chi(\tau + \alpha)\, dV_1(\alpha) = \int_0^\infty \chi(\tau + \alpha)\, dV_2(\alpha), \qquad \tau > 0. \tag{2.2.7}$$

As per the definition, $\chi(\tau)$ is a function continuous and bounded on $(0, \infty)$; $\chi(\tau) \to \phi(0)$ as $\tau \to \infty$; V_1 and V_2 are nondecreasing functions satisfying, according to (2.1.11), the condition

$$\int_0^\infty e^{\delta\alpha}\, d[V_1(\alpha) + V_2(\alpha)] < \infty \tag{2.2.8}$$

for some $\delta > 0$.

We apply the Laplace transform to (2.2.7). Then, for complex z lying in the half-plane $\operatorname{Re} z < \delta$,

$$\int_0^\infty dV_1(\alpha) \int_\alpha^\infty e^{-z(\tau-\alpha)}\chi(\tau)\, d\tau = \int_0^\infty dV_2(\alpha) \int_\alpha^\infty e^{-z(\tau-\alpha)}\chi(\tau)\, d\tau. \tag{2.2.9}$$

If z is chosen in the strip $0 < \operatorname{Re} z < \delta$, then, in view of (2.2.8), we may rewrite (2.2.9) in the form

$$\int_0^\infty e^{z\alpha}\, dV_1(\alpha) \int_0^\infty e^{-z\tau}\chi(\tau)\, d\tau - \int_0^\infty e^{z\alpha}\, dV_1(\alpha) \int_0^\alpha e^{-z\tau}\chi(\tau)\, d\tau$$

$$= \int_0^\infty e^{z\alpha}\, dV_2(\alpha) \int_0^\infty e^{-z\tau}\chi(\tau)\, d\tau - \int_0^\infty e^{z\alpha}\, dV_2(\alpha) \int_0^\alpha e^{-z\tau}\chi(\tau)\, d\tau. \tag{2.2.10}$$

We introduce the notation

$$\sigma(z) = \int_0^\infty e^{z\alpha}\, d\{V_2(\alpha) - V_1(\alpha)\}$$

$$\mathscr{X}(z) = \int_0^\infty e^{-z\tau}\chi(\tau)\, d\tau$$

$$K(z) = \int_0^\infty e^{z\alpha}\, dV_1(\alpha) \int_0^\alpha e^{-z\tau}\chi(\tau)\, d\tau - \int_0^\infty e^{z\alpha}\, dV_2(\alpha) \int_0^\alpha e^{-z\tau}\chi(\tau)\, d\tau. \tag{2.2.11}$$

* In view of the multiplicative character of the expressions in (2.1.9), it would have been natural to apply the Mellin transform (cf. [44]), but it is less convenient in computations.

In terms of these, (2.2.10) can be rewritten in the form

$$\sigma(z)\mathscr{X}(z) + K(z) = 0, \qquad 0 < \operatorname{Re} z < \delta. \tag{2.2.12}$$

From the definitions of the functions in (2.2.11), we can obtain some of their properties enumerated below.

(1) $\sigma(z)$ in the half-plane $\operatorname{Re} z < \delta$ represents an analytic function which is almost periodic on every vertical line $\operatorname{Re} z = \text{constant}$ lying in this half-plane.

(2) For any $\gamma > 0$ and $\varepsilon > 0$, we can indicate an $m(\gamma, \varepsilon)$ such that

$$|\sigma(z)| > m(\gamma, \varepsilon) \tag{2.2.13}$$

for z lying in the strip $-\gamma + \varepsilon \leqq \operatorname{Re} z \leqq \delta - \varepsilon$, but outside discs of radius ε with centres at the zeros of $\sigma(z)$.

(3) For any positive γ and ε, the number of zeros of $\sigma(z)$ lying in the rectangle $\{-\gamma + \varepsilon \leqq \operatorname{Re} z \leqq \delta - \varepsilon, y < \operatorname{Im} z < y + 1\}$ is bounded by a number $k(\gamma, \varepsilon)$ independent of y.

(4) $K(z)$ represents a function analytic in the half-plane $\operatorname{Re} z < \delta$ and bounded in every half-plane $\operatorname{Re} z \leqq \gamma' < \delta$.

(5) $\mathscr{X}(z)$ is analytic in the half-plane $\operatorname{Re} z > 0$ and satisfies there the inequality

$$|\mathscr{X}(z)| \leqq c_2/(\operatorname{Re} z) \tag{2.2.14}$$

and, in the strip $0 < \operatorname{Re} z < \delta$, admits the representation

$$\mathscr{X}(z) = -K(z)/\sigma(z). \tag{2.2.15}$$

Property (1) follows immediately from the definition of $\sigma(z)$ in view of condition (2.1.11) and the definition of almost periodic functions (cf. for instance [76], p. 342).

Properties (2) and (3) are immediate consequences of the results on almost periodic functions analytic in a strip. We cite these properties here without proof in the two lemmas below (cf. [76], p. 347).

LEMMA 2.2.1. Let $f(z)$, $z = x + iy$, be an analytic almost periodic function (not identically zero) in the strip $|y| < \text{h}$. Then, to any $\delta > 0$, there corresponds an $m(\delta) > 0$ such that $|f(z)| \geqq m(\delta)$ for z lying in the strip $|y| \leqq h - \delta$ and outside discs of radius δ with centers at the zeros of $f(z)$.

LEMMA 2.2.2. The number of zeros lying in the rectangle $t < x < t + 1$, $|y| \leqq h - \delta$, of a function which is analytic and almost periodic in the strip $|y| < h$, is bounded by a number $k(\delta)$ not depending on t.

For the proof of property (5), it suffices to note that the analyticity of $\mathscr{X}(z)$ and estimate (2.2.14) follow from the definition in view of the boundedness of $\chi(\tau)$, and the validity of representation (2.2.15) follows from (2.2.12).

We now turn to the derivation of the expansion in terms of the zeros of $\sigma(z)$ for the solution of (2.1.9).

We shall use here the inversion formula for the Laplace transform (cf. for instance [24], p. 107), whose applicability in our case is obvious.

$$\int_0^\tau \chi(t)\,dt = \frac{1}{2\pi i}\,\text{P.V.}\int_{x'-i\infty}^{x'+i\infty} e^{\tau z}\,\frac{\mathscr{X}(z)}{z}\,dz, \qquad x'>0$$

where P.V. denotes the principal value of the integral.

Taking an arbitrary $\Lambda > 0$, we can also see that

$$\int_0^\tau e^{-\Lambda t}\chi(t)\,dt = \frac{1}{2\pi i}\,\text{P.V.}\int_{x'-i\infty}^{x'+i\infty} e^{\tau z}\,\frac{\mathscr{X}(z+\Lambda)}{z}\,dz, \qquad x' > -\Lambda \tag{2.2.16}$$

$$\int_\tau^\infty e^{-\Lambda t}\chi(t)\,dt = \frac{1}{2\pi i}\,\text{P.V.}\int_{x'-i\infty}^{x'+i\infty} e^{\tau z}\,\frac{\mathscr{X}(\Lambda)-\mathscr{X}(\Lambda+z)}{z}\,dz, \qquad x' > -\Lambda$$

$$\int_\tau^\infty e^{-\Lambda t}\chi(t)\,dt = -\frac{1}{2\pi i}\,\text{P.V.}\int_{x'-i\infty}^{x'-i\infty} e^{\tau(z-\Lambda)}\,\frac{\mathscr{X}(z)}{z-\Lambda}\,dz, \qquad 0 < x' < \Lambda.$$

If here we subject x' also to the condition $x' < \delta$, then, according to (2.2.15), we have

$$\int_\tau^\infty e^{-\Lambda t}\chi(t)\,dt = \frac{1}{2\pi i}\,\text{P.V.}\int_{x'-i\infty}^{x'+i\infty} e^{\tau(z-\Lambda)}\,\frac{K(z)}{(z-\Lambda)\sigma(z)}\,dz \tag{2.2.17}$$

$$0 < x' < \min(\delta, \Lambda).$$

Similarly, we can obtain the relation

$$\int_\tau^\infty (v-\tau)e^{-\Lambda v}\chi(v)\,dv = \frac{1}{2\pi i}\int_{x'-i\infty}^{x'+i\infty} e^{\tau(z-\Lambda)}\,\frac{K(z)}{(z-\Lambda)^2\sigma(z)}\,dz$$

$$0 < x' < \min(\delta, \Lambda). \tag{2.2.18}$$

We note that the integral on the right side of (2.2.18) converges absolutely.

The singularities of the integrand on the right side of (2.2.18) are located at those zeros of $\sigma(z)$ which, according to (2.2.15), lie in the half-plane $\operatorname{Re} z \leqq 0$.

We take arbitrary $\gamma > 0$, $\varepsilon > 0$, and an arbitrarily small ε'; $0 < \varepsilon' < \varepsilon$. Consider the strip $-\gamma - \varepsilon' < \operatorname{Re} z < \varepsilon'$. By property (3), in every rectangle $\{-\gamma-\varepsilon' < \operatorname{Re} z < \varepsilon',\ n < \operatorname{Im} z < n+1\}$, $n = 0, 1, \ldots,$ the number of zeros of $\sigma(z)$ counting multiplicities is bounded by a number $m(\gamma, \varepsilon')$ not depending on n. We can indicate a small enough $\varepsilon'' > 0$ which permits, for $n = 1, 2, \ldots,$ the choice of a y_n in $(n-1, n)$ such that all points of the interval $\{\operatorname{Im} z = y_n,\ -\gamma-\varepsilon' < \operatorname{Re} z < \varepsilon'\}$ are at a distance of at least ε'' from any of the zeros of $\sigma(z)$. We further set $y_0 = 0$.

Similarly, we shall choose $x_n \in (\gamma, \gamma + \varepsilon')$, $\varepsilon_n \in (0, \varepsilon')$ such that the segments $\{\operatorname{Re} z = x_n,\ y_n \leqq \operatorname{Im} z < y_{n+1}\}$ and $\{\operatorname{Re} z = \varepsilon_n,\ y_n \leqq \operatorname{Im} z < y_{n+1}\}$ do not contain points situated at a distance of $\leqq \varepsilon'$ from the zeros of $\sigma(z)$.

Let us denote by l_n the rectangular contour

$$\{-x_n < \operatorname{Re} z < \varepsilon_n,\ y_n < \operatorname{Im} z < y_{n+1}\}, \qquad n = 1, 2, \ldots$$

and by l_{-n} the mirror reflection in the real axis of the contour l_n. We denote by l the rectangular contour

$$\{-x_0 < \operatorname{Re} z < \varepsilon',\ -y_0 < \operatorname{Im} z < y_0\}.$$

Since the complex zeros of $\sigma(z)$ occur in conjugate pairs, all the points of the contours $l_{\pm n}$, $n = 1, 2, \ldots$ are situated, by our construction, at a distance of at least ε'' from the zeros of $\sigma(z)$.

We rewrite (2.2.18) in the form

$$\int_\tau^\infty (v - \tau)e^{-\Lambda v}\chi(v)\,dv = \frac{1}{2\pi i}\lim_{n\to\infty}\int_{x'-iy_{n+1}}^{x'+iy_{n+1}} e^{\tau(z-\Lambda)}\frac{K(z)}{(z-\Lambda)^2\sigma(z)}\,dz,$$

$$0 < x' < \min(\varepsilon', \delta, \Lambda). \tag{2.2.19}$$

$$\frac{1}{2\pi i}\int_{x'-iy_{n+1}}^{x'+iy_{n+1}} e^{\tau(z-\Lambda)}\frac{K(z)}{(z-\Lambda)^2\sigma(z)}\,dz = S(\tau) + \sum_{j=1}^n S_j(\tau) + R_{1n}(\tau) + R_{2n}(\tau)$$

where we have used the notation

$$S_n(\tau) = \frac{1}{2\pi i}\oint_{l_n \cup l_{-n}} e^{\tau(z-\Lambda)}\frac{K(z)}{(z-\Lambda)^2\sigma(z)}\,dz \tag{2.2.20}$$

$$S(\tau) = \frac{1}{2\pi i}\oint_l e^{\tau(z-\Lambda)}\frac{K(z)}{(z-\Lambda)^2\sigma(z)}\,dz \tag{2.2.21}$$

$$R_{1n}(\tau) = \frac{1}{2\pi i}\left\{\int_{-x_n}^{\varepsilon_n} e^{\tau(x-\Lambda+iy_{n+1})}\frac{K(x+iy_{n+1})}{(x-\Lambda+iy_{n+1})^2\sigma(x+iy_{n+1})}\,dx\right.$$
$$\left.+\int_{-x_n}^{\varepsilon_n} e^{\tau(x-\Lambda-iy_{n+1})}\frac{K(x-iy_{n+1})}{(x-\Lambda-iy_{n+1})^2\sigma(x-iy_{n+1})}\,dx\right\} \tag{2.2.22}$$

$$R_{2n}(\tau) = \frac{1}{2\pi i}\int_{\hat{l}_n} e^{\tau(z-\Lambda)}\frac{K(z)}{(z-\Lambda)^2\sigma(z)}\,dz \tag{2.2.23}$$

where $\hat{l}_n$ is the broken line having its vertices at the sequence of points

$$-x_n - iy_{n+1},\ -x_n - iy_n,\ \ldots,\ -x_0 - iy_1,\ -x_0 + iy_1,\ \ldots,\ -x_n + iy_{n+1}.$$

We shall now find estimates for (2.2.20)–(2.2.23). Let $z \in l_n$. Since z is at a distance $\geqq \varepsilon''$ from all the zeros of $\sigma(z)$, we have $|\sigma(z)| > m(\gamma, \varepsilon'')$, in view of the properties of $\sigma(z)$—item (2) above. Also, by item (4) above, $|K(z)| < c_3$. Since $|z - \Lambda| > |n - \Lambda|$, and a similar estimate holds for $z \in l_{-n}$, we easily see that (using the inequality $x > A$)

$$|R_{1n}(\tau)| \leqq \frac{c_4}{(n-\Lambda)^2} e^{\tau(\varepsilon' - \Lambda)}, \tag{2.2.24}$$

$$|S_n(\tau)| \leqq \frac{c_5}{(n-\Lambda)^2} e^{\tau(\varepsilon' - \Lambda)}. \tag{2.2.25}$$

Since Re $z < -\gamma$ for $z \in l_n$, a similar argument gives

$$|R_{2n}(\tau)| \leqq c_6 e^{-\tau(\gamma + \Lambda)}. \tag{2.2.26}$$

The constants c_4, c_5, c_6 do not depend on n.

Further, it is obvious that $R_2(\tau) = \lim_{n\to\infty} R_{2n}(\tau)$ exists, the convergence being uniform in the interval $(0, \infty)$.

We now let $n \to \infty$ in (2.2.19), and find

$$\int_\tau^\infty (v - \tau)e^{-\Lambda v}\chi(v)\, dv = S(\tau) + \sum_{n=1}^{\infty} S_n(\tau) + R_2(\tau). \tag{2.2.27}$$

The series in (2.2.27) converges absolutely and uniformly in $(0, \infty)$, in view of (2.2.25). For $R_2(\tau)$, the following estimate holds

$$|R_2(\tau)| \leqq c_6 e^{-\tau(\gamma + \Lambda)}. \tag{2.2.28}$$

For evaluating $S_n(\tau)$, as is obvious from (2.2.20), it is necessary to obtain the residues of the integrand at the zeros of $\sigma(z)$ located inside the contour l_n. As we noted above (item (3) of the list of properties), the number of such zeros, counting multiplicities, is bounded.

If $z = \eta$ is any zero of $\sigma(z)$, with, say, multiplicity $k_\eta + 1$, then, evidently

$$\text{Res}\left\{e^{\tau(z-\Lambda)} \frac{K(z)}{(z-\Lambda)^2\sigma(z)}\right\}_{z=\eta} = P_{-\eta}(\tau)e^{\tau(\eta - \Lambda)}, \tag{2.2.29}$$

where $P_{-\eta}(\tau)$ is a polynomial of degree K_η; its coefficients clearly depend on Λ.

If $z_{n1}, \ldots, z_{nk_n}$ denote all the real zeros of $\sigma(z)$ which lie inside $l_n (n = 1, 2, \ldots)$, then, on the basis of (2.2.20), we can see that

$$S_n(\tau) = \sum_{\substack{k=1 \\ \text{Re}\, z_{nk} < 0}}^{k_n} \{P_{-z_{nk}}(\tau)e^{\tau(z_{nk} - \Lambda)} + \overline{P_{-z_{nk}}(\tau)e^{\tau(z_{nk} - \Lambda)}}\}. \tag{2.2.30}$$

For uniformity, we denote by l_0 the rectangular contour $\{-x_0 < \text{Re}\, z < \varepsilon_0;\ 0 < \text{Im}\, z < y_1\}$ and by $z_{01}, \ldots, z_{0k_0}$ the zeros of $\sigma(z)$ lying inside l_0.

Let an enumeration of the real zeros of $\sigma(z)$ lying in $(-x_0, 0]$ be: $-\gamma_1, \ldots, -\gamma_{\bar{m}(\gamma,\varepsilon)}$ so that

$$\gamma_j \in [0, \gamma], \qquad 1 \leqq j \leqq \bar{m}; \qquad \gamma_j \in (\gamma, x_0), \qquad \bar{m} < j \leqq \bar{m}(\gamma, \varepsilon).$$

Then we have

$$S(\tau) = S_0(\tau) + \pi(\tau) + \pi_1(\tau)$$

where

$$S_0(\tau) = \sum_{k=0}^{k_0} \{P_{-z_{0k}}(\tau)e^{\tau(z_{0k}-\Lambda)} + \overline{P_{-z_{0k}}(\tau)e^{\tau(z_{0k}-\Lambda)}}\} \tag{2.2.31}$$

$$\pi(\tau) = \sum_{j=1}^{\bar{m}} P_{\gamma_j}(\tau)e^{\tau(\gamma_j-\Lambda)}; \qquad \pi_1(\tau) = \sum_{j=\bar{m}+1}^{\bar{m}(\gamma,\varepsilon)} P_{\gamma_j}(\tau)e^{\tau(\gamma_j-\Lambda)}. \tag{2.2.32}$$

If we now return to the original notation in (2.2.27), taking (2.2.31) into account, and further setting

$$R(\tau) = R_2(\tau) + \pi_1(\tau) \tag{2.2.33}$$

we obtain the desired expansion.

For $r(t) = R(\ln t)$, (2.2.5) holds, according to (2.2.28) and (2.2.32). The estimates obtained in the process of the above argument guarantee the stated character of convergence.∎

We may note that, generally speaking, the requirement of continuity made on the solution $\phi(t)$ is stringent. For our purposes, however, relaxing this condition is not of importance since we are interested only in solutions of the type (2.1.8) continuous in some neighborhood of the origin.

In constructing the expansion (2.2.1), we need in fact take into account only those zeros of $\sigma(z)$ which correspond to nonzero residues in (2.2.29). These residues, as we noted above, in general depend on Λ. From (2.2.29) it is easily seen that the degree of the polynomial $P_{-\eta}(\tau)$ does not depend on Λ.

We make a definition (cf. [81]).

DEFINITION 2.2.1. Those zeros of $\sigma(z)$ which correspond to nonzero residues in formula (2.2.29) will be called *active*. A point z_0 will be called an *active exponent* if $-z_0$ is an active zero. The degree of the polynomial corresponding to it in (2.2.29) *plus* unity will be called the *multiplicity* of the active exponent.

Let us denote by S the set of all active exponents, by S' the set of all active complex exponents lying in the half-plane $\operatorname{Im} z > 0$. We also define

$$\sigma_1 = \inf\{\operatorname{Re} \xi : \xi \in S\}, \qquad \sigma_3 = \inf\{\operatorname{Re} \xi : \xi \in S'\}. \tag{2.2.34}$$

In case all the active exponents are located in a strip $0 \leqq \operatorname{Re} z \leqq \gamma_0$, we similarly define

$$\sigma_2 = \sup\{\operatorname{Re} \xi : \xi \in S\}, \qquad \sigma_4 = \sup\{\operatorname{Re} \xi : \xi \in S'\}. \tag{2.2.35}$$

In terms of the concepts introduced above, we can derive a corollary from the theorem proved above, which will be very useful in what follows.

Corollary. Under the conditions of Theorem 2.2.1,

$$\sigma_1 \geqq 0. \tag{2.2.36}$$

For any positive Λ, ε, as $t \to 0+$,

$$\int_0^t u^\Lambda \ln (t/u)\phi(u)\, du = O(t^{\sigma_1 + \Lambda - \varepsilon}), \tag{2.2.37}$$

and in the case when S' is not empty, as $t \to 0+$,

$$\int_0^t u^\Lambda \ln (t/u)\phi(u)\, du = \pi(t) + O(t^{\sigma_3 + \Lambda - \varepsilon}) \tag{2.2.38}$$

where $\pi(t)$ has the form (2.2.4) and contains terms corresponding to the real active exponents not exceeding σ_3.

We shall not pause to prove the above simple fact, but merely remark that (2.2.36) follows from the analyticity of $\mathscr{X}(z)$ in the half-plane $\operatorname{Re} z > 0$; also, when S' is not empty, taking $\gamma > \sigma_3$, the right walls of the contours $l_n (n = 0, 1, \ldots)$ can be chosen along the vertical lines $\operatorname{Re} z = -\sigma_3 + \varepsilon_n (0 < \varepsilon_n < \varepsilon')$, which evidently leads to the estimate as $t \to 0+$:

$$W(t) = \sum_{n=0}^{\infty} S_n(t) = O(t^{\sigma_3 + \Lambda - \varepsilon}).$$

In some cases, the representation (2.2.1) can be somewhat simplified. An expansion under conditions analogous to those of Lemma 2.1.1 can serve as an example.

THEOREM 2.2.2. Let the real-valued function $\phi(t)$ be defined and continuous on $[0, \infty)$ and satisfy the equations (2.1.9) and (2.1.11) for all $t \geqq 0$, and let, as $t \to \infty$,

$$\int_0^t |\phi(u)|\, du = O(t^A) \tag{2.2.39}$$

where $A > 0$. Then all the active exponents of $\phi(t)$ lie in the strip $0 \leqq \operatorname{Re} z \leqq A$ and for any $\Lambda > 0$ the following expansion holds:

$$\int_0^t u^{\Lambda - 1} \ln (t/u)\phi(u)\, du = \pi(t) + \sum_{n=0}^{\infty} \mathrm{S}_n(t); \qquad 0 < t < \infty \tag{2.2.40}$$

where $\pi(t)$ is defined in the same way as in Theorem 2.2.1, but the summation runs only over real exponents lying on $[0, A]$; the $S_n(t)$, $n = 0, 1, \ldots,$ are defined in the same way as in Theorem 2.2.1, with Λ being taken to be $(\gamma + \varepsilon)$. The series in (2.2.40) converges absolutely and uniformly in every interval $0 < t \leqq t_1$, where $0 < t_1 < \infty$.

Proof: In the case under consideration, $\chi(\tau) = \phi(e^{-\tau})$ satisfies (2.2.7) for all real τ.

For $\tau \geqq \tau_0$ (arbitrary but fixed), Theorem 2.2.1 can be applied. A similar construction can be effected for $\tau \leqq \tau_0$. For definiteness, we take $\tau_0 = 0$ and consider (2.2.7).

We set, formally at first,

$$\tilde{\mathscr{X}}(z) = \int_0^\infty e^{z\tau}\chi(-\tau)\,d\tau. \tag{2.2.41}$$

In view of condition (2.2.39), $\tilde{\mathscr{X}}(z)$ is defined and analytic in the half-plane $\operatorname{Re} z > A$, and using arguments and notations similar to those used in the derivation of (2.2.12), we obtain the relation

$$\sigma(-z)\tilde{\mathscr{X}}(z) - K(-z) = 0, \qquad \operatorname{Re} z > A. \tag{2.2.42}$$

It follows from (2.2.41) that $K(z)/\sigma(z)$ cannot have singularities in $\operatorname{Re} z < -A$, and this means that there are no active exponents in the half-plane $\operatorname{Re} z > A$.

Further, for $\gamma > A$, we have

$$\begin{aligned} R_{2n}(\tau) &= \int_{\hat{l}_n} e^{\tau(z-\Lambda)} \frac{K(z)}{(z-\Lambda)^2\sigma(z)}\,dz \\ &= \frac{1}{2\pi}\int_{-y_{n+1}}^{y_{n+1}} e^{\tau(-N+iy-\Lambda)} \frac{K(-N+iy)}{(-N+iy-\Lambda)^2\sigma(-N+iy)}\,dy \\ &\quad + \frac{1}{2\pi i}\int_{-N}^{-x_n} e^{\tau(x+iy_n-\Lambda)} \frac{K(x+iy_n)}{(x+iy_n-\Lambda)^2\sigma(x+iy_n)}\,dx \\ &\quad + \frac{1}{2\pi i}\int_{-x_n}^{-N} e^{\tau(x-iy_n-\Lambda)} \frac{K(x-iy_n)}{(x-iy_n-\Lambda)^2\sigma(x-iy_n)}\,dx. \end{aligned} \tag{2.2.43}$$

Here $N > A + \varepsilon$, and the rest of the notation is as in Theorem 2.2.1. From (2.2.39) it follows immediately that in the half-plane $\operatorname{Re} z \geqq A + \varepsilon$ ($\varepsilon > 0$)

$$|\tilde{\mathscr{X}}(z)| < c_7. \tag{2.2.44}$$

Taking (2.2.42) and (2.2.44) into account, and letting first $N \to \infty$ and then $n \to \infty$, we find that as $n \to \infty$, in (2.2.43),

$$R_{2n}(\tau) \to 0$$

which enables us to justify the expansion (2.2.40).

To conclude the proof, it is obviously sufficient to note that we can choose $t_1 = e^{-\tau_1}$ arbitrarily, and the character of convergence of the series in (2.2.40) for any t_1 is the same as in Theorem 2.2.1. ∎

We may remark that in situations similar to those considered in Theorem 2.2.2, the expansion (2.2.40) can be still further simplified somewhat by using (2.2.17) rather than (2.2.18). For example, in [81], such expansions are given for the solutions of the equation (2.1.3). In the case considered there, $\sigma(z)$ has the form

$$\sigma(z) = \sum_{j=1}^{r} \{|a_j|^z - |b_j|^z\}, \qquad a_j \neq 0,\ b_j \neq 0. \tag{2.2.45}$$

All the zeros of (2.2.45) are located in a finite strip (cf. for instance [76], p. 350). The same is true of the equation considered by Ramachandran and Rao (cf. Chapter 5). The series obtained in these expansions may however be nonabsolutely convergent. In the case considered in Theorem 2.2.1, we turned to (2.2.18), rather than to (2.2.17), mainly with the proof of the remainder term in view.

Let us now consider what effect the existence of derivatives for the solutions has on the location of the active exponents. We shall establish the following theorem (cf. [81], Lemma IX).

THEOREM 2.2.3. Let, under the conditions of Theorem 2.2.1, $\phi(t)$ have the derivative $\phi^{(n)}(t)$ of order n for $0 < t < t'$, and let $\lim_{t\to 0+} \phi^{(n)}(t) = \phi_+^{(n)}(0)$ exist finitely. Then, if there exist real active exponents which are not simultaneously integers and simple active exponents, then every one of them is $\geqq n$; if there exist active exponents, then every one of them has its real part $\geqq n$, and $\sigma_3 \geqq n$.

***Proof*:** Take the $\sigma(z)$ corresponding to equation (2.2.9), and let z_0 ($\operatorname{Re} z_0 < \delta$) be an arbitrary zero thereof, with multiplicity m_0. If we consider the function

$$\sigma(z \mid z_0) = \frac{\sigma(z)}{(z - z_0)^{m_0}} \tag{2.2.46}$$

analytic in the half-plane $\operatorname{Re} z < \delta$, then, in view of (2.2.11), it is obviously possible to represent it in the form

$$\sigma(z \mid z_0) = \int_0^\infty e^{z\beta} \mathscr{V}(\beta \mid z_0)\, d\beta,$$

$$\mathscr{V}(\beta \mid z_0) = \frac{e^{-z_0\beta}}{(m-1)!} \int_\beta^\infty e^{z_0\alpha} (\alpha - \beta)^{m-1}\, d[V_1(\alpha) - V_2(\alpha)] \tag{2.2.47}$$

so that

$$\int_0^\infty e^{\delta'\beta} |\mathscr{V}(\beta \mid z_0)|\, d\beta < \infty \qquad \text{for } \delta' < \delta. \tag{2.2.48}$$

Suppose the theorem is false and that, in the strip $-n \leqq \operatorname{Re} z \leqq 0$, there is at least one active zero of $\sigma(z)$ which is not simultaneously an integer and a simple active zero. Denote the (or one such) zero by z_1 and the polynomial corresponding to it in the expansion (2.2.1) by P_{-z_1} and the degree of this polynomial by m_1.

We apply Taylor's formula to $\phi(t)$. For sufficiently small $t(t \to 0+)$

$$\phi(t) = \sum_{k=0}^{n} b_k t^k + o(t^n). \tag{2.2.49}$$

We take (2.2.27) with $\gamma = n + 1$. Here $\chi(t)$ is defined in accordance with (2.2.6). Correspondingly to (2.2.49), we shall consider (2.2.27) for sufficiently large $\tau(\tau \to \infty)$.

Replace τ by $\tau + \beta$ $(\beta \geqq 0)$ in (2.2.27), multiply both sides of the resulting relation by $\mathscr{V}(\beta | z_1)$ and integrate over $(0, \infty)$ with respect to β, the series on the right in (2.2.27) is obviously integrable term by term. Then, in view of the definition of $\mathscr{V}(\beta | z_1)$, we obtain for $\tau \to \infty$

$$\sum_{k=0}^{n} \tilde{b}_k e^{-k\tau} = \int_0^\infty P_{z_1}(\tau + \beta) e^{z_1(\tau+\beta)} \mathscr{V}(\beta | z_1)\, d\beta + o(e^{-n\tau}) \tag{2.2.50}$$

or,

$$\sum_{k=0}^{n} \tilde{b}_k e^{-k\tau} = e^{z_1 \tau} \sum_{j=0}^{m_1} d_j P_{z_1}^{(j)}(\tau) + o(e^{-n\tau}) \tag{2.2.51}$$

where we have denoted

$$\frac{1}{j!} \int_0^\infty \beta^j e^{z_1 \beta} \mathscr{V}(\beta | z_1)\, d\beta$$

by d_j. Noting that by our construction

$$d_0 = \int_0^\infty e^{z_1 \beta} \mathscr{V}(\beta | z_1)\, d\beta \tag{2.2.52}$$

and comparing the two sides of (2.2.51) as $\tau \to \infty$ for the case when $-z_1$ is a complex active exponent, and then in the case when $-z_1$ is a real active exponent which is not simultaneously an integer and a simple active exponent, we obtain a contradiction.▮

2.3 ACTIVE EXPONENTS OF PROBABILISTIC SOLUTIONS

In the preceding section, we obtained the basic expansion for continuous solutions of the equations under consideration, and examined some of their properties. We shall continue the study of such expansions here, with the determination of the special nature of the expansions corresponding to probabilistic solutions as our principal aim.

With this end in view, we sharpen the expansions obtained in the statements of Theorems 2.2.1 and 2.2.2 under the supplementary requirements that the solution vanishes at the origin and is nonnegative near it. As is easily seen, these conditions are necessary if the solution is to be of the form $\phi(t) = -\ln|f(t)|$, where f is a c.f. Basically, the procedure will be the same as in [81].

THEOREM 2.3.1. If under the conditions of Theorem 2.2.1, $\phi(t) \geqq 0$ for all $0 < t < t'' \leqq t'$, $\phi(0) = 0$, then σ_1 (cf. (2.2.34)) is an active exponent, and

$$\sigma_1 > 0 \tag{2.3.1}$$

$$(-1)^{m_{\sigma_1}} \cdot a_{0\sigma_1} > 0 \tag{2.3.2}$$

where $a_{0\sigma_1}$ is the leading coefficient of the polynomial P_{σ_1} corresponding to σ_1 in the expansion (2.2.1), and m_{σ_1} is the degree of that polynomial.

***Proof*:** Without loss of generality, we may take $t'' = t' = 1$. We shall first show that σ_1 is an active exponent and then establish (2.3.1) and (2.3.2).

Let us suppose that σ_1 is not an active exponent. Then a $\Lambda > 0$ and a complex active zero z_Λ exist such that, on the circle with centre at Λ and passing through z_Λ, there are no active zeros other than z_Λ and $\bar{z}_\Lambda$, and there are no active zeros inside the circle. Let

$$|z_\Lambda - \Lambda| = d_\Lambda > 0. \tag{2.3.3}$$

By our construction, for all active zeros z other than z_Λ, $\bar{z}_\Lambda$,

$$|z - \Lambda| > d_\Lambda. \tag{2.3.4}$$

We choose and fix $\gamma > 0$ subject to the condition

$$\gamma + \Lambda > 2d_\Lambda. \tag{2.3.5}$$

Then we can find an integer n_0 such that, for $n > n_0$,

$$\inf_{z \in l_n} |z - \Lambda| > 2d_\Lambda \tag{2.3.6}$$

where l_n is the contour defined in the construction of the expansion (2.2.1).

Denote by $\mathscr{A}$ the set of all active zeros lying inside the contours $l_{-n_0}, \ldots, l_{-1}, l, l_1, \ldots, l_{n_0}$, other than the zeros z_Λ and $\bar{z}_\Lambda$. Such active zeros are finite in number; let $z_1, \ldots, z_{\tilde{k}}$ be an enumeration of them in some order. Turning to the expansion (2.2.1), we choose some integer $s > 0$ and apply the operation of an s-fold integration.

$$(\mathcal{T}_s g)(\tau) = \frac{1}{s!}\int_\tau^\infty (x-\tau)^{s-1} g(x)\,dx.$$

We apply $\mathcal{T}_s$ to (2.2.27), which obviously belongs to its domain of definition. As a result, we find

$$\frac{1}{(s+1)!}\int_\tau^\infty (x-\tau)^{s+1} e^{-\Lambda x}\chi(x)\,dx$$
$$= \sum_{n=1}^{\infty} \frac{1}{2\pi i}\left\{\int_{l_n} e^{\tau(z-\Lambda)} \frac{K(z)}{(z-\Lambda)^{s+2}\sigma(z)}\,dz + \int_{l_{-n}} e^{\tau(z-\Lambda)} \frac{K(z)}{(z-\Lambda)^{s+2}\sigma(z)}\,dz\right\}$$
$$+ \frac{1}{2\pi i}\int_l e^{\tau(z-\Lambda)} \frac{K(z)}{(z-\Lambda)^{s+2}\sigma(z)}\,dz + \mathcal{T}_s(R(\tau)) \tag{2.3.7}$$

which we rewrite as

$$\frac{1}{(s+1)!}\int_\tau^\infty (x-\tau)^{s+1} e^{-\Lambda x}\chi(x)\,dx = 2\,\mathrm{Re}\left\{\mathrm{res}\left[e^{\tau(z-\Lambda)} \frac{K(z)}{(z-\Lambda)^{s+2}\sigma(z)}\right]_{z=z_\Lambda}\right\}$$
$$+ \frac{1}{2\pi i}\sum_{j=1}^{\check{k}} \int_{c_{\rho_j}} e^{\tau(z-\Lambda)} \frac{K(z)}{(z-\Lambda)^{s+2}\sigma(z)}\,dz$$
$$+ 2\,\mathrm{Re}\sum_{n>n_0} \frac{1}{2\pi i}\int_{l_n} e^{\tau(z-\Lambda)} \frac{K(z)}{(z-\Lambda)^{s+2}\sigma(z)}\,dz$$
$$+ \mathcal{T}_s(R(\tau)). \tag{2.3.8}$$

Here c_{ρ_j} is a circle with centre z_j and of a radius ρ_j so small that it does not contain any other points of $\mathcal{A}$.

We shall estimate each of the members of the relation (2.3.8) for large s and suitably chosen τ.

According to the conditions of the theorem, $\chi(\tau) = \phi(e^{-\tau}) \geqq 0$ for $\tau \geqq 0$ ($t'' = t' = 1$.) Hence we see that

$$\frac{1}{(s+1)!}\int_\tau^\infty (x-\tau)^{s+1} e^{-\Lambda x}\chi(x)\,dx \geqq 0, \qquad \tau > \tau'. \tag{2.3.9}$$

We shall estimate the right member of (2.3.8) for large s. Taking the first term and writing

$$P_{z_\Lambda}(\tau) = a_0\tau^m + a_1\tau^{m-1} + \cdots + a_m, \qquad a_0 \neq 0, \qquad m \geqq 0$$

we easily see that*

$$\mathscr{T}_s\{P_{z_\Lambda}(\tau)e^{\tau(z_\Lambda-\Lambda)}\}$$

$$= \begin{cases} (-1)^s a_0 \dfrac{e^{(z_\Lambda-\Lambda)\tau}}{(z_\Lambda-\Lambda)^s} & \text{for } m=0 \\ (-1)^{m+s}s(s+1)\cdots(s+m-1)\dfrac{e^{(z_\Lambda-\Lambda)\tau}}{(z_\Lambda-\Lambda)^{s+m}}[1+R(\tau,s)]a_0 & \\ \qquad \text{for } m>0 & \end{cases} \tag{2.3.10}$$

where, for $\tau > \varepsilon > 0$

$$|R(\tau, s)| \leqq c_8 \frac{\tau^m}{s}. \tag{2.3.11}$$

By the definition of $\mathscr{A}$, in view of (2.3.4), there exists a $\beta > 0$ so small that

$$|z-\Lambda| \geqq d_\Lambda + \beta, \qquad z \in \mathscr{A}. \tag{2.3.12}$$

Taking this into account and $\rho < \beta/2$, we find

$$\left|\int_{c_{\rho_j}} e^{\tau(z-\Lambda)} \frac{K(z)}{(z-\Lambda)^{s+2}\sigma(z)}\,dz\right| \leqq \frac{c_9}{(d_\Lambda + (\beta/2))^s} \tag{2.3.13}$$

$$\tau \geqq 0, j = 1, 2, \ldots.$$

Similarly, taking (2.3.6) and (2.2.25) into account,

$$\left|\sum_{n>n_0}\int_{l_n} e^{\tau(z-\Lambda)} \frac{K(z)}{(z-\Lambda)^{s+2}\sigma(z)}\,dz\right| \leqq \frac{c_{10}}{(2d_\Lambda)^s} \tag{2.3.14}$$

and taking (2.3.5) and (2.2.5) into account ($R(\tau) = r(e^{-\tau})$),

$$|\mathscr{T}_s(R(\tau))| \leqq \frac{c_{10}}{(2d_\Lambda)^s}. \tag{2.3.15}$$

Since $\operatorname{Im} z_\Lambda > 0$ by assumption, we have the representation

$$z_\Lambda - \Lambda = d_\Lambda e^{i\theta}, \qquad 0 < \theta < \pi. \tag{2.3.16}$$

Also, let

$$a_0 = |a_0| e^{i\theta_0}. \tag{2.3.17}$$

* $\mathscr{T}_s((\tau^n e^{\Delta\tau}) = (-1)^n \dfrac{e^{\Delta\tau}}{\Delta}\left\{\dfrac{\tau^n}{\Delta^s} + \sum_{j=1}^{n}\dbinom{n}{j}(-1)^j s(s+1)\cdots(s+j-1)\dfrac{\tau^{n-j}}{\Delta^{s+j}}\right\}$

for positive integers n and $\operatorname{Re}\Delta < 0$ (cf. [15], p. 333).

With these notations,

$$(-1)^{s+m}\frac{e^{(z_\Lambda-\Lambda)\tau}}{(z_\Lambda-\Lambda)^{s+m}}a_0=|a_0|\frac{1}{d_\Lambda^{s+m}}e^{\tau d_\Lambda\cos\theta+i(\tau d\sin\theta+\pi(s+m)+\theta_0)} \tag{2.3.18}$$

For any s, τ can obviously be chosen in the segment $[(2\pi/d_\Lambda \sin\theta), (4\pi/d_\Lambda \sin\theta)]$ such that

$$\tau d_\Lambda \sin\theta+\pi(s+m)+\theta_0=(2k+1)\pi \tag{2.3.19}$$

for a suitable integer k. We choose and fix a sufficiently large s such that the right members of each of the relations (2.3.13)–(2.3.15) does not exceed

$$\frac{1}{12}|a_0|\,d_\Lambda^{-(s+m)}e^{4\pi\cot\theta}. \tag{2.3.20}$$

If $m \geqq 1$, then we subject the choice of s to the additional restriction

$$c_8\left[\frac{4\pi}{d_\Lambda\sin\theta}\right]^m\frac{1}{s}<\frac{1}{4}. \tag{2.3.21}$$

Choosing and fixing s subject to these conditions in the interval $[(2\pi/d_\Lambda \sin\theta), (4\pi/d_\Lambda \sin\theta)]$, we correspondingly choose $\tau=\tau_1$ such that (2.3.19) holds. Under such a choice of s and $\tau=\tau_1$, we find, in view of (2.3.10) and (2.3.18) that

$$\operatorname{Re}\mathscr{T}_s\{P_{-z_\Lambda}(\tau)e^{\tau(z_\Lambda-\Lambda)}\}_{\tau=\tau_1}<-\frac{1}{2}|a_0|\frac{e^{4\pi\cot\theta}}{d_\Lambda^{s+m}}. \tag{2.3.22}$$

Taking (2.3.13)–(2.3.15), (2.3.20), and (2.3.22) into account, from (2.3.8) we finally obtain for (2.3.7)

$$\mathscr{T}_s\left\{\int_\tau^\infty(v-\tau)e^{-\Lambda v}\chi(v)\,dv\right\}<-\frac{1}{2}|a_0|\frac{e^{4\pi\cot\theta}}{d_\Lambda^{s+m}}. \tag{2.3.23}$$

A comparison of (2.3.23) and (2.3.9) leads us to a contradiction, thus showing that σ_1 is an active exponent.

We shall now show that $\sigma_1>0$.

By the Corollary to Theorem 2.1.2, $\sigma_1<0$ is not possible. Thus, it only remains to show that $\sigma_1=0$ is also not possible. To prove this, we make use of the fact that $\phi(0)=0$. Since ϕ is continuous, in terms of (2.2.6) this fact means that as $\tau\to\infty$

$$\chi(\tau)\to 0. \tag{2.3.24}$$

We choose some small $\varepsilon>0$; then there exists $\tau_\varepsilon>0$ such that

$$\chi(\tau)<\varepsilon \qquad \text{for} \qquad \tau>\tau_\varepsilon. \tag{2.3.25}$$

Using (2.3.25), we can find an estimate for the left member of (2.3.7) for $\tau > \tau_\varepsilon$

$$\frac{1}{s!}\int_\tau^\infty (v-\tau)^s e^{-\Lambda v}\chi(v)\,dv \leqq \varepsilon\,\frac{e^{-\Lambda\tau}}{\Lambda^{s+1}}. \tag{2.3.26}$$

On the other hand, under the assumption that $\sigma_1 = 0$, using arguments similar to those applied for deriving (2.3.14), to estimate the right member of (2.3.7), we easily obtain that, for fixed $\tau > \tau_\varepsilon$ and sufficiently large $s(\Lambda > 0)$

$$\frac{1}{s!}\int_\tau^\infty (v-\tau)^s e^{-\Lambda v}\chi(v)\,dv > \frac{1}{2}\,|a_0|\,\frac{e^{-\Lambda\tau}}{\Lambda^{s+1}} \tag{2.3.27}$$

where a_0 is the leading coefficient of the polynomial P_{σ_1} corresponding to the active exponent σ_1 in the expansion (2.2.1).

If we choose $\varepsilon < |a_0|/2$, then (2.3.26) and (2.3.27) are in contradiction, which shows that σ_1 has to be positive.

If m_{σ_1} be the degree of the polynomial P_{σ_1}, then applying the same arguments again as for the derivation of (2.3.25) leads in this case to the relation, valid for suitably chosen τ and s $(\Lambda > 0)$

$$\frac{1}{s!}\int_\tau^\infty (v-\tau)^s e^{-\Lambda v}\chi(v)\,dv < \frac{1}{2}\,|a_0|(-1)^{m_{\sigma_1}}\,\frac{e^{-(\Lambda+\sigma_1)\tau}}{(\Lambda+\sigma_1)^{s+1}} \tag{2.3.28}$$

which concludes the argument. ▌

From the theorem just proved, a simple corollary, useful in the sequel, follows immediately.

Corollary. Under the conditions of Theorem 2.3.1, we may take $\Lambda = 0$ in the expansion (2.2.1); also, for any $\varepsilon > 0$, the following estimate holds as $t \to 0+$.

$$\int_0^t \frac{\phi(u)}{u}\,du = O(t^{\sigma_1-\varepsilon}) \tag{2.3.29}$$

***Proof*:** Since $\sigma_1 > 0$, we may take $x' < 0$ in (2.2.18), which permits us to take $\Lambda = 0$ in the expansion (2.2.1). For proving (2.3.29), we note that, taking $\Lambda = 0$, $\gamma = \sigma_1 - \varepsilon$ $(\varepsilon > 0)$ in (2.2.37), we have as $t \to 0+$

$$\int_0^t \ln\left(\frac{t}{u}\right)\phi(u)\,\frac{du}{u} = O(t^{\sigma_1-\varepsilon}) \tag{2.3.30}$$

whence, using the nonnegativity of ϕ and taking any $0 < \lambda < 1$,

$$\int_0^{\lambda t} \phi(u)\,\frac{du}{u} = O(t^{\sigma_1-\varepsilon})$$

which leads to (2.3.29), proving the corollary. ▌

We may note that in the above case, (2.3.30) can also be obtained by starting from the analyticity of $\mathscr{X}(z)$ in the half-plane Re $z < -\sigma_1 < 0$.

We also note that under the conditions of Theorem 2.3.1, if the set of complex active exponents is non-empty, we can obtain a corresponding estimate for $W(t) = \sum_{n=0}^{\infty} S_n(t)$, in (2.2.1), taking $\gamma > \sigma_3$; as $t \to 0+$,

$$W(t) = O(t^{\sigma_3 - \varepsilon}) \qquad (\varepsilon > 0). \tag{2.3.31}$$

In the case where ϕ satisfies (2.1.9) throughout the half-line $[0, \infty)$, for instance, we can obtain for the upper bound of the real parts of the active exponents a result similar to Theorem 2.3.1, under the conditions of Theorem 2.2.2. In fact, we have

THEOREM 2.3.2. If, under the conditions of Theorem 2.2.2, $\phi(t) \geqq 0$ for all $0 \leqq t' < t < \infty$, then σ_2 is finite (cf. (2.2.35)), is an active exponent, and further

$$a_{0\sigma_2} > 0 \tag{2.3.32}$$

where $a_{0\sigma_2}$ is the leading coefficient of the polynomial P_{σ_2} corresponding to σ_2 in the expansion (2.2.1).

We shall not pause to prove this theorem, since the proof proceeds along the same lines as that of Theorem 2.3.1. We also have a Corollary similar to that of Theorem 2.3.1, but with these changes: analogues of (2.3.29) and (2.3.31) hold, not as $t \to 0+$ but as $t \to \infty$, and with σ_1 replaced by σ_2, and σ_3 by σ_4.

We now turn to consider the properties of probabilistic solutions of the original equations. We first take the family $\mathscr{F}(V_1, V_2, \delta, t')$. Here, the following theorem obtains (cf. [81], Lemma X).

THEOREM 2.3.3. Let $F \in \mathscr{F}(V_1, V_2, \delta, t')$ ([cf. (2.1.1)]). If f is its c.f., then

(1) for $\phi = -\ln |f|$, (2.2.1) holds, with σ_1(cf. (2.2.34)) an active exponent, and

$$0 < \sigma_1 \leqq 2 \tag{2.3.33}$$

$$(-1)^{m_{\sigma_1}} \cdot a_{0\sigma_1} > 0 \tag{2.3.34}$$

where $a_{0\sigma_1}$ is the leading coefficient, and m_{σ_1} the degree of the polynomial $P_{\sigma_1}(t)$ corresponding to σ_1 in the expansion.

(2) if $F \in \mathscr{F}(V_1, V_2, \delta, t')$ for every $t' > 0$ such that in the interval $[0, t')$, $f(t) \neq 0$, and condition (2) of Lemma 2.1.1 is satisfied ,then, for all $0 \leqq t < \infty$, $f(t) \neq 0$, $\phi(t) = -\ln |f(t)|$ exists, and (2.1.40) holds for ϕ on $(0, \infty)$; and in addition to our assertions in (1) above, σ_2 exists finitely (cf. (2.2.35)), is an active exponent, and

$$a_{0\sigma_2} > 0 \tag{2.3.35}$$

where $a_{0\sigma_2}$ is the leading coefficient of the polynomial P_{σ_2} corresponding to σ_2 in the expansion (2.2.1).

Proof:

(1) Under the conditions of (1), it is easy to see that ϕ satisfies the conditions of Theorem 2.3.1. We have only to prove the inequality on the right in (2.3.33) to complete the proof. Suppose the inequality does not hold. Then let

$$\sigma_1 = 2 + \Delta, \qquad \Delta > 0. \tag{2.3.36}$$

Then we have from (2.3.29) that, as $t \to 0+$,

$$\int_0^t \phi(u) \frac{du}{u} = O(t^{2+(\Delta/2)}). \tag{2.3.37}$$

Hence using arguments usual in such cases, we easily see that F is necessarily degenerate. Indeed, from (2.3.37), it follows that as $t \to 0+$

$$\int_{t/2}^t \phi(u) \frac{du}{u} = O(t^{2+(\Delta/2)}). \tag{2.3.38}$$

Applying the mean value theorem in (2.3.38), we see that a sequence of positive constants $\{t_\nu\}$ exists such that, as $\nu \to \infty$,

$$t_\nu \to 0, \qquad \phi(t_\nu) = O(t_\nu^2). \tag{2.3.39}$$

If we denote by F_1 the convolution of F with its conjugate $\tilde{F}$, then

$$1 - |f(t)|^2 = 2 \int \sin^2 \frac{tx}{2} \, dF_1(x) \tag{2.3.40}$$

whence

$$\frac{1 - |f(t)|^2}{t^2} \geqq \frac{1}{\pi^2} \int_{-\pi/t}^{\pi/t} x^2 \, dF_1(x), \qquad t > 0. \tag{2.3.41}$$

Since, for sufficiently small t, $|f(t)|^2 = \exp[-2\phi(t)]$, we see from (2.3.39) and (2.3.41) that

$$\int_{-\pi/t_\nu}^{\pi/t_\nu} x^2 dF_1(x) = O(1) \text{ as } \nu \to \infty, \tag{2.3.42}$$

contradicting (2.3.36).

(2) Under the conditions of (2), Lemma 2.1.1 is valid and permits us to apply Theorem 2.3.2. ∎

Theorem 2.3.3 essentially exhausts the general properties of the classes $\mathscr{F}(V_1, V_2, \delta, t')$, although it is based on comparatively few properties of c.f.'s. In [81] for example, it is shown that for $0 < \sigma_1 \leqq 1$, one can construct a solution of a highly arbitrary nature for (2.1.2) on a finite interval, which is extensible to the whole real line as a c.f. (cf. [81], Theorem V). For such a purpose, the well-known theorem* of Polya (cf. [115]) can be applied, according to which every convex function f on $[0, \infty)$, satisfying the requirements that $f(0) = 1, f(t) \to 0$ as $t \to \infty$, and defined for negative t by symmetry is a c.f.

In case (2) of Theorem 2.3.3 also, it is possible to construct probabilistic solutions of a highly arbitrary type. Examples of such are constructed in [81], Theorem IV. Following the same procedure, one can construct c.f.'s having a more general type of spectrum of active exponents.

We now pause to dwell on nonsymmetric probabilistic solutions of (2.1.7). We shall mainly concern ourselves with the connection between the spectra of active exponents of the real and of the imaginary parts of such solutions.

We first establish a lemma relating to properties of c.f.'s.

LEMMA 2.3.1. Let f be the c.f. of some law F on the real line; then, for any real t_1, t_2, and t_3,

$$\left| \sum_{1 \leqq j<k \leqq 3} \operatorname{Im} f(t_j - t_k) \right| \leqq \frac{1}{\sqrt{3}} \sum_{1 \leq j<k \leq 3} \{1 - \operatorname{Re} f(t_j - t_k)\}. \tag{2.3.43}$$

***Proof*:** In view of the positive definiteness of c.f.'s, we have for any real t_1, t_2, t_3 and any complex z_1, z_2, z_3 such that $z_1 + z_2 + z_3 = 0$ that

$$\sum_{1 \leq j<k \leq 3} \{f(t_j - t_k) - 1\} z_j \bar{z}_k \geqq 0. \tag{2.3.44}$$

Taking $z_1 = \exp(i\pi/3)$, $z_2 = \exp(-i\pi/3)$, and $z_3 = \exp(i\pi)$, we find

$$\sum_{1 \leq j<k \leq 3} \operatorname{Im} f(t_j - t_k) + \frac{1}{\sqrt{3}} \sum_{1 \leq j<k \leq 3} \{1 - \operatorname{Re} f(t_j - t_k)\} \geqq 0. \tag{2.3.45}$$

The relation (2.3.43) follows hence immediately on noting that $\operatorname{Re} f$ is even, $\operatorname{Im} f$ is odd, and $|f(t)| \leqq 1$.∎

THEOREM 2.3.4. If $F \in \mathscr{F}_1(V_{\mathrm{I}}, V_{\mathrm{II}}, \delta, t')$ and f is its c.f., then

(1) Theorem 2.3.3 holds for F;

(2) for $\psi = \arg f$, the expansion (2.2.1) of Theorem 2.2.1 holds; further, if $\mathscr{N}$, the set of active exponents for this expansion, contains elements other than a possible simple active exponent $\xi = 1$, then

$$\mu_1 = \inf_{\xi \in \mathscr{N}} \operatorname{Re} \xi \geqq \sigma_1. \tag{2.3.46}$$

***Proof*:** Only (2.3.46) requires to be proved, the other assertions following immediately from the earlier discussion.

* It is indicated in [2], p. 263, that this fact was known even in the year 1939 to M. G. Krein.

The argument will be divided into two stages. First we establish that, as $t \to 0+$,

$$\int_0^t \frac{\psi^2(u)}{u}\, du = O(t^{\sigma_1 - \varepsilon}) \tag{2.3.47}$$

where $\varepsilon > 0$ is arbitrary; then we shall prove (2.3.46).

For sufficiently small t we may write

$$\begin{aligned} \int \cos tx\, dF(x) &= e^{-\phi(t)} \cdot \cos \psi(t) \\ \int \sin tx\, dF(x) &= e^{-\phi(t)} \cdot \sin \psi(t) \end{aligned} \tag{2.3.48}$$

and using the well-known Cauchy-Schwarz-Bunyakowski inequality,

$$\begin{aligned} e^{-2\phi(t)} \sin^2 \psi(t) &\leqq \int \sin^2 tx\, dF(x) \\ &= \frac{1 - e^{-\phi(2t)}}{2} + e^{-\phi(2t)} \sin^2 \left[\frac{\psi(zt)}{2}\right]. \end{aligned} \tag{2.3.49}$$

Hence

$$\sin^2 \psi(t) \leqq \frac{1}{4} \psi^2(2t) + 2\phi(t) + \frac{1}{2} \phi(2t). \tag{2.3.50}$$

We take an arbitrary $\varepsilon > 0$ and write

$$\alpha = 1 - 2^{-(1/2)(2 - \sigma_1 + \varepsilon)}. \tag{2.3.51}$$

Obviously $0 < \alpha < 1$. We choose $t_1 > 0$ so small that for $0 \leqq t < t_1$

$$\sin^2 \psi(t) \geqq (1 - \alpha)\psi^2(t). \tag{2.3.52}$$

From (2.3.50)–(2.3.52), we have for $0 \leqq t < t_1$:

$$\psi^2(t) \leqq \frac{1 - \alpha}{2^{\sigma_1 - \varepsilon}} \psi^2(2t) + \frac{1}{1 - \alpha} \left\{2\phi(t) + \frac{1}{2} \phi(2t)\right\}. \tag{2.3.53}$$

We write

$$\lambda(t) = \sup_{t \leq x \leq t_1} \frac{1}{x^{\sigma_1 - \varepsilon}} \int_{x/2}^{x} \psi^2(u) \frac{du}{u}, \qquad 0 < t < t_1. \tag{2.3.54}$$

By definition, λ is a nondecreasing function. We shall show that it is bounded for $0 < t < t_1$. From (2.3.53) and (2.3.29), it follows that

$$\lambda(t/2) \leqq (1 - \alpha)\lambda(t) + c_{12} \tag{2.3.55}$$

We set $t = t_1/2^\nu$, $\nu = 1, 2, \ldots$, successively in (2.3.55):

$$\lambda(t_1/2^{\nu+1}) \leqq (1-\alpha)\lambda(t/2^\nu) + c_{12} \tag{2.3.56}$$

whence

$$\lambda(t_1/2^{\nu+1}) \leqq (1-\alpha)^\nu \lambda(t_1/2) + c_{12}\frac{1-(1-\alpha)^\nu}{\alpha} \tag{2.3.57}$$

and since λ is nondecreasing, it follows from (2.3.57) on letting $\nu \to \infty$ that

$$\lambda(t) \leqq c_{13}, \qquad 0 < t < t_1. \tag{2.3.58}$$

Relation (2.3.47) follows immediately from this on taking

$$\int_0^t \psi^2(u)\frac{du}{u} = \sum_{\nu=0}^{\infty} \int_{t/2^{\nu+1}}^{t/2^\nu} \psi^2(u)\frac{du}{u}.$$

Let us now turn to the proof of (2.3.46).

We shall assume that the set $\mathcal{N}$ of active exponents for $\psi(t)$, ignoring the element $\xi = 1$ in the case when it happens to be a simple active exponent for the corresponding expansion, is nonempty. Suppose then that (2.3.46) is not true, so that a $\Delta > 0$ can be found along with at least one exponent $\xi_0 \in \mathcal{N}$ such that

$$\operatorname{Re} \xi_0 \leqq \sigma_1 - \Delta. \tag{2.3.59}$$

We consider the functions $\psi_1(t) = \psi(t) - 2\psi(t/2)$ and $\psi_2(t) = \psi(t) - 3\psi(t/3)$. Expansions of the form (2.2.1) hold for them; denoting the sets of active exponents corresponding to them by $\mathcal{N}_1$ and $\mathcal{N}_2$, we easily see that

$$\mathcal{N}_j \subset \mathcal{N}, \qquad \mathcal{N} - \mathcal{N}_j \subset \tilde{\mathcal{N}}_j, \qquad j = 1, 2 \tag{2.3.60}$$

where $\tilde{\mathcal{N}}_1$ is the set of zeros of the function $2^z - 2$, and $\tilde{\mathcal{N}}_2$ is the set of zeros of $3^z - 3$, i.e.,

$$\tilde{\mathcal{N}}_1 = \left\{1 + \frac{2\pi i}{\ln 2}\nu;\ \nu = 0, \pm 1, \pm 2, \ldots\right\}$$

$$\tilde{\mathcal{N}}_2 = \left\{1 + \frac{2\pi i}{\ln 3}\nu;\ \nu = 0, \pm 1, \pm 2, \ldots\right\}.$$

We note that a member of $\tilde{\mathcal{N}}_1$ does not belong to $\mathcal{N}_1$ when it belongs to $\mathcal{N}$ as a simple exponent, and similarly in the dual case.

We now use Lemma 2.3.1. Setting $t_1 - t_2 = t_2 - t_3 = -t/2$ $(t_3 - t_1 = t)$, where $t > 0$ is sufficiently small, we find from (2.3.43)

$$|e^{-2\phi(t)}\sin\psi(2t) - 2e^{-\phi(t)}\sin\psi(t)|$$
$$\leqq \frac{1}{\sqrt{3}}\{2(1 - e^{-\phi(t)}\cos\psi(t)) + (1 - e^{-\phi(2t)}\cos\psi(2t))\}. \tag{2.3.61}$$

Hence it follows as $t \to 0+$ that

$$\psi(t) - 2\psi\left(\frac{t}{2}\right) = O\left[\phi\left(\frac{t}{2}\right) + \phi(t) + \psi^2\left(\frac{t}{2}\right) + \psi^2(t)\right]. \tag{2.3.62}$$

From (2.3.62), on taking (2.3.47) and (2.3.29) into account, it follows that as $t \to 0+$

$$\int_0^t \left|\psi(t) - 2\psi\left(\frac{t}{2}\right)\right| \frac{dt}{t} = O(t^{\sigma_1 - \varepsilon}) \qquad \text{for any } \varepsilon > 0. \tag{2.3.63}$$

Similarly, setting $t_1 - t_2 = -t/3, t_2 - t_3 = -2t/3, (t_3 - t_1 = t)$, where $t > 0$ is sufficiently small, we can see that as $t \to 0+$,

$$\psi(t) - \psi\left(\frac{2t}{3}\right) - \psi\left(\frac{t}{3}\right) = O\left[\phi\left(\frac{t}{3}\right) + \phi\left(\frac{2t}{3}\right) + \phi(t) + \psi^2\left(\frac{2t}{3}\right) + \psi^2(t)\right] \tag{2.3.64}$$

which, together with (2.3.29) and (2.3.47), leads to the relation

$$\int_0^t \left|\psi(t) - 3\psi\left(\frac{t}{3}\right)\right| \frac{dt}{t} = O(t^{\sigma_1 - \varepsilon}), \qquad t \to 0+. \tag{2.3.65}$$

From (2.3.63) and (2.3.65) respectively, it obviously follows in view of the arbitrariness of $\varepsilon > 0$ that

$$\operatorname{Re} \xi \geqq \sigma_1 \qquad \text{for} \qquad \xi \in \mathcal{N}_j, j = 1, 2$$

which immediately leads to a contradiction with (2.3.59), since, by our construction, ξ_0 belongs at least to one of the two sets $\mathcal{N}_1, \mathcal{N}_2$. ∎

2.4 CHARACTERIZATION OF THE NORMAL LAW

In the present section, we consider the location of the active exponents in the case $\sigma_1 = 2$ and apply the results obtained to a characterization of the normal law on the basis of equations of the form (2.1.7) and (2.1.8). Thus we shall obtain conditions under which the family $\mathcal{F}(V_1, V_2)$ consists of normal laws alone. In [81], this problem was studied for the pair of statistics (2.1.1) in the presence of condition (2.1.11). In [36], Zinger studied the same problem under the more general conditions of Theorem 2.3.3.

The introduction of certain additional considerations permits a considerable simplification of the presentation and the refinement of some intermediate results (cf. [81], Lemma XI, and also Theorem 2.4.1 below).

It is necessary to note that in view of Cramer's theorem on the decomposition of the normal law (cf. [70], p. 66), we may confine our attention to equation (2.1.9) alone, because it sufficies to show that $|f(t)|^2 = f(t)f(-t)$ is the c.f. of a normal law, on the strength of the above theorem.

We start with a detailed study of the character and location of the active exponents in the case; $\sigma_1 = 2$ is a simple active exponent. The following simple lemma will prove very useful.

LEMMA 2.4.1. Let g be the c.f. of a symmetric law. If the law G has finite moment of order $2k$, but not of order $2k + 2$ for some positive integer k, then $\ln g(t)$ exists and has a continuous derivative of order $2k$ in some neighborhood of the origin, and

$$(-1)^{k+1}\{[\ln g(t)]^{(2k)} - \kappa_{2k}\} \geqq 0 \tag{2.4.1}$$

where $\kappa_{2k} = (\ln g(t))^{(2k)}|_{t=0}$ is the semi-invariant of order $2k$.

***Proof*:** From the properties of c.f.s, it is well-known that the following two assertions are equivalent:

(1) the symmetric law G has finite moment of order $2k$, and

(2) g, the c.f. of G, has a finite derivative of order $2k$ at the origin.

Either of the above equivalent conditions then implies the existence throughout the real axis of bounded and uniformly continuous derivatives for g: $g^{(2j)}(t), j = 1, \ldots, k$, defined by the relations

$$g^{(2j)}(t) = (-1)^j \int_{-\infty}^{+\infty} (\cos tx)x^{2j}\, dG(x). \tag{2.4.2}$$

Under the given conditions, for $t > 0$,

$$(-1)^{k+1}\frac{g^{(2k)}(t) - g^{(2k)}(0)}{t^2} \geqq \frac{2}{\pi^2}\int_{-\pi/t}^{\pi/t} x^{2k+2}\, dG(x) \tag{2.4.3}$$

where the right member tends to ∞ as $t \to 0+$, and consequently so does the left member. We take $t_1 > 0$ such that $g(t) > 0$ for $|t| < t_1$. Then $\gamma(t) = -\ln g(t)$ exists in that interval, possesses a continuous derivative of order $2k$, which is of the form

$$-\gamma^{(2k)}(t) = \frac{g^{(2k)}(t)}{g(t)} - (2k-1)\frac{g^{(2k-1)}(t)}{g(t)}\frac{g'(t)}{g(t)} + u(t) \tag{2.4.4}$$

where $u(t)$ is a polynomial in the functions $g^{(j)}/g$: more precisely, it is a linear combination of summands of the form

$$\frac{g^{(\alpha_1)}(t)}{g(t)} \cdots \frac{g^{(\alpha_l)}(t)}{g(t)}$$

where $\alpha_1 + \cdots + \alpha_l = 2k$, $0 \leqq \alpha_j \leqq 2k - 2$. We have from (2.4.4)

$$-\frac{\gamma^{(2k)}(t) - \gamma^{(2k)}(0)}{t^2} = \frac{g^{(2k)}(t) - g^{(2k)}(0)}{t^2 g(t)} + g^{(2k)}(0)\frac{1 - g(t)}{t^2 g(t)}$$
$$- (2k-1)\frac{g^{(2k-1)}(t)}{tg(t)}\frac{g'(t)}{tg(t)} + \frac{u(t) - u(0)}{t^2}. \tag{2.4.5}$$

In the right member of (2.4.5), all the summands other than the first, converge to finite limits as $t \to 0+$, and the first summand, as we have already noted, diverges to ∞, preserving its sign given by (2.4.3). Hence (2.4.1) follows immediately.▌

We now study the nature of the locations of the active exponents under the conditions of Theorem 2.3.1, when $\sigma_1 = 2$. The following theorem holds.

THEOREM 2.4.1. Let $\sigma_1 = 2$ under the conditions of Theorem 2.3.1, and let it further be a simple active exponent. Then, if in the expansion (2.2.1), complex active exponents are present, there must exist a real active exponent s, which is not simultaneously an even integer and a simple active exponent, satisfying the condition

$$2 < s \leqq \sigma_3. \tag{2.4.6}$$

If s_1 is the smallest among such exponents (their number is finite), then in addition to (2.4.6) we have

$$(-1)^{q+m_{s_1}} \cdot a_{0s_1} > 0. \tag{2.4.7}$$

Here, as before, σ_3 is as defined in (2.2.34), a_{0s_1} is the leading coefficient of the polynomial P_{s_1} corresponding to s_1 in the expansion (2.2.1), m_{s_1} is its degree, and

$$q = \left[\frac{s_1}{2}\right] - \left[1 + \left[\frac{s_1}{2}\right] - \frac{s_1}{2}\right].$$

***Proof*:** We first establish the existence of a finite second moment for F. In (2.2.1), we take some $\gamma > 2$, and if there exist complex active exponents, then we also subject γ to the restriction $\gamma > \sigma_3 + 1$ ($\varepsilon = 1$), so that

$$\int_0^t \phi(u) \ln \frac{t}{u} \frac{du}{u} = a_0 t^2 + \pi(t) + W(t) + R(t). \tag{2.4.8}$$

Here $\pi(t)$ corresponds to the real active exponents—they are situated in the interval $(2, \gamma + 1)$; $W(t)$ to the complex active exponents (if any) which lie in the strip $\sigma_3 \leqq \operatorname{Re} z < \gamma + 1$;

$$W(t) = \sum_{n=0}^{\infty} S_n(t). \tag{2.4.9}$$

For $R(t)$ and $W(t)$, the following estimates hold as $t \to 0+$ (cf. the corollary to Theorem 2.3.1)

$$R(t) = O(t^{\gamma}) \tag{2.4.10}$$

$$W(t) = O(t^{\sigma_3 - \varepsilon}) \tag{2.4.11}$$

where $\varepsilon > 0$ is arbitrary.

If $W(t) \not\equiv 0$ in some neighborhood of the origin, then, as $t \to 0+$, it must change sign infinitely often, since otherwise we can apply the arguments of Theorem 2.3.1 to establish that $W(t)$ must contain a term corresponding to the real active exponent σ_3, which contradicts the definition of $W(t)$. Thus there must exist a sequence of positive numbers $\{t_n\}$ such that

$$t_n \to 0 \quad \text{as} \quad n \to \infty, \quad W(t_n) \leqq 0, \qquad n = 1, 2, \ldots. \tag{2.4.12}$$

From (2.4.8), taking (2.4.10) and (2.4.12) into account as well as the obvious $\pi(t) = O(t^2)$, we have as $n \to \infty$, in view of the nonnegativity of ϕ,

$$\int_0^{t_n} \ln \frac{t_n}{u} \phi(u) \frac{du}{u} = O(t_n^2). \tag{2.4.13}$$

Hence, in view of the continuity of ϕ (in a neighborhood of the origin), and applying the mean value theorem for example to

$$\int_{(1/2)t_n}^{t_n} \ln \frac{t_n}{u} \phi(u) \frac{du}{u},$$

we can indicate a sequence $\{\tilde{t}_n\}$ of positive numbers such that as $n \to \infty$

$$\tilde{t}_n \to 0, \qquad \phi(\tilde{t}_n) = O(\tilde{t}_n^2). \tag{2.4.14}$$

We see at once from (2.4.14) that F has finite variance, applying for instance the relation (2.3.41) already used by us in Theorem 2.3.3 (also cf. (2.4.3)), whence it follows in particular that ϕ has a continuous second derivative in some neighborhood of the origin.

We emphasize that here and in what follows—in particular, in applications of Lemma 2.4.1—we rely on these properties of c.f.'s: $|f|^2$ is the c.f. of the symmetric law $F_1 = F * \tilde{F}$ ($\tilde{F}$ is the conjugate d.f. of F), and either *both* F and F_1 have, or *neither* of them has, the absolute moment of a given order.

We now turn to proving (2.4.6). We shall assume that complex active exponents exist. First consider the case

$$2 \leqq \sigma_3 < 4. \tag{2.4.15}$$

In this case, F does not have a finite fourth moment, since the existence of the latter would imply that of a continuous fourth derivative in some neighborhood of the origin for ϕ, which in turn, by Theorem 2.2.3, would lead to $\sigma_3 \geqq 4$. Therefore, applying Lemma 2.4.1, in some neighborhood of $t = 0$,

$$\phi''(t) - \phi''(0) \leqq 0. \tag{2.4.16}$$

In (2.4.8), we take $\phi(t)$ in the form

$$\phi(t) = \frac{t^2}{2} \phi''(0) + \int_0^t (t - u)\{\phi''(u) - \phi''(0)\}\, du. \tag{2.4.17}$$

Then we have, as $t \to 0+$,

$$\int_0^t \ln \frac{t}{x} \frac{dx}{x} \int_0^x (x-u)\{\phi''(u) - \phi''(0)\}\, du = \pi_1(t) + W(t) + O(t^\gamma). \quad (2.4.18)$$

Since $\phi''(t) - \phi''(0) = O(1)$, the left side of (2.4.18) is obviously $O(t^2)$. Further, taking (2.4.16) into account as well, we can apply the same arguments as in the proof of Theorem 2.3.1 and see that all active exponents on the right side of (2.4.18) have real parts greater than 2; moreover, among these exponents, there is necessarily at least one real exponent $\leqq \sigma_3$. Since this exponent is greater than 2, (2.4.16) is established, if (2.4.15) holds.

We turn to the general case, to which we apply the same considerations. As before, we assume that active complex exponents exist. We take an integer $p > 1$ such that

$$2p \leqq \sigma_3 < 2p + 2. \quad (2.4.19)$$

Now there are two possibilities: if there exists a real active exponent s, which is not simultaneously an even integer and a simple exponent, satisfying the condition $s \leqq 2p$, then we have (2.4.6); we therefore have only to consider the case where such an s does not exist. In that case, setting $\gamma > 2p + 2$, we can rewrite (2.4.8) in the form

$$\int_0^t \ln \frac{t}{u}\, \phi(u) \frac{du}{u} = P(t) + \pi_2(t) + W(t) + O(t^{2p+2}). \quad (2.4.20)$$

Here P is an even polynomial of degree $\leqq 2p$, $\pi_2(t)$ corresponds to the real active exponents lying in the interval $(2p, \gamma + 1)$, $W(t)$ corresponds to the complex active exponents; their abscissae lie in the half-closed interval $[\sigma_3, \gamma + 1)$. For $\pi_2(t)$ in particular, the following estimate holds for small t.

$$\pi_2(t) = O(t^{2p}). \quad (2.4.21)$$

We shall show that in this case ϕ has continuous derivatives of orders up to and including $2p$ in some neighborhood of the origin. Suppose not; then there exists an integer $k > 0$ such that $k \leqq p - 1$, and ϕ has (for all sufficiently small t) a continuous derivative of order $2k$ but not a finite derivative of order $2k + 2 \leqq 2p$, at least at the origin. Because of the existence of $\phi^{(2k)}$, we have (for small t)

$$\phi(t) = \frac{t^2}{2!}\phi''(0) + \cdots + \frac{t^{2k}}{(2k)!}\phi^{(2k)}(0)$$
$$+ \frac{1}{(2k-1)!}\int_0^t (t-u)^{2k-1}\{\phi^{(2k)}(u) - \phi^{(2k)}(0)\}\, du. \quad (2.4.22)$$

Taking $\phi(t)$ in the form (2.4.22) and substituting it in (2.4.20), we find as $t \to 0+$ that

$$\int_0^t \ln\frac{t}{x} \cdot \frac{dx}{x} \int_0^x (x-u)^{2k-1}\{\phi^{(2k)}(u) - \phi^{(2k)}(0)\}\,du$$
$$= P_1(t^2) + \pi_2(t) + W(t) + O(t^{2p}) \qquad (2.4.23)$$

where P_1 is a polynomial of degree $\leqq p$. From (2.4.23), it is easily seen that, in view of the continuity of $\phi^{(2k)}$ at the origin, the left side of (2.4.19) is $O(t^{2k})$, and because of (2.4.21) and (2.4.11), this polynomial does not contain terms of degrees smaller than $k + 1$.

Further, in view of the definition of k, we can apply Lemma 2.4.1 and see that for sufficiently small t

$$(-1)^{k+1}\{\phi^{(2k)}(t) - \phi^{(2k)}(0)\} \geqq 0. \qquad (2.4.24)$$

Using the same arguments as for deriving (2.4.12), we can then assert that there exists a sequence $\{t_{n_k}, n = 1, 2, \ldots\}$ of positive constants such that

$$t_{n_k} \to 0 \quad \text{as} \quad n \to \infty, \qquad (-1)^k W(t_{nk}) \leqq 0, \qquad n = 1, 2, \ldots.$$

Applying then the same arguments as for the derivation of (2.4.14) from (2.4.8), we can establish the existence of a sequence of positive numbers $\{\tilde{t}_{nk}, n = 1, 2, \ldots\}$ such that as $n \to \infty$

$$\tilde{t}_{nk} \to 0, \qquad \phi^{(2k)}(\tilde{t}_{nk}) - \phi^{(2k)}(0) = O(\tilde{t}_{nk}^2). \qquad (2.4.25)$$

Using (2.4.25) similarly to (2.4.14), we can see that F has finite moment of order $2k + 2$. Since this contradicts the conditions of the choice of k, it follows that F has finite moment of order $2p$.

In view of (2.4.18), the law F then cannot have finite moment of order $2p + 2$ (cf. Theorem 2.2.3) and consequently Lemma 2.4.1 can be applied again. We take (2.4.22) with $k = 2p$ and substitute in (2.4.20).

$$\int_0^t \frac{dt}{t} \int_0^t \frac{dt}{t} \int_0^t (t-x)^{2p-1}\{\phi^{(2p)}(x) - \phi^{(2p)}(0)\}\,dx$$
$$= at^{2p} + \pi_1(t) + W(t) + O(t^{2p+2}). \qquad (2.4.26)$$

Since $\phi^{(2p)}(t) - \phi^{(2p)}(0) = O(1)$ as $t \to 0+$, we can establish similarly to (2.4.22) that the left member of (2.4.26) is $O(t^{2p})$ as $t \to 0+$. Further, we can conclude by Lemma 2.4.1 that for sufficiently small $t > 0$

$$(-1)^p\{\phi^{(2p)}(t) - \phi^{(2p)}(0)\} \geqq 0. \qquad (2.4.27)$$

All this permits us here, as in the case $p = 1$, to apply the same arguments as for the proof of Theorem 2.3.1 and conclude that there necessarily exists a real active exponent $\geqq 2p$ and $\leqq \sigma_3$. Thus it is established that, if complex active exponents are present, then there is necessarily a real active exponent which is not simultaneously an even integer and a simple exponent, and satisfies (2.4.6). We denote by s_1 the smallest of such exponents. Also, let $q \geqq 1$ be an integer such that

$$2q < s_1 \leqq 2q + 2. \tag{2.4.28}$$

Since $\gamma > 2q + 2$, we can rewrite (2.4.20) in the form

$$\int_0^t \frac{dt}{t} \int_0^t \frac{dt}{t} \phi(t) = P(t) + \pi(t) + W(t) + O(t^{2q+2}) \tag{2.4.29}$$

where P is an even polynomial of degree at most $2q$, and $\pi(t)$ and $W(t)$ are of the order $O(t^{2q})$ as $t \to 0$ in view of (2.4.6), (2.4.28), and (2.4.9). Using (2.4.29) in the same way as we did (2.4.20) in the foregoing argument permits the conclusion that F must have finite moment of order $2q$. On the other hand, F cannot have finite moment of order $2q + 2$, since its existence would be contradictory to (2.4.28) (cf. Theorem 2.2.3). This circumstance permits the repeated use of Lemma 2.4.1. If in (2.4.26) we replace p by q and consider the corresponding relation (2.4.27), then arguing in the same way as in the concluding stages of the proof of Theorem 2.4.1 for deriving (2.3.2), we obtain

$$(-1)^{m_{s_1}+q} \cdot a_{0s_1} > 0 \tag{2.4.30}$$

where a_{0s_1} is the leading coefficient and m_{s_1} the degree of the polynomial P_{s_1} corresponding to the active exponent s_1 in the expansion (2.4.29). Since q is defined by (2.4.28), (2.4.30) coincides with (2.4.7) as required.∎

We shall use the above theorem to obtain conditions under which probabilistic solutions of (2.1.7) and (2.1.9) can only be normal laws. For characterizing (2.1.2) in the presence of (2.1.12), this problem was considered in [81]. Under the more general conditions of Section 2, Theorem 2.3.3, this was done by Zinger [36]; here we follow his presentation.

THEOREM 2.4.2. In order that, for some probability law F, the statements:

(I) F is a normal law; and

(II) $F \in \mathscr{F}(V_{\mathrm{I}}, V_{\mathrm{II}}, \delta, t')$

are equivalent, it is necessary and sufficient that the following conditions be satisfied:

(1) $\sigma(-2) = 0$, where $\sigma(z)$ corresponds to (2.1.9) according to formula (2.2.3);

(2) all the positive zeros of $\sigma(-z)$ which are even integers and multiples of 4 must be simple zeros;

(3) all the positive zeros of $\sigma(-z)$ which are even integers $\equiv 2 (\mathrm{mod}\ 4)$ must have multiplicity at most two; if such a zero has multiplicity two, then it must be unique and the largest among the positive zeros;

(4) if $\sigma(-z)$ has a positive zero s, which is not an even integer, then it must be unique, simple, and the largest among the positive zeros, and $[s/2]$ must be odd.

***Proof*:** We first establish the sufficiency of the conditions. Let then conditions (1)–(4) be satisfied. By (1), a normal law satisfies the conditions of Theorem 2.3.3. We shall show that if (2)–(4) are satisfied, there are no laws other than normal which satisfy the conditions of that theorem.

On the basis of Theorem 2.4.1 and (2)–(4), we may conclude that complex active exponents are not present in the expansion (2.2.1) for F. A comparison of Theorems 2.3.3 and 2.4.1 shows that active exponents which are not simultaneously even integers and simple exponents also must be absent, since otherwise relations (2.4.7) and (2.3.34) which must be simultaneously satisfied, would be in contradiction. Thus active exponents can only be even integers which, again, must all be simple, and they are finite in number. Thus $\phi(t) = P(t^2)$ for all t, where P is a polynomial, and

$$|f(t)|^2 = \exp[-2P(t^2)]. \tag{2.4.31}$$

Applying the well-known lemma of Marcinkiewicz (cf. Section 1.4 or [101]) and Cramer's theorem mentioned earlier, we conclude from (2.4.31) that F is a normal law. This proves the sufficiency of the conditions of the theorem.

We turn to the necessity part. For establishing it, we may use a lemma proved in [81].

LEMMA 2.4.2. (cf. [81], Lemma 15). Let two numbers γ_1 and γ_2 be given such that $2 \leqq \gamma_1 < \gamma_2$, and also a polynomial of degree $n \geqq 0$ with real coefficients

$$P(u) = a_0 u^n + a_1 u^{n-1} + \cdots + a_n, \qquad a_0 < 0$$

subject further to the following condition: if γ_1 is an even integer, then $n \geqq 1$. Then, for sufficiently large $A > 0$, either

(1) f_1 given by

$$f_1(t) = \exp\{-At^2 - A_1|t|^{\gamma_1}(\ln|t|)^m\} \tag{2.4.32}$$

with $0 \leqq m \leqq n$ and $A_1 > 0$ an integer, or

(2) f_2 given by

$$f_2(t) = \exp\{-At^2 - |t|^{\gamma_1}P(\ln|t|) - |t|^{\gamma_2}\} \tag{2.4.33}$$

will be a c.f.

Proof of lemma 2.4.2: We first take the function f_1 with A and m arbitrary numbers satisfying the conditions given above. Set

$$y_1(x) = (1/2\pi)\int e^{iux} f_1(u)\, du. \tag{2.4.34}$$

It is clear from (2.4.32) that the integral in (2.4.34) converges absolutely and that

$$y_1(x) = (1/\pi)\,\mathrm{Re}\int_0^\infty e^{iux} f_1(u)\,du. \tag{2.4.35}$$

Write, for $u > 0$,

$$f_1(u) = \exp\,[-\,Au^2 - A_1 u^{\gamma_1}(\ln u)^m]. \tag{2.4.36}$$

We note that f_1 is analytic in the complex plane $w = u + iv$ with a cut along the positive semiaxis and continuous up to the cut. We shall show that for any $\Lambda = \Lambda(x) > 0$

$$\int_0^\infty e^{iux} f_1(u)\,du = \int_{L_1} e^{iux} f_1(u)\,du + \int_{L_2} e^{iux} f_1(u)\,du \tag{2.4.37}$$

where L_1 is the segment $[u = 0,\ 0 \leqq v \leqq \Lambda]$ of the imaginary axis, and L_2 is the half-line $[v = \Lambda,\ 0 \leqq u < \infty]$.

To prove the above, we apply Cauchy's theorem to the rectangular contour having two of its sides along the axes of coordinates, with a cut given by

$$w = \rho e^{i\phi}, \qquad \rho \to 0, \qquad 0 \leqq \phi \leqq \pi/2.$$

We need only show that, as $u \to \infty$,

$$\int_0^\Lambda e^{i(u+iv)x} f_1(u + iv)\,dv \to 0. \tag{2.4.38}$$

Setting $u + iv = \mathrm{R}e^{i\phi}$, we see that, for $0 < v < \Lambda$ and $u \to \infty$, $\phi \to 0$. Hence $(u + iv)^{\gamma_1}[\ln\,(u + iv)]^m$ will have its argument as small as desired, and

$$\mathrm{Re}\,\{(u + iv)^{\gamma_1}[\ln\,(u + iv)]^m\} > 2^{-1}\rho^{\gamma_1}. \tag{2.4.39}$$

Hence it follows that $f_1(u + iv) \to 0$ uniformly in v and the integral in (2.4.38) has the estimate

$$\int_0^\Lambda e^{-v}|\,f_1(u + iv)|\,dv \to 0.$$

Thus (2.4.37) stands proved. In what follows, choosing a suitable $\Lambda = \Lambda(x)$, we first find an asymptotic expression for $y_1(x)$ and show that $y_1(x) > 0$ for $x \geqq x_0(A)$. Then we show that, for sufficiently large A, $y_1(x) > 0$ also for $0 \leqq x < x_0(A)$.

Let

$$x > k_0\,A^{(1/2)+\eta_0}, \ A \geqq 1 \tag{2.4.40}$$

where k_0, η_0 are positive constants; the value of k_0 is to be fixed later. Let

$$\Lambda(x) = (\ln x)^2/x. \tag{2.4.41}$$

We have

$$I_2 = \int_{L_2} e^{iwx} f_1(w)\, dw = \int_0^{\infty} e^{i(u+i\Lambda(x))} f_1(u + i\Lambda(x))\, du \tag{2.4.42}$$

whence

$$|I_2| \leqq e^{-x\Lambda(x)} \int_0^{\infty} |f_1(u + i\Lambda(x))|\, du. \tag{2.4.43}$$

The integral on the right exists finitely in view of the obvious inequality

$$\ln |f_1(u + i\Lambda(x))| < C' - C'' u^{\gamma_1}. \tag{2.4.44}$$

Hence, by our choice (2.4.41) of the function $\Lambda(x)$,

$$I_2 = O(e^{-(\ln x)^2}). \tag{2.4.45}$$

We shall now calculate

$$I_1 = \int_{L_1} e^{iwx} f_1(w)\, dw = i \int_0^{\Lambda(x)} e^{-vx} \exp\left[Av^2 - A_1 e^{i\pi 2^{-1}\gamma_1} v^{\gamma_1}\left(\ln v + \frac{i\pi}{2}\right)^m\right] dv. \tag{2.4.46}$$

For k_0 sufficiently large we have, independently of A,

$$Av^2 < A\,\frac{(\ln x)^4}{x^2} < \frac{[\ln (k_0 A^{(1/2)+\eta_0})]^4}{k_0^2 A^{2\eta_0}} < \frac{1}{4},$$

$$\left| v^{\gamma_1}\left(\ln v + \frac{i\pi}{2}\right)^m \right| < v^{\gamma_1/2} < \frac{1}{x} < \frac{1}{4}$$

with our notations. Hence in (2.4.46), the integrand can be rewritten as*

$$ie^{-vx} e^{Av^2}\left[1 - A_1 e^{(i\pi\gamma_1/2)} v^{\gamma_1}\left(\ln v + \frac{i\pi}{2}\right)^m + Bv^{2\gamma_1}(\ln v)^{2m}\right].$$

The real part of this function will be

$$\operatorname{Re}\left\{e^{-vx}\, i\left[-A_1 e^{i\pi\gamma_1/2} v^{\gamma_1}\left(\ln v + \frac{i\pi}{2}\right)^m \right.\right.$$
$$\left.\left. + A_1 \cdot A \cdot e^{i\pi\gamma_1/2} v^{\gamma_1+2}\left(\ln v + \frac{i\pi}{2}\right)^m + BA^2 v^{2\gamma_1+2}(\ln v)^{2m}\right]\right\}. \tag{2.4.47}$$

* Hereafter in the proof of the lemma, we shall denote by B any quantity bounded in absolute value, and not always one and the same.

We also have for $c > 0$, $\mu > 0$,

$$\int_0^\infty e^{-vx} v^c (\ln v)^\mu \, dv = \int_0^\infty e^{-t} \left(\frac{t}{x}\right)^c (\ln t - \ln x)^\mu \frac{dt}{x}$$

$$= \frac{1}{x^{c+1}} (-\ln x)^\mu \int_0^\infty e^{-t} t^c \, dt + B \frac{(\ln x)^{\mu-1}}{x^{c+1}} \tag{2.4.48}$$

$$= \frac{\Gamma(c+1)}{x^{c+1}} (-\ln x)^\mu + B \frac{(\ln x)^{\mu-1}}{x^{c+1}}.$$

The integral of (2.4.47) over the interval $[\Lambda(x), \infty)$ gives a term $B \exp(-(\ln x)^2)$; then, using (2.4.46) and (2.4.48), we have

$$\operatorname{Re}(I_1) = \operatorname{Re}\left[-iA_1 e^{i\pi\gamma_1/2} \int_0^\infty e^{-vx} v^{\gamma_1} \left(\ln v + \frac{i\pi}{2}\right)^m dv\right]$$

$$+ BA \frac{(\ln x)^m}{x^{\gamma_1+3}} + BA^2 \frac{(\ln x)^{2m}}{x^{2\gamma_1+3}}. \tag{2.4.49}$$

For the indicated values of x, the last two terms in (2.4.49) give

$$B \frac{A}{x^2} \frac{(\ln x)^m}{x^{\gamma_1+3}} \tag{2.4.50}$$

where $A/x^2 < 1$. We take the first term in (2.4.49)

$$T(x) = \operatorname{Re}\left[-iA_1 e^{i\pi\gamma_1/2} \int_0^\infty e^{-vx} v^{\gamma_1} \left(\ln v + \frac{i\pi}{2}\right)^m dv\right]. \tag{2.4.51}$$

Let us first take the case γ_1 is not an even integer, and $[\gamma_1/2]$ is even. Then, in the expression (2.4.32) for $f_1(u)$, we set $m = 0$ and obtain ($\sin(\pi\gamma_1/2) > 0$)

$$T(x) = A_1 \sin \frac{\pi\gamma_1}{2} \cdot \frac{\Gamma(\gamma_1 + 1)}{x^{\gamma_1+1}} \tag{2.4.52}$$

and in view of the estimates above

$$y_1(x) > \frac{1}{2} A_1 \sin \frac{(\pi\gamma_1/2)}{x^{\gamma_1+1}} \tag{2.4.53}$$

for $x \geqq k_0 A^{1/2+\eta_0}$.

We then show that $y_1(x) > 0$ for $0 \leqq x < k_0 A^{1/2+\eta_0}$, if A is sufficiently large. We set $x = \xi A^{1/2}$ and in (2.4.35) take $u = tA^{-1/2}$. Then (2.4.35) can be rewritten in the form

$$y_1(x) = \frac{1}{\pi\sqrt{A}} \operatorname{Re} \int_0^\infty \exp(i\xi t) \exp\left(-t^2 - A_1 \left(\frac{t}{\sqrt{A}}\right)^{\gamma_1} \left(\ln \frac{t}{\sqrt{A}}\right)^m\right) dt. \tag{2.4.54}$$

We set $t_1 = \ln A$; then

$$y_1(x) = \frac{1}{\pi\sqrt{A}} \operatorname{Re} \int_0^{t_1} \exp(i\xi t) \exp\left(-t^2 - A_1\left(\frac{t}{\sqrt{A}}\right)^{\gamma_1}\left(\ln\frac{t}{\sqrt{A}}\right)^m\right) dt$$
$$+ B\exp(-(\ln A)^2). \quad (2.4.55)$$

We shall take A to be so large that for $t \leqq t_1$

$$A_1\left(\frac{t}{\sqrt{A}}\right)^{\gamma_1}\left(\ln\frac{t}{\sqrt{A}}\right)^m < 1. \quad (2.4.56)$$

Then expanding in series the exponent corresponding to this term, we easily see that

$$y_1(x) = \frac{1}{\pi\sqrt{A}} \operatorname{Re} \int_0^{t_1} e^{i\xi t} e^{-t^{\sigma_1}}\, dt + \frac{B}{\sqrt{A}} A^{-(1/2)\gamma_1}\left(\ln\frac{1}{\sqrt{A}}\right)^m \quad (2.4.57)$$

or, in view of the definition of t_1, for sufficiently large A,

$$y_1(x) = \frac{1}{\pi\sqrt{A}} \operatorname{Re} \int_0^{\infty} e^{i\xi t - t^2}\, dt + \frac{B}{\sqrt{A}} A^{-(1/2)\gamma_1}$$

or

$$y_1(x) = \frac{1}{\pi\sqrt{A}} e^{-\xi^2/4} + \frac{B}{\sqrt{A}} A^{-(1/2)\gamma_1} \quad (2.4.58)$$

for sufficiently large A and for $0 \leqq x < k_0 A^{1/2+\eta_0}$. Thus, we have established that $y_1(x) > 0$ for all real x.

Let now γ_1 be not an even integer and $[\gamma_1/2]$ be odd. If the degree n of the polynomial $P(z)$ be $\geqq 1$, then we set $m = 1$, $A_1 = 1$. Then, in view of (2.4.48), (2.4.50) gives

$$T(x) = \sin\left(\frac{\pi\gamma_1}{2}\right) \cdot \left(-\frac{\Gamma(\gamma_1 + 1)}{x^{\gamma_1+1}} \ln x\right) + B \cdot \frac{1}{x^{\gamma_1+1}}$$

where $-\sin[(1/2)\pi\gamma_1] = |\sin[(1/2)\pi\gamma_1]| > 0$. Under the conditions (2.4.40), we obtain $y_1(x) > 0$. If under the same conditions the degree $n = 0$, then $P(z) = a_0 < 0$, and in this case we set up the function (2.4.33) which takes the form

$$f_2(u) = \exp(-Au^2 - a_0|u|^{\gamma_1} - |u|^{\gamma_2}).$$

Arguing wholly similarly to the above, we can see that for $x > k_0 A^{(1/2)+\eta_0}$

$$y_2(x) = \frac{1}{\pi} \operatorname{Re} \int_0^{\infty} e^{iux} f_2(u)\, du > \frac{1}{2} \frac{|a_0 \sin[(1/2)\pi\gamma_1]|}{x^{\gamma_1+1}}. \quad (2.4.59)$$

We have yet to consider the case when γ_1 is an even integer. Under the conditions, we must have $n \geqq 1$.

We shall distinguish two cases: $\gamma_1 \equiv 0 (\text{mod } 4)$ and $\gamma_1 \equiv 2 (\text{mod } 4)$. In the first case, turning to (2.4.49) and (2.4.50), we will have $\exp[(1/2)\pi\gamma_1 i] = 1$, and with $A_1 = 1$ and taking (2.4.48) into account,

$$T(x) = \text{Re}\left[-i\int_0^\infty e^{-vx}v^{\gamma_1}\left(\ln v + \frac{i\pi}{2}\right)^m dv\right]$$
$$= \frac{\pi m}{2}\frac{\Gamma(\gamma_1+1)(-\ln x)^{m-1}}{x^{\gamma_1+1}} + B\cdot\frac{(\ln x)^{m-2}}{x^{\gamma_1+1}}. \tag{2.4.60}$$

For odd m, $T(x)$ will be positive and

$$y_2(x) > \frac{1}{2}\pi m\Gamma(\gamma_1+1)\frac{(\ln x)^{m-1}}{x^{\gamma_1+1}}. \tag{2.4.61}$$

We may take $m = 1$ in the above.

Let, finally, $\gamma_1 \equiv 2 (\text{mod } 4)$. Then $\exp[(1/2)\pi\gamma_1 i] = -1$, and

$$T(x) = \text{Re}\left[i\int_0^\infty e^{-vx}v^{\gamma_1}\left(\ln v + \frac{i\pi}{2}\right)^m dv\right]$$
$$= -\frac{\pi m}{2}\frac{\Gamma(\gamma_1+1)}{x^{\gamma_1+1}}(-\ln x)^{m-1} + B\frac{(\ln x)^{m-1}}{x^{\gamma_1+1}}. \tag{2.4.62}$$

If $m = 2$, then

$$y_2(x) > \frac{1}{2}\pi\Gamma(\gamma_1+1)\frac{\ln x}{x^{\gamma_1+1}}. \tag{2.4.63}$$

If further $n = 1$ and we cannot choose $m = 2$, then

$$P(\ln|u|) = a_0 \ln|u| + a_1, \qquad a_0 < 0.$$

In this case, we shall set up our function in the form

$$f_2(u) = \exp(-Au^2 - |u|^{\gamma_1}(a_0\ln|u| + a_1) - |u|^{\gamma_2}). \tag{2.4.64}$$

In view of the foregoing, we obtain

$$y_2(x) > \frac{\pi\Gamma(\gamma_1+1)}{4x^{\gamma_1+1}}.$$

To conclude the proof of Lemma 2.4.2, it now remains to show that in all those cases where we have proved that $y_j(x) > 0$ $(j = 1, 2)$ for $x > k_0 A^{(1/2)+\eta_0}$, we will also have $y_j(x) > 0$ for

$$0 \leqq x < k_0 A^{1/2}. \tag{2.4.65}$$

For this purpose, consider

$$y_j(x) = \frac{1}{\pi} \operatorname{Re} \int_0^\infty e^{iux} f_j(u)\, du \tag{2.4.66}$$

under the conditions (2.4.65). Then $f_j(u)$ being chosen as above,

$$f_j(u) = \exp\left(-Au^2 \pm |u|^{\gamma_1} Q(\ln|u|) - b|u|^{\gamma_2}\right) \tag{2.4.67}$$

where Q is a polynomial of degree $\leqq n$, and $b = 0$ or 1. We take (2.4.66) and set $\xi = x/\sqrt{A}$; $u = t/\sqrt{A}$. Then

$$y_j(x) = \frac{1}{\pi\sqrt{A}} \operatorname{Re} \int_0^\infty e^{i\xi t} \exp\left[-t^2 + \left(\frac{t}{\sqrt{A}}\right)^2 Q\left(\ln\left|\frac{t}{\sqrt{A}}\right|\right) - b\left|\frac{t}{\sqrt{A}}\right|^{\gamma_2}\right] dt.$$

In view of the foregoing, such a substitution gives us (the role of A is played by 1)

$$y_j(x) > 0 \qquad \text{for} \qquad \xi > k_0.$$

Setting $t_1 = \ln A$ as before, we find

$$y_j(x) = \frac{1}{\pi\sqrt{A}} \operatorname{Re} \int_0^{t_1} e^{i\xi t - t^2}\, dt + B \frac{A^{-\gamma_1/2}}{\sqrt{A}} (\ln A)^n$$

or

$$y_j(x) = \frac{1}{\pi\sqrt{A}} \operatorname{Re} \int_0^\infty e^{i\xi t - t^2}\, dt + \frac{B}{\sqrt{A}} A^{-\gamma_1/2} (\ln A)^n$$

$$= \frac{1}{\sqrt{2\pi A}} e^{-\xi^2/4} + \frac{B}{\sqrt{A}} A^{-\gamma_1/2} (\ln A)^n. \tag{2.4.68}$$

For $\xi < k_0$ and sufficiently large A, $y_j(x) > 0$. Thus Lemma 2.4.2 stands proved completely.∎

From the lemma just proved, it follows that if one of the conditions (2), (3), and (4) is violated, then it is possible to construct a c.f. of the form (2.4.32) or (2.4.33), other than the c.f.'s of the normal law, satisfying the conditions of Theorem 2.3.3. If condition (1) is also violated, then a normal law cannot satisfy the conditions of the theorem. Thus the theorem 2.4.2 is proved.∎

We remark that conditions characterizing the normal law can be formulated in terms of the simultaneous satisfaction of both the relations (2.1.7) and (2.1.9). In such a case, it is obviously sufficient to add to the conditions of Theorem 2.4.2 a condition (1′) guaranteeing together with (1), in the presence of (2.1.7), a solution of the form $w(t) = iat - bt^2$, where a and b are real constants, and assume that (2.1.9) has been obtained from (2.1.7) by separating out the real parts.

We also take note of a particular case when certain assumptions about the law F permit the simplification of the conditions for the normality of F. This is the case when the law concerned possesses finite moments of a sufficiently high order. Here the following proposition holds (cf. [81], Theorem II′).

THEOREM 2.4.3. If F satisfies the conditions of Theorem 2.3.3, and the least upper bound σ_2 of the abscissae of the active exponents corresponding to the expansion (2.2.1) exists finitely, and F has finite moment of order $2[(\sigma_2/2) + 1]$, then F is normal.

An immediate consequence of Theorem 2.4.3 is a theorem with more stringent conditions, also established in Linnik [81].

THEOREM 2.4.3′. Let the two linear forms given by (2.1.1) be identically distributed, and let the condition (2.1.12) be satisfied:

$$\max(|a_1|, \ldots, |a_n|) \neq \max(|b_1|, \ldots, |b_n|).$$

Let γ be the maximal real zero of the determining function $\sigma(z)$. If F has a finite moment of order $2m$, where $m = [(\gamma/2) + 1]$, then F is normal.

Proof of Theorem 2.4.3. In this case, in view of Theorem 2.2.3, the real active exponents corresponding to $\phi = -\ln|f|$, where f is the c.f. of F, can only be even integers, and the exponents are also simple. According to Theorem 2.2.3, there are no complex active exponents. Thus, in some neighborhood of the origin, the relation

$$|f(t)|^2 = \exp[P(t^2)] \tag{2.4.69}$$

holds. Then we can apply the following lemma ([81], Lemma XII).

LEMMA 2.4.3. Let the c.f. g of some random variable have the form

$$g(u) = \exp Q(u) \tag{2.4.70}$$

in some neighborhood $|u| < \delta$ of the origin, where Q is a polynomial. Then

$$Q(u) = Au^2 + iCu \tag{2.4.71}$$

where $A \leq 0$ and C are real constants, and relation (2.4.71) holds for all real u.

Proof: g has derivatives of all orders at $u = 0$. Since g is a c.f., if G be the corresponding d.f., then

$$g(u) = \int e^{iux}\, dG(x) \tag{2.4.72}$$

and (cf. (viii) of section 1.1.1) G has moments of all orders. If these be α_n, then $|\alpha_n| = |g^{(n)}(0)|$. We estimate the growth of α_n in terms of n. We have by Cauchy's integral formula

$$g^{(n)}(0) = \frac{d^n}{du^n} e^{Q(u)}\bigg|_{u=0} = \frac{n!}{2\pi i} \oint_{C_R} \frac{e^{Q(u)}}{u^{n+1}}\, du \tag{2.4.73}$$

where C_R is the circle $|u| = R$. But $|Q(u)| \leqq k|u|^q$ for $|u| > 1$. Hence, for any $R > 0$,

$$|g^{(n)}(0)| \leqq \frac{n!}{2\pi} \cdot 2\pi \cdot \frac{1}{R^n} e^{kR^q} \tag{2.4.74}$$

We choose $R = n^{1/q}$; then, for $n \geqq n_0$, (2.4.74) gives

$$|g^{(n)}(0)| \leq n!\, n^{-n/q} e^{kn} = n!\, n^{-n/2q} e^{kn-(n/2)\ln(n^{1/q})} \leq n!\, n^{-n/2q}. \tag{2.4.75}$$

Hence, for even $n = 2m > n_0$, we have $\alpha_{2m} < (2m)!/(2m)^{m/q}$, and $\gamma_{2m-1} < \alpha_{2m}$ where γ_{2m-1} is the absolute $(2m-1)$-th moment. Hence, for any $r > 0$,

$$\int_{-\infty}^{\infty} e^{rx}\, dG(x) = \sum_{n=0}^{\infty} \frac{r^n}{n!} \int_{-\infty}^{\infty} x^n\, dG(x)$$

$$\leqq \sum_{n=0}^{n_0} \frac{r^n}{n!} \int_{-\infty}^{\infty} x^n\, dG(x) + 2 \sum_{n_0+1}^{\infty} \left(\frac{r}{n^{(1/2q)}}\right)^n < \infty. \tag{2.4.76}$$

Thus g is an entire function, and since it coincides for $|u| < \delta$ with the entire function $\exp Q$, $g(u) = \exp Q(u)$ everywhere, and in particular for real u. By the lemma of Marcinkiewicz mentioned earlier, g is therefore the c.f. of a normal law, so that $Q(u) = Au^2 + iCu$, $A \leqq 0$, C real.∎

In various problems concerning characterizations of the normal law, a very important role is played by factorizations of the form (cf. for instance [85], p. 78)

$$f(t) = f_1^{\alpha_1}(t) \cdots f_s^{\alpha_s}(t), \qquad |t| < t' \tag{2.4.77}$$

where $\alpha_1, \ldots, \alpha_s$ are some positive numbers, and the f_j are characteristic functions all of which are nonvanishing for $|t| < t'$. $f_j^{\alpha_j}(t)$ is defined of course as $\exp[\alpha_j \ln f_j(t)]$, $j = 1, \ldots, s$. Analytical properties of such functions have been considered in detail in the work [85] in connection with the theory of α-decompositions.

Functions representable in the form (2.4.77) are in general not c.f.s, but have many properties in common with them. For our applications to characterization problems connected with the normal law, the following generalization of the decomposition theorem of Cramer's, indicated in [38], is essential.

The results obtained in section 2.4 can in particular be carried over to functions of the form (2.4.77). For this purpose, the role played by Lemma 2.4.1 in our earlier considerations is now taken over by the following generalization thereof.

LEMMA 2.4.1′. Let $g_1, \ldots, g_s$ be c.f.s of symmetric laws $G_1, \ldots, G_s$, be nonvanishing for $|t| < t'$ and let

$$g(t) = g_1^{\alpha_1}(t) \cdots g_s^{\alpha_s}(t) \tag{2.4.78}$$

where the α_j are positive constants. If each of the G_j has finite moment of order $2k$, for some integer $k > 0$, and at least one of them does not have finite moment of order $2k + 2$, then there exists $t'' \in (0, t']$ such that, for all $|t| < t''$, $\ln g(t)$ exists, has continuous derivative of order $2k$, and

$$(-1)^{k+1}\{[\ln g(t)]^{(2k)} - ([\ln g(t)]^{(2k)}|_{t=0})\} \geqq 0, \qquad |t| < t''. \tag{2.4.79}$$

The problem of identical distribution of pairs of linear statistics (L_1, L_2) and (L_3, L_4) is considered in Chapter 10, p. 317. The problem is the same if the conditional distribution of L_1 given L_2 is identical with the conditional distribution of L_3 given L_4. In general $L_1, \ldots, L_4$ may be linearly dependent.

2.5 IDENTICALLY DISTRIBUTED LINEAR VECTOR STATISTICS

The problem of describing the class of laws admitting identically distributed linear statistics can be extended to multivariate distributions.

Let $\mathbf{X}$ be a random vector in n-dimensional Euclidean space, having the d.f. $F(y_1, \ldots, y_n)$. We shall take vectors as column vectors. We take $\mathbf{X}^1, \ldots, \mathbf{X}^r$ as r mutually independent vectors, identically distributed according to the above law F. Let us form the vector statistics

$$\mathbf{L}_1 = \sum_1^r \mathbf{A}_j \mathbf{X}^j \quad \text{and} \quad \mathbf{L}_2 = \sum_1^r \mathbf{B}_j \mathbf{X}^j \tag{2.5.1}$$

where the $\mathbf{A}_j$ and $\mathbf{B}_j$ are real square $(n \times n)$ matrices. Here, as in the univariate case, one may raise the problem of describing the class $\Omega^{(n)}_{L_1, L_2}$ of all laws F for which the random vectors $\mathbf{L}_1$ and $\mathbf{L}_2$ are identically distributed.

If we introduce $f(\mathbf{u})$, the c.f. of the distribution of the random vector $\mathbf{X}$, denoting as usual (the symbol T denotes the operation of transposing)

$$f(\mathbf{u}) = E \exp i\mathbf{u}^T\mathbf{x} \tag{2.5.2}$$

then the condition $F \in \Omega^{(n)}_{L_1, L_2}$ can be rewritten in the equivalent form

$$\prod_{j=1}^r f(\mathbf{A}_j^T \mathbf{u}) = \prod_{j=1}^r f(\mathbf{B}_j^T \mathbf{u}), \qquad \mathbf{u} \in R^n. \tag{2.5.3}$$

In some neighborhood of the origin in R^n, $w(\mathbf{u}) = \ln f(\mathbf{u})$ exists. If we take a neighborhood where $w(\mathbf{A}_j^T\mathbf{u})$ and $w(\mathbf{B}_j^T\mathbf{u})$ exist, then we have from (2.5.3)

$$\sum_{j=1}^{r} w(\mathbf{A}_j^T\mathbf{u}) = \sum_{j=1}^{r} w(\mathbf{B}_j^T\mathbf{u}). \tag{2.5.4}$$

The above equation can be generalized in the spirit of (2.1.7). The construction and study of general solutions of (2.5.4) or of its linear generalizations, and the singling out therefrom of the probabilistic solutions, encounter a whole series of specific difficulties, connected primarily with the significantly more complicated character of particular solutions of (2.5.4). This circumstance can be seen, for example, by a comparison of multivariate stable distributions studied by Sakovich (cf. [135]) and the univariate stable laws. At the present time, only isolated results of a specialized nature are known, relating to probabilistic solutions of equations of the type (2.5.4) (cf. [47]).

We shall briefly touch here on just one problem relating to vectors with independent components, which admits reduction to a case studied above.

If we consider vectors with independent components, then a natural generalization of the univariate problem (section 2.1) consists in the assumption that the corresponding components (those with the same index) of the vectors $\mathbf{L}_1$ and $\mathbf{L}_2$ are identically distributed.

Let $\mathbf{X} = (X_1, \ldots, X_n)$ be a random vector in R^n, whose components are independent. We write

$$\begin{aligned} &f_j(t) = E\exp(itX_j), \qquad t \text{ real} \\ &\mathbf{A}_j = ((a_{kl}^j))_1^n; \qquad \mathbf{B}_j = ((b_{kl}^j))_1^n; \qquad j = 1, \ldots, r. \end{aligned} \tag{2.5.5}$$

If $\mathbf{X}^1, \ldots, \mathbf{X}^r$ are random vectors in R^n, mutually independent and all having the same distribution as $\mathbf{X}$, then the condition of identical distribution of components of the vectors $\mathbf{L}_1$ and $\mathbf{L}_2$ with the same suffix is equivalent to the condition that the corresponding c.f.s satisfy the system of equations

$$\prod_{l=1}^{n}\prod_{j=1}^{r} f_l(a_{kl}^j t) = \prod_{l=1}^{n}\prod_{j=1}^{r} f_l(b_{kl}^j t) \qquad \text{for all } t, \tag{2.5.6}$$
$$k = 1, 2, \ldots, n.$$

In some interval $|t| < t_0$, $f_l(t) \neq 0$ and $w_l(t) = \ln f_l(t)$ exists for $l = 1, 2, \ldots, n$. Choosing $t_1 > 0$ such that for $|t| < t_1$

$$f_l(a_{kl}^j t) \neq 0, \qquad f_l(b_{kl}^j t) \neq 0; \qquad k, l = 1, 2, \ldots, n; \qquad j = 1, 2, \ldots, r.$$

We have from (2.5.6)

$$\sum_{l=1}^{n}\sum_{j=1}^{r} w_l(a_{kl}^j t) = \sum_{l=1}^{n}\sum_{j=1}^{r} w_l(b_{kl}^j t), \qquad |t| < t_1, \qquad 1 \leq k \leq n. \tag{2.5.7}$$

The system (2.5.7) or (2.5.6) can be reduced, by eliminating all the c.f.s concerned excepting one, to a system of equations, one for each of the c.f.s, of the type (2.1.2), respectively (2.1.3). In the cases where these equations are found to be nondegenerate, the foregoing theory can be applied to them.

We take the case $n = 2$ for illustrative purposes and formulate conditions under which a characterization of the normal law obtains.

Rewriting (2.5.7) for $n = 2$ in the form ($\varepsilon_{kl}^j = \pm 1$)

$$\sum_{j=1}^{2r} \{\varepsilon_{k1}^j w_1(e_{k1}^j t) + \varepsilon_{k2}^j w_2(e_{k2}^j t)\} = 0 \tag{2.5.8}$$

where $e_{kl}^j = a_{kl}^j$, $e_{kl}^{r+j} = b_{kl}^j$ for $j = 1, \ldots, r$, we obtain by eliminating one of the functions from (2.5.7), for sufficiently small t

$$\sum_{j'=1}^{2r} \sum_{j''=1}^{2r} \{\varepsilon_{11}^{j'} \varepsilon_{22}^{j''} w(e_{11}^{j'} e_{22}^{j''} t) + \varepsilon_{21}^{j'} \varepsilon_{12}^{j''} w(e_{21}^{j'} e_{12}^{j''} t)\} = 0. \tag{2.5.9}$$

Each of the functions w_1 and w_2 satisfies (2.5.9) for sufficiently small t.

We introduce a condition corresponding to (2.1.12)

$$\begin{aligned} &\max\,[\{|e_{11}^{j'_1} e_{22}^{j''_1}|,\ |e_{21}^{j'_2} e_{12}^{j''_2}|\}: \{j'_1, j''_1; j'_2, j''_2: \varepsilon_{11}^{j'_1}\varepsilon_{22}^{j''_1} = \varepsilon_{21}^{j'_2}\varepsilon_{12}^{j''_2} = 1,\\ &\qquad (j) = 1, 2, \ldots, 2r\}]\\ &\neq \max\,[\{|e_{11}^{j'_1} e_{22}^{j''_1}|,\ |e_{21}^{j'_2} e_{12}^{j''_2}|\}: \{j'_1, j''_1, j'_2, j''_2: \varepsilon_{11}^{j'_1}\varepsilon_{22}^{j''_1} = \varepsilon_{21}^{j'_2}\varepsilon_{12}^{j''_2} = -1,\\ &\qquad (j) = 1, 2, \ldots, 2r\}] \end{aligned} \tag{2.5.10}$$

and the defining function $\sigma(z)$

$$\sigma(z) = \sum \{\varepsilon_{11}^{j'} \varepsilon_{22}^{j''} |e_{11}^{j'} e_{22}^{j''}|^z + \varepsilon_{21}^{j'} \varepsilon_{12}^{j''} |e_{21}^{j'} e_{12}^{j''}|^z\} \tag{2.5.11}$$

where the summation is taken over $e_{11}^{j'} \cdot e_{22}^{j''}$, $e_{21}^{j'} \cdot e_{12}^{j''}$ different from zero. In terms of (2.5.10) and (2.5.11), we can rewrite Theorem 1 of [81] (Theorem 2.4.2 above) for the system (2.5.6).

THEOREM 2.5.1. In order that the c.f.s f_1 and f_2 satisfying (2.5.6) in the presence of (2.5.10)* correspond to normal laws, it is necessary and sufficient that $\sigma(z)$ defined by (2.5.11) satisfy the conditions (1)–(4) of Theorem 2.4.2. For $n > 2$, similar results hold.

The system (2.5.7) admits generalization in the form

$$\sum_{l=1}^{n} \int_{-a'}^{a'} w_l(at)\, dV_{kl}^{\mathrm{I}}(a) = \sum_{l=1}^{n} \int_{-a'}^{a'} w_l(at)\, dV_{kl}^{\mathrm{II}}(a) \tag{2.5.12}$$

* For simplicity of presentation, we have omitted statements of conditions corresponding to normal solutions of (2.5.6) for $n > 2$.

where the functions $V_{kl}^{\mathrm{I(II)}}(a)$ have the same character as the corresponding functions in (2.1.7). To carry out a study of (2.5.12) similar to that of (2.1.7), we may use the Laplace transform, whose application permits the reduction of the problem to linear algebraic systems. If we divide the probabilistic solutions of the system (2.5.12) into two classes, corresponding to the real and imaginary parts of the unknown functions, then, in those cases where the system of equations corresponding to the real parts is nondegenerate, results wholly similar to Theorems 2.3.3 and 2.4.2 hold; and if the systems corresponding to the real and the imaginary parts are both nondegenerate, then Theorem 2.3.4 has its complete analogue.

We also note that even in some cases where the systems of equations become degenerate, we may characterize the normal solutions. This happens, for example, in cases where the discussion can be reduced to the consideration of linear combinations with positive coefficients of the functions $\ln w_j(t)$, $j = 1, 2, \ldots, n$, and to the application of generalizations of the theorems obtained in section 2.4.

For further remarks on the problem of identical distribution of linear vector statistics, see Addendum A at the end of the book.

2.6 SUPPLEMENTARY REMARKS. ORDER STATISTICS

We begin by making some supplementary remarks. As already stated in the introduction, the substitution $X_i = H(Y_i)$, where H is some measurable function, in (2.1.1), transforms linear functions of the X_i into linear functions of the $H(Y_i)$, and their identical distribution yields theorems characterizing the Y_i, corresponding to each of the above theorems characterizing the $H(Y_i)$ (for instance, those guaranteeing normality).

Of greater interest in this respect is the application of algebraic addition theorems, in particular "relativistic linear forms" (cf. section 3.6). Applications to statistical physics are also possible (cf. section 3.7).

We also note that in view of the well-known duality between c.f.'s and d.f.'s for addition of independent random variables and the formation of order statistics, we may consider, by analogy with identically distributed linear statistics, identically distributed order statistics of a random sample (cf. [80]). The operation of addition of random variables, corresponding to which the c.f.s are multiplied, will correspond to the operation of taking the maximal term of a variational series, corresponding to which the integral laws are multiplied. The linear statistic $a_1 X_1 + \cdots + a_n X_n$ will correspond to a statistic of the form

$$\max\,[(X_1 - a)/A_1, \ldots, (X_n - a)/A_n] \tag{2.6.1}$$

where the A_i are positive, and a is some real number. We shall call them *M-statistics.*

The following "dual" to Theorem 2.4.3 holds.

THEOREM 2.6.1. Let $M_1 = \max\,[(X_1 - a)/A_1, \ldots, (X_n - a)/A_n]$ and $M_2 = \max\,[(X_1 - a)/B_1, \ldots, (X_n - a)/B_n]$ be two M-statistics which are identically distributed. Let further

$$\min\,(A_1, \ldots, A_n) \neq \min\,(B_1, \ldots, B_n).$$

We set up the "determining function"

$$\sigma(z) = A_1^{-z} + \cdots + A_n^{-z} - B_1^{-z} - \cdots - B_n^{-z}.$$

Let γ be the maximum modulus of its real zeros, and $m = [(1/2)\gamma + 1]$. If the integral law F vanishes for $x < a$, and, for some $\delta > 0$,

$$\frac{d^{2m}}{dx^{2m}}\,[(x - a)^{2m} \ln F(x)] \qquad \text{exists for} \qquad x \in (a, a + \delta)$$

and converges to a finite limit as $x \to a + 0$, then

$$F(x) = \exp\left[-\frac{E_1}{(x - a)} - \frac{E_2}{(x - a)^2} - \cdots - \frac{E_{2m}}{(x - a)^{2m}}\right] \tag{2.6.2}$$

where the E_j are indeterminate constants, not necessarily positive.

We shall not pause to prove this and similar theorems given in the Note [80]. The proof is "dual" in character to those of the earlier sections of this chapter. In particular, the condition of the existence of the limit of

$$\frac{d^{2m}}{dx^{2m}}\,[(x - a)^{2m} \ln F(x)] \text{ as } x \to a + 0$$

is the dual, in a well-known sense, of the condition of the existence of the moment of order $2m$ in Theorem 2.4.3′.

We note that (2.6.2) yields a set of d.f.'s with support on the half-line $x > a$, for any values of the parameters $E_1, \ldots, E_n$ for which F turns out to be a d.f.

CHAPTER 3

Independent Linear Statistics. Modifications. Relativistic Linear Statistics

3.1 THE SKITOVICH-DARMOIS THEOREM

As we have seen earlier (cf. Chapter 2), the phenomenon of independence of statistics is closely connected with that of their identical distribution. The most thoroughly studied phenomenon of independence is that concerning linear statistics. We first consider independent scalar random variables $X_1, \ldots, X_n$ (not necessarily identically distributed) and two linear statistics

$$L_1 = \sum \alpha_i X_i, \qquad L_2 = \sum \beta_i X_i \tag{3.1.1}$$

where the α_i, β_i are constant coefficients (in what follows, we shall also consider random coefficients).

As it happens, the independence of two linear statistics essentially characterizes the normality of the variables X_j. To be precise, the following assertion, due to Skitovich [138, 139] and Darmois [22] holds.

THEOREM 3.1.1. Let L_1 and L_2 given by (3.1.1) be independent. Then the random variables X_j for which $\alpha_j \beta_j \neq 0$ are all normal.

Remark. We may take note of the almost trivial converse proposition in the following form: if $\sum_1^n \sigma_j^2 \alpha_j \beta_j = 0$ and those X_j for which $\alpha_j \beta_j \neq 0$ are normal, then L_1 and L_2 are independent. Indeed, $\sum \sigma_j^2 \alpha_j \beta_j = 0$ is the condition of uncorrelatedness of our linear forms when the variables X_j entering into both the forms are normal.

Theorem 3.1.1 has an obvious geometrical interpretation. Let, for simplicity, all the $\alpha_j \cdot \beta_j$ be nonzero. Then this theorem can be formulated in geometrical language as follows: let $(X_1, \ldots, X_n)$ be a random vector in the Euclidean space R^n and a system of n axes $\boldsymbol{\delta}_1, \boldsymbol{\delta}_2, \ldots, \boldsymbol{\delta}_n$ be such that the projections of $(X_1, \ldots, X_n)$ on these axes are mutually independent; let further two axes, $\boldsymbol{\alpha}$ and $\boldsymbol{\beta}$ be such that they do not depend on subsets of

$\boldsymbol{\delta}_1, \ldots, \boldsymbol{\delta}_n$. Let the projections of the vector $(X_1, \ldots, X_n)$ on the $\boldsymbol{\alpha}$ and $\boldsymbol{\beta}$ axes be stochastically independent. Then the vector $(X_1, \ldots, X_n)$ has n-variate normal distribution.

Proof of the Theorem: Since L_1 and L_2 are stochastically independent, we have for all real u and v

$$E \exp \{i(uL_1 + vL_2)\} = E(\exp (iuL_1))E(\exp (ivL_2)). \tag{3.1.2}$$

For those j for which $\alpha_j \beta_j \neq 0$, we write $\beta_j X_j = Y_j$; $\beta_j/\alpha_j = b_j$. For those j for which $\beta_j \neq 0$, $\alpha_j = 0$, we write $\beta_j X_j = Y_j$; we denote the c.f.'s of the Y_j by ϕ_j.

We have

$$uL_1 + vL_2 = \sum (\alpha_j u + \beta_j v)X_j. \tag{3.1.3}$$

We consider such a neighborhood of the (u, v) origin, $|u| < \delta_0$, $|v| < \delta_0$, that for all j,

$$E \exp \{i(\alpha_j u + \beta_j v)X_j\} \neq 0; \qquad E \exp \{iu\alpha_j X_j\} \neq 0; \qquad E \exp \{iv\beta_j X_j\} \neq 0. \tag{3.1.4}$$

The above conditions are clearly trivial for $\alpha_j = 0$ or $\beta_j = 0$. From (3.1.2) and (3.1.3), we obtain for u and v in the neighborhood of the origin indicated above

$$\prod{}' \phi_j(u + b_j v) = \prod{}' \phi_j(u)\phi_j(b_j v) \tag{3.1.5}$$

where the product is taken over all those j for which $\alpha_j \beta_j \neq 0$. The other factors entering into (3.1.2) on the left and on the right cancel out. Changing the enumeration of the variables if necessary, we rewrite (3.1.5) in the form

$$\prod_1^m \phi_j(u + b_j v) = \prod_1^m \phi_j(u)\phi_j(b_j v). \tag{3.1.6}$$

In view of (3.1.4), we set in the above neighborhood of the origin $\psi_j(t) = \ln \phi_j(t)$, where $\psi_j(0) = 0$, and write (3.1.6) in the form

$$\sum_1^m \psi_j(u + b_j v) = A(u) + B(v). \tag{3.1.7}$$

We now apply the remark to Lemma 1.5.1, according to which $A(u)$ is a polynomial of degree at most m in some neighborhood of the origin. Thus, in that neighborhood, $\prod \phi_j$ has the form $\exp A$, where A is a polynomial of degree at most m. By Lemma 1.2.1, the ϕ_j are therefore c.f.'s of normal laws.

It is clear that under the independence of the linear forms (3.2.1), the relation $\sum \sigma_j^2 \alpha_j \beta_j = 0$, where $\sigma_j^2 = \text{Var}(X_j)$, is automatically satisfied.∎

Concerning special cases of Theorem 3.1.1 and their connection with Cramer's theorem on the decomposition of the normal law, cf. [85], Chapter VI. Also given there are generalizations to the case of "quasirandom" variables (having "distribution functions" which are not necessarily non-decreasing).

We remark that for characterizing normality, instead of the independence of L_1 and L_2, we may require the relation of constancy of regression of the type $E(L_2|L_1) = \text{constant}$. Concerning such phenomena, *vide* Chapter 5 on problems of regression connected with linear statistics.

3.2 MULTIVARIATE GENERALIZATIONS

Multivariate analogues of Theorem 3.1.1 were indicated by Skitovich [139]. Ghurye and Olkin [19] generalized these analogues. Zinger gave a new and relatively simple account of their results in his dissertation [37]. We shall follow his presentation below.

Let $\mathbf{X}^1, \ldots, \mathbf{X}^r$ be mutually independent random vectors in R^n. We consider the random vectors

$$\mathbf{L}_1 = \sum \mathbf{A}_j \mathbf{X}^j \quad \text{and} \quad \mathbf{L}_2 = \sum \mathbf{B}_j \mathbf{X}^j \tag{3.2.1}$$

where the $\mathbf{A}_j$ and $\mathbf{B}_j$ are real nonsingular $n \times n$ matrices.

THEOREM 3.2.1. If the linear statistics (3.2.1) are independent, then each of the vectors $\mathbf{X}^j (j = 1, 2, \ldots, r)$ is normally distributed.

***Proof*:** We write the condition of independence of $\mathbf{L}_1$ and $\mathbf{L}_2$ in terms of c.f.'s, in the form of the relation

$$\prod_1^r f_j(\mathbf{A}_j^T \mathbf{u} + \mathbf{B}_j^T \mathbf{v}) = \prod_1^r f_j(\mathbf{A}_j^T \mathbf{u}) f_j(\mathbf{B}_j^T \mathbf{v}) \tag{3.2.2}$$

where $\mathbf{u}$ and $\mathbf{v}$ are arbitrary column vectors in R^n: $\mathbf{u}^T = (u_1, \ldots, u_n)$ and $\mathbf{v}^T = (v_1, \ldots, v_n)$. The f_j are the c.f.'s of the $\mathbf{X}_j$.

In view of the nonsingularity of the $\mathbf{A}_j$, we may take $\mathbf{A}_j = \mathbf{I}_n$, the identity matrix, by suitably transforming the variables u_j. Then (3.2.2) takes the form

$$\prod_1^r f_j(\mathbf{u} + \mathbf{B}_j^T \mathbf{v}) = \prod_1^r f_j(\mathbf{u}) f_j(\mathbf{B}_j^T \mathbf{v}). \tag{3.2.3}$$

In view of the properties of c.f.'s, a $\rho_0 > 0$ can be found such that for all $\mathbf{u}$ and $\mathbf{v}$ such that $|\mathbf{u}| < \rho_0, |\mathbf{v}| < \rho_0$, the functions $f_j(\mathbf{u})$, $f_j(\mathbf{B}_j^T\ \mathbf{v})$, and $f_j(\mathbf{u} + \mathbf{B}_j^T\ \mathbf{v})$ do not vanish. Then we find from (3.2.3)

$$\sum_1^r \phi_j(\mathbf{u} + \mathbf{B}_j^T \mathbf{v}) = \sum_1^r \phi_j(\mathbf{u}) + \sum_1^r \phi_j(\mathbf{B}_j^T \mathbf{v}) \tag{3.2.4}$$

where $\phi_j = -\ln|f_j|$ and $\phi_j(\mathbf{0}) = 0$. The ϕ_j are obviously nonnegative.

From the remark following Lemma 1.5.9, we find that $\sum \phi_j(\mathbf{u})$ is a polynomial, say $P(\mathbf{u})$, so that the c.f. of $\mathbf{L}_1$ is of the form $\exp(P(\mathbf{u}))$. Then by Marcinkiewicz's theorem $\mathbf{L}_1$ has n-variate normal distribution. But $\mathbf{L}_1$ is a linear combination of independent vector variables, and by Cramer's theorem each component of $\mathbf{L}_1$ is an n-variate normal variable.

We shall present a different proof due to Zinger [37], using the properties of the right member of (3.2.4). We shall attempt to make (3.2.4) meaningful for arbitrary $\mathbf{u}$ and $\mathbf{v}$ in R^n.

We introduce the usual norm of the matrix $\mathbf{B}$ as the norm of the operator: $\|\mathbf{B}\| = \sup_{|\mathbf{u}| \leqq 1} |\mathbf{Bu}|$, and set

$$\alpha = \max_{1 \leq j, h \leq r} (\|\mathbf{B}_j^T(\mathbf{B}_h^T)^{-1}\|, \|(\mathbf{B}_j^T)^{-1}\|. \tag{3.2.5}$$

From the definition (3.2.5), it is seen that

$$\alpha \geqq \|\mathbf{B}_j^T(\mathbf{B}_j^T)^{-1}\| = 1.$$

For $0 < \rho < \rho_0$, we write

$$M_j(\rho) = \max_{|u| \leqq \rho} |\phi_j(\mathbf{u})|, \qquad j = 1, \ldots, r \tag{3.2.6}$$

$$M(\rho) = \max_{1 \leqq j \leqq r} M_j(\rho). \tag{3.2.7}$$

We take any j, say $j = j_0$. From (3.2.4), since ϕ_j is nonnegative, we find

$$\phi_{j_0}(\mathbf{u} + \mathbf{B}_{j_0}^T\mathbf{v}) \leqq \sum_1^r \phi_j(\mathbf{u}) + \sum_1^r \phi_j(\mathbf{B}_j^T\mathbf{v}). \tag{3.2.8}$$

We take $\mathbf{u}$ such that $|\mathbf{u}| \leqq \rho < \rho_0$, and set $\mathbf{v} = (1/\alpha)(\mathbf{B}_{j_0}^T)^{-1}\mathbf{u}$. From (3.2.8), we then find

$$M_{j_0}\left[\left(1 + \frac{1}{\alpha}\right)\rho\right] \leqq 2rM(\rho).$$

Since j_0 is arbitrary among the numbers $1, 2, \ldots, r$, we find

$$M\left[\left(1 + \frac{1}{\alpha}\right)\rho\right] \leqq 2rM(\rho). \tag{3.2.9}$$

Hence we find that if $\phi_j(\mathbf{u})$ is finite for $|\mathbf{u}| \leqq \rho$, then it is finite for $|\mathbf{u}| \leqq (1 + 1/\alpha)\rho$, i.e., if f_j does not vanish for $|\mathbf{u}| \leqq \rho$, then it does not vanish for $|\mathbf{u}| \leqq (1 + 1/\alpha)\rho$. Thus, successively enlarging ρ, we see that f_j is nonvanishing in any bounded subset of R^n for any j; a finite $M(\rho)$ is defined for all values of $\rho > 0$ by formula (3.2.7), and (3.2.9) holds for all ρ. In view of the nondecreasing nature of M, we deduce by elementary arguments from (3.2.8) that

$$M(\rho) = O(\rho^k) \tag{3.2.10}$$

where $k > 0$ is a suitable constant.

Let us now turn to equation (3.2.4). We introduce the normal density

$$g(\mathbf{v}) = (2\pi)^{-n/2} \exp(-(1/2)\mathbf{v}^T\mathbf{v}), \qquad \mathbf{v} \in R^n$$

and integrate (3.2.4) after multiplying both sides by $g(\mathbf{v})$ over the whole of R^n, this being possible in view of the estimate (3.2.10). As a result, we obtain

$$\sum_1^r \int \phi_j(\mathbf{u} + \mathbf{B}_j^T\mathbf{v})g(\mathbf{v})\, d\mathbf{v} = \sum_1^r \phi_j(\mathbf{u}) + \sum_1^r \int \phi_j(\mathbf{B}_j^T\mathbf{v})g(\mathbf{v})\, d\mathbf{v}. \qquad (3.2.11)$$

We take any one of the summands on the left; it can be written in the form

$$\int \phi_j(\mathbf{u} + \mathbf{B}_j^T\mathbf{v})g(\mathbf{v})\, d\mathbf{v} = \int \phi_j(\mathbf{B}_j^T\mathbf{v})g(\mathbf{v} - (\mathbf{B}_j^T)^{-1}\mathbf{u})\, d\mathbf{v}. \qquad (3.2.12)$$

In view of the elementary equality

$$(\mathbf{v} - (\mathbf{B}_j^T)^{-1}\mathbf{u})^T(\mathbf{v} - (\mathbf{B}_j^T)^{-1}\mathbf{u}) = \mathbf{v}^T\mathbf{v} + ((\mathbf{B}_j^T)^{-1}\mathbf{u})^T((\mathbf{B}_j^T)^{-1}\mathbf{u}) - 2\mathbf{v}^T(\mathbf{B}_j^T)^{-1}\mathbf{u}$$

we can rewrite in a new form the expression for $g(\mathbf{v} - (\mathbf{B}_j^T)^{-1}\mathbf{u})$ in the formula (3.2.12); after this, we obtain

$$\int \phi_j(\mathbf{u} + \mathbf{B}_j^T\mathbf{v})g(\mathbf{v})\, d\mathbf{v} = g((\mathbf{B}_j^T)^{-1}\mathbf{u}) \int \phi_j(\mathbf{B}_j^T\mathbf{v}) \exp\{\mathbf{v}^T(\mathbf{B}_j^T)^{-1}\mathbf{u} - (1/2)\mathbf{v}^T\mathbf{v}\}\, d\mathbf{v}. \qquad (3.2.13)$$

We now apply the well-known theorem of Cramer and Wold, which permits us to reduce the consideration of multivariate distributions to that of univariate ones. We set $\mathbf{u} = \lambda\mathbf{u}_0$, where $\mathbf{u}_0 \in R^n$, $|\mathbf{u}_0| = 1$ and λ is a real number; then $f_j(\lambda\mathbf{u}_0)$ will be the c.f. of a scalar random variable. Then, from (3.2.11) and (3.2.13) it follows, after taking exponentials and raising to the second power

$$\prod_1^r |f_j(\lambda\mathbf{u}_0)|^2 = \exp\left[2C - 2\sum_1^r g((\mathbf{B}_j^T)^{-1}\lambda\mathbf{u}_0) \times \int \phi_j(\mathbf{B}_j^T\mathbf{v}) \exp\{\mathbf{v}^T(\mathbf{B}_j^T)^{-1}\lambda\mathbf{u}_0 - (1/2)\mathbf{v}^T\mathbf{v}\}\, d\mathbf{v}\right] \qquad (3.2.14)$$

where

$$C = \sum_1^r \int \phi_j(\mathbf{B}_j^T\mathbf{v})g(\mathbf{v})\, d\mathbf{v}.$$

In view of the estimate (3.2.10), the right side of (3.2.14), as a function of $\mathbf{u}_0$, can be extended throughout the complex plane as an entire function of λ, not having any zeros. Further, the squared modulus of a c.f. also being a c.f., the functions $|f_j(\lambda\mathbf{u}_0)|^2$ can also be similarly extended. Choosing λ to be a purely imaginary number, we have that for any h in R^n

$$E \exp(\mathbf{h}^T\mathbf{X}^j) < \infty.$$

Let us write $\xi_j(\mathbf{h}) = \ln E \exp(\mathbf{h}^T \mathbf{X}^j)$; then we obtain corresponding to (3,2.4)

$$\sum_1^r \xi_j(\mathbf{u} + \mathbf{B}_j^T \mathbf{v}) = \sum_1^r \xi_j(\mathbf{u}) + \sum_1^r \xi_j(\mathbf{B}_j^T \mathbf{v}). \tag{3.2.15}$$

Then $\xi_j(\mathbf{h}) \geqq 0$ for $\mathbf{h}$ in R^n, for all j. Therefore, we can apply to $\xi_j(\mathbf{h})$ the same arguments as in the derivation of the estimates (3.2.10), so that $\xi_j(\mathbf{h})$ will have the same estimate as $|\mathbf{h}| \to \infty$, $\mathbf{h} \in R^n$,

$$\xi_j(\mathbf{h}) = O(|\mathbf{h}|^k) \tag{3.2.16}$$

where k is a positive constant. Hence, for the c.f. $|f_j(\lambda \mathbf{u}_0)|^2$, we find for any $\mathbf{u}_0$ in R^n, by the "ridge property" of c.f.'s

$$f_j(\lambda \mathbf{u}_0) f_j(-\lambda \mathbf{u}_0) = O(|\lambda \mathbf{u}_0|^k) \tag{3.2.17}$$

for any complex value of λ. Thus, the above c.f. is of finite order and non-vanishing. By Marcinkiewicz' theorem (Lemma 1.4.2) we have that it is the c.f. of a normal law, and by Cramer's theorem, $f_j(\lambda \mathbf{u}_0)$ is a normal c.f. Finally, in view of the arbitrariness of $\mathbf{u}_0$ in R^n, we obtain from the Cramer-Wold theorem that the f_j are c.f.'s of n-dimensional normal laws, proving the theorem.▌

3.3 LINEAR FORMS WITH A DENUMERABLE NUMBER OF VARIABLES

The phenomenon of independence of linear forms in an infinite number of variables was first studied by Mamai in [99]. The most general results were obtained by Ramachandran [120] in 1967. We present his result below, following his note [120] (also cf. [121]).

THEOREM 3.3.1. Let $\{X_j\}$ be a sequence of independent r.v.'s and $\{a_j\}$, $\{b_j\}$ be two sequences of real constants such that

(I) the sequences $\{a_j/b_j : a_j b_j \neq 0\}$ and $\{b_j/a_j : a_j b_j \neq 0\}$ are both bounded;

(II) $\sum a_j X_j$ and $\sum b_j X_j$ converge with probability one to random variables U and V respectively; and

(III) U and V are independent.

Then, for every j such that $a_j b_j \neq 0$, X_j is normally distributed.

***Proof*:** Let f_j be the c.f. of X_j. Then, for all real u and v, we have

$$E \exp i(uU + vV) = E \exp(iuU) E \exp(ivV)$$

which yields the relation

$$\prod_1^\infty f_j(a_j u + b_j v) = \prod_1^\infty f_j(a_j u) \cdot \prod_1^\infty f_j(b_j v). \tag{3.3.1}$$

Setting $g_j(t) = f_j(t)f_j(-t) = |f_j(t)|^2$, for real t, we have

$$\prod g_j(a_j u + b_j v) = \prod g_j(a_j u) \cdot \prod g_j(b_j v). \tag{3.3.2}$$

There is an interval around the origin in which $\Pi g_j(a_j u)$, the c.f. of $\sum a_j X_j$ is nonvanishing. It follows that $g_j(a_j u) \neq 0$ for u in an interval I', for all j; similarly, $g_j(b_j v) \neq 0$ for v in an interval I'', for all j. In view of (3.3.2), for u and v in some interval I around the origin, $g_j(a_j u + b_j v) \neq 0$ and taking the logarithms of both sides of (3.3.2) for such u and v, and setting $h_j = \ln g_j$, we have

$$\sum h_j(a_j u + b_j v) = \sum h_j(a_j u) + \sum h_j(b_j v). \tag{3.3.3}$$

All the three series in (3.3.3) are uniformly convergent series of negative terms for u and v in I, so that any subseries of any of them also converges uniformly. Since the relation $h_j(a_j u + b_j v) = h_j(a_j u) + h_j(b_j v)$ is automatically satisfied when $a_j b_j = 0$, we can subtract from the three series in (3.3.3) the subseries respectively corresponding to those indices for which $a_j b_j = 0$; the remaining series, where the summation in the case of each of the three runs over all those j for which $a_j b_j \neq 0$, continue to be uniformly convergent in I.

Let $\psi_j(u) = h_j(a_j u)$. Then in view of the boundedness of $c_j = b_j/a_j$ and of a_j/b_j, the equation (3.3.3) becomes

$$\sum \psi_j(u + c_j v) = \sum \psi_j(u) + \sum \psi_j(c_j v) = A(u) + B(v) \tag{3.3.4}$$

where A and B are continuous functions of u and v on I, and the series on the left converges uniformly for u and v in I. Since the conditions of Lemma 1.5.8 are satisfied, this leads to the conclusion

$$\sum c_j^2 \psi_j(x) = a + bx + cx^2 \qquad \text{for } x \in I.$$

Since the ψ_j are nonnegative even functions and $\psi_j(0) = 0$, we must have $a = b = 0$ and $c \leqq 0$; it follows from Lemma 1.2.1 that the g_j are normal c.f.'s. Then it follows as usual from Cramer's theorem that for every j such that $a_j b_j \neq 0$, f_j is a normal c.f.∎

3.4 MODIFICATIONS. "RELATIVISTIC LINEAR STATISTICS"

As we have already pointed out in the introduction, starting from characterizations of some distribution, say, the normal, through the property of independence of some statistics, we can obtain in a trivial manner characterizations of a wider class of distributions in each case.

For instance, let $(X_1, \ldots, X_n)$ be a random sample, and $L_1 = \sum_1^n \alpha_j X_j$ and $L_2 = \sum_1^n \beta_j X_j$ be linear statistics where $\alpha_j \beta_j \neq 0$ for all j, and $\sum \alpha_j \beta_j = 0$. Then, by Theorem 3.1.1, for the normality of the X_j, it is necessary and sufficient that L_1 and L_2 be independent.

Let now $H(x)$ be some Lebesgue-measurable function on the real line. Consider the statistics $S_1 = \sum \alpha_i H(X_i)$ and $S_2 = \sum \beta_i H(X_i)$. For the independence of S_1 and S_2, it is necessary and sufficient that the $H(X_i)$ be normally distributed.

The condition $H(X_i)$ is normal, clearly, cannot always be satisfied. It can possibly be, for instance, if $H(x) = x^3$, but cannot be if $H(x) = x^2$ or x^4 (so that in these cases the statistics S_1 and S_2 are necessarily dependent). But the class of random variables for which it is possible to pick out a measurable function H subject to the condition that $H(X_i)$ is normal for given H, constitute in a well-known sense a "two-parameter" family characterized by the independence of the statistics S_1 and S_2.

Modifications of Theorem 3.1.1 other than such relatively trivial ones as the above are also possible. In particular, some can be obtained using functions having algebraic addition properties.

A function f of a real variable x will be said to have an algebraic addition property if there exists an algebraic function F such that $f(x + y) = F(f(x), f(y))$.

The following theorem is well-known (cf. Aczel [3], p. 61).

Every single-valued analytic function with an algebraic addition property is either a rational function of x, or a rational function of e^{cx}, or a doubly periodic function, i.e., a rational function or a Weierstrass function

$$\zeta(z) = \frac{1}{z^2} + \sum{}' \left[\frac{1}{(z - 2m\omega - 2n\omega')^2} - \frac{1}{(2m\omega + 2n\omega')^2}\right]$$

where z is a complex variable, ω and ω' are periods, and the summation extends over all integers m and n other than the pair (0, 0). The special case of a rational function of e^{cx} will be of interest to us, namely, $\tanh x = (e^x - e^{-x})/(e^x + e^{-x})$. We see that $|\tanh x| \leqq 1$ and that the following addition theorem holds.

$$\tanh(x + y) = \frac{\tanh x + \tanh y}{1 + \tanh x \tanh y}$$

As is well-known, the last formula has interesting physical significance, as first pointed out by A. Sommerfeld (vide [154]), in regard to relativistic addition of collinear velocities. If v_1, v_2 are two collinear velocities in the Einsteinian universe, then the "sum" of the velocities: $v = v_1 \oplus v_2$ will be

$$v = \frac{v_1 + v_2}{1 + (v_1 v_2/c^2)} \tag{3.4.1}$$

where c is the speed of light. Measuring all velocities in terms of that of light (taking $c = 1$), we obtain the addition formula corresponding to the hyperbolic tangent

$$v_1 \oplus v_2 = \frac{v_1 + v_2}{1 + v_1 v_2} \tag{3.4.2}$$

Thus, under the Einsteinian addition of collinear velocities, the quantities

$$\tanh^{-1} v = \frac{1}{2} \ln \frac{1 - v}{1 + v} \tag{3.4.3}$$

will be added in the usual manner. As is well-known (cf. for instance [154]), the addition (3.4.2) corresponds to the addition of collinear segments in a Lobachevsky space, and is noncommutative.

From a general point of view, the addition (3.4.2) and the more general cases of noncommutative addition indicated above, constitute special cases of algebraic structures, where a fairly developed theory of addition of random variables and limit theorems exists (cf. [16] and also [149]).

Let $v_1, \ldots, v_n$ be collinear, independent, random, and not necessarily identically distributed velocities in the Einsteinian universe, with the velocity of light $c = 1$. Thus $|v_j| \leqq 1$ with probability one for all j. Let $\mu_1, \ldots, \mu_n$ and $\nu_1, \ldots, \nu_n$ be nonzero integers. We define $\mu_j v_j$ as the addition

$$v_j \oplus v_j \oplus \cdots \oplus v_j \qquad (\mu_j \text{ times}) \text{ if } \mu_j > 0$$

and as the expression

$$-(v_j \oplus v_j \oplus \cdots \oplus v_j) \qquad (-\mu_j \text{ times}) \text{ if } \mu_j < 0$$

(the addition is carried out $-\mu_j$ times if $\mu_j < 0$). We set up the "relativistic integral linear forms"

$$\begin{aligned} L_1 &= \mu_1 v_1 \oplus \mu_2 v_2 \oplus \cdots \oplus \mu_n v_n \\ L_2 &= \nu_1 v_1 \oplus \nu_2 v_2 \oplus \cdots \oplus \nu_n v_n . \end{aligned} \tag{3.4.4}$$

We can assume a relation of the form

$$\sum_{j=1}^{r} \sigma_j^2 \mu_j \nu_j = 0 \tag{3.4.5}$$

where $\sigma_j^2 > 0$ are some given numbers.

The relativistic linear forms will in fact be fairly complicated rational expressions in $v_1, \ldots, v_n$. If, for example, $\mu_1 = 2$, $\mu_2 = -2$; $\nu_1 = \nu_2 = 2$, then

$$L_1 = 2v_1 \oplus (-2v_2) = 2 \frac{v_1(1 + v_2^2) - v_2(1 + v_1^2)}{(1 + v_1^2)(1 + v_2^2) - 4v_1 v_2}$$

$$L_2 = 2v_1 \oplus 2v_2 = 2 \frac{v_1(1 + v_2^2) + v_2(1 + v_1^2)}{(1 + v_1^2)(1 + v_2^2) + 4v_1 v_2} .$$

Suppose that the two linear forms (3.4.4), subject to some such condition as (3.4.5) are independent, i.e., the rational fractions of the type indicated above are independent. Then the following theorem holds.

THEOREM 3.4.1. In order that two relativistic integral linear forms given by (3.4.4), subject to (3.4.5), be stochastically independent, it is necessary and sufficient that

$$v_j = \tanh X_j \tag{3.4.6}$$

where the X_j are normally distributed random variables.

Proof: Let L_1 and L_2 be independent. Under the relativistic addition of velocities (3.4.2), we have $|L_j| \leqq 1$ for $j = 1, 2$. We set, for $j = 1, 2$,

$$L'_j = \tanh^{-1}(L_j) = \frac{1}{2}\ln\left(\frac{1 - L_j}{1 + L_j}\right); \qquad X_j = \tanh^{-1} v_j = \frac{1}{2}\ln\left(\frac{1 - v_j}{1 + v_j}\right).$$

Then we have $L'_1 = \sum \mu_j X_j$ and $L'_2 = \sum \nu_j X_j$ with the usual addition. Further, L'_1 and L'_2 are independent. Hence the X_j are normal with variance σ_j^2 and $v_j = \tanh X_j$. ∎

From an analytical point of view, we thus obtain a complete description of the phenomenon of the independence of certain special rational statistics of independent but not necessarily identically distributed random variables v_j subject to the condition $|v_j| \leqq 1$. Here the families of laws (3.4.6) are "two-parameter" ones.

Applying other addition theorems (for example the theorems connected with the Weierstrass function $\zeta(z)$), we can obtain descriptions of the phenomenon of independence of fairly complicated algebraic functions of independent and not necessarily identically distributed random variables. However, we shall not pause to dwell on them here.

If we consider identically distributed as well as independent r.v.'s, in other words, a random sample, then theorems on "relativistic integral linear forms" of the type (3.4.4) can be obtained, corresponding to the theorems of Chapter 2, in the manner described above.

The characterization of normality given by Theorem 2.4.2 yields a theorem characterizing families of the type $v = \tanh X$ where X is normal, through the identical distribution of special rational statistics of a random sample $(v_1, \ldots, v_n)$. The conditions of Theorem 2.4.2 will be formulated in terms of the integral coefficients μ_j and ν_j.

For further comments on real random variables satisfying an addition theorem see [178].

3.5 APPLICATIONS TO STATISTICAL PHYSICS

We present below some elementary applications of the above theorems as well as of those of Chapter 2 to statistical physics, following the note [79] and the book [85], somewhat generalizing the conditions given there.

Theorem 3.1.1 permits us to make a simple derivation of the well-known Maxwell's law on the distribution of velocities of motion of molecules (under conditions of "Newtonian chaos") under hypotheses weaker than those used in classical statistical physics (cf. for example [156]).

Let $\mathbf{X} = (X_1, X_2, X_3)$ be a three-dimensional random vector, representing the velocities of motion of a molecule in three-dimensional space. We make two assumptions:

(1) There exists a system of orthogonal axes such that the projections X_1, X_2, X_3 of our vector $\mathbf{X}$ on these axes are stochastically independent and identically distributed,

(2) There also exist two axes, not orthogonal to any of the above three axes, such that the projections of $\mathbf{X}$ on them are independent.

From these two assumptions, it follows from Theorem 3.1.1 that $\mathbf{X}$ is normal, with all its components X_1, X_2, X_3 having the same variance. This is the classical Maxwell law.

If we drop the assumption of identical distribution from (1) above, then $\mathbf{X}$ will again be normally distributed, but the variances of the components need not be equal to one another; this is the ideal "ellipsoidal" law, corresponding to Schwarzchild's law for the distribution of the velocities of stars (cf. for instance [107]).

Theorem 4.4.1 of the next chapter permits us to weaken condition (2) also; one of the two auxiliary axes of condition (2) can be chosen at random. Let a constant axis $\mathbf{l}_2$ with direction cosines $\cos\alpha_1$, $\cos\alpha_2$, $\cos\alpha_3$, none zero, be chosen, and let us consider the projection of $\mathbf{X}$ on the axis $\mathbf{l}_2$; $L_2 = X_1 \cos\alpha_1 + X_2 \cos\alpha_2 + X_3 \cos\alpha_3$. We take another axis $\mathbf{l}_1$ at random so that the projection of $\mathbf{X}$ thereon is $L_1 = u_1' X_1 + u_2' X_2 + u_3' X_3$, where (u_1', u_2', u_3') is a random vector which we subject to the condition

$$u_1' \cos\alpha_1 + u_2' \cos\alpha_2 + u_3' \cos\alpha_3 = 0 \text{ with probability one.}$$

We set $u_j = u_j'/\cos\alpha_j$, $j = 1, 2, 3$. If the new random vector (u_1, u_2, u_3) satisfies the conditions (4.4.1)–(4.4.3), then the vector (X_1, X_2, X_3) is normal.

We see that the random choice of the axis $\mathbf{l}_1$ weakens condition (2), but then the imposition of a requirement stronger than (4.4.1) is needed. The problem of replacing condition (2) by one where both the axes $\mathbf{l}_1$ and $\mathbf{l}_2$ are chosen at random has not been studied so far.

Theorems 2.4.3 and 2.4.3′ enable us to give yet another proof of Maxwell's law. Retaining condition (1), we introduce in place of (2) the condition:

(2′) There exist two axes, not necessarily orthogonal, such that the projections of the speeds on them

$$L_1 = a_1 X_1 + a_2 X_2 + a_3 X_3, \qquad L_2 = b_1 X_1 + b_2 X_2 + b_3 X_3$$

are identically distributed.

Let the coefficients a_j, b_j be such that the conditions of Theorem 2.4.3 are satisfied. Then, if the d.f. of the velocities satisfies the conditions of Theorem 2.4.3′ and has finite moment of order $2[\sigma_2/2 + 1]$, where σ_2 is as defined there, then the velocities are normally distributed. The normality of the velocities can also be obtained under more stringent but more easily formulated conditions (cf. [79]). Let the axis $\mathbf{l}_2$ have the direction of the unit vector $a_1\mathbf{i} + a_2\mathbf{j} + a_3\mathbf{k}$, situated in the positive octant, with $a_1 > a_2$, $a_1 > a_3$. Let another axis with the direction of the unit vector $b_1\mathbf{i} + b_2\mathbf{j} + b_3\mathbf{k}$ make a small angle ϕ with the former. If ϕ is small enough, we will have $b_j > 0$ for all j, $b_1 > b_2$ and $b_1 > b_3$. Let further $a_1 = b_1 + \xi$, where $\xi > 0$. We have (cf. Chapter 2) $\sigma(z) = \sum_1^3 (a_j^z - b_j^z)$. For $x > 0$, we obtain

$$\sigma(x) > a_1^x - 3b_1^x = b_1^x[(1 + \xi/b_1)^x - 3] > b_1^x(x\xi/2b_1 - 2)$$

for sufficiently small ϕ. For $x > 4b_1/\xi$, $\sigma(x) \neq 0$. Hence the largest positive zero of the function $\sigma(z)$ does not exceed $4b_1/\xi$. We set $m = [2b_1/\xi + 2]$.

If the moment of order $2m$ of the speed of the molecule is finite, then Maxwell's law holds.

CHAPTER 4

Independent Nonlinear Statistics. Random Linear Statistics

4.1 STATEMENT OF THE PROBLEM. DEFINITIONS

In Chapter 3, the phenomenon of independence of linear statistics with constant coefficients was studied. In particular, theorems characterizing normality were obtained through functional transformations of two-parameter families of d.f.'s for the X_i such that $Y_i = H(X_i)$ were distributed as $N(a, \sigma^2)$.

As we shall see below, random linear forms connected with nonlinear statistics are of a fairly common nature. In this chapter, we shall study the phenomenon of independence of nonlinear statistics and present some results obtained in this area. We first introduce some definitions, following [83] and [31].

Let $\mathbf{X} = \mathbf{X}^{(n)} = (X_1, \ldots, X_n)$ be a random vector with independent components. The statistic $S(\mathbf{X})$ will be called a *quasipolynomial* if there exists a continuous function ϕ and two nonnegative polynomials r and R of the same degree such that for all $\mathbf{x} \in R^n$

$$r(\mathbf{x}) \leqq \phi[S(\mathbf{x})] \leqq R(\mathbf{x}). \tag{4.1.1}$$

A polynomial $P(x_1, \ldots, x_n)$ will be called *admissible with respect to* x_j, if the coefficient of x_j^m in its irreducible form is nonzero, where m is the degree of the polynomial.

A quasipolynomial statistic will be called *admissible with respect to* x_j if its "lower polynomial" r is so.

Let a statistic $S(\mathbf{X})$ be given. If there exists a nondegenerate affine transformation $\mathbf{X} \to \mathbf{X}'$ of the sample space, under which $S(\mathbf{X})$ becomes $S_1(\mathbf{X}')$, where the latter is a function of the "observations" $X_1', \ldots, X_k'$ with $k \leqq n$, and the number k of observations on which the statistic $S_1(\mathbf{X}')$ depends cannot be decreased by taking any other transformation of the type indicated, then the number k will be called the *dimension* of the statistic $S(\mathbf{X})$.

It is easy to see that in the above definition we may speak of orthogonal transformations of the sample space instead of nondegenerate affine transformations, without affecting the definition.

A statistic S will be called a *tube statistic* if it is continuous, nonnegative, homogeneous with a positive degree of homogeneity, and its level surface $S = 1$ consists of one connected continuous surface. If the level surface $S = 1$ consists of a finite number of connected continuous surfaces, S will be called *multitubular*.

If a tubular or multitubular statistic has dimension $k \leqq n$, and vanishes only for $x_1' = \cdots = x_k' = 0$, then it will be called *definite*. Its level surfaces will be bounded.

Examples

(1) The linear form $L = a_1 X_1 + \cdots + a_n X_n$ where the coefficients are all nonzero will be a definite tube statistic of dimension 1.

(2) Statistics in the form of the even central moments of a random sample $m_{2k} = (1/n) \sum_1^n (X_i - \bar{X})^{2k}$ will be definite tube statistics of dimension $(n - 1)$.

(3) Let $\mathbf{A} = \mathbf{A}_{nn} = ((a_{jk}))$ be a matrix of rank $r \leqq n$, and $\rho > 0$ be a positive constant. The statistic

$$S = \sum_{j=1}^{n} \left| \sum_{k=1}^{n} a_{jk} X_k \right|^{\rho} \tag{4.1.2}$$

will be a definite tube statistic of dimension r.

Two tube statistics will be called *unlinkable* if there exists an orthogonal transformation which transforms them into functions of two distinct sets of variables.

The results to be presented below will relate mainly to a random sample vector and two tube statistics S_1 and S_2 of dimensions k_1 and k_2 such that

$$k_1 + k_2 = n. \tag{4.1.3}$$

Some results concerning random linear forms will also cover the case of a general sample vector.

It is fairly well known that, for the independence of two statistics S_1 and S_2, it is necessary and sufficient that, for any continuous function of two real variables $F(x, y)$, the two statistics $F(S_1(\mathbf{X}), S_2(\mathbf{X}))$, and $F(S_1(\mathbf{X}), S_2(\mathbf{Y}))$ should be identically distributed, where $\mathbf{Y}$ is a random vector having its distribution identical with, but independent of, that of $\mathbf{X}$.

For studying the independence of two statistics, we shall apply mainly two methods: the method of differential equations, and that of integral equations with shifts in the argument.

The method of differential equations was first applied in the year 1942 by Lukacs [92] for studying the condition of independence of the sample mean and the sample variance. It can also be applied to study the independence of polynomial tube statistics one of which is linear.

The method of integral equations with shifted argument (integro-functional equations) will be applied to a more general class of pairs of statistics, but, generally speaking, requires that they be definite tube statistics; the omission of the definiteness requirement requires the introduction of very stringent conditions, for characterization problems, as we shall see in section 4.11.

4.2 THE CONDITION OF INDEPENDENCE OF $\bar{X}$ AND S^2: GENERALIZATIONS

Let $\mathbf{X}^{(n)} = (X_1, \ldots, X_n)$ be a random sample vector with $n \geqq 2$; $\bar{X} = (1/n) \sum_1^n X_i$ and $S^2 = (1/n) \sum_1^n (X_i - \bar{X})^2$ are the first two sample moments. The condition of independence of $\bar{X}$ and S^2 has been fully studied (cf. [49] and [29]).

THEOREM 4.2.1. For $n \geqq 2$, for the independence of $\bar{X}$ and S^2, it is necessary and sufficient that the sample be normal.

Remark. The proof we give does not depend on the existence of the moments of X_i, and is a little involved though direct. In section 4.3 we prove a general result (Theorem 4.3.1) which shows that the existence of moments is implied by the independence of $\bar{X}$ and S^2. Then the simpler arguments of Theorem 4.2.2 or of Theorems 6.3.1 and 6.3.2 in Chapter 6, which use the condition $E(S^2/\bar{X}) =$ constant can be applied.

For the proof by the method of differential equations, we shall follow the presentation in [29].

***Proof*:** The sufficiency of the condition is trivial, since, as is well known, the (degenerate) normal vector $(X_1 - \bar{X}, \ldots, X_n - \bar{X})$ is independent of $\bar{X}$. For proving the necessity, we invoke the well-known identity

$$\sum X_j^2 = n\bar{X}^2 + nS^2 \tag{4.2.1}$$

and introduce the function f given by

$$f(t) = E \exp(itX_1 - X_1^2).$$

Obviously, f is an entire function of t. Consider the expression

$$E\left[nS^2 \exp \sum_{j=1}^{n} (itX_j - X_j^2)\right].$$

In view of the independence of $\bar{X}$ and S^2, and of the identity (4.2.1), we may write

$$E\left(nS^2 \exp \sum_{j=1}^{n} (itX_j - X_j^2)\right) = E(nS^2 e^{-nS^2}) \cdot E \exp\left(it\left(\sum_1^n X_j\right) - n\bar{X}^2\right)$$

and also

$$E(nS^2 e^{-nS^2}) E\left(\exp\left(it \sum_1^n X_j - n\bar{X}^2\right)\right) = a \cdot E \exp \sum_{j=1}^{n} (itX_j - X_j^2) \quad (4.2.2)$$

where we have set

$$a = \frac{E(nS^2 e^{-nS^2})}{E(e^{-nS^2})} > 0.$$

Further, in view of the independence of the X_j, we have

$$E \exp \sum_{j=1}^{n} (itX_j - X_j^2) = [f(t)]^n.$$

In view of the identity (4.2.1), we then have

$$E\left[nS^2 \exp\left(\sum_{j=1}^{n} (itX_j - X_j^2)\right)\right] = E \sum_{j=1}^{n} X_j^2 \exp \sum_{j=1}^{n} (itX_j - X_j^2)$$
$$- E\left(n\bar{X}^2 \exp \sum_{j=1}^{n} (itX_j - X_j^2)\right).$$

Then we compute

$$E\left[nS^2 \exp\left(\sum_{j=1}^{n} (itX_j - X_j^2)\right)\right] = n(f(t))^{n-1} E(X_1^2 \exp(itX_1 - X_1^2))$$
$$- \frac{1}{n} E\left(\left(\sum_{j=1}^{n} X_j^2\right) \exp\left(\sum_{j=1}^{n} (itX_j - X_j^2)\right)\right);$$

$$E(X_1^2 \exp(itX_1 - X_1^2)) = -f''(t);$$

$$E\left(\left(\sum_{j=1}^{n} X_j\right)^2 \exp\left(it \sum_{j=1}^{n} X_j - \sum_{j=1}^{n} X_j^2\right)\right) = -[(f(t))^n]''.$$

Thereafter, (4.2.2) can be rewritten in the form

$$-n(f(t))^{n-1} f''(t) + (f(t))^{n-1} f''(t) + (n-1)(f(t))^{n-2}(f'(t))^2 = a(f(t))^n$$

or

$$(f(t))^{n-1} f''(t) - (f(t))^{n-2}(f'(t))^2 = a_1 (f(t))^n.$$

Here, $a_1 = a/(n-1) > 0$; f is an entire function, with

$$f(0) = \int_{-\infty}^{\infty} e^{-u^2} dF(u) > 0.$$

Hence, in some neighborhood of the origin, we have

$$\frac{f''(t)}{f(t)} - \left[\frac{f'(t)}{f(t)}\right]^2 = -a_1$$

or, $(\ln f(t))'' = -a_1$. The solution of this equation will be

$$f(t) = k \exp(i\gamma t - (1/2)a_1 t^2) \tag{4.2.3}$$

where $k = f(0)$, and $\gamma = -if'(0)$. Since f is an entire function, it coincides with the right-hand side of (4.2.3) throughout the real axis, if it does so in a neighborhood of the origin.

We shall now show that F is a normal d.f. We set

$$G(x) = \int_{-\infty}^{x} e^{-u^2}\, dF(u).$$

But, $f = k \cdot f_0$ where f_0 is the c.f. of a normal law. Hence, $G(x)$ can be represented in the form

$$G(x) = \frac{k}{\sqrt{2\pi a_1}} \int_{-\infty}^{x} \exp\left(-\frac{(x-\gamma)^2}{2a_1}\right) dx$$

so that, for any x,

$$\int_{-\infty}^{x} e^{-u^2}\, dF(u) = \frac{k}{\sqrt{2\pi a_1}} \int_{-\infty}^{x} \exp\left[\frac{-(u-\gamma)^2}{2a_1}\right] du. \tag{4.2.4}$$

We shall show that F is everywhere differentiable and calculate its derivative. Let, for instance, $x_0 > 0$, $h > 0$. We have from (4.2.4)

$$\frac{k}{\sqrt{2\pi a_1}} \frac{e^{x_0^2}}{h} \int_{x_0}^{x_0+h} \exp\left[-\frac{(x-\gamma)^2}{2a_1}\right] dx < \frac{F(x_0+h) - F(x_0)}{h}$$

$$< \frac{k}{\sqrt{2\pi a_1}} \frac{e^{(x_0+h)^2}}{h} \cdot \int_{x_0}^{x_0+h} \exp\left[-\frac{(x-\gamma)^2}{2a_1}\right] dx$$

and similarly for $h < 0$. Hence it follows that

$$F'(x_0) = \frac{k}{\sqrt{2\pi a_1}} \exp\left(x_0^2 - \frac{(x_0-\gamma)^2}{2a_1}\right)$$

so that F is a normal d.f. ∎

Theorem 4.2.1 admits the following generalization.

Let $\mathbf{X}^{(n)} = (X_1, \ldots, X_n)$ be a random sample vector from the distribution F. We form the linear statistic

$$\tilde{X} = a_1 X_1 + \cdots + a_n X_n \tag{4.2.5}$$

where $\sum_1^n a_i^2 = 1$, and the statistic

$$\tilde{S}^2 = \sum_1^n X_i^2 - \left(\sum_1^n a_i X_i\right)^2.$$

THEOREM 4.2.1′ For the independence of $\tilde{X}$ and $\tilde{S}^2$, it is necessary and sufficient that the random sample be normal.

***Proof*:** If the X_i are normal, an orthogonal transformation with its first row equal to $(a_1, \ldots, a_n)$ will unlink $\tilde{X}$ and $\tilde{S}^2$ in the sense of section 4.1, and thus they are independent. We shall merely indicate the proof of the necessity part. Again, we introduce

$$f(t) = \int e^{itx - x^2}\, dF(x)$$

and using the same argument as above arrive at the equation

$$\sum_{j=1}^n \frac{f''(a_j t)}{f(a_j t)}(1 - a_j^2) - \sum_{j \neq k} a_j a_k \frac{f'(a_j t) f'(a_k t)}{f(a_j t) f(a_k t)} = -a < 0.$$

This equation is investigated in the same way as before, and leads to the normality of the sample.∎

More general results on the independence of $\bar{X}$ and certain quadratic forms, in the case of a random sample vector, under the assumption that the second moment exists, were obtained by Laha [75], whose presentation we follow below.

Let $(X_1, \ldots, X_n)$ be a random sample where $EX_i = m$ and $\text{Var}(X_i) = \sigma^2$ exist. Let $((a_{ij}))$ be an $n \times n$ matrix. Under what conditions on the distribution of the X_i will the quadratic form $Q = \sum_{i,j} a_{ij} X_i X_j$ be independent of $\bar{X}$?

This problem also can be studied by the method of differential equations.

Let ϕ be the c.f. of the X_j and $L = n\bar{X}$. The independence of Q and $\bar{X}$ is clearly equivalent to that of Q and L. We then have

$$Ee^{itL + iuQ} = Ee^{itL} \cdot Ee^{iuQ}. \tag{4.2.6}$$

In view of the existence of $\text{Var}(X_j) = \sigma^2$, we may differentiate both sides of (4.2.6) once with respect to u, and then set $u = 0$, so that we have

$$E(Qe^{itL}) = E(e^{itL}) \cdot E(Q). \tag{4.2.7}$$

Substituting in (4.2.7) the expressions for L and Q, we obtain after some algebraic simplifications, the relation

$$\left(\sum_{i=1}^n a_{ii}\right)\phi''(t)(\phi(t))^{n-1} + \left(\sum_{i \neq j} a_{ij}\right)(\phi'(t))^2(\phi(t))^{n-2} = -E(Q) \cdot (\phi(t))^n. \tag{4.2.8}$$

In a sufficiently small neighborhood of the origin, $|t| < \delta$, where $\phi(t) \neq 0$, we may divide through by $(\phi(t))^n$ to obtain

$$\left(\sum_{i=1}^{n} a_{ii}\right) \frac{\phi''(t)}{\phi(t)} + \left(\sum_{i \neq j} a_{ij}\right) \left[\frac{\phi'(t)}{\phi(t)}\right]^2 = -E(Q). \tag{4.2.9}$$

Setting $\theta(t) = \ln \phi(t)$ in the above neighborhood, we see that (4.2.9) leads to

$$A\theta'' + B(\theta')^2 = C$$

where

$$A = \sum_{i=1}^{n} a_{ii}, \qquad B = \sum_{i,j} a_{ij}, \qquad C = -EQ = -A\sigma^2 + Bm^2. \tag{4.2.10}$$

We shall suppose first that $C \neq 0$. Then we shall distinguish three cases:
(I) $A \neq 0$, $B = 0$;
(II) $A = 0$, $B \neq 0$;
(III) $A \neq 0$, $B \neq 0$.
In Case (I), the equation (4.2.10) takes the form

$$\theta'' = -\sigma^2$$

so that we obtain a normal distribution, with $EQ = A\sigma^2$. Further, according to a well-known theorem of Craig [65], if $X_1, \ldots, X_n$ are a random sample from a normal population, then, for the independence of $\bar{X}$ and Q, it is necessary and sufficient that

$$\sum_j a_{ij} = 0, \qquad i = 1, 2, \ldots, n. \tag{4.2.11}$$

Hence we have

THEOREM 4.2.2. Under the above conditions, if $A \neq 0$, $B = 0$, then the condition (4.2.11) together with the normality of the variables X_j is necessary and sufficient for the independence of $\bar{X}$ and Q.

In Case (II), we obtain the equation

$$(\theta')^2 = -m^2. \tag{4.2.12}$$

This corresponds to a degenerate distribution, in which case $\bar{X}$ and Q are trivially independent.

In Case (III), we again have the possibility of a trivial solution, $\theta' =$ constant, leading to a degenerate distribution as in Case (II). Otherwise, let $\theta' = w$; we obtain

$$Aw' + Bw^2 = C. \tag{4.2.13}$$

We have for $t = 0$, $w(0) = im$, so that the solution for (4.2.13) under such initial conditions will be

$$w(t) = \sqrt{\frac{C}{B}}\,\frac{C_0 \exp(t\sqrt{BC}/A) - \exp(-t\sqrt{BC}/A)}{C_0 \exp(t\sqrt{BC}/A) + \exp(-t\sqrt{BC}/A)} \tag{4.2.14}$$

where

$$C_0 = \frac{\sqrt{C/B} + im}{\sqrt{C/B} - im}.$$

Then, since $\theta'(t) = w(t)$ and $\theta(0) = 0$, we have

$$\theta(t) = (A/B)\ln(\lambda_1 e^{\alpha t} + \lambda_2 e^{-\alpha t})$$

or,

$$\phi(t) = (\lambda_1 e^{\alpha t} + \lambda_2 e^{-\alpha t})^{\beta} \tag{4.2.15}$$

where

$$\lambda_1 = C_0/(C_0 + 1) \qquad \text{and} \qquad \lambda_2 = 1/(C_0 + 1)$$

(so that $\lambda_1 + \lambda_2 = 1$)

$$\beta = A/B = \left(\sum_i a_{ii}\right)\bigg/\left(\sum_{i,j} a_{ij}\right); \qquad \alpha = \sqrt{BC}/A = (1/\beta)(m^2 + \beta\sigma^2)^{1/2}.$$

Hence we find

$$\lambda_1 = \frac{1}{2}\left(1 + \frac{im}{\sqrt{-(m^2 + \beta\sigma^2)}}\right); \qquad \lambda_2 = \frac{1}{2}\left(1 - \frac{im}{\sqrt{-(m^2 + \beta\sigma^2)}}\right).$$

Depending on the sign of β, two cases arise:
(a) $m^2 + \beta\sigma^2 > 0$,
(b) $m^2 + \beta\sigma^2 < 0$.
Case (b), fully studied in the work of Laha [75], pp. 29–31, does not lead to any possible distribution for the X_i yielding the independence of $\overline{X}$ and Q, and therefore we shall not present it here. Case (a), which leads to the characterization of some simple discrete distributions, will be dealt with here; in that case, ϕ reduces to the form

$$\phi(t) = (\lambda_1 e^{ict} + \lambda_2 e^{-ict})^{\beta} \tag{4.2.16}$$

where λ_1, λ_2, and c are real numbers. For $\beta > 0$, λ_1 and λ_2 are obviously positive for any value for the sign of the radical. If then β is an integer, then (4.2.16) corresponds to a distribution which is a simple transformation of the Binomial, namely that given by

$$P[X = (2r - \beta)c] = \frac{\beta(\beta - 1)\cdots(\beta - r + 1)}{r!}\lambda_1^r \lambda_2^{\beta - r} \tag{4.2.17}$$

for $r = 0, 1, \ldots, \beta$. If $\beta > 0$ is not an integer, then an expression similar to (4.2.17) does not correspond to a probability distribution, since negative quantities appear in it. If $\beta = -\beta_0 < 0$, then, depending on the choice of the sign for the radical, either $\lambda_2 < 0$ or $\lambda_1 < 0$, the other being positive in view of $\lambda_1 + \lambda_2 = 1$.

In the case $\lambda_1 < 0$, the c.f. (4.2.16) leads to a distribution of the negative binomial type.

$$P[X = c(2r + \beta_0)] = \frac{\beta_0(\beta_0 + 1)\cdots(\beta_0 + r - 1)}{r!}(-\lambda_1)^r \lambda_2^{-\beta_0 - r} \quad (4.2.18)$$

Thus, the independence of $\bar{X}$ and Q can lead in the case under consideration only to discrete distributions of the type (4.2.17) and (4.2.18).

We now investigate whether, for the above discrete distributions, statistics Q independent of $\bar{X}$ exist. In the case (4.2.17) as well as (4.2.18), the random variable concerned can take extremal values (smallest or largest). In both cases,

$$P[X = -c\beta] = \lambda_2^{\beta}.$$

Let L_e be the corresponding extremal value of $L = n\bar{X}$. Obviously, $L_e = -nc\beta$, and is taken when and only when each X_i takes the value L_e/n, and then the statistic Q must take a definite value Q_e. Thus,

$$P(Q = Q_e | L = L_e) = 1.$$

From the condition of independence of Q and L, it then follows that

$$P(Q = Q_e) = P(Q = Q_e | L = L_e) = 1.$$

Thus, the distribution of Q is degenerate, with $Q = Q_e$ as the point of support. If the X_i take discrete values with the finite or denumerable set of points $\{a_i,\ i = 1, 2, \ldots\}$ as the support of the distribution, then we must obviously have

$$a_i^2 B = a_j^2 B, \qquad i, j = 1, 2, \ldots$$

so that $a_i = \pm a$ $(i = 1, 2, \ldots)$, where a is a constant. $a \neq 0$ since the X_i are nondegenerate. Without loss of generality, we may assume that $a > 0$, and the independent random variables X_i take the values $-a$ and a with probabilities p and $1 - p$, $p \in (0, 1)$. We shall show that in this case, the coefficients a_{ij} for $i \neq j$ in the quadratic form Q must be absent, so that

$$Q = \sum_i a_{ii} X_i^2. \quad (4.2.19)$$

Indeed, when the X_i take the values $\pm a \neq 0$ independently of one another, Q remains constant. Thus the quadratic form $\sum a_{ij} x_i x_j$ takes a constant value also when $x_i = \pm 1$; let q be this constant value. Then we have that for all the 2^n possible choices $x_i = \pm 1$, $Q(x_1, \ldots, x_n) = q$, so that

$$q = 2^{-n} \sum_{x_i = \pm 1} Q(x_1, \ldots, x_n) = A + 2^{-n} \sum_{i \neq j} a_{ij} \sum_{x_i, x_j = \pm 1} x_i x_j = A$$

since

$$\sum_{x_i, x_j = \pm 1} x_i x_j = 0 \qquad \text{for } i \neq j.$$

Thus, $q = A$. We now write for $x_i = \pm 1$,

$$Q = Q(x_1, \ldots, x_n) = A + \sum_{i \neq j} a_{ij} x_i x_j .$$

If, for the indicated values of the x_i, $Q = q$, then for such values of the x_i,

$$\sum_{i \neq j} a_{ij} x_i x_j = 0. \tag{4.2.20}$$

We shall show that this implies that $a_{ij} = 0$ for $i \neq j$. We proceed by induction. If $n = 2$, then this is obvious. Suppose it true for sets of variables $\leqq n$ in number; we shall prove it for the case of $(n + 1)$ variables. We have

$$\sum_{\substack{i \neq j \\ i, j \leqq n+1}} a_{ij} x_i x_j = \sum_{i=1}^{n} a_{i,n+1} x_i x_{n+1} + \sum_{\substack{i \neq j \\ i, j \leqq n}} a_{ij} x_i x_j = 0$$

for $x_i = \pm 1$. Let $x_1, \ldots, x_n$ take any of the 2^n sets of values, where each of them $= \pm 1$. In the equality obtained, we set $x_{n+1} = 1$ and $x_{n+1} = -1$ in succession, and add up the two equalities obtained. Then we obtain

$$2 \sum_{\substack{i \neq j \\ i, j \leqq n}} a_{ij} x_i x_j = 0$$

for any of the above-stated values of the x_i. By the induction assumption, we have hence $a_{ij} = 0$ $(i, j = 1, \ldots, n;\ i \neq j)$. Hence we obtain

$$\sum_{i=1}^{n} a_{i,n+1} x_i = 0 \tag{4.2.21}$$

for $x_i = \pm 1$. Taking $x_2, \ldots, x_n$ fixed, and $x_1 = +1$, and then $x_1 = -1$, and subtracting the relations obtained from one another, we see that $a_{1,n+1} = 0$. Proceeding similarly, we see that $a_{2,n+1} = a_{3,n+1} = \cdots = a_{n,n+1} = 0$, which proves our assertion. Thus Q has the form (4.2.19).

As a result of what has been just presented, we obtain, using the same notations as before, the following

THEOREM 4.2.3. Under the conditions

$$\sum_i a_{ii} \neq 0, \qquad \sum_{i,j} a_{ij} \neq 0, \qquad m^2 + \sigma^2 (\sum a_{ii}) / (\sum a_{ij}) > 0$$

for the independence of $\bar{X}$ and the quadratic form Q, it is necessary and sufficient that

(1) the X_i have a discrete distribution given by

$$P[X_i = a] = p, \qquad P[X_i = -a] = 1 - p; \qquad a \text{ real}, p \in (0, 1)$$

and

(2) $Q = \sum a_{ii} X_i^2$ (i.e., $a_{ij} = 0$ for $i \neq j$).

The above theorem yields a characterization of the discrete distributions indicated above, among the class of all distributions having finite variance. Similar results hold for certain more complicated polynomial statistics defined in terms of the cumulants (cf. the summary in the review [93] of E. Lukacs). More general results are given in section 4.7, but the problem of independence of $\bar{X}$ and polynomial statistics in its general form is still unsolved.

4.3 INDEPENDENCE OF QUASIPOLYNOMIAL STATISTICS

We now return to the consideration of a vector $\mathbf{X} = (X_1, \ldots, X_n)$ with independent (but not necessarily identically distributed as well) components.

THEOREM 4.3.1. Let $S_1(\mathbf{X})$ and $S_2(\mathbf{X})$ be two quasipolynomial statistics admissible with respect to all the n variables. If they are independent, then the following relation holds for any of the components of $\mathbf{X}$:

$$P[|X_j| \geqq x] = O(e^{-x^a}), \qquad j = 1, 2, \ldots, n \tag{4.3.1}$$

where $a > 0$ is some constant, and $x > 0$.

An obvious deduction from Theorem 4.3.1 is the following corollary.

Corollary. Under the conditions of Theorem 4.3.1, every component of $\mathbf{X}$ has moments of all orders.

We remark that a proposition of the above kind for independent polynomial statistics was studied by Chanda [162].

The proof of Theorem 4.3.1 is very similar to that of Theorem 3.3.1 on random linear forms, and is based on similar considerations. We shall again invoke the lemmas of Chapter 1. We shall follow Zinger [31] in our presentation.

***Proof*:** Without loss of generality, we may assume that our statistics $S_1(\mathbf{X})$ and $S_2(\mathbf{X})$ satisfy the inequalities (4.1.1), for suitable lower polynomials r_1, r_2, and upper polynomials R_1, R_2. Let $F_j(x) = P[X_j < x], j = 1, \ldots n$, be the distribution functions of the components of X.

In order not to make special qualifying statements in what follows, we shall assume that none of the components has its distribution concentrated on a finite interval. If such exist, the theorem holds trivially for them, and for the remaining, we may argue in the same way as we shall below.

We first consider the component X_1. There exists a bounded $(n-1)$-dimensional set G such that

$$\int_G \cdots \int dF_2(x_2) \cdots dF_n(x_n) = \gamma > 0. \tag{4.3.2}$$

Then a sufficiently large C_1 exists such that, for $|x_1| > C_1$ and $(x_2, \ldots, x_n)$ in G, the folllowing inequalities hold:

$$r_1(\mathbf{x}) \geqq C_2 x_1^{2k}; \qquad r_2(\mathbf{x}) \geqq C_2 x_1^{2m} \tag{4.3.3}$$

where $2k$ and $2m$ are the degrees of the polynomials $r_1(x)$ and $r_2(x)$ respectively (the evenness of the degrees of the polynomials follows from their non-negativity).

For the upper polynomials, we have the obvious inequalities.

$$\begin{aligned} R_1(\mathbf{x}) &\leqq C_3(|x_1| + \cdots + |x_n|)^{2k} + C_4 \\ R_2(\mathbf{x}) &\leq C_3(|x_1| + \cdots + |x_n|)^{2m} + C_4 \end{aligned} \tag{4.3.4}$$

We choose two numbers $\lambda > 0$ and $\mu > 0$ subject to the conditions

$$\lambda(R_1(\mathbf{x}))^{1/2k}, \qquad \mu(R_2(\mathbf{x}))^{1/2m} \leqq (1/n) \sum_1^n |x_j| + 1. \tag{4.3.5}$$

This is possible, in view of (4.3.4). We write

$$\tilde{S}_1(\mathbf{x}) = \lambda(S_1(\mathbf{x}))^{1/2k} - 1; \qquad \tilde{S}_2(\mathbf{x}) = \mu(S_2(\mathbf{x}))^{1/2m} - 1. \tag{4.3.6}$$

From the definition of quasipolynomials, and from (4.3.3) and (4.3.5), it follows that there exists $\alpha_1 > 0$ such that

$$\alpha_1 |x_1| \leqq \tilde{S}_i(\mathbf{x}) \leqq (1/n) \sum |x_j|; \qquad i = 1, 2 \tag{4.3.7}$$

where the right-hand relation is valid for all x, and the left one only for $|x_1| > C_5$ and $(x_2, \ldots, x_n) \in G$ (cf. (4.3.3) and (4.3.4)). It also follows from (4.3.7) that $\alpha_1 \leqq 1/n$.

We now consider the function

$$\Omega(x) = 1/\sum_{j=1}^n P[|X_j| \geqq x]. \tag{4.3.8}$$

According to the assumption made at the beginning, this function is finite for all real x, and is nondecreasing. If we introduce a random variable X with the distribution function

$$\phi(x) = (1/n) \sum_1^n F_j(x) \tag{4.3.9}$$

then $\Omega(x)$ can be written in terms of this X as

$$\Omega(x) = 1/\{nP[|X| \geqq x]\}. \tag{4.3.10}$$

Since $S_1(\mathbf{X})$ and $S_2(\mathbf{X})$ are independent, we may write (only formally at first)

$$E\{\Omega^\delta(\tilde{S}_1(\mathbf{X}))\Omega^\delta(\tilde{S}_2(\mathbf{X}))\} = E\{\Omega^\delta(\tilde{S}_1(\mathbf{X}))\} \cdot E\{\Omega^\delta(\tilde{S}_2(\mathbf{X}))\}. \tag{4.3.11}$$

We choose δ in the interval (1/2, 1). To establish the validity of (4.3.11), it suffices to show that the mathematical expectations entering into either one of the two sides of the relation exist finitely.

We consider $E[\Omega^\delta(\tilde{S}_1(\mathbf{X}))]$. In view of (4.3.7),

$$E\{\Omega^\delta(\tilde{S}_1(\mathbf{X}))\} \leqq E\{\Omega^\delta(n^{-1} \sum |X_j| + 1)\}. \tag{4.3.12}$$

Consider the n subsets of the sample space given by

$$E_j = \{(X_1, \ldots, X_n): |X_j| \geqq |X_i|, i = 1, \ldots, n\}, \qquad j = 1, \ldots, n. \tag{4.3.13}$$

Each point $(X_1, \ldots, X_n)$ belongs to at least one of the E_j, so that the sample space $= \bigcup_j E_j$, and it follows from (4.3.12) that

$$E\{\Omega^\delta(\tilde{S}_1(\mathbf{X}))\} \leqq \sum_{j=1}^{n} E\{\Omega^\delta(|X_j|)\} \tag{4.3.14}$$

or, in terms of the d.f. of the random variable X of (4.3.9),

$$E\{\Omega^\delta(\tilde{S}_1(\mathbf{X}))\} \leqq n \int_{-\infty}^{\infty} \Omega^\delta(|x| + 1)\, d\phi(x). \tag{4.3.15}$$

Since $\Omega(x) = 1/\{nP[|X| \geq x]\}$, and $0 < \delta < 1$, applying Lemma 1.1.8 to the variable $|X|$, we obtain from (4.3.15) that

$$E[\Omega^\delta(\tilde{S}_1(\mathbf{X}))] < \infty. \tag{4.3.16}$$

In exactly the same way, can establish that

$$E[\Omega^\delta(\tilde{S}_2(\mathbf{X}))] < \infty. \tag{4.3.17}$$

Thus, the validity of (4.3.11) may be taken as proved.

If, on the left side of (4.3.11), the integration is carried out over the set

$$\{(x_1, \ldots, x_n): |x_1| > C_1, (x_2, \ldots, x_n) \in G\} \tag{4.3.18}$$

instead of over the whole space, then, from the monotonicity of $\Omega(x)$, the left relation in (4.3.7), and (4.3.11), it follows that

$$E\{\Omega^{2\delta}(\alpha_1 |X_1|)\} < \infty. \tag{4.3.19}$$

Precisely the same argument applies to the other X_j. Hence we can indicate an α, $0 < \alpha \leqq 1/n$, such that, for all j,

$$E[\Omega^{2\delta}(\alpha |X_j|)] < \infty. \tag{4.3.20}$$

Hence it immediately follows that

$$E\{\Omega^{2\delta}(\alpha|X|)\} < \infty \tag{4.3.21}$$

where X is a random variable with distribution law (4.3.9).

We had chosen $1/2 < \delta < 1$. Hence $2\delta > 1$. Hence, in view of (4.3.21), we may apply Lemma 1.1.9 to the variable X, to obtain for any $x > 0$, for some $\beta > 0$,

$$P[|X| \geqq x] = O(e^{-x\beta}) \tag{4.3.22}$$

and consequently

$$P[|X_j| \geqq x] = O(e^{-x\beta}) \qquad \text{for all } j. \tag{4.3.23}$$

Theorem 4.3.1 stands proved.∎

Remark. For the proof above, we have made essential use of the admissibility of both the statistics $S_1(\mathbf{X})$ and $S_2(\mathbf{X})$. If one of them is not admissible, then the existence of the moments can be proved under the supplementary requirement: for some integer $s > 0$, we should have

$$E \log(\cdots 1 + \log(1 + \log(1 + |X_j|))\cdots)) < \infty \qquad \text{for all } j. \tag{4.3.24}$$

We also note that if the components of the vector $\mathbf{X}$ are identically distributed, then all the conditions need be imposed on only one component.

4.4 A QUASIPOLYNOMIAL STATISTIC AND A LINEAR FORM

If in the above, one of the quasipolynomial statistics be a linear form, then Theorem 4.3.1 can be considerably sharpened. The admissibility of the linear statistic obviously means merely that all its coefficients be nonzero.

THEOREM 4.4.1. If $S_1(\mathbf{X})$ be a quasipolynomial statistic, admissible with respect to all the variables, and $S_2(\mathbf{X})$ a linear form with nonzero coefficients, independent of $S_1(X)$, then

(1) for any $t > 0$, and for all j,

$$E \exp(t|X_j|) < \infty; \tag{4.4.1}$$

(2) the following estimate holds for some $A \geqq 1$,

$$E \exp(t|X_j|) = O(\exp t^A) \tag{4.4.2}$$

We also have an obvious corollary as follows.

Corollary. Under the conditions of Theorem 4.4.1, the c.f.'s of the components of $\mathbf{X}$ are entire functions of finite order.

Proof: The proof of Theorem 4.4.1 falls naturally into three parts: we first prove (1) for small t, then for all $t > 0$, and finally prove (2).

We start by establishing (1) for small $t > 0$.

First of all, we note that the statistic $S_2(\mathbf{X}) = \sum a_i X_i$ may be taken to be $\sum X_i$, since it is always to possible effect the transformation $X_j' = a_j X_j$ in view of $a_j \neq 0$ for any j.

As in the proof of Theorem 4.3.1, we choose a sufficiently large $C_1 > 0$ and an $(n-1)$-dimensional bounded set G such that

$$\int_G \cdots \int dF_2(x_2) \cdots dF_n(x_n) = \gamma > 0 \tag{4.4.3}$$

and for $|x_1| > C_1$ and $(x_2, \ldots, x_n)$ in G

$$S_1(\mathbf{x}) \geqq r_1(\mathbf{x}) \geqq C_2 x_1^{2k} \tag{4.4.4}$$

where $r_1(\mathbf{X})$ is the lower polynomial for $S_1(X)$, and $2k$ its degree.

We introduce the statistic

$$\bar{S}_1(\mathbf{X}) = [S_1(X)]^{m/2k} \tag{4.4.5}$$

where $m > 0$ is some number. Since $\bar{S}_1(\mathbf{X})$ and $\sum X_i$ are independent, we may write

$$E\{\bar{S}_1(\mathbf{X})|X_1 + \cdots + X_n|^N\} = E\{\bar{S}_1(\mathbf{X})\} \cdot E\{|X_1 + \cdots + X_n|^N\}, \qquad N = 0, 1, \ldots . \tag{4.4.6}$$

The existence of the above expectations is obviously guaranteed by Theorem 4.3.1. We obtain from (4.4.6), on using (4.4.3) and (4.4.4),

$$C_3 \int_{|x_1| > C_1} |x_1|^m \, dF_1(x_1) \int_G \cdots \int |x_1 + \cdots + x_n|^N \, dF_2(x_2) \cdots dF_n(x_n)$$
$$\leqq C \int \cdots \int |x_1 + x_2 + \cdots + x_n|^N \, dF_1(x_1) \cdots dF_n(x_n) \tag{4.4.7}$$

where $C = E\{\bar{S}_1(\mathbf{X})\}$.

Using the obvious inequalities

$$|x_1| - |x_2 + \cdots + x_n| \leqq |x_1 + x_2 + \cdots + x_n| \leqq |x_1| + |x_2| + \cdots + |x_n|$$

we obtain from (4.4.7)

$$\int_{|x_1| > C_1} |x_1|^m \, dF_1(x_1) \int_G \cdots \int (|x_1| - |x_2| + \cdots + x_n|)^N \, dF_2(x_2) \cdots dF_n(x_n)$$
$$\leqq C_4 \int \cdots \int (|x_1| + |x_2| + \cdots + |x_n|)^N \, dF_1(x_1) \cdots dF_n(x_n). \tag{4.4.8}$$

We add to both sides of (4.4.8) the quantity

$$\gamma I_1 = \int_{|x_1| \leqq C_1} |x_1|^m \, dF_1(x_1) \int \underset{G}{\cdots} \int (|x_1| - |x_2 + \cdots + x_n|)^N \, dF_2(x_2) \cdots dF_n(x_n)$$

and introduce the notations

$$\begin{aligned} a_k^j &= E|X_j|^k = \int |x|^k \, dF_j(x) \\ b_k^1 &= \frac{1}{\gamma} \int \underset{G}{\cdots} \int |x_2 + \cdots + x_n|^k \, dF_2(x_2) \cdots dF_n(x_n). \end{aligned} \tag{4.4.9}$$

Then, we obtain from (4.4.8)

$$\int |x_1|^m \, dF_1(x_1) \int \underset{G}{\cdots} \int (|x_1| - |x_2 + \cdots + x_n|)^N \, dF_2(x_2) \cdots dF_n(x_n)$$

$$\leqq \gamma I_1 + C_4 \int \cdots \int (|x_1| + |x_2| + \cdots + |x_n|)^N \, dF_1(x_1) \cdots dF_n(x_n). \tag{4.4.10}$$

Expanding the expressions in parentheses on both sides of (4.4.10), and using the above definitions, we obtain the inequality (with $C_5 = C_4/\gamma$)

$$\sum_{k=0}^{N} (-1)^k \binom{N}{k} a_{N+m-k}^1 b_k^1 \leqq I_1 + C_5 \sum{}^* \frac{N!}{i_1! \cdots i_n!} a_{i_1}^1 \cdots a_{i_n}^n$$

where $\sum^*$ indicates summation over $i_1 + \cdots + i_n = N$. On carrying over to the right side from the left all the terms except the one corresponding to $k = 0$, we find

$$a_{N+m}^1 \leqq \sum_{k=1}^{N} \binom{N}{k} a_{N+m-k}^1 b_k^1 + I_1 + C_5 \sum{}^* \frac{N!}{i_1! \cdots i_n!} a_{i_1}^1 \cdots a_{i_n}^n. \tag{4.4.11}$$

An inequality wholly similar to (4.4.11) can be obtained for $j = 2, 3, \ldots, n$, on proceeding similarly for each X_j and introducing the quantities b_k^j and I_j.

$$a_{N+m}^j \leqq \sum_{k=1}^{N} \binom{N}{k} a_{N+m-k}^j b_k^j + I_j + C_5 \sum{}^* \frac{N!}{i_1! \cdots i_n!} a_{i_1}^1 \cdots a_{i_n}^n \tag{4.4.12}$$

$$j = 1, 2, \ldots, n; \qquad N = 0, 1, 2, \ldots.$$

To establish the first part of the proof, it suffices to obtain an estimate of the form

$$a_k^i \leqq k! \, M^k \tag{4.4.13}$$

for $i = 1, 2, \ldots, n$; $k = 0, 1, \ldots$, for some constant $M > 0$. We shall obtain this estimate by induction, using (4.4.12).

We now choose and fix two integers m and m_0 such that

$$n - 1 \leqq m < m_0. \tag{4.4.14}$$

We choose the number M subject to the condition that the estimate (4.4.13) holds for all $k \leqq m_0$ (we shall impose further restrictions on M later). This condition will be the basis for our induction argument. Suppose (4.4.13) is proved for $k \leqq N + m - 1$; we shall prove it for $k = N + m$.

On the basis of the induction assumption, we estimate the right-hand side of (4.4.12). We take

$$\sum_{k=1}^{N} \binom{N}{k} a_{N+m-k}^1 b_k^1. \tag{4.4.15}$$

Since the set G is bounded, $|x_2 + \cdots + x_n|$ is bounded there. Hence there exists $b > 0$ such that

$$b_k^1 = \frac{1}{\gamma} \int \underset{G}{\cdots} \int |x_2 + \cdots + x_n|^k \, dF_2(x_2) \cdots dF_n(x_n) \leqq b^k, \qquad k = 0, 1, \ldots. \tag{4.4.16}$$

Then, using the induction assumption, we obtain for (4.4.15) the estimate

$$\begin{aligned} \sum_{k=1}^{N} \binom{N}{k} a_{N+m-k}^1 b_k^1 &\leqq \sum_{k=1}^{N} \frac{N(N-1)\cdots(N-k+1)}{k!} M^{N+m-k} b^k \\ &\leqq M^{N+m}(N+m)! \sum_{k=1}^{N} \frac{(b/M)^k}{k!}. \end{aligned} \tag{4.4.17}$$

The above inequality holds *a fortiori* if we extend the summation in the rightmost member to infinity, to obtain

$$\sum_{k=1}^{N} \binom{N}{k} a_{N+m-k}^1 b_k^1 \leqq (N+m)!\, M^{N+m}(e^{b/M} - 1). \tag{4.4.18}$$

We now impose on the number M the supplementary restriction

$$M > \frac{b}{\ln 3 - \ln 2}$$

which is obviously equivalent to

$$e^{b/M} - 1 < \frac{1}{2}$$

which enables us to assert, on the basis of (4.4.18)

$$\sum_{k=1}^{N} \binom{N}{k} a_{N+m-k}^1 b_k^1 < \frac{1}{2}(N+m)!\, M^{N+m}. \tag{4.4.19}$$

Considerations similar to those used for deriving (4.4.16) show that a $B > 0$ exists such that

$$I_1 \leqq B^N \qquad \text{for all } N. \tag{4.4.20}$$

We turn to the last sum in (4.4.12).

$$C_6 \sum{}^* \frac{N!}{i_1! \cdots i_n!} a_{i_1}^1 \cdots a_{i_n}^n \leqq C_7 N! M^N \sum{}^* 1. \tag{4.4.21}$$

The sum on the right side of (4.4.21) is the number of solutions in non-negative integers of the equation $x_1 + \cdots + x_n = N$ and it is easy to see that it does not exceed $(N+1)^{n-1}$, since each of the x_i except one can take at most $(N+1)$ values $\left(\text{the actual value of the sum is } \binom{N+n-1}{N}\right)$. Hence the left member of (4.4.21) can be estimated by

$$C_7 N!\, M^N (N+1)^{n-1} \leqq C_7\, (N+n-1)!\, M^N. \tag{4.4.22}$$

According to our choice, $m \geqq n-1$. We subject M again to the restriction

$$M \geqq \max\{4B, 4C_7).$$

Then it follows from (4.4.22) that the left member of (4.4.21) is less than $(1/4)(N+m)!\, M^{N+m}$; further, the same estimate also holds for I_1. From (4.4.11), it then follows in view of (4.4.19), (4.4.20), and (4.4.22) that

$$a_{N+m}^1 \leqq (N+m)!\, M^{N+m}. \tag{4.4.23}$$

Similar arguments and estimates hold for the a_{N+m}^k, $k = 2, \ldots, n$; it may be assumed that b and B are chosen in common for all k. Thus, (4.4.13) stands proved.

We pass on to the second part of the proof. It follows from (4.4.13) that, for $0 \leqq t < 1/M$,

$$E \exp\, (t|X_j|) < \infty. \tag{4.4.24}$$

We shall show that the above relation holds for all $t > 0$. Suppose t_0, the supremum of values of t for which (4.4.24) holds, is finite. Obviously, $t_0 \geqq 1/M$. We choose $\mu > 0$ such that for all $\mathbf{x}$, ($2k$ is the degree of the lower polynomial),

$$\tilde{S}_1(\mathbf{x}) + \mu[S_1(\mathbf{x})]^{1/2k} < \frac{1}{n} \sum |x_j| + 1. \tag{4.4.25}$$

Then, as in (4.4.3) and (4.4.4), a $\beta > 0$, $C_1 > 0$, and a set G exist such that, for $|x_1| > C_1$ and $(x_2, \ldots, x_n)$ in G,

$$\tilde{S}_1(\mathbf{x}) \geqq \beta |x_1|. \tag{4.4.26}$$

Take

$$0 < t < t_0 \tag{4.4.27}$$

and consider the relation

$$E \exp \left\{\tilde{S}_1(\mathbf{X}) + t \left|\sum_1^n X_j\right|\right\} = E \exp \tilde{S}_1(\mathbf{X}) \cdot E \exp \left(t \left|\sum_1^n X_j\right|\right). \tag{4.4.28}$$

The existence of the above expectations is guaranteed by (4.4.24) and (4.4.25). It follows from (4.4.28) that

$$\int_{|x_1|>C_1} dF_1(x_1) \int \underset{G}{\cdots} \int dF_2(x_2) \cdots dF_n(x_n) \exp\{\tilde{S}_1(\mathbf{x}) + t|\sum \mathbf{x}_j|\} < \infty \quad (4.4.29)$$

and consequently

$$\int_{|x_1|>C_1} e^{(t+\beta)|x_1|}\, dF_1(x_1) < \infty$$

so that

$$E \exp\{(t+\beta)|X_1|\} < \infty \qquad \text{for } 0 < t < t_0. \quad (4.4.30)$$

Similar relations can obviously be obtained for the other variables X_j, so that

$$E \exp\{(t+\beta)|X_j|\} < \infty, \qquad 0 < t < t_0 \qquad (j = 1, 2, \ldots, n). \quad (4.4.31)$$

We choose τ from the interval $(t_0 - \beta,\ t_0)$, and at once see that (4.4.31) contradicts the definition of t_0. Hence a finite t_0 cannot exist, and (4.4.24) is valid for all $t > 0$. This proves the assertion (1) of Theorem 4.4.1.

We turn to the proof of assertion (2) of the theorem. We introduce the statistic

$$S_1^*(\mathbf{X}) = \lambda|[S_1(\mathbf{X})]^{1/2k}$$

where $2k$ is the degree of the lower (and upper) polynomials for the statistic $S_1(\mathbf{X})$. The number λ is chosen according to the condition (cf. (4.3.5))

$$S_1^*(x) \leqq |x_1| + |x_2| + \cdots + |x_n| + 1. \quad (4.4.32)$$

As before, we choose an $(n-1)$-dimensional set G and a sufficiently large C_1, such that for $|x_1| > C_1$ and $(x_2, \ldots, x_n) \in G$,

$$\int \underset{G}{\cdots} \int dF_2(x_2) \cdots dF_n(x_n) = \gamma > 0,$$

$$S_1^*(x) \geqq C_8|x_1|, \qquad |x_1 + x_2 + \cdots + x_n| \geqq (1 - (1/2)C_8)|x_1|.$$

In view of the independence of $S_1^*(\mathbf{X})$ and $X_1 + X_2 + \cdots X_n$, we may write

$$E \exp[tS_1^*(\mathbf{X}) + t(X_1 + X_2 + \cdots + X_n)]$$
$$= E \exp[tS_1^*(\mathbf{X})] \cdot E \exp[t(X_1 + X_2 + \cdots + X_n)]. \quad (4.4.33)$$

We shall take t to be positive. The existence of the above expectations for all $t > 0$ is guaranteed by assertion (1) of Theorem 4.4.1, in view of (4.4.32). We have

$$E[\exp tS_1^*(\mathbf{X})] \leqq E[\exp t(|X_1| + \cdots + |X_n|)] \cdot e^t,$$
$$E \exp t(X_1 + \cdots + X_n) \leqq E \exp t(|X_1| + \cdots + |X_n|). \quad (4.4.34)$$

From (4.4.33) and (4.4.34), on carrying out the integration on the left side of (4.4.33) only over the set

$$\{(x_1, \ldots, x_n): |x_1| > C_1, (x_2, \ldots, x_n) \in G\}$$

we obtain

$$\int_{|x_1|>C_1} dF_1(x_1) \int_G \cdots \int dF_2(x_2) \cdots dF_n(x_n) \exp t(S_1^* + |x_1 + \cdots + x_n|)$$
$$\leqq e^t \cdot [E \exp (t(|X_1| + \cdots + |X_n|))]^2. \quad (4.4.35)$$

In the set under consideration, we have

$$S_1^*(\mathbf{x}) + |x_1 + \cdots + x_n| > C_8|x_1| + (1 - (1/2)C_8)|x_1| = (1 + (1/2)C_8)|x_1|. \quad (4.4.36)$$

If we apply the above estimate in the left side of (4.4.35), we obtain *a fortiori*

$$\int_{|x_1|>C_1} dF_1(x_1) \int_G \cdots \int dF_2(x_2) \cdots dF_n(x_n) \exp t(1 + 1/2)C_8)|x_1|$$
$$\leqq e^t \cdot [E \exp t(|X_1| + \cdots + |X_n|)]^2$$

or,

$$\gamma \int_{|x_1|>C_6} \exp t(1 + (1/2)C_8)|x_1| \, dF_1(x_1) \leqq e^t \cdot [E \exp t(|X_1| + \cdots + |X_n|)]^2.$$

Adding to both sides of the above relation the quantity

$$\gamma \int_{|x_1|>C_6} \exp (t(1 + (1/2)C_8)|x_1|) \, dF_1(x_1) \leqq \gamma \exp (C_9 t),$$

writing $\phi_j(t) = E \exp (t|X_j|)$, and using the obvious identity

$$E \exp t(|X_1| + \cdots + |X_n|) = \prod_{j=1}^{n} \phi_j(t)$$

we obtain

$$\phi_1[(1 + (1/2)C_8)t] \leqq \exp (C_9 t) + (1/\gamma)e^t \prod_{j=1}^{n} \phi_j^2(t).$$

A wholly similar relation holds for the other ϕ_j as well. Thus, positive numbers $\Delta > 1$, C_{10}, and γ exist such that, for all j.

$$\phi_j(\Delta t) \leqq \exp (C_{10} t) + (1/\gamma)e^t \prod_{j=1}^{n} \phi_j^2(t). \quad (4.4.37)$$

Writing

$$w(t) = \sup_{\substack{0 \leqq t' \leqq t \\ 1 \leqq j \leqq n}} \{\phi_j(t'), \exp (C_{10} t'/2n), 1 + (e^{t'}/\gamma)\} \quad (4.4.38)$$

we see from (4.4.37) that for any $0 \leqq t' \leqq t$,

$$\phi_j(\Delta t') \leqq w^{2n}(t) + (e^{t'}/\gamma)w^{2n}(t) \leqq w^{2n+1}(t). \tag{4.4.39}$$

Since $t' \leqq t$ is arbitrary, it follows that

$$w(\Delta t) \leqq w^{2n+1}(t). \tag{4.4.40}$$

Proceeding in the same way as in the derivation of Lemma 1.1.9, we obtain from (4.4.40) a relation of the form

$$w(t) = O(\exp t^A)$$

where A is some positive constant.▮

THEOREM 4.4.2. Under the conditions of Theorem 4.4.1, those components of the vector **X** whose c.f.'s do not have zeros (in the complex plane) are normally distributed.

***Proof*:** This is an immediate consequence of the Corollary to Theorem 4.4.1 and Lemma 1.4.2. Since the c.f. of such a component of the vector **X** is an entire function of finite order, not having any zeros by assumption, it is the c.f. of a normal law.▮

Thus it is seen that if there exists a quasipolynomial statistic $S_1(\mathbf{X})$ independent of the linear form $\sum X_i$ for a random vector **X** with a nonnormal distribution, then the c.f.'s of the nonnormal components of **X** must have zeros (in the complex plane, to which they are extensible as entire functions of finite order). We cite some examples confirming this fact.

Let $\mathbf{X} = (X_1, \ldots, X_n)$ be a random sample vector, with X_1 having a discrete distribution with N points of increase, and $P(X_1 = a_k) = p_k > 0$, $\sum_1^N p_k = 1$, $N > 1$. Let $\{b_1, \ldots, b_m\}$ be the set of values which can be assumed by the differences $a_i - a_j (i, j = 1, \ldots, N)$. Consider the polynomial statistic

$$Q(\mathbf{X}) = \prod_{k=1}^{n} \prod_{1 \leqq i \leqq j \leqq k} (X_i - X_j - b_k).$$

$Q(\mathbf{X})$ takes the constant value zero with probability one, and thus, trivially, is distributed independently of $\sum X_i$. The c.f. of X_1 given by $\sum_{k=1}^N p_k e^{ita_k}$ has infinitely many zeros in the complex plane.

We consider one more example, of two polynomial statistics admissible with respect to all the components of the random vector. Let $\mathbf{X} = (X_1, X_2)$ be a random sample vector with $X_j \sim N(0, 1)$ for $j = 1, 2$. Then, the linear forms $L_1 = X_1 + X_2$ and $L_2 = X_1 - X_2$ are independent. Following the remarks in section 3.5, we introduce a transformation of the sample variables:

$X_j = Y_j^{2m+1}$. Consider the random vector $\mathbf{Y} = (Y_1, Y_2)$ with independent components. The statistics

$$S_1(\mathbf{Y}) = Y_1^{2m+1} + Y_2^{2m+1}$$

$$S_2(\mathbf{Y}) = Y_1^{2m+1} - Y_2^{2m+1}$$

are independent, and the c.f. of either component of $\mathbf{Y}$ is

$$\frac{1}{\sqrt{2\pi}} \int_{-\infty}^{\infty} \exp\,(itx^{1/(2m+1)} - x^2/2)\, dx$$

$$= \frac{2m+1}{\sqrt{2\pi}} \int_{-\infty}^{\infty} \exp\,(ity - (1/2)y^{4m+2})\, y^{2m}\, dy.$$

This is an entire function of finite order, which accords with what has been said above. Further, according to the same, it must have zeros in the complex t-plane. The existence of an infinity of zeros for functions of a similar type was proved by Polya [116].

4.5 RANDOM LINEAR FORMS

Recently, theorems on independent linear forms have been generalized to linear forms with random coefficients [43]. The consideration of such random linear functionals proved useful in studying the phenomena of independence of many kinds of nonlinear statistics; further, the apparatus used in the discussion makes significant use of the inequalities given in section 4.4. Hence we have thought it fit to deal with investigations of the phenomena of independence of random linear forms in the present section.

Suppose a pair of linear forms

$$L_1 = u_1 X_1 + \cdots + u_n X_n, \qquad \text{and} \qquad L_2 = u_{n+1} X_1 + \cdots + u_{2n} X_n \tag{4.5.1}$$

are given, where the vector of coefficients $(u_1, \ldots, u_{2n})$ is a random vector in $R^{(2n)}$, independent of $(X_1, \ldots, X_n)$. We denote the corresponding column vectors by $\mathbf{u}^{(2n)}$ and $\mathbf{X}^{(n)}$ respectively.

Suppose that the distribution of $\mathbf{u}^{(2n)}$ in $R^{(2n)}$ has the following properties:

(1) It has bounded support, so that, for some $R_0 > 0$,

$$P[|\mathbf{u}^{(2n)}| < R_0] = 1. \tag{4.5.2}$$

(2) For any j, $1 \leqq j \leqq n$, the pair (u_n, u_{n+j}) can take values away from zero componentwise with positive probability; precisely, there exists an $\varepsilon > 0$ such that

$$P[|u_j| > \varepsilon,\ |u_{n+j}| > \varepsilon] > 0, \qquad j = 1, \ldots, n. \tag{4.5.3}$$

If the coefficients u_j of the forms are constants, then we are back to the case considered in section 2.1; these constants, however, must be nonzero if (4.5.3) is to be satisfied.

THEOREM 4.5.1. Let, in the above situation, the random linear forms L_1 and L_2 be independent. Then,

(1) $E|X_j|^N < \infty$ for all $N > 0$, for all j, and (4.5.4)

(2) if one of the forms has constant nonzero coefficients, then the c.f.'s of all the components of $\mathbf{X}^{(n)}$ can be extended to the complex plane as entire functions of finite order.

Corollary. Under the conditions of Part (2) above, those components of $\mathbf{X}^{(n)}$, whose c.f.'s do not vanish anywhere in the complex plane, are normal.

***Proof*:** We shall use mainly the same arguments as in section 4.4.

Let F_j and ϕ_j denote respectively the d.f. and the c.f. of X_j. If all the components X_j of $\mathbf{X}^{(n)}$ have distributions with compact support, then the relation (4.5.4) and assertion (2) above are trivial. Hence we may assume that the d.f. of at least one component is not concentrated on a finite segment. Then consider the function

$$\Omega(x) = 1/\sum_{j=1}^{n} P[|X_j| \geqq x], \qquad x \text{ real.} \tag{4.5.5}$$

In view of what we have said above, Ω is a nonnegative, nondecreasing function. We prove a lemma.

LEMMA 4.5.1. For some $\tilde{\alpha} > 0$ and any $\delta \in (0, 1)$, the following relation holds.

$$E\{[\Omega(\tilde{\alpha}|X_j|)]^{2\delta}\} < \infty \qquad \text{for all } j. \tag{4.5.6}$$

***Proof*:** We note that the independence of L_1 and L_2 implies the formal relation below for any $\alpha > 0$.

$$E\{[\Omega(\alpha|L_1|)]^{\delta}\,[\Omega(\alpha|L_2|)]^{\delta}\} = E[\Omega(\alpha|L_1|)]^{\delta} \cdot E[\Omega(\alpha|L_2|)]^{\delta}. \tag{4.5.7}$$

We show below that this relation is meaningful and establish its validity. We first show that the expectations on the right side of (4.5.7) exist. We choose $\alpha > 0$ so small that

$$\alpha|L_j| \leqq (1/n)\sum_{1}^{n} |X_j|, \qquad j = 1, 2, \text{ with probability one.} \tag{4.5.8}$$

Such a choice of α is possible in view of the condition (4.5.2) imposed on the vector of random coefficients, $\mathbf{u}^{(2n)}$.

We now define n subsets of $R^{(n)}$ given by

$$E_j = \{\mathbf{x}^{(n)} : |x_k| \leqq |x_j| \text{ for all } k\}; \qquad j = 1, \ldots, n. \tag{4.5.9}$$

These sets are measurable, and further, obviously,

$$R^{(n)} = \bigcup_{1}^{n} E_j. \tag{4.5.10}$$

We now apply Lemma 1.1.8. Let S_r be the sphere in $R^{(n)}$ given by $x_1^2 + \cdots + x_n^2 \leqq r^2$, $r > 0$, and $I(S_r)$ the indicator function of S_r. Then, in view of the nondecreasing nature of Ω, we may write

$$\begin{aligned}
E\{I(S_r)[\Omega(\alpha|L_j|)]^\delta\} &\leqq E\left\{I(S_r)\left[\Omega\left(\frac{1}{n}\sum_{j=1}^{n}|X_j|\right)\right]^\delta\right\} \\
&\leqq \sum_{j=1}^{n}\int_{E_j} I(S_r)[\Omega(|x_j|)]^\delta\, dF_1(x_1)\cdots dF_n(x_n) \\
&\leqq \sum_{j=1}^{n}\int_{-\infty}^{\infty}\frac{dF_j(x)}{\left\{\sum_{j=1}^{n}P[|X_j| \geqq x]\right\}^\delta} \\
&\leqq \sum_{j=1}^{n}\int_{-\infty}^{\infty}\frac{dF_j(x)}{\{P[|X_j| \geqq x]\}^\delta} < \infty,
\end{aligned} \tag{4.5.11}$$

by Lemma 1.1.8, where, for values of x for which the denominator and hence the numerator vanishes, we set the corresponding part of the integral equal to zero, as indicated earlier, this being permissible in view of the condition that $\delta \in (0, 1)$.

Letting $r \to \infty$, the existence of $E[\Omega(\alpha|L_j|)]^\delta$ follows for sufficiently small $\alpha > 0$ and all $\delta \in (0, 1)$. This proves (4.5.7) for such α.

Changing our enumeration of the components if necessary, let $X_1, \ldots, X_{n_0}$ be those components not having compact support, and the remaining have such support. As pointed out earlier, it suffices to study only the former type of components. Consider X_1. We take a bounded $(n-1)$-dimensional set G_1 such that

$$\int_{G_1}\cdots\int dF_2(x_2)\cdots dF_n(x_n) = \gamma > 0. \tag{4.5.12}$$

We then choose $C_1 > 0$ so large that, if

$$|x_1| > C_1, \qquad (x_2, \ldots, x_n) \in G_1, \qquad |u_1| \geqq \varepsilon, \qquad |u_{n+1}| \geqq \varepsilon,$$

the following inequality is satisfied for sufficiently small $\alpha_1 > 0$.

$$|L_1| \geqq \alpha_1|x_1|, \qquad |L_2| \geqq \alpha_1|x_1|. \tag{4.5.13}$$

We turn to (4.5.7). The left member of this relation is not smaller than the corresponding integral taken over the following subset of $R^{(n)} \times R^{(2n)}$.

$$\{(\mathbf{x}^{(n)}, \mathbf{u}^{(2n)}): |x_1| > C_1, (x_2, \ldots, x_n) \in G_1;\ |u_1| > \varepsilon,\ |u_{n+1}| > \varepsilon\}.$$

Since Ω is nondecreasing, from the above circumstance and also from (4.5.12), (4.5.13), and the finiteness of the right-hand side of (4.5.7), we obtain for sufficiently small $\alpha_2 > 0$

$$E[\Omega(\alpha_2 |X_1|)]^{2\delta} < \infty.$$

The same line of reasoning holds for the other X_j as well. Thus we can choose $\tilde{\alpha} > 0$ such that

$$E[\Omega(\tilde{\alpha} |X_j|)]^{2\delta} < \infty \qquad \text{for all } j. \tag{4.5.14}$$

Choosing some $\delta \in (1/2, 1)$, so that $\delta_1 = 2\delta > 1$, we have then

$$\int_{-\infty}^{\infty} \frac{dF_j(x)}{\{P[|X_j| \geqq \tilde{\alpha}x]\}^{\delta_1}} < \infty. \tag{4.5.15}$$

Then, Lemma 1.1.9 implies that $P[|X_j| \geqq x] = O\,(\exp(-x^{\beta}))$, where $\beta = \beta(\tilde{\alpha}, \delta_1) > 0$. Hence it follows that, for every j, X_j has moments of all orders, thus proving Part (1) of Theorem 4.5.1.

We turn to the proof of Part (2). Without loss of generality, we may assume that it is the linear form L_2 that has the constant nonzero coefficients, and then again we may take it to be of the form

$$L_2 = X_1 + \cdots + X_n. \tag{4.5.16}$$

Clearly, Part (1) of the theorem holds for the new X_j. From the independence of L_1 and L_2, we deduce in view of (4.5.4) that

$$E|L_1|^m |L_2|^N = E|L_1|^m \cdot E|L_2|^N. \tag{4.5.17}$$

Here N is any nonnegative integer, and m a positive integer, whose value we shall specify later. Since the arguments below are in many ways similar to those of section 4.4, we shall present them here in a more concise fashion. We write

$$a_k^j = E|X_j|^k, \qquad k = 1, 2, \ldots; \qquad j = 1, \ldots, n.$$

As in the derivation of (4.5.14), we consider a set of the form

$$\{(\mathbf{x}^{(n)}, \mathbf{u}^{(2n)}) : |x_1| > C_1, (x_2, \ldots, x_n) \in G_1, |u_1| > \varepsilon\}$$

recalling that by assumption, $u_{n+1} = u_{n+2} = \cdots = u_{2n} = 1$. We carry out the integration on the left-hand side of (4.5.17) over this set. Arguing as for deriving (4.4.12), we obtain the estimate

$$a_{N+m}^j \leqq A_1^N + A_2 \sum_{i_1 + \cdots + i_n = N} \frac{N!}{i_1! \cdots i_n!} a_{i_1}^1 \cdots a_{i_n}^n + \sum_{k=1}^{N} \binom{N}{k} a_{N+m-k}^j \cdot A_3^k \tag{4.5.18}$$

for the value of $j = 1$; $A_1, A_2, A_3, \ldots$, here and in what follows, will denote positive constants. Replacing $j = 1$ in our derivation above by $j = 2, \ldots, n$, we obtain (4.5.18) for all j.

We now fix the number m and another integer m_0 such that

$$n - 1 \leqq m < m_0. \tag{4.5.19}$$

We also choose $M > 0$ such that

$$a_k^j \leqq k!\, M^k \qquad \text{for } j = 1, \ldots, n; \qquad k = 1, \ldots, m. \tag{4.5.20}$$

Starting from (4.5.17), we shall show that for a suitable $M > 0$, the relation (4.5.20) will be satisfied for all k. The proof is made by induction, starting with (4.5.20). Let this relation be valid for $k \leqq N + m - 1$; we shall prove that it holds for $k = N + m$. We estimate the right member of (4.5.18). According to our induction assumption

$$\sum_{k=1}^{N} \binom{N}{k} a_{N+m-k}^j A_3^k \leqq \sum_{k=1}^{N} \frac{N(N-1)\cdots(N-k+1)}{k!} (N+m-k)!\, M^{N+m-k} A_3^k$$

$$\leqq (N+m)!\, M^{N+m} \sum_{k=1}^{N} \frac{(A_3/M)^k}{k!} < (N+m)!\, M^{N+m}(e^{A_3/M} - 1).$$

Let M be so large that $\exp(A_3/M) \leqq 3/2$. Then we have

$$\sum_{k=1}^{N} \binom{N}{k} a_{N+m-k}^j A_3^k \leqq \frac{1}{2}(N+m)!\, M^{N+m}. \tag{4.5.21}$$

Then, noting as before that the number of solutions in nonnegative integers of the equation $i_1 + \cdots + i_n = N$ does not exceed $(N+1)^{n-1}$, we similarly obtain the estimate

$$\sum \frac{N!}{i_1! \cdots i_n!} a_{i_1}^1 \cdots a_{i_n}^n \leqq N!\, M^N \sum 1 \tag{4.5.22}$$

the summations being taken over all $(i_1, \ldots, i_n)$ such that $i_1 + \cdots + i_n = N$, the i_k being nonnegative integers,

$$\leqq N!\,(N+1)^{n-1} M^N \leqq (N+n-1)!\, M^N.$$

We set further $M > \max(4A_2, 4A_3)$; then we find from (4.5.18) together with (4.5.21) and (4.5.22) that

$$a_{N+m}^j \leqq (N+m)!\, M^{N+m}$$

as required. Hence we have that for $t \in (0, 1/M)$,

$$E(\exp t|X_j|) < \infty. \tag{4.5.23}$$

Thus, the c.f. of X_j is analytic in some horizontal strip. We shall show that it is an entire function, i.e., that (4.5.23) holds for all positive t.

Suppose not, and let t_0 be the supremum of values of t for which (4.5.23) holds. Then, $1/M \leqq t_0 < \infty$. We take some t in $(0, t_0)$ and choose $\alpha \in (0, 1)$ such that $\alpha|L_1| \leqq \sum_1^n |X_j|$ with probability one (as we have seen already, such an α exists). Then, in view of what has been proved,

$$E \exp(t\{|L_2| + \alpha|L_1|\}) = E(\exp t|L_2|) \cdot E(\exp \alpha t|L_1|). \quad (4.5.24)$$

We can then proceed as in the derivation of (4.5.14) and (4.5.18) and see that for some $\tilde{\alpha} > 0$,

$$E \exp[t(1 + \tilde{\alpha})|X_j|] < \infty.$$

If here we choose $t \in (t_0/(1 + \tilde{\alpha}), t_0)$, then we arrive at a contradiction to the definition of t_0, and hence (4.5.23) has to be valid for all $t > 0$.

Now we can consider (4.5.24) for any t. Let $w_j(t) = E(\exp t|X_j|)$. Applying the same method as for the derivation of (4.5.14) and (4.5.18)—integrating over a suitable subset of the type indicated earlier—we easily arrive at the estimate

$$w_j(t(1 + \tilde{\alpha})) \leqq A_4 \exp(A_5 t) + \prod_{j=1}^{n} (w_j(t))^2 \quad (4.5.25)$$

for all positive t and all j. We shall deduce hence the relation

$$E(\exp t|X_j|) = O(\exp t^A) \quad (4.5.26)$$

where $A > 0$ is some constant.

We set $\Delta = 1 + \tilde{\alpha} > 1$, and introduce the monotone function

$$\theta(t) = \sup_{\substack{1 \leqq j \leqq n \\ 0 \leqq t' \leqq t}} [(A_4 + 1) \exp(A_5 t'), w(t'), 2].$$

Then, as in section 4.4,

$$\theta(\Delta t) \leqq [\theta(t)]^{2n+1}$$

and as before, in view of the monotonicity of θ, we easily obtain (4.5.26). Thus the c.f. of X_j is extensible to the complex plane as an entire function of finite order. This proves Part (2) of Theorem 4.5.1, which thus stands completely proved.∎

The Corollary to Theorem 4.5.1, on conditions for the normality of X_j, follows from Marcinkiewicz' theorem (Lemma 1.4.2).

4.6 GENERALIZATION OF THE SKITOVICH-DARMOIS THEOREM: ONE OF THE FORMS RANDOM

Let the form L_2 have nonzero constant coefficients, so that, without loss of generality, we may take

$$L_2 = X_1 + \cdots + X_n.$$

The conditions (4.5.2) and (4.5.3) take the forms

$$P[|\mathbf{u}^{(n)}| < R_0] = 1 \qquad \text{for some } R_0 > 0 \tag{4.6.1}$$

and

$$P[|u_j| > \varepsilon] > 0 \qquad \text{for all } j, \qquad \text{for some } \varepsilon > 0. \tag{4.6.2}$$

$\mathbf{u}^{(n)}$ of course denotes $(u_1, \ldots, u_n)$, where $L_1 = u_1 X_1 + \cdots + u_n X_n$. Corresponding to $\mathbf{u}^{(n)}$, we set up a family of "defining polynomials" in the real variable ξ given by

$$Q_{m_1, \ldots, m_n}(\xi) = E[(1 + u_1\xi)^{m_1} \cdots (1 + u_n \xi)^{m_n}]$$

where $(m_1, \ldots, m_n)$ is a multiple index ranging over all possible choices of nonnegative integer-values for the m_i. Then the following theorem holds.

THEOREM 4.6.1. Let (4.6.1) and (4.6.2) be satisfied, and let, for every multiple index $(m_1, \ldots, m_n)$ with $\sum_1^n m_i \neq 0$ (i.e., with not all the m_i equal to zero),

$$Q_{m_1, \ldots, m_n}(\xi) \neq \text{constant}. \tag{4.6.3}$$

Then the independence of L_1 and L_2 involves the normality of $\mathbf{X}^{(n)}$.

Remark. We note that this theorem is a generalization of the Skitovich-Darmois theorem (Theorem 3.1.1) in the case of nonzero coefficients in both forms. In fact, in such a case, the form L_2 can be written in the form (4.5.16), i.e., as above, and, for L_1, the random vector $\mathbf{u}^{(n)}$ has a degenerate distribution concentrated at the point $(u_1, \ldots, u_n)$ with $u_j \neq 0$ for any j. Let us set up the polynomial Q as per the given formula; in this case, it takes the form

$$Q_{m_1, \ldots, m_n}(\xi) = (1 + u_1\xi)^{m_1} \cdots (1 + u_n \xi)^{m_n}.$$

If at least one of the m_i is positive (for instance, let $m_1 > 0$; then $Q(0) = 1$, while $Q(-1/u_1) = 0$, so that) Q is not constant-valued. It would follow from Theorem 4.6.1 that $\mathbf{X}^{(n)}$ is normal.

***Proof*:** For any complex s and t, the following relation holds, in view of Theorem 4.5.1.

$$E \exp(tL_2 + sL_1) = E \exp(tL_2) \cdot E \exp(sL_1). \tag{4.6.4}$$

Let π be the probability measure corresponding to the d.f. of the vector $\mathbf{u}^{(n)}$. If, as before, ϕ_j denote the c.f. of X_j, then (4.6.4) can be rewritten in the form

$$\int_{R^n} \phi_1(t + su_1) \cdots \phi_n(t + su_n)\pi(d\mathbf{u}) = \phi_1(t) \cdots \phi_n(t) \int_{R^n} \phi_1(su_1) \cdots \phi_n(su_n)\pi(d\mathbf{u}). \tag{4.6.5}$$

We shall show that ϕ_j never vanishes, for any j. Suppose, for some j_0 and t_0, $\phi_{j_0}(t_0) = 0$. If m_j be the multiplicity of the zero at t_0 for ϕ_j, then

$$\phi_j(t) = (t - t_0)^{m_j} \cdot \theta_j(t), \qquad m_j \geqq 0, \qquad m_{j_0} > 0 \tag{4.6.6}$$

where θ_j is an entire function with $\theta_j(t_0) \neq 0$.

We set $s = \xi(t - t_0)$ in (4.6.5) and substitute $\phi_j(t)$ in the form (4.6.6) in it, to obtain

$$(t - t_0)^{\Sigma_1^n m_j} \int_{R^n} \prod_1^n (1 + u_j \xi)^{m_j} \theta_j(t + \xi(t - t_0)u_j)\pi(d\mathbf{u})$$

$$= (t - t_0)^{\Sigma_1^n m_j} \theta_1(t) \cdots \theta_n(t) \int_{R^n} \prod_1^n \phi_j((t - t_0)u_j)\pi(d\mathbf{u}).$$

We divide out by $(t - t_0)^{\Sigma_1^n m_j}$ and let $t \to t_0$. Since, in view of (4.6.1), the r.v.'s u_j are bounded by R_0 with probability one, we have

$$\int_{R^n} (1 + u_1\xi)^{m_1} \cdots (1 + u_n \xi)^{m_n} \pi(du) \cdot \prod_1^n \theta_j(t_0) = \prod_1^n \theta_j(t_0) \cdot \prod_1^n \phi_j(0) = \prod_1^n \theta_j(t_0).$$

Dividing out by $\prod_1^n \theta_j(t_0)$, we find

$$\int_{R^n} (1 + u_1\xi)^{m_1} \cdots (1 + u_n \xi)^{m_n} \pi(d\mathbf{u}) = 1 \tag{4.6.7}$$

where at least one of the m_j, namely m_{j_0}, is positive. Equation (4.6.7) contradicts the conditions of Theorem 4.6.1. Hence, none of the ϕ_j can vanish anywhere in the complex plane. By the corollary to Theorem 4.5.1, $\mathbf{X}^{(n)}$ must be a normal vector. This concludes the proof.∎

The condition (4.6.3) imposed on the defining polynomials $Q(m_1, \ldots, m_n)$ will not be needed if, instead, some conditions are imposed on the components $X_1, \ldots, X_n$ themselves.

THEOREM 4.6.2. Suppose (4.6.1) is satisfied, (4.6.2) is replaced by the stronger condition that, for some $\varepsilon > 0$,

$$P[|u_j| > \varepsilon, j = 1, \ldots, n] > 0 \tag{4.6.8}$$

and the components X_j of $\mathbf{X}^{(n)}$ are representable in the form

$$X_j = X_j' \pm X_j'' \qquad \text{for all } j$$

where X_j' and X_j'' are i.i.d.r.v.'s. Then $\mathbf{X}^{(n)}$ is normal.

***Proof*:** We argue as in the proof of Theorem 4.6.1 down to formula (4.6.6). Then we note that if g_j be the c.f. of X_j', then $\phi_j(t) = g_j(t)g_j(\pm t)$. Since ϕ_j is an entire function, as is well-known (cf. [85]), so are $g_j(\pm t)$. If t_0 is a zero of $\phi_j(t)$, in the complex plane, then it is a zero of $g_j(t)$ or of $g_j(-t)$, and, by the Hermitian property of c.f.'s, it is also a zero of $g_j(-t)$ or of $g_j(t)$ respectively. Thus any zero in the complex plane of ϕ_j is of even multiplicity. In view of this, (4.6.7) reduces to the form (writing $m_j = 2q_j$)

$$\int_{R^n} (1 + u_1\xi)^{2q_1} \cdots (1 + u_n\xi)^{2q_n}\pi(d\mathbf{u}) = 1 \tag{4.6.9}$$

with the condition: $\sum_1^n q_i > 0$. We show below that this cannot be. The coefficient of the highest power of ξ in the polynomial on the left side of (4.6.9) is

$$E(u_1^{2q_1} \cdots u_n^{2q_n}).$$

In view of (4.6.8), this coefficient is obviously nonzero, and hence (4.6.9) is impossible.∎

From the foregoing, it is clear that while Theorem 3.1.1 is comparatively simple to prove, its generalization to the case where a random linear form appears (Theorem 4.6.1) is significantly more difficult to establish.

In general, very little information is available concerning the independence of two random linear forms in nonidentically distributed r.v.'s; characterizations of distributions through such phenomena are yet to be obtained. In the identically distributed case, however, it is possible to say significantly more; as we shall see later, independence of random linear forms can be related in such a case to that of nonlinear statistics.

4.7 INDEPENDENCE OF THE SAMPLE MEAN AND A NONSINGULAR POLYNOMIAL STATISTIC

It is possible to indicate some conditions to be imposed on the polynomial statistic $P(\mathbf{X})$ independent of the linear from $L_2 = \sum_1^n X_i$ under which the c.f.'s of the components of $\mathbf{X}$ will not have zeros in the complex plane and therefore the vector $\mathbf{X}$ will be normal.

In the polynomial $P(\mathbf{x})$, we single out the homogeneous polynomial $P_0(\mathbf{x})$ of the highest degree, and in each term of this polynomial, we replace x_j^k by $m_j(m_j - 1)\cdots(m_j - k + 1)$; as a result, we get a polynomial in the m_j, which we denote by $\pi(m_1, \ldots, m_n)$, and call the *defining polynomial* of $P(\mathbf{x})$.

We can give another, equivalent definition of the above polynomial. In $P_0(\mathbf{x})$, we replace each monomial $x_1^{k_1} \ldots x_n^{k_n}$ by the differential operator

$$\frac{\partial^{k_1}}{(\partial\xi_1)^{k_1}} \cdots \frac{\partial^{k_n}}{(\partial\xi_n)^{k_n}}$$

obtaining as a result a differential operator $P_0(\partial/\partial\xi)$. Then,

$$\pi(m_1, \ldots, m_n) = P_0\left(\frac{\partial}{\partial\xi}\right)\xi_1^{m_1} \cdots \xi_n^{m_n}\Big|_{\xi_1 = \cdots = \xi_n = 1}.$$

The statistic $P(\mathbf{X})$ will be called *nonsingular* if, for any choice whatever of the nonnegative integers $m_1, \ldots, m_n$,

$$\pi(m_1, \ldots, m_n) \neq 0 \tag{4.7.1}$$

provided at least one of them is positive, i.e.,

$$m_1 + \cdots + m_n > 0. \tag{4.7.2}$$

We note the resemblance of (4.7.2) to the condition (4.6.3) in the study of random linear forms made in section 4.6.

For the case of a random sample $(X_1, \ldots, X_n)$ and a polynomial statistic based on it, we introduce the notion of *diagonal nonsingularity*. In such a situation, a statistic $P(\mathbf{X})$ will be called *diagonally nonsingular* if its defining polynomial is such that $\pi(m, \ldots, m) \neq 0$ for $m = 1, 2, \ldots$.

THEOREM 4.7.1. Let $\mathbf{X}$ be a random sample vector, and the polynomial statistic $P(\mathbf{X})$ admissible with respect to (one of the, and hence with respect to) all the variables, and further diagonally nonsingular. If $P(\mathbf{X})$ is independent of $\mathbf{X}$, then the sample is normal.

***Proof*:** We start by establishing a lemma.

LEMMA 4.7.1. Let $\mathbf{X} = (X_1, \ldots, X_n)$ be a random vector with independent components. If the polynomial statistic $P(\mathbf{X})$, admissible with respect to all the components, and $L = \sum_1^n X_i$ are independent, then, in order that at the same point of the complex plane, the c.f.'s of the X_j have zeros of multiplicity respectively $m_j (j = 1, \ldots, n)$, it is necessary that $\pi(m_1, \ldots, m_n) = 0$, π being the defining polynomial.

We remark that this condition is not sufficient for the existence of zeros for the c.f.'s. As an example, we take the case of a random sample (X_1, X_2), and $P(\mathbf{X}) = X_1 - X_2$, $L(\mathbf{X}) = X_1 + X_2$. Then the defining polynomial $\pi(m_1, m_2) = m_1 - m_2$ vanishes for $m_1 = m_2 = 1, 2, \ldots$. If (X_1, X_2) is a normal sample, then the c.f.'s of X_1 and X_2 are nonvanishing, while P and L are independent.

Proof of the Lemma: The lemma and its proof are some-what strengthened versions of the results of section 6 of [83]. In view of Theorem 4.4.1, we have for any complex z

$$E(Pe^{zL}) = E(P) \cdot E(e^{zL}). \tag{4.7.3}$$

Now, $E(e^{zL}) = f_1(z) \cdots f_n(z)$, where the f_j are the c.f.'s of the X_j, and $E(P) = A$, where A can possibly be zero. Equation (4.7.3) yields a relation between the functions f_j and their derivatives, which we write in the form

$$D(f_1, \ldots, f_n) = A \prod_1^n f_j. \tag{4.7.4}$$

Consider the formation of the left member of (4.7.4). We have

$$P(\mathbf{x}) = \sum a_{k_1, \ldots, k_n} x_1^{k_1} \cdots x_n^{k_n}, \qquad 0 \leqq k_i \leqq m$$

where m is the degree of $P(\mathbf{x})$. Further,

$$E[X_1^{k_1} \cdots X_n^{k_n} \exp z(X_1 + \cdots + X_n)] = f_1^{(k_1)}(z) \cdots f_n^{(k_n)}(z). \tag{4.7.5}$$

Thus, the left side of (4.7.4) consists of sums of monomials in n factors taken from among the $f_j^{(k)}$, $1 \leqq j \leqq n$, $0 \leqq k \leqq m$. If P_0 is the homogeneous polynomial of degree m contained in P, then P_0 gives rise to a differential expression of order m on the left side of (4.7.4), i.e., in each monomial thereof; const. $f_1^{(k_1)} \cdots f_n^{(k_n)}$, we have $k_1 + \cdots + k_n = m$.

Let, if possible, f_j have a zero of order $m_j \geqq 0$ at z_0, for $1 \leqq j \leqq n$, with

$$m_1 + \cdots + m_n > 0. \tag{4.7.6}$$

Let $D' = \{0 < |z - z_0| < r\}$ denote any deleted neighborhood of z_0 in which none of the f_j vanishes. Let ϕ_j denote any continuous branch of $\ln f_j$ in, say, an open angle of the form $D' = \{|\arg(z - z_0)| < \pi\}$; then, $\phi_j' = f_j'/f_j$ is uniquely defined in the sense of being independent of the branch of $\ln f_j$ that ϕ_j represents, is analytic in D', and has a simple pole at z_0. Further in, D', $f_j^{(k)}/f_j$ is a polynomial in $\phi_j', \phi_j'', \ldots, \phi_j^{(k)}$, which has the property that, in any monomial thereof, the sum of the orders of the derivatives involved (taking multiplicity into account) is k. For instance, $f_j''/f_j = \phi_j'' + (\phi_j')^2$. For $z \in D'$, dividing out both sides of (4.7.4) by $\prod_1^n f_j(z)$, we obtain a relation of the form

$$\sum a_{k_1, \ldots, k_n} \frac{f_1^{(k_1)}(z) \cdots f_n^{(k_n)}(z)}{f_1(z) \cdots f_n(z)} = A. \tag{4.7.7}$$

Using the fact that, in D', $\phi_j'(z) = g_j(z) + m_j/(z - z_0)$ where g_j is analytic in the open disc $D = D' \cup \{z_0\}$, the terms in (4.7.7) corresponding to the polynomial P_0 give rise to an expression of the form

$$\frac{c_m}{(z - z_0)^m} + \cdots + \frac{c_1}{(z - z_0)} + G_1(z) \tag{4.7.8}$$

and those corresponding to the other terms in P, to an expression of the form

$$\frac{d_{m-1}}{(z - z_0)^{m-1}} + \cdots + \frac{d_1}{(z - z_0)} + G_2(z) \tag{4.7.9}$$

where the c_j, d_j are constants, and G_1 and G_2 are analytic in D. It follows from (4.7.7)–(4.7.9) that we must have

$$c_m = (c_{m-1} + d_{m-1} = \cdots = c_1 + d_1) = 0.$$

It is easily seen that, if

$$P_0(\mathbf{x}) = \sum a_{k_1,\ldots,k_n} x_1^{k_1} \cdots x_n^{k_n},\ k_1 + \cdots + k_n = m$$

then

$$c_m = \sum a_{k_1,\ldots,k_n} m_1^{(k_1)} \ldots m_n^{(k_n)} \tag{4.7.10}$$

where $m_j^{(k_j)} = m_j(m_j - 1)\cdots(m_j - k_j + 1)$. This proves the lemma.▌

Theorem 4.7.1 concerns the identically distributed case, so that $f_j = f$ (say) for all j; from the proof of the above lemma, it follows that f cannot have any zeros in the complex plane if, for every positive integer m, $\pi(m, \ldots, m) \neq 0$; the normality of the X_j is then an immediate consequence of Theorem 4.4.2.

Lemma 4.7.1 also enables us to easily derive Theorem 3.1.1. For simplicity, we confine ourselves to the case where both linear forms have nonzero coefficients (for all j).

Let the linear form $L_1 = \sum_1^n A_j X_j$ be independent of $L = \sum_1^n X_j$, the X_j being themselves independent. The admissibility of L_1 is equivalent to none of the A_j being zero. The defining polynomials for the linear form L_1 and for quadratic form L_1^2 are respectively

$$\pi_1(m_1, \ldots, m_n) = \sum A_j m_j, \text{ and}$$

$$\pi_2(m_1, \ldots, m_n) = \sum_{j=1}^{n} A_j^2 m_j(m_j - 1) + \sum_{i \neq j} A_i A_j m_i m_j$$

$$= (\pi_1(m_1, \ldots, m_n))^2 - \sum_{j=1}^{n} A_j^2 m_j.$$

If at some point z_0, the f_j were to have zeros of orders m_j respectively, with at least one m_j positive, then both $\pi_1(m_1, \ldots, m_n)$ and $\pi_2(m_1, \ldots, m_n)$ must be zero, so that we must have $\sum A_j^2 m_j = 0$, which is impossible. Hence, the f_j cannot have zeros, and since they are of finite order, by Theorem 4.4.1, it follows from Lemma 1.4.2 that the X_j are normal.

We can derive some further consequences for a vector $\mathbf{X}$ with independent components, from Lemma 4.7.1.

We call a polynomial statistic $P(\mathbf{X})$ *weakly nonsingular* if its defining polynomial $\pi(m_1, \ldots, m_n) \neq 0$ for any n-tuple of nonnegative integers $(m_1, \ldots, m_n)$ such that $m_1 + \cdots + m_n \geqq 2$.

THEOREM 4.7.2. Let $P(\mathbf{X})$ be admissible with respect to all components, weakly nonsingular, and independent of $L = X_1 + \cdots + X_n$. For the normality of $\mathbf{X}$, it is sufficient that at least one of the following conditions be satisfied.

(a) The components X_j can be split up into pairs of identically distributed ones.

(b) For all j, X_j admits the representation: $X_j = X_j' \pm X_j''$ where X_j' and X_j'' are independent and identically distributed.

***Proof*:** Let $f_j(z) = E \exp(zX_j)$ be the c.f. of X_j, and let z_0 be a zero of the f_j, the multiplicity of the zero being respectively $m_1, \ldots, m_n$. Let $\mu = m_1 + \cdots + m_n$. If $\mu > 0$, then $\mu \geqq 2$, under condition (a) or (b), since the f_j will either coincide in pairs or have zeros of even multiplicity. In view of the condition of weak nonsingularity, we must have $\mu = 0$.∎

We remark that instead of $L = \sum_1^n X_i$, we may also take an arbitrary linear form $a_1 X_1 + \cdots + a_n X_n$ with nonzero coefficients, setting $Y_j = a_j X_j$, we then accordingly change the condition of nonsingularity or weak nonsingularity of $P(\mathbf{X})$.

Also, as shown by Theorem 4.2.3, cases of c.f.'s which vanish do occur.

We consider some applications of Theorem 4.7.1.

Examples. Let $P(\mathbf{X}) = (1/n) \sum_1^n (X_i - \bar{X})^3 = m_3$ be the third sample central moment. Suppose m_3 is independent of $\bar{X}$. The polynomial $P(\mathbf{X})$ is obviously admissible for $n \geqq 3$. Straightforward calculation shows* that the corresponding defining polynomial has the form

$$\pi(m, \ldots, m) = 2m\left(1 - \frac{1}{n}\right)\left(1 - \frac{2}{n}\right) \tag{4.7.11}$$

so that it does not vanish for $m = 1, 2, \ldots$, if $n \geqq 3$. Thus $P(\mathbf{X})$ is diagonally nonsingular, and thus the independence of m_3 and $\bar{X}$ turns out to be necessary and sufficient for the normality of $\mathbf{X}$, without any supplementary condition.

For the fourth, fifth, and seventh sample central moments, we have respectively* the following expressions for $\pi(m, \ldots, m)$

$$\frac{3m(n-1)}{n^3}[mn(n-1) - 2(n^2 - 3n + 3)];$$

$$\frac{4m(n-1)(n-2)}{n^4}[-5mn(n-1) + 6(n^2 - 2n + 2)];$$

and

$$\frac{6m(n-1)(n-2)}{n^5}[35m^2n^2(n-1)^2 - 14mn(n-1)(11n^2 - 27n + 27) + 120(n^2 - n + 1)(n^2 - 3n + 36)].$$

These polynomials do not vanish for $n \geqq 4$, $n \geqq 5$, and $n \geqq 7$ respectively. Thus the normal law is characterized for such n by the independence of $\bar{X}$ and the sample moments indicated above.

* These calculations were carried out by N. M. Khalfin and V. N. Chugujeva of the Leningrad Division of the Mathematical Institute of the U.S.S.R. Academy of Sciences.

It should be noted that sample moments of even orders are definite tube statistics of dimension $(n-1)$ (cf. the definition in section 4.1) so that the independence of such a moment and $\bar{X}$ implies the normality of the sample under the supplementary assumption that a continuous probability density exists for the X_j, by Theorem 4.9.1. No such considerations obtain for odd order moments; hence they are of special interest in the sense indicated above.

In the examples above, a distinctive condition has been imposed on the sample size n *vis-à-vis* the order of the sample moment, namely, that $n \geqq k$ (for $n = 3, 4, 5, 7$). If, generally, we consider sample sizes n which are sufficiently large compared to k, the order of the sample moment m_k, then, following the note [88], we can establish the following theorem.

THEOREM 4.7.3. Let $k \geqq 5$ be a square-free integer, and, if even, not be of the form $2(2^s + 1)$. Let the sample moment m_k be independent of $\bar{X}$, and the sample size n satisfy the condition

$$n \geqq 2(2^k\, k! + 1)^{k+1}. \tag{4.7.12}$$

Then the sample is normal.

***Proof*:** Our proof will have a number-theoretic character. It will be based on Theorem 4.7.1. We note that the polynomial concerned is admissible with respect to all the variables. To establish its diagonal nonsingularity, it is obviously necessary to consider its defining polynomial $\pi(m, \ldots, m)$. Setting $\bar{\partial}/\partial\xi = (1/n) \sum_1^n (\partial/\partial\xi_i)$, following the definition at the beginning of this section, we have

$$\pi(m, \ldots, m) = \frac{1}{n} \sum_1^n \left(\frac{\partial}{\partial\xi_i} - \frac{\bar{\partial}}{\partial\xi}\right)^k (\xi_1 \cdots \xi_n)^m \Big|_{\xi_1 = \cdots = \xi_n = 1}.$$

Introducing the simpler expression

$$\pi_1(m) = \left(\frac{\partial}{\partial\xi_1} - \frac{\bar{\partial}}{\partial\xi}\right)^k (\xi_1 \cdots \xi_n)^m \Big|_{\xi_1 = \cdots = \xi_n = 1} \tag{4.7.13}$$

we note that by considerations of symmetry, $\pi_1(m) = \pi(m, \ldots, m)$, and hence we shall study the zeros of $\Pi_1(m)$ among the natural numbers. The expression (4.7.13) can be rewritten in the form

$$\left\{\left(\frac{\partial}{\partial\xi_1}\right)^k - \binom{k}{1}\left(\frac{\partial}{\partial\xi_1}\right)^{k-1}\left(\frac{\bar{\partial}}{\partial\xi}\right) + \cdots + (-1)^{k-1}\binom{k}{k-1}\left(\frac{\partial}{\partial\xi_1}\right)\left(\frac{\bar{\partial}}{\partial\xi}\right)^{k-1} + (-1)^k\left(\frac{\bar{\partial}}{\partial\xi}\right)^k\right\}(\xi_1 \cdots \xi_n)^m \Big|_{\xi_1 = \xi_2 = \cdots = \xi_n = 1}. \tag{4.7.14}$$

First of all, we shall show that for $n \geqq 2k$, $\pi_1(1) \neq 0$. In fact, setting $m = 1$ in (4.7.14), we find

$$\begin{aligned}(-1)^{k-1} n^k \pi_1(1) &= kn(n-1)\cdots(n-(k-2)) - n(n-1)\cdots(n-(k-1)) \\ &= n(n-1)\cdots(n-(k-2))(2k-1-n) \neq 0 \qquad \text{for } n \geqq 2k.\end{aligned}$$

Further, we see from (4.7.14) that, for given n, $\pi_1(m)$ is a polynomial in m of degree at most k. It is not identically zero, since $\pi_1(1) \neq 0$. If it is constant, then this constant is not zero, and $\pi_1(m) \neq 0$ for $m = 1, 2, \ldots$, which proves our theorem. Hence we need only consider the case where the degree d of $\Pi_1(m)$ is positive. The coefficients of the polynomial $\pi_1(m)$, as easily seen, do not exceed $2^k k!$ Hence, in view of the well-known MacLaurin theorem on the zeros of a polynomial, we have for the zeros of $\pi_1(m)$ the estimate

$$m \leqq 2^k k! + 1. \tag{4.7.15}$$

We now write (4.7.14) in the form

$$m(m-1)\cdots(m-k+1) - \binom{k}{1} m^2(m-1)\cdots(m-k+2) + \cdots + (-1)^{k-1}\binom{k}{1} m^k + (-1)^k m^k + \theta 2^k k! \frac{m^k}{n} \tag{4.7.16}$$

where $\theta = \theta(m, n, k)$ and $|\theta| \leqq 1$. We have seen that $m > 1$. We divide the expression (4.7.16) by $(m - 1)$ and equate the resulting expression to zero. We obtain the relation:

$$\frac{(k-1)m^k}{m-1} + z = \theta \frac{2^k k!\, m^k}{n(m-1)} \tag{4.7.17}$$

where z is an integer. We shall show that under the condition (4.7.12), $(m - 1)$ divides $(k - 1)$, i.e.

$$(m-1)\,|\,(k-1). \tag{4.7.18}$$

In fact, the numbers m and $(m - 1)$ are mutually prime.

If (4.7.18) does not hold, then the left member of (4.7.17) can be written in the form $z_1 + a/(m - 1)$, where z_1 is an integer, and $1 \leqq a \leqq m - 2$. Under the conditions (4.7.12) and (4.7.15), (4.7.17) would then yield a contradiction. Thus, (4.7.18) must be true.

We shall now show that, for $n \geqq 2$, it follows from (4.7.12) and (4.7.15) that

$$n\,|\,m. \tag{4.7.19}$$

The first term in the expression (4.7.16), namely, the product $m(m - 1)\cdots(m - k + 1)$ is obviously divisible by k, since the factors constitute a full system of residues modulo k. Then follow terms of the form

$$\binom{k}{r} m^{r+1}(m-1)\cdots(m-(k-r-1)). \tag{4.7.20}$$

We shall show that, for r such that $1 \leqq r \leqq k-1$, these expressions are divisible by k. In fact, the above expression

$$= m^r \frac{k(k-1)\cdots(k-(k-r-1))}{(k-r)!} m(m-1)\cdots(m-(k-r-1))$$

$$= m^r k \frac{(k-1)\cdots(k-1-(k-r-2))}{(k-r-1)!} \frac{m(m-1)\cdots(m-(k-r-1))}{(k-r)}.$$

Of these factors, the first, m^r, is an integer, the third equals $\binom{k-1}{k-r-1}$ and hence is an integer, and the numerator in the fourth factor is the product of factors which constitute a complete system of residues modulo $(k-r)$ and hence is divisible by $(k-r)$, so that the fourth factor above is also an integer. Thus, the expression (4.7.20) is divisible by k. On dividing by k, and equating the result to zero, (4.7.16) yields the relation $z - m^k/k + \theta(2^k k!\, m^k/nk) = 0$, where z is an integer. In view of (4.7.18), we see that $m \leqq k$, so that the above relation can be rewritten.

$$z - \frac{m^k}{k} + \theta_1 \frac{2^k\, k!\, k^k}{nk} = 0, \qquad |\theta_1| \leqq 1$$

If the zero m is not divisible by k, which is square-free, then m^k cannot be divisible by k either, so that the left member above takes the form

$$z_1 + \frac{a}{k} + \theta_1 \frac{2^k k!\, k^k}{nk} = 0$$

where $1 \leqq a \leqq k-1$ and z_1 is an integer. This contradicts (4.7.12), and hence (4.7.19) is proved. We deduce immediately from (4.7.18) and (4.7.19) that $m = k$. Now (4.7.16) can be rewritten as

$$k(k-1)\cdots 2\cdot 1 - \binom{k}{1} k^2(k-1)\cdots 2 + \cdots + (-1)^{k-2}\binom{k}{2} k^{k-1}(k-1)$$
$$+ (-1)^{k-1}\binom{k}{1} k^k + (-1)^k \cdot k^k + \theta\, 2^k k!\, \frac{k^k}{n}. \qquad (4.7.21)$$

All the terms of this expression excepting the last four are divisible by $(k-1)(k-2)$. Let us consider the first three of the last four terms. Their sum is

$$(-1)^k k^k (k-1)\frac{k-3}{2}. \qquad (4.7.22)$$

We shall show that this is not divisible by $(k-1)(k-2)$. Suppose that it is so divisible. First consider the case when k is odd. Then $k^k(k-3)/2$ is an integer and it must be divisible by $(k-2)$. But $(k, k-2) = 1$ so that $(k-3)/2$

should be divisible by $(k-2)$, which is impossible. Then consider the case when k is even. Then, by our assumption that $k \neq 2(2^s+1)$, we have $(k-2)$ is not a power of 2. Since $k \geqq 5$, $(k-2)$ must have an odd prime factor, p. Since $(k, k-2) = 2$, p does not divide k. Further, $k^k(k-3)/2$ is an integer which must be divisible by $(k-2)$. It follows hence that $(k-3)$ must be divisible by p. But, since $k \geqq 5$ and even, we have $k \geqq 6$, or $(k-3) \geqq 3$. Further, $(k-3)$ is relatively prime to $(k-2)$ and hence cannot be divisible by p which divides $(k-2)$. Hence our assertion stands proved.

Dividing out (4.7.21) by $(k-1)(k-2)$ and equating the resulting expression to zero, we obtain

$$z + \frac{a}{(k-1)(k-2)} = \theta \frac{2^k k!\, k^k}{n(k-1)(k-2)}, \qquad z \text{ integer.}$$

This contradicts the relation (4.7.12). Thus, $k = m$ cannot be a zero for $\pi_1(m)$, and $\pi_1(m) = \pi(m, \ldots, m)$ cannot have zeros for $m = 1, 2, 3, \ldots$, which proves our theorem.▌

Theorem 4.7.3 admits some generalization.
Consider the statistic

$$\chi_k = \sum_{i=1}^{r} b_i(X_i - \bar{X})^k + P_{k-1}(X_i - \bar{X}) \tag{4.7.23}$$

where k is assumed to be a square-free integer, and, in case it is even, not to be of the form $2(2^s+1)$; $b_i \neq 0$ are real constants, $r \leqq n$, and P_{k-1} is a polynomial of degree at most $k-1$; further, we assume that

$$\sum_1^r b_i \neq 0. \tag{4.7.24}$$

In particular, if $P_{k-1} \equiv 0$, $r = n$, $b_i = 1/n$ for all i, we have $\chi_k = m_k$.

The defining polynomial for the statistic χ_k will be $\pi(m, \ldots, m) \sum_1^r b_i$, where $\pi(m, \ldots, m)$ is the defining polynomial for m_k, studied in the proof of the preceding theorem. Thus, in view of (4.7.24), it does not vanish for $m = 1, 2, \ldots$, under condition (4.7.12). Let us then consider the condition of admissibility of χ_k with respect to all the components. The "leading" polynomial in χ_k, $\sum_1^r b_i(X_i - \bar{X})^k$, can be written in the form

$$\sum_1^r b_i\left(\frac{n-1}{n} X_i - \sum_{s \neq i} \frac{X_s}{n}\right)^k.$$

Thus, the coefficients for the powers X_i^k will have the form

$$b_i\left(1 - \frac{1}{n}\right)^k - \left(\frac{1}{n}\right)^k \sum_{j \neq i} b_j \qquad \text{for } i \leqq r$$

$$-\left(\frac{1}{n}\right)^k \sum_{j=1}^r b_j \qquad \text{for } i > r.$$

Thus we see that the coefficients of the X_i^k do not vanish for $i > r$, in view of (4.7.24), and for $i \leqq r$, they will not vanish under the condition

$$n \geqq 3\left(\frac{\sum_{j=1}^{r} |b_j|}{\min_j |b_j|}\right)^{1/k}. \tag{4.7.25}$$

Hence we have the theorem.

THEOREM 4.7.4. Let $k \geqq 5$ be a square-free integer and, if even, not be of the form $2(2^s + 1)$. Let the polynomial statistic χ_k given by (4.7.23) be independent of $\bar{X}$, and let the sample size satisfy the conditions (4.7.22) and (4.7.25). Then the sample is normal.

4.8 THE METHOD OF INTEGRO-FUNCTIONAL EQUATIONS

We consider the problem of characterizing the normal distribution through the independence of pairs of definite tube statistics (see section 4.1 for the definitions). Let the linear form $L = \sum X_i$ be independent of the definite tube statistic S(**X**) of dimension $(n-1)$; let the components of **X** have distributions with continuous densities $y_j(x)$, $j = 1, 2, \ldots, n$. Then the integro-functional equation below holds.

$$\int_\Phi \prod_{j=1}^{n} y_j(t + s\sigma_j(\phi))T_1(\phi)\, d\phi = c y_1(t) \cdots y_n(t) \int_\Phi \prod_{1}^{n} y_j(s\sigma_j(\phi))T_1(\phi)\, d\phi \tag{4.8.1}$$

for all real t and positive s; $\phi = (\phi_1, \ldots, \phi_{n-2})$ is a point on the $(n-2)$-dimensional torus Φ: $\{0 \leqq \phi_j \leqq \pi$ for $j = 1, 2, \ldots, n-3$; $0 \leqq \phi_{n-2} \leqq 2\pi\}$; $c > 0$ is a constant; $T_1(\phi) \geqq 0$ is a continuous function with $\int_\phi T_1(\phi)\, d\phi < \infty$; $\sigma_j(\phi)$ are continuous functions with

$$\sum_{1}^{n} \sigma_j(\phi) = 0. \tag{4.8.2}$$

We shall now derive the relation (4.8.1) below.

We make an orthogonal transformation

$$(X_1, \ldots, X_n) \to (X_1', \ldots, X_n')$$

which unlinks the statistics L and $S(\mathbf{X})$. Then we may set

$$X_n' = \sqrt{n}\,\bar{X} = \left(\sum_{1}^{n} X_j\right) \Big/ \sqrt{n}. \tag{4.8.3}$$

The inverse of an orthogonal matrix being its transpose, the inverse transformation can be written in the form

$$X_j = \sum_{k=1}^{n-1} a_{jk} X_k' + (1/\sqrt{n}) X_n'; \qquad j = 1, 2, \ldots, n. \tag{4.8.4}$$

Under our transformation, L becomes $\sqrt{n}X_n'$ and $S(\mathbf{X})$ must become $S_1(\mathbf{X}')$ depending only on $X_1', \ldots, X_{n-1}'$.

Instead of the variables $(X_1', \ldots, X_{n-1}')$, we introduce spherical coordinates: $(X_1', \ldots, X_{n-1}') \to (\rho, \phi)$; $0 \leqq \rho < \infty$; $\phi = (\phi_1, \ldots, \phi_{n-2}) \in \Phi$; we also set $X_n' = \sqrt{n}v$ (so that $v = \bar{X}$). Consider the transformation $(X_1, \ldots, X_n) \to (v, \rho, \phi)$; in view of (4.8.4), we may write

$$X_j = v + \rho z_j; \qquad j = 1, \ldots, n \tag{4.8.5}$$

where z_j depend only on ϕ; in view of the orthogonality of the vectors $(a_{1k}, a_{2k}, \ldots, a_{nk})$ to $(1/\sqrt{n}, \ldots, 1/\sqrt{n})$,

$$\sum_1^n z_j = \sum_1^n \sum_{k=1}^{n-1} a_{jk} X_k' = \sum_{k=1}^{n-1} X_k' \sum_{j=1}^n a_{jk} = 0$$

and

$$\rho^2 \sum_1^n z_j^2 = \sum_{j=1}^n \left(\sum_{k=1}^{n-1} a_{jk} X_k' \right)^2 = \sum_{k=1}^{n-1} (X_k')^2$$

in view of the orthogonality of the columns of the matrix $((a_{jk}))$. Hence $\sum_1^n z_j^2 = 1$. We display the relations obtained for the z_j.

$$\sum_1^n z_j = 0; \qquad \sum_1^n z_j^2 = 1. \tag{4.8.6}$$

By our assumption, we must then have $S(\mathbf{X}) = S_1(\mathbf{X}')$, $S_2(\rho z_1, \ldots, \rho z_n) = S_2(X_1 - \bar{X}, \ldots, X_n - \bar{X})$. Thus the original statistic depends only on $X_j - \bar{X}$, and hence—cf. (4.8.5)

$$S(\mathbf{X}) = S(v + \rho z_1, \ldots, v + \rho z_n) = S(\rho z_1, \ldots, \rho z_n).$$

Since it is always possible to consider S as homogeneous and of the first degree of homogeneity (see last paragraph of this section), it then follows that

$$S(\mathbf{X}) = \rho S(z_1, \ldots, z_n) = \rho S(\phi). \tag{4.8.7}$$

In view of the definiteness of the tube statistic S, $S(\phi)$ takes on, on Φ, a smallest positive value. The independence of S and $\bar{X}$ implies that for any t, $s \geqq 0$, $\Delta t \geqq 0$, $\Delta s \geqq 0$,

$$\begin{aligned} P[t \leqq \bar{X} < t + \Delta t,\ s \leqq S(\mathbf{X}) < s + \Delta s] \\ = P[t \leqq \bar{X} < t + \Delta t] \cdot P[s \leqq S(\mathbf{X}) < s + \Delta s]. \end{aligned} \tag{4.8.8}$$

We apply the transformation of coordinates (4.8.5) to the left side of (4.8.8); using (4.8.7) also, we find the following expression for it.

$$\Delta P = \int_\Phi T(\phi)\, d\phi \int_t^{t+\Delta t} dv \int_{r/S(\phi)}^{(r+\Delta s)/S(\phi)} \rho^{n-2} y_1(v + \rho z_1) \cdots y_n(v + \rho z_n)\, d\rho \tag{4.8.9}$$

where $T(\phi)$ is the factor corresponding to the Jacobian of the transformation from rectangular coordinates to spherical ones. We divide (4.8.9) through by $\Delta t \Delta s$ and let Δt and Δs tend to zero. We remark that the definiteness of the tube statistic S is very important here, in that it implies that $S(\phi)$ has a finite maximum on Φ. In view of this, we easily see that

$$\lim_{\substack{\Delta t \to 0 \\ \Delta s \to 0}} (\Delta P / \Delta t \Delta s) = \sqrt{n} s^{n-2} \int_\Phi T_1(\phi)\, d\phi \prod_1^n y_j(t + s\sigma(\phi) z_j) \qquad (4.8.10)$$

where $\sigma(\phi) = 1/S(\phi)$ and $T_1(\phi) = T(\phi)/(S(\phi))^{n-1}$.

The expression (4.8.10) represents the density of the joint distribution of $\bar{X}$ and $S(\mathbf{X})$. By assumption, it must be equal to the product of the densities $P_{\bar{X}}(t)$ and $P_S(s)$ of the statistics $\bar{X}$ and S. Hence we deduce that

$$\sqrt{n} s^{n-2} \int_\Phi T_1(\phi) \prod_1^n y_j(t + s\sigma_j(\phi))\, d\phi = P_{\bar{\mathbf{X}}}(t) P_S(s) \qquad (4.8.11)$$

for t real and $s \in (0, \infty)$; $\sigma_j(\phi) = z_j \sigma(\phi)$ for all j. It follows from (4.8.6) that

$$\sum_1^n \sigma_j(\phi) = 0 \qquad \text{and} \qquad \sum_1^n \sigma_j^2(\phi) = \sigma^2(\phi) \qquad \text{for } \phi \in \Phi. \qquad (4.8.12)$$

We see that $P_{\bar{\mathbf{X}}}(0) \neq 0$, since otherwise the left side of (4.8.11) would vanish for all s, which is impossible. Let us set $t = 0$ in (4.8.11). Then we obtain the expression for $P_s(s)$ from (4.8.11) which we substitute in (4.8.10). Dividing both sides of the equation thus obtained by s^{n-2}, we find

$$\int_\Phi \prod_1^n y_j(t + s\sigma_j(\phi)) T_1(\phi)\, d\phi = c_1 P_{\bar{X}}(t) \int_\Phi \prod_1^n y_j(s\sigma_j(\phi)) T_1(\phi)\, d\phi. \qquad (4.8.13)$$

As usual, $c_1, c_2, \ldots$ will denote positive constants.

Setting $s = 0$ in the above, we obtain the expression for $P_{\bar{X}}(t)$

$$P_{\bar{X}}(t) = c_2 y_1(t) \cdots y_n(t). \qquad (4.8.14)$$

Substitution of this expression in (4.8.13) leads to (4.8.1).∎

We note that, for multitubular statistics, as follows from their definition, we obtain an expression of the form (4.8.1), where, in place of $\int_\Phi$ there will be a finite sum of such integrals. It is possible to take it as $\int_\Phi$ when ϕ ranges over a certain number of identical copies of the torus Φ.

We now consider the construction of an equation of the form (4.8.1) for two independent definite tube statistics of degree 1 and dimensions n_1 and n_2 respectively ($n_1 + n_2 = n$). We shall assume that S_1 and S_2 are unlinkable. Then we shall proceed as in the preceding argument.

Let

$$(X_1, \ldots, X_n) \to (X_1', \ldots, X_n')$$

be an orthogonal transformation which unlinks the statistics S_1 and S_2. Then we may assume that $S_1(\mathbf{X})$ is a function of $X_1', \ldots, X_{n_1}'$ alone, and $S_2(\mathbf{X})$ of the variables $X_{n_1+1}', \ldots, X_n'$ alone. We write down the inverse transformation

$$X_j = \sum_{k=1}^{n_1} a_{jk} X_k' + \sum_{m=n_1+1}^{n} a_{jm} X_m'; \qquad j = 1, 2, \ldots, n. \tag{4.8.15}$$

Instead of the variables $X_1', \ldots, X_{n_1}'$, we introduce spherical coordinates

$$(X_1', \ldots, X_{n_1}') \to (\rho_1, \phi), \qquad 0 \leqq \rho_1 < \infty, \qquad \phi \in \Phi$$

where Φ is a $(n_1 - 1)$-dimensional torus. Similarly, instead of $X_{n_1+1}', \ldots, X_n'$, we introduce the spherical coordinates

$$(X_{n_1+1}', \ldots, X_n') \to (\rho_2, \psi), \qquad 0 \leqq \rho_2 < \infty, \qquad \psi \in \Psi$$

where Ψ is a $(n_2 - 1)$-dimensional torus. We set

$$\rho_1 \sigma_j(\phi) = \sum_{k=1}^{n_1} a_{jk} X_k', \qquad j = 1, \ldots, n$$

$$\rho_2 \kappa_j(\psi) = \sum_{m=n_1+1}^{n} a_{jm} X_m', \qquad j = 1, \ldots, n.$$

Then, from (4.8.15) and the orthogonality of our transformation, we find

$$\sum_1^n \sigma_j(\phi)\kappa_j(\psi) = 0 \qquad \text{for } \phi \in \Phi, \qquad \psi \in \Psi. \tag{4.8.16}$$

As per our assumptions, $S_1(\mathbf{X})$ has the form

$$\begin{aligned} S_1(\mathbf{X}) &= S_1'(X_1', \ldots, X_{n_1}') = S_1''(\rho_1\sigma_1(\phi), \ldots, \rho_1\sigma_n(\phi)) \\ &= \rho_1 S_1''(\sigma_1(\phi), \ldots, \sigma_n(\phi)); \qquad 0 \leqq \rho_1 < \infty \end{aligned}$$

and similarly

$$S_2(\mathbf{X}) = \rho_2 S_2''(\kappa_1(\psi), \ldots, \kappa_n(\psi)); \qquad 0 \leqq \rho_2 < \infty.$$

Here we have used the fact that the ranks of the rectangular matrices $((a_{jk}))$ in the relations (4.8.16) are respectively n_1 and n_2, so that these two matrices together constitute an orthogonal matrix (this is easily deduced, for Laplace's theorem on determinants for instance).

After this, arguments entirely similar to those leading to (4.8.1) above, easily lead to the basic integro-functional equation

$$\begin{aligned} &\int_\Phi \int_\Psi \prod_{j=1}^n y_j(t\sigma_j(\phi) + s\kappa_j(\psi)) T_1(\phi) T_2(\psi)\, d\phi\, d\psi \\ &\quad = c \int_\Phi \prod_{j=1}^n y_j(t\sigma_j(\phi)) T_1(\phi)\, d\phi \cdot \int_\Psi \prod_{j=1}^n y_j(s\kappa_j(\psi)) T_2(\psi)\, d\psi \end{aligned} \tag{4.8.17}$$

where the notation has been changed to $\rho_1 \to t \in [0, \infty)$; $\rho_2 \to s \in [0, \infty)$; $c > 0$ constant; T_1 and T_2 are the Jacobian factors corresponding to the transformation from rectangular to spherical coordinates; further, condition (4.8.16) is satisfied.

If we restrict ourselves only to values of $t \geqq 0$, then (4.8.1) is a special case of (4.8.17). In fact, the role of one of the statistics is played by the definite tube statistic $|\bar{X}|$ of degree 1 and dimension 1. The second statistic S is also a tube statistic, of degree 1 and dimension $(n - 1)$. The torus Φ degenerates to a point, and the factor $y_1(t) \cdots y_n(t)$ is obtained in place of the integral. The torus Ψ is denoted in (4.8.1) by Φ.

We remark that, for definite independent tube statistics S_1 and S_2, it is always possible to assume, without loss of generality, that they are of degree 1: if these degrees are in fact respectively k_1 and k_2, then we may consider the statistics S_1^{1/k_1} and S_2^{1/k_2} which will be independent and of degree 1. Thus, equations (4.8.1) and (4.8.17) will have the same form if the statistics S, S_1, and S_2 which they involve are of any arbitrary positive degrees.

Further, if S_1 and S_2 are multitubular definite statistics, then, instead of the integrals in (4.8.1) and (4.8.17), there will appear finite sums of integrals, requiring only nonessential changes in the preceding argument.

4.9 ANOSOV'S THEOREM

We consider a random sample vector $\mathbf{X} = (X_1, \ldots, X_n)$ and the phenomenon of independence of $\bar{X}$ and a definite tube statistic $S(\mathbf{X})$ of dimension $n - 1$. The integro-functional equation (4.8.1) can be written, with $y_j(t) = y(t)$ for all j,

$$\int_\Phi \prod_{j=1}^n y(t + s\sigma_j(\phi))T_1(\phi)\, d\phi = c(y(t))^n \int_\Phi \prod_{j=1}^n y(s\sigma_j(\phi))T_1(\phi)\, d\phi. \quad (4.9.1)$$

If $y(t)$ is a normal density, then it is clear that the equation (4.9.1) turns into an identity; this is easily verified immediately, taking relation (4.8.2) into account.

It is natural to ask: under what conditions on $y(t)$ will the normal densities be the only solutions of (4.9.1)? We can establish that such will be the case if $y(t)$ is continuously twice-differentiable: for, differentiate (4.9.1) twice with respect to s and set $s = 0$, the operations involved being valid in view of the conditions imposed on $y(t)$ and the boundedness of the continuous functions $\sigma_j(\phi)$. We obtain as a result, in view of (4.8.2),

$$y''y^{n-1} - (y')^2y^{n-2} = c_1 y^n \qquad \text{for all } t$$

where $c_1 > 0$ is a constant. It easily follows that y is a normal density.

We note that we have not exploited above the probabilistic character of y, namely, $y(t) \geqq 0$.

Thus, for continuously twice-differentiable densities $y(t)$, a characterization of normality easily follows from (4.9.1). But, for a long time, a relaxation of these conditions on $y(t)$ was not found possible, till they were removed in 1963 by Anosov [1]. We now present his theorem, following [1].

THEOREM 4.9.1. Let $\mathbf{X} = (X_1, \ldots, X_n)$ be a random sample vector from a population having a probability density function $y(t)$. Let $\bar{X}$ be independent of the tube statistic $S(\mathbf{X}) = S(X_1 - \bar{X}, \ldots, X_n - \bar{X})$ of dimension $n - 1$. If $y(t)$ is continuous, then X is normal.

Remark. Thus, in Anosov's theorem, the condition that $y(t)$ be continuously twice-differentiable is replaced by the mere continuity of $y(t)$.

Examples.

(1) $S(\mathbf{X}) = m_{2k}$, the sample central moment of order $2k$: on the contrary, m_3 is not suitable as an example, though, as shown at the end of section 4.7, the independence of m_3 and $\bar{X}$ implies the normality of the sample; for m_3, the method of differential equations applies, but not that of integro-functional equations.

(2) $$S = \sum_{j=1}^{n} \left| \sum_{k=1}^{n} a_{jk} X_k \right|^{\rho}, \rho > 0$$

cf. (4.1.2). We assume that the rank of $((a_{jk}))$ is $n - 1$.

***Proof*:** We turn to the integro-functional equation (4.9.1). For simplicity of presentation, we take $n = 3$, the general case being treated exactly in the same way. The torus Φ will have dimension 1 and (4.9.1) can be rewritten as

$$\int_0^{2\pi} y(t + s\sigma_1(\phi))y(t + s\sigma_2(\phi))y(t + s\sigma_3(\phi))\, \Delta(\phi)\, d\phi = f(s)y^3(t) \quad (4.9.2)$$

where $\Delta(\phi) \geqq 0, f(s)$ is the last factor in the right-hand side of (4.9.1) with $n = 3$, and the $\sigma_j(\phi)$ are continuous functions, with

$$\sum_1^3 \sigma_j(\phi) \equiv 0, \qquad T = \int_0^{2\pi} \sum_1^3 \sigma_j(\phi)\, \Delta(\phi)\, d\phi > 0.$$

We shall not be using here the probabilistic character of $y(t)$, namely, the conditions $y(t) \geqq 0$, $\int_{-\infty}^{\infty} y(t)\, dt = 1$.

LEMMA 4.9.1. Let $y(t)$ be defined and continuous on the segment $[t_1, t_2]$, nonvanishing there, and satisfy (4.9.2) for any t in (t_1, t_2) and any $s > 0$ for which the integral in (4.9.1) is defined. Then, throughout the segment $[t_1, t_2]$, $y(t)$ has the form

$$y(t) = \pm \exp(At^2 + Bt + C). \quad (4.9.3)$$

Corollary. Let $y(t)$ be defined and continuous on the entire real line, be not identically zero, and satisfy (4.9.2) for all t. Then $y(t)$ has the form (4.9.3) for all t.

In fact, to prove the corollary, it suffices to show that $y(t) \neq 0$ for any t. By the lemma, $y(t)$ must be representable in the form (4.9.3) in any segment where it does not vanish; but, if there exists a segment at the end points of which $y(t)$ vanishes, then this would contradict the representation (4.9.3) for $y(t)$ in the segment in view of the continuity of $y(t)$.

Proof of the Lemma. We choose a segment $[t_3, t_4] \subset (t_1, t_2)$. Since $y(t)$ does not vanish, we may assume that $y(t) > 0$ on (t_1, t_2), since, together with $y(t)$, the function $-y(t)$ also satisfies (4.9.2).

We set $u(t) = \ln y(t)$, which is defined and continuous on (t_1, t_2). We set

$$L_{s,\phi} u(t) = u(t + s\sigma_1(\phi)) + u(t + s\sigma_2(\phi)) + u(t + s\sigma_3(\phi)) - 3u(t).$$

Then $L_{s,\phi}$ is a linear operator, and (4.9.2) can be rewritten as

$$\int_0^{2\pi} \exp\,(L_{s,\phi} u(t))\Delta(\phi)\, d\phi = f(s). \tag{4.9.4}$$

Further, $L_{0,\phi} u(t) = 0$, and hence, for $s = 0$, we obtain

$$\int_0^{2\pi} \Delta(\phi)\, d\phi = f(0).$$

Let $\theta(x) = e^x - 1 - x$. Then we may rewrite (4.9.4) as

$$\int_0^{2\pi} L_{s,\phi} u(t)\Delta(\phi)\, d\phi + \int_0^{2\pi} \theta(L_{s,\phi} u(t))\Delta(\phi)\, d\phi = f(s) - f(0). \tag{4.9.5}$$

We require the following properties of $\theta(x)$:

$$\theta(x) \geqq 0 \text{ for all } x \text{ and } \theta(x) \leqq x^2 \text{ for } |x| \leqq 1 \tag{4.9.6}$$

(the latter property follows from the fact that, for $|x| \leqq 1$,

$$\theta(x) = \sum_{n=2}^{\infty} x^r/r! \leqq \sum_{n=2}^{\infty} x^2/r! = (e-2)x^2 \leqq x^2).$$

We shall establish some simple auxiliary results.

LEMMA 4.9.2. If $w(t)$ is continuously twice-differentiable on (t_1, t_2), then

$$\lim_{s\to 0} [L_{s,\phi} w(t)/s^2] = (1/2)w''(t) \sum_1^3 (\sigma_j(\phi))^2$$

uniformly for t in $[t_3, t_4]$, $\phi \in [0, 2\pi]$, and

$$\lim_{s\to 0} (1/s^2) \int_0^{2\pi} L_{s,\phi} w(t)\Delta(\phi)\, d\phi = (1/2)w''(t)T$$

uniformly for t in $[t_3, t_4]$.

The lemma follows immediately from the definition of the operator $L_{s,\phi}$

LEMMA 4.9.3. Let the function w be defined and continuous on the segment $[t_1, t_2]$. Then, either it is convex downwards or there exists a linear function $l(t)$ such that the difference: $v = w - l$ has a local maximum in (t_1, t_2).

Proof: If w is not convex downwards, then there exist τ_1, τ_2, and $\lambda \in [0, 1]$ such that

$$w(\lambda\tau_1 + (1-\lambda)\tau_2) > \lambda w(\tau_1) + (1-\lambda)w(\tau_2).$$

Consider the linear function

$$l(t) = w(\tau_1) + \frac{w(\tau_2) - w(\tau_1)}{\tau_2 - \tau_1}(t - \tau_1).$$

Note that $l(\tau_1) = w(\tau_1)$, $l(\tau_2) = w(\tau_2)$ and

$$l(\lambda\tau_1 + (1-\lambda)\tau_2) = \lambda w(\tau_1) + (1-\lambda)w(\tau_2).$$

Then, for the difference $v = w - l$, we have

$$v(\tau_1) = v(\tau_2) = 0, \qquad v(\lambda\tau_1 + (1-\lambda)\tau_2) > 0,$$

so that v has a local maximum on $(\tau_1, \tau_2) \subset (t_1, t_2)$.▮

LEMMA 4.9.4. If there exists a linear function $l(t)$ such that $v(t) = u(t) - l(t)$ has a local maximum, then, for sufficiently small s, $f(s) \leqq f(0)$.

Proof: Let t_0 be a local maximum point for v. We have $L_{s,\phi}\, u(t_0) = L_{s,\phi}\, v(t_0) + L_{s,\phi}\, l(t_0) = L_{s,\phi}\, v(t_0)$ since $L_{s,\phi}\, l(t) = 0$ in view of the relation $\sum_1^3 \sigma_j(\phi) \equiv 0$. But, for sufficiently small s, $v(t_0 + s\sigma_j(\phi)) \leqq v(t_0)$, and consequently $L_{s,\phi}\, v(t_0) \leqq 0$. Hence

$$f(s) = \int_0^{2\pi} \exp\,(L_{s,\phi}\, v(t_0))\Delta(\phi)\, d\phi < \int_0^{2\pi} 1 \cdot \Delta(\phi)\, d\phi = f(0).▮$$

LEMMA 4.9.5. If, for all sufficiently small s, $f(s) \leqq f(0)$, then u is convex upwards.

Proof: In view of (4.9.5) and the fact that $\theta(x) \geqq 0$,

$$\int_0^{2\pi} L_{s,\phi}\, u(t)\, \Delta(\phi)\, d\phi \leqq f(s) - f(0) \leqq 0.$$

We now introduce an averaging operation which transforms $u(t)$ into

$$u_\varepsilon(t) = \int_{-\infty}^{\infty} K_\varepsilon(t - \tau)u(\tau)\, d\tau$$

where the kernel K_ε is chosen such that it has the following properties: it is nonnegative and twice continuously differentiable on $(-\varepsilon, \varepsilon)$, zero outside, and $\int K_\varepsilon(t)\,dt = 1$. Such an averaging leaves constants invariant, and commutes with integration with respect to ϕ and with the operator $L_{s,\phi}$. Therefore,

$$\int_0^{2\pi} L_{s,\phi} u_\varepsilon(t)\Delta(\phi)\,d\phi \leqq 0$$

under the conditions of Lemma 4.9.5. Dividing this relation through by s^2 and letting $s \to 0$, we obtain in view of Lemma 4.9.2 that $u_\varepsilon''(t) \leqq 0$, i.e., $u_\varepsilon(t)$ is a convex-upwards function. As $\varepsilon \to 0$, $u_\varepsilon(t) \to u(t)$ uniformly for all t in $[t_3, t_4]$, and the limit of convex-upwards functions is itself such a function. ∎

LEMMA 4.9.6. The function $u(t)$ is either convex downwards or convex upwards.

***Proof*:** This follows from Lemmas 4.9.3, 4.9.4, and 4.9.5. ∎

LEMMA 4.9.7. If w is convex downwards, then $L_{s,\phi} w(t) \geqq 0$, and if w is convex upwards, then $L_{s,\phi} w(t) \leqq 0$.

***Proof*:** Since $\sum_1^3 \sigma_j(\phi) \equiv 0$, the center of mass of identical masses placed at the points

$$(t + s\sigma_j(\phi),\ w(t + s\sigma_j(\phi)), \qquad j = 1, 2, 3$$

is the point:

$$(t,\ (1/3)[w(t + s\sigma_1(\phi)) + w(t + s\sigma_2(\phi)) + w(t + s\sigma_3(\phi))]).$$

In view of the convexity downwards (upwards), the center of mass must lie above (below) the graph of w, i.e., we must have, respectively,

$$w(t) \leq (\geq)(1/3)[w(t + s\sigma_1(\phi)) + w(t + s\sigma_2(\phi)) + w(t + s\sigma_3(\phi))]$$

and this is equivalent to the inequality (respectively)

$$L_{s,\phi} w(t) \leq (\geq) 0. \blacksquare$$

LEMMA 4.9.8. For all t in $[t_3, t_4]$, we have

$$u(t) = At^2 + Bt + C.$$

***Proof*:** For sufficiently small s, $L_{s,\phi} u(t) \leqq 1$ for $t \in [t_3, t_4]$. From Lemmas 4.9.6 and 4.9.7, we derive

$$L_{s,\phi} u(t) \leqq (L_{s,\phi} u(t))^2 \leqq \pm \sup_{\substack{0 \leqq \phi \leqq 2\pi \\ t_3 \leqq \tau \leqq t_4}} |L_{s,\phi} u(\tau)| L_{s,\phi} u(t)$$

where the sign $+$ is taken in the convex-downwards case and the sign $-$ in the other. We then deduce from (4.9.5)

$$\begin{aligned}\int_0^{2\pi} L_{s,\phi} u(t)\Delta(\phi)\,d\phi &\leqq f(s) - f(0) \\ &\leqq \int_0^{2\pi} L_{s,\phi} u(t)\Delta(\phi)\,d\phi \pm \sup_{\substack{0 \leqq \phi \leqq 2\pi \\ t_3 \leqq \tau \leqq t_4}} |L_{s,\phi} u(\tau)| \cdot \int_0^{2\pi} L_{s,\phi} u(t)\Delta(\phi)\,d\phi.\end{aligned}$$

We then carry out the averaging operation described earlier and divide through by s^2, to obtain

$$\frac{1}{s^2}\int_0^{2\pi} L_{s,\phi}u_\varepsilon(t)\Delta(\phi)\,d\phi \leqq \frac{f(s)-f(0)}{s^2}$$

$$\leqq \left(1 \pm \sup_{\substack{0\leqq\phi\leqq 2\pi\\ t_3\leqq\tau\leqq t_4}} |L_{s,\phi}u(\tau)|\right)\cdot\frac{1}{s^2}\int_0^{2\pi} L_{s,\phi}\,u_\varepsilon(t)\Delta(\phi)\,d\phi.$$

As $s \to 0$, the integral terms tend to $(1/2)u_\varepsilon''(t)T$ in view of Lemma 4.9.1; also, $\sup_{\substack{0\leq\phi\leq 2\pi\\ t_3\leq\tau\leq t_4}} |L_{s,\phi}u(\tau)| \to 0$. Hence it follows that the limit below exists

$$\lim_{s\to 0}\frac{f(s)-f(0)}{s^2} = A$$

and that $u_\varepsilon''(t) = 2A$, so that $u_\varepsilon(t) = At^2 + B_\varepsilon t + C_\varepsilon$. As $\varepsilon \to 0$, $u_\varepsilon(t) \to u(t)$ uniformly for t in $[t_3, t_4]$, and the limit of quadratic polynomials is again one such. This proves Lemma 4.9.8, and consequently Lemma 4.9.1 and Theorem 4.9.1.∎

4.10 TWO NONLINEAR DEFINITE TUBE STATISTICS. CONNECTION WITH RANDOM LINEAR FORMS

We now consider the problem of characterizing normality through the independence of two nonlinear tube statistics. For studying it, we need to impose on the distributions concerned conditions which are considerably more stringent than in the case where one of the statistics is linear.

THEOREM 4.10.1. Let $\mathbf{X} = (X_1, \ldots, X_n)$ be a random sample vector from a distribution with a probability density function $y(x)$ which is an even function and is analytic in some neighborhood of the real axis. If two definite tube statistics $S_1(\mathbf{X})$ and $S_2(\mathbf{X})$, the sum of whose dimensions is n, are independent, then $\mathbf{X}$ is a normal vector.

***Proof*:** We shall use the equation (4.8.17) in our proof. Since $\mathbf{X}$ is a random sample vector, we have $y_j(t) = y(t)$ for $j = 1, 2, \ldots, n$. In view of (4.8.16), the function $y(t) = \exp(At^2)$, A a constant, will satisfy our equation. We shall show that under the conditions of the present theorem, all the solutions of our equation will differ from the above only by a constant multiplier; it is clear that if $y(t)$ is to be a probability density, then $A < 0$ and this constant multiplier can be determined through the normalizing condition.

For simplicity of presentation, we take $n = 3$; the general case is discussed in exactly the same way.

We differentiate (4.8.17) twice with respect to s, and set $s = 0$. Then, taking (4.8.16) into account, we find

$$\int_{\Phi} \{A_1 y''(t\sigma_1(\phi))y(t\sigma_2(\phi))y(t\sigma_3(\phi)) + A_2 y(t\sigma_1(\phi))y''(t\sigma_2(\phi))y(t\sigma_3(\phi)) + A_3 y(t\sigma_1(\phi))y(t\sigma_2(\phi))y''(t\sigma_3(\phi))\}T_1(\phi)\, d\phi = C_1 \int_{\Phi} y(t\sigma_1(\phi))y(t\sigma_2(\phi))y(t\sigma_3(\phi))T_1(\phi)\, d\phi \quad (4.10.1)$$

where

$$A_j = \int_{\Psi} \kappa_j^2(\psi)T_2(\psi)\, d\psi,\ j = 1, 2, 3.$$

We introduce the further notation

$$A_{jk} = \int_{\Phi} (\sigma_j(\phi))^{2k} T_1(\phi)\, d\phi; \qquad j = 1, 2, 3; \qquad k = 1, 2, \ldots$$

$$M_k = \sum_{j=1}^{3} A_j A_{jk} > 0.$$

We differentiate (4.10.1) $2k$ times and set $t = 0$, to obtain

$$M_k y^{(2k+2)}(0)y^2(0) + S_{2k} = 0. \quad (4.10.2)$$

Here S_{2k} contains terms involving derivatives at the origin of $y(t)$, of orders at most $2k$. Now, setting $t = 0$ in (4.8.17), we see that $y(0) \neq 0$. Hence it follows that (4.10.2) determines, for an even function $y(t)$ analytic at the origin and satisfying (4.8.17), the sequence of even order derivatives at the origin. Let $y_\sigma(t) = (1/\sigma\sqrt{2\pi}) \exp(-t^2/2\sigma^2)$, where $\sigma > 0$ is arbitrary. $y_\sigma(t)$ is even, analytic, and satisfies (4.8.17). Further, $y_\sigma(0) = 1/\sigma\sqrt{2\pi}$. From what we have seen above, it follows that if another probability density $y(t)$ is given, satisfying (4.6.17), then $y(0) \neq 0$ and we may select σ such that $y(0) = y_\sigma(0)$. Then we obtain $y_\sigma^{(n)}(0) = y^{(n)}(0)$ for all n, since the even order derivatives coincide and the odd order ones vanish, for both functions. If $y(t)$ is analytic in the neighborhood of the real axis, then it must coincide with $y_\sigma(t)$, which proves the theorem.∎

For a general vector $\mathbf{X} = (X_1, \ldots, X_n)$, with independent components, a study of the phenomenon of independence of two nonlinear statistics has not so far been successful. We shall consider cases where one of the independent statistics is linear, with nonzero coefficients. Without loss of generality, we may denote it by $X_1 + \cdots + X_n$. In such a case, a connection has been found between this phenomenon and the case of random linear forms. For studying this connection, we turn to section 4.6.

We set up the random vector

$$\mathbf{X}^{(n+m)} = (X_1, \ldots, X_n, X_{n+1}, \ldots, X_{n+m})$$

with independent components, of which the first n coincide with the components of our $\mathbf{X}$ and the rest are arbitrary, and a random vector $\mathbf{U} = (U_1, \ldots, U_n)$ with Lebesgue-measurable components $U_j = U_j(X_{n+1}, \ldots, X_{n+m})$ for $j = 1, 2, \ldots, n$. We subject this vector to the conditions of Theorem 4.6.2:

(1) $P[\sum_1^n U_j^2 \leqq R^2] = 1$ for some $R > 0$;

(2) $P[|U_1| > \varepsilon, \ldots, |U_n| > \varepsilon] > 0$ for some $\varepsilon > 0$. Further, we require that condition (4.6.3) be satisfied.

$$Q_{m_1, \ldots, m_n}(\xi) = E[(1 + U_1\xi)^{m_1} \cdots (1 + U_n\xi)^{m_n}] \neq \text{constant}.$$

We set up the statistic

$$S = \sum_{j=1}^{n} X_j U_j(X_{n+1}, \ldots, X_{n+m}). \tag{4.10.3}$$

This nonlinear statistic is of "quasilinear" type; it is linear in $X_1, \ldots, X_n$. If it is independent of $L = X_1 + \cdots + X_n$, then the vector $\mathbf{X}$ is normal. In order for it to be true and conversely, we impose the following conditions. Let $\mathbf{X}^{(n+m)}$ be a normal vector, and $EX_j = a_j$, Var $(X_j) = \sigma_j^2$, $j = 1, 2, \ldots, n + m$. We require that the following relations be satisfied identically.

$$\sum_1^n \sigma_j U_j = 0, \qquad \sum_1^n a_j U_j = \text{constant}, \qquad \text{and} \qquad \sum_1^n \sigma_j^2 U_j^2 = \text{constant}.$$

Then, for fixed $X_{n+1}, \ldots, X_{n+m}$, the conditional distribution of the vector $\mathbf{Y}^{(n+m)} = (X_1 - a_1, \ldots, X_{n+m} - a_{n+m})$ and $S^* = S - \sum_1^n a_j U_j$ will be the product of the conditional normal density of S^* : $p_{S^*}(x_1, \ldots, x_n)$, not depending on $X_{n+1}, \ldots, X_{n+m}$, and the conditional density $p_{\mathbf{X}}(x_1, \ldots, x_n)$, also not depending on $X_{n+1}, \ldots, X_{n+m}$. Multiplying this product by the joint density of $X_{n+1}, \ldots, X_{n+m}$, and integrating, we obtain the product of the (unconditional) densities of $\mathbf{X}$ and S^*, which coincide with the conditional densities $p_{\mathbf{X}}$ and p_{S^*}. We see that the use of random linear forms leads to results which are apparently not amenable to study by the methods of differential and integro-functional equations described above. In fact, quasilinear statistics of the form (4.10.3) are rather unusual, but we shall indicate below more natural examples of connections between nonlinear statistics and random linear forms.

THEOREM 4.10.2. Let $\mathbf{X} = (X_1, \ldots, X_n)$ have independent components with densities $y_j(x)$, $j = 1, 2, \ldots, n$, such that $(y_j(x))^{1/2}$ is a positive-definite function. If the statistic $L = X_1 + \cdots + X_n$ is independent of the $(n-1)$-dimensional definite tube statistic $S(X_1 - \overline{X}, \ldots, X_n - \overline{X})$, then $\mathbf{X}$ is a normal vector.

***Proof*:** We use the connection with the theorems of section 4.8. We may interpret the basic equation (4.8.1) as the equation (4.6.5) for c.f.'s of random linear forms (cf. section 4.6).

$$\int \phi_1(t+su_1)\,\phi_2(t+su_2)\cdots\phi_n(t+su_n)\pi\,(d\mathbf{u})$$
$$= \phi_1(t)\phi_2(t)\cdots\phi_n(t)\int \phi_1(su_1)\phi_2(su_2)\cdots\phi_n(su_n)\pi\,(d\mathbf{u})$$

Here π denotes the probability measure corresponding to the vector $\mathbf{U} = (U_1, \ldots, U_n)$. We may set $U_j = \sigma_j(\phi)$ and assume that ϕ is distributed on the torus Φ with density proportional to $T_1(\phi)$. Then, $y_j(t) = (\sqrt{y_j(t)})^2$ will be positive definite functions, and may be taken to be proportional to the c.f.'s of the sums $X_j' + X_j''$ of two identically distributed independent random variables. Such an interpretation makes it possible for us to consider Theorem 4.10.2 as a special case of Theorem 4.6.2.∎

We note that in this manner we establish a duality between tube statistics and random linear forms. A comparison of tube statistics to linear forms with random coefficients can also be useful in tackling other problems (in particular, in problems of characterization through the identical distribution of pairs of tube statistics, but we shall not consider such problems here.

We also note that, in terms of the correspondence between tube statistics and random linear forms, we can formulate a condition corresponding to (4.6.3), namely,

$$Q_{m_1, \ldots, m_n}(\xi) \neq \text{constant}$$

so that Theorem 4.6.1 can be paraphrased. Indeed, for every multiple index $(m_1, \ldots, m_n)$ of nonnegative integers m_j such that $m_1 + m_2 + \cdots + m_k > 0$, we construct a polynomial

$$\tilde{Q}_{m_1, \ldots, m_n}(\xi) = \int_\Phi \prod_1^n (1 + \xi\sigma_j(\phi))^{m_j} T_1(\phi)\, d\phi. \tag{4.10.4}$$

Then the following result is obtained.

THEOREM 4.10.3. Let, under the conditions of Theorem 4.10.2, the condition of the positive definiteness of $\sqrt{y_j(x)}$ be replaced by the weaker one of the positive definiteness of $y_j(x)$, $j = 1, 2, \ldots, n$, and let the polynomials corresponding to the statistic S according to the formula (4.10.4) be such that

$$\tilde{Q}_{m_1, \ldots, m_n}(\xi) \neq \text{constant}$$

for all multiple indices $(m_1, \ldots, m_n)$ such that $m_1 + \cdots + m_n > 0$. Then $\mathbf{X}$ is a normal vector.

4.11 INDEPENDENCE OF TUBE STATISTICS, ONE OF THEM INDEFINITE

We turn to the question of the independence of a linear form $L = X_1 + \cdots + X_n$ and a tube statistic $S = S(X_1 - \bar{X}, \ldots, X_n - \bar{X})$, which we shall not assume to be definite. In such a case, the function $\sigma(\phi)$ corresponding to it will, in general, be unbounded on Φ. This circumstance implies that, for instance, in the case of the independence of $\bar{X}$ and such a statistic $S(\mathbf{X}^{(n)})$ which is unlinkable from $\bar{X}$, a relation of the form (4.8.1) may not hold (the corresponding distributions do not possess densities), but even in cases where such a relation can be set up, the methods developed in the preceding sections for its solution turn out to be unsuitable since they depend essentially on the boundedness of $\sigma(\phi)$.

In such a case, a characterization of the normal law can be successfully effected, using an asymptotic analysis of the relation of independence, under fairly stringent restrictions on the tails of the densities y_j, $1 \leqq j \leqq n$, and, in fact, without requiring the positive definiteness of these densities.

According to the conditions of the following theorem, the density $y_j(x)$ is assumed representable in the form

$$y_j(x) = \exp(-px^2 + a_j x)\,\theta_j(x) \tag{4.11.1}$$

where $p > 0$, the a_j are constants, and

$$[\theta_j(x \pm y)/\theta_j(\pm y)] \to 1 \qquad \text{as} \qquad y \to \infty. \tag{4.11.2}$$

For the functions θ_j, we can establish the following representations on the positive and negative semiaxes (cf. [24])

$$\theta_j(\pm x) = g_j^{\pm}(x) \exp\left(\int_0^x h_j^{\pm}(u)\,du\right), \qquad x \geqq 0 \tag{4.11.3}$$

where, as $x \to \infty$, $g_j^{\pm}(x) \to b_j^{\pm} \neq 0$, $h_j^{\pm}(x) \to 0$. For the θ_j in (4.11.1), it is easy to establish under the condition (4.11.2) that, for any $\varepsilon > 0$, there exists an $x' > 0$ such that for $|x| > x'$,

$$c \exp(-\varepsilon|x|) \leqq \theta_j(x) \leqq c \exp(\varepsilon|x|). \tag{4.11.4}$$

THEOREM 4.11.1. Let $\mathbf{X} = (X_1, \ldots, X_n)$ be a random vector with independent components, which have probability densities y_j representable in the form (4.11.1), with the θ_j satisfying (4.11.2). Let $S = S(X_1 - \bar{X}, \ldots, X_n - \bar{X})$ be a tube statistic, not necessarily definite, but satisfying the additional condition that, on the hypersphere $(x_1 - \bar{x})^2 + \cdots + (x_n - \bar{x})^2 = 1$, it attains its maximum value at a unique point such that at that point $x_j - \bar{x} \neq 0$, $j = 1, \ldots, n$, and that, in some neighborhood of this maximum point, it satisfies the Lipschitz condition for some $\alpha > 0$. Let us further assume that the relation (4.8.1) can be set up in the present case and that all the integrals concerned are absolutely convergent. Under such conditions, if S is independent of $L = X_1 + \cdots + X_n$, then $\mathbf{X}$ is a normal vector.

***Proof*:** We take equation (4.8.1), where the integrals concerned are assumed to be absolutely convergent.

We substitute the $y_j(x)$ in the form (4.11.1) into equation (4.8.1). First consider the right-hand side.

$$\int_\Phi T_1(\phi) \prod_1^n y_j(s\sigma(\phi)z_j)\, d\phi$$
$$= \int_\Phi T_1(\phi) \prod_1^n \theta_j(s\sigma(\phi)z_j) \cdot \exp\left(-ps^2\sigma^2(\phi) + s\sigma(\phi)\sum_1^n a_j z_j\right) d\phi. \quad (4.11.5)$$

We divide up this integral into two parts I_1 and I_2. If ϕ_0 be the point at which $\sigma(\phi)$ attains its minimum, the first integral is taken over R_δ, a δ-neighborhood of ϕ_0, and the second integral over the complementary set $\Phi - R_\delta$. If we write

$$\min_\Phi \sigma^2(\phi) = \Lambda > 0 \quad (4.11.6)$$

then, obviously, for $\delta > 0$ sufficiently small, there exists a $\Delta > 0$ such that

$$\sigma^2(\phi) \geqq \Lambda + \Delta \qquad \text{for } \phi \in \Phi - R_\delta. \quad (4.11.7)$$

We shall show that, as $s \to \infty$,

$$I_2/I_1 \to 0. \quad (4.11.8)$$

From (4.11.1), (4.11.4), and (4.11.7), it is easily seen that for $s > c_1$, sufficiently large,

$$I_2 = \int_{\Phi - R_\delta} T_1(\phi) \prod_1^n \theta_j(s\sigma(\phi)z_j) \cdot \exp\left(-ps^2\sigma^2(\phi) + s\sigma(\phi)\sum_1^n a_j z_j\right) d\phi$$
$$\leqq c_2 \exp\left(-ps^2(\Lambda + \Delta) + c_3 s\right). \quad (4.11.9)$$

On the other hand, for $s > c_4$,

$$I_1 = \int_{R_\delta} T_1(\phi) \prod_1^n \theta_j(s\sigma(\phi)z_j) \exp\left(-ps^2\sigma^2(\phi) + s\sigma(\phi)\sum_1^n a_j z_j\right) d\phi$$
$$\geqq c_5 (\exp c_6 s) \int_0^\delta \rho^{n-3} \exp\left[-ps^2(\Lambda + c_7 \rho^\alpha)\right] d\rho$$
$$\geqq c_8 e^{-p\Lambda s^2} \cdot s^{-2(n-2)/\alpha}. \quad (4.11.10)$$

Here we assumed δ so small that, in accordance with the conditions of the theorem, the Lipschitz condition of order α is satisfied in R_δ. Combining (4.11.9) and (4.11.10) leads immediately to (4.11.8).

We then consider the left member of (4.8.1).

$$
\begin{aligned}
\int_{\Phi} T_1(\phi) \prod_1^n y_j(t + s\sigma(\phi)z_j)\, d\phi
&= \exp\left(-pnt^2 + t\sum_1^n a_j\right) \\
&\quad \times \left\{ \int_{\Phi - R_\delta} T_1(\phi) \prod_1^n \theta_j(t + s\sigma(\phi)z_j) \exp\left(-ps^2\sigma^2(\phi) + s\sigma(\phi)\sum_1^n a_j z_j\right) d\phi \right. \\
&\quad \left. + \int_{R_\delta} T_1(\phi) \prod_1^n \theta_j(t + s\sigma(\phi)z_j) \exp\left(-ps^2\sigma^2(\phi) + s\sigma(\phi)\sum_1^n a_j z_j\right) d\phi \right\}.
\end{aligned} \tag{4.11.11}
$$

Let us denote, for brevity, the first and second integrals appearing above within the curly brackets by I_3 and I_4 respectively. Proceeding here also as in the derivation of (4.11.8), we can easily verify that

$$
I_3/I_4 \to 0 \qquad \text{as} \qquad s \to \infty. \tag{4.11.12}
$$

Further, we can see that

$$
I_4/I_1 \to 1 \qquad \text{as} \qquad s \to \infty. \tag{4.11.13}
$$

We now divide both sides of (4.8.1) by I_1 and let $s \to \infty$. In view of (4.11.8), (4.11.11), (4.11.12), and (4.11.13), we then find

$$
p_{\bar{X}}(t) = c_9 \exp\left(-pnt^2 + t\sum_1^n a_j\right)
$$

whence it follows that $\bar{X}$ has a normal distribution. Then it follows from Cramer's theorem that all the $X_j, j = 1, \ldots, n$, also have normal distributions. ∎

We may remark that the case of the independence of two tube statistics, for instance equation (4.8.17), can also be dealt with similarly. There, however, in the final stage, we would obtain that one of the statistics has the same distribution as it would if $\mathbf{X}^{(n)}$ were a normal vector. The question whether the normality of $\mathbf{X}^{(n)}$ would follow, and for what statistics, still remains open. The linear case is an exception, where the question is answered in the affirmative with the help of the theorem of Cramer's which we have just invoked.

CHAPTER 5

Regression Problems Connected with Linear Statistics

5.1 INTRODUCTION

In Chapter 3, we studied characterizations of distributions through the property of independence of two linear forms in independent random variables, and in Chapter 2, characterizations of distributions through the property of identical distribution of two linear statistics based on a random sample. The problems studied in the present chapter concern characterizations of distributions through the property of constancy of regression of one statistic on another. We also consider the general case of the problem, where the regression of each of p given linear functions on q other linear functions is constant.

The starting point of investigations in this area was the following theorem [56] obtained by the authors of the present work in 1965.

Kagan-Linnik-Rao Theorem: If $X_1, \ldots, X_n$, $n \geqq 3$, are independent and identically distributed random variables with finite expectation $EX_i = 0$, and

$$E(\overline{X} \mid X_1 - \overline{X}, \ldots, X_n - \overline{X}) = 0$$

where $\overline{X} = (X_1 + \cdots + X_n)/n$, then the X_i are normally distributed.

The note [56] was followed by a series of papers: Rao [130], Ramachandran and Rao [122, 123], Khatri and Rao [73, 74], Pathak and Pillai [110], Pillai [112] and Shimizu [165]. We sum up the results of these studies here. We shall use some lemmas from Chapter 1 (sections 1.1 and 1.2) and some results of Chapter 2.

5.2 SOME GENERALIZATIONS OF THE KAGAN-LINNIK-RAO THEOREM

Let $X_1, \ldots, X_n$ be independent random variables, not necessarily identically distributed. Consider the linear function $a_1 X_1 + \cdots + a_n X_n$ with nonzero coefficients, which under a suitable change of scale can be written as

$$L = X_1 + \cdots \; X_n. \tag{5.2.1}$$

Further, let

$$b_{i1} X_1 + \cdots + b_{in} X_n, \qquad i = 1, \ldots, p \geqq 2 \tag{5.2.2}$$

be p linearly independent linear functions, which, after a suitable transformation can be represented in the canonical form (renaming the variables if necessary).

$$\begin{aligned} M_1 &= X_1 + c_{11} X_{p+1} + \cdots + c_{1n} X_n \\ &\cdots \qquad \cdots \qquad \cdots \\ M_p &= X_p + c_{p1} X_{p+1} + \cdots + c_{pn} X_n \end{aligned} \tag{5.2.3}$$

We denote the matrix of coefficients c_{ij} by $\mathbf{C}$ (its order is $p \times (n-p)$), and its ith column by $\mathbf{C}_i$.

THEOREM 5.2.1. Let $p \geqq 2$ and let no column of $\mathbf{C}$ be proportional to any other or to any column of the identity matrix I_p of order $p \times p$. If $EX_i = 0$, then, the condition

$$E(L \mid M_1, \ldots, M_p) = 0 \tag{5.2.4}$$

implies that the X_j are normally distributed.

***Proof*:** By Lemma 1.1.1, the condition (5.2.4) is equivalent to

$$E\{Le^{i(t_1 M_1 + \cdots + t_p M_p)}\} = 0 \tag{5.2.5}$$

which leads to the functional equation

$$\phi_1(t_1) + \cdots + \phi_p(t_p) + \phi_{p+1}(\mathbf{C}_1' \mathbf{t}) + \cdots + \phi_n(\mathbf{C}_{n-p}' \mathbf{t}) = 0 \tag{5.2.6}$$

valid for $|t_i| < \delta$, $i = 1, \ldots, p$, for sufficiently small $\delta > 0$; here $\mathbf{t}' = (t_1, \ldots, t_p)$, $\phi_i = f_i'/f_i$, and f_i is the c.f. of X_i. It follows from Lemma 1.5.3 that ϕ_i are polynomials and hence the X_i are normal.∎

Corollary 1. Let $L_1, \ldots, L_n$ be linearly independent linear functions of the r.v.'s $X_1, \ldots, X_n$, where all the coefficients in the form L_1 are different from zero. Suppose further that $EX_i = 0$ for all i. Then the condition

$$E(L_1 \mid L_2, \ldots, L_n) = 0 \tag{5.2.7}$$

guarantees that the X_i are normal r.v.'s if $n \geqq 3$.

The original theorem of Kagan-Linnik-Rao, cited in section 5.1, is a special case of this corollary.

Corollary 2. Suppose that the column $\mathbf{C}_i$ is proportional to $\mathbf{C}_j$, $\mathbf{C}_k, \ldots$ and the s-th column e_s of the identity matrix. Let

$$\mathbf{C}_j = \lambda_j \mathbf{C}_i, \qquad \mathbf{C}_k = \lambda_k \mathbf{C}_i, \ldots, \qquad \mathbf{e}_s = \lambda_s \mathbf{C}_i.$$

If all the numbers $\lambda_j, \lambda_k, \ldots, \lambda_s$ are of the same sign and at least one column of $\mathbf{C}$ contains two nonzero elements, then (5.2.4) implies the normality of the r.v.'s $X_i, X_j, X_k, \ldots$.

Corollary 2 is proved with the help of Lemma 1.5.3 and Lemma 1.2.1.∎

We now consider two sets of q and $p \geqq 2$ linear functions, each set being assumed to consist of linearly independent ones. We can then write them in a canonical form.

$$\begin{aligned} L_j &= \sum_{u=1}^{p} a_{ju} X_u + \sum_{i=p+1}^{n} b_{j,i-p} X_i, \qquad j = 1, \ldots, q, \\ M_i &= X_i + \sum_{u=p+1}^{n} c_{i,u-p} X_u, \qquad i = 1, \ldots, p \geqq 2. \end{aligned} \tag{5.2.8}$$

THEOREM 5.2.2. Let $X_1, \ldots, X_n$ be independent r.v.'s with $EX_i = 0$ for all i. Consider the linear functions $L_1, \ldots, L_q$ and $M_1, \ldots, M_p$ defined by (5.2.8). Consider the matrices $\mathbf{A} = ((a_{ij}))$, $\mathbf{B} = ((b_{ij}))$, and $\mathbf{C} = ((c_{ij}))$ of orders $q \times p$, $q \times (n-p)$, and $p \times (n-p)$ respectively. If (i) each column of the matrices $\mathbf{A}$ and $\mathbf{B}$ has at least one nonzero element, and (ii) no column of $\mathbf{C}$ is proportional to any other or to any column of the identity matrix $\mathbf{I}_p$, then

$$E(L_i | M_1, \ldots, M_p) = 0, \qquad i = 1, \ldots, q \tag{5.2.9}$$

implies the normality of the X_i.

***Proof*:** We have from (5.2.9)

$$E\{L_i e^{i(t_1 M_1 + \cdots + t_p M_p)}\} = 0, \qquad i = 1, \ldots, q \tag{5.2.10}$$

which yields the functional equation (using the same notation as before)

$$\sum_{u=1}^{p} a_{ju} \phi_u(\mathbf{e}'_u \mathbf{t}) + \sum_{i=p+1}^{n} b_{j,i-p} \phi_i(\mathbf{t}'\mathbf{C}_{i-p}) = 0, \qquad j = 1, \ldots, q \tag{5.2.11}$$

valid for $|t_i| < \delta$, $i = 1, \ldots, p$. Since the conditions of Lemma 1.5.3 are satisfied, the ϕ_i must be polynomials and hence the X_i must be normal.∎

Before formulating a corollary to Theorem 5.2.2, we introduce a definition. Let $L_1, \dots, L_n$ be linearly independent linear functions of the r.v.'s $X_1, \dots, X_n$, which we write in the canonical form (after renaming the variables if necessary)

$$L_i = X_i + \sum_{u=q+1}^{n} w_{i,\,u-q} X_u, \qquad i = 1, \dots, q, \tag{5.2.12}$$

$$L_j = X_j + \sum_{i=1}^{q} u_{j-q,\,i} X_i, \qquad j = q+1, \dots, n. \tag{5.2.13}$$

We shall say that the r.v.'s X_{q+i} and X_{q+j} are *linked* in (5.2.12) if both of them appear with nonzero coefficients in at least one of the equations (5.2.12).

Corollary. If, in (5.2.12), the r.v. X_{q+i} is linked to at least one of the r.v.'s X_{q+j}, and further if, in representation (5.2.13), none of the columns of coefficients u_{ji} is the null vector, then

$$E(L_i \mid L_{q+1}, \dots, L_n) = 0, \qquad i = 1, \dots, q \tag{5.2.14}$$

implies the normality of the r.v.'s X_j if $n - q \geqq 2$.

5.3 CONSTANCY OF REGRESSION OF ONE LINEAR STATISTIC ON ANOTHER (THE CASE OF TWO R.V.'S)

Let X_1 and X_2 be two independent and identically distributed (i.i.d.) r.v.'s with $EX_1 = 0$. There are given two linear functions $a_1 X_1 + a_2 X_2$ and $b_1 X_1 + b_2 X_2$, with all the coefficients nonzero, such that

$$E(a_1 X_1 + a_2 X_2 \mid b_1 X_1 + b_2 X_2) = 0. \tag{5.3.1}$$

Without loss of generality, the above condition may be rewritten.

$$E(X_1 - \alpha X_2 \mid X_1 + \beta X_2) = 0 \qquad \text{for } |\beta| \leqq 1. \tag{5.3.2}$$

We shall consider characterizations of the distribution of the r.v. X_1 through the relation (5.3.2) under various conditions on α and β.

THEOREM 5.3.1. Let X_1 and X_2 be i.i.d.r.v.'s with $EX_1 = 0$, satisfying the relation (5.3.2). Then

(i) if $\alpha\beta < 0$, then $X_1 = 0$ with probability one;

(ii) if $\alpha\beta > 0$ and $|\beta| = 1$, then $X_1 = 0$ with probability one if $|\alpha| \neq 1$, has an arbitrary distribution if $\alpha = 1$, and an arbitrary symmetric distribution if $\alpha = -1$;

(iii) if $\alpha\beta > 0$ and $|\beta| < 1$, and λ is the unique real number such that $|\alpha|\,|\beta|^{\lambda-1} = 1$, then

(a) $X_1 = 0$ with probability one, if $\lambda \leqq 1$ or $\lambda > 2$;

(b) X_1 is a normal (possibly degenerate) r.v. if $\lambda = 2$; and

(c) if $1 < \lambda < 2$, then X_1 is infinitely divisible (cf. (1.1.7))* and $L(\mu, \sigma^2, M, N)$ be the Levy representation for the logarithm of the c.f. of X_1, then $\sigma = 0$, and, if $\beta > 0$,

$$M(u) = \xi(\log |u|)/|u|^\lambda, \qquad N(u) = -\eta(\log u)/u^\lambda \tag{5.3.3}$$

where ξ and η are nonnegative right-continuous functions with period $(-\log \beta)$; while, if $\beta < 0$,

$$M(u) = \xi(\log |u|)/|u|^\lambda, \qquad N(u) = -\xi(\log u - \log |\beta|)/u^\lambda \tag{5.3.4}$$

where ξ is a nonnegative, right-continuous function with period $-2\log|\beta|$.

***Proof*:** From (5.3.2), we have for all real t,

$$E(X_1 - \alpha X_2)e^{it(X_1+\beta X_2)} = 0. \tag{5.3.5}$$

Let f be the c.f. of X_1 and I an interval containing the origin in which f does not vanish. Then we obtain from (5.3.5)

$$\phi(t) = \alpha\phi(\beta t), \qquad t \in I \tag{5.3.6}$$

where $\phi(t) = f'(t)/f(t)$. On integrating both sides, we have

$$f(t) = [f(\beta t)]^\gamma \qquad \text{for } t \in I \text{ and } \gamma = \alpha/\beta. \tag{5.3.7}$$

(*i*) Let $\alpha\beta < 0$, i.e., $\gamma < 0$, then, since $|f(\beta t)| \leqq 1$, we have from (5.3.7) that $|f(t)| \geqq 1$ and hence $|f(t)| = 1$, so that $|f(t)| \equiv 1$ for all $t \in I$, and hence $X_1 = EX_1 = 0$ with probability one. This proves assertion (i) of the theorem.

(ii) Let $\alpha\beta > 0$ and $|\beta| = 1$. If $\alpha = 1$, then $\beta = 1$, and in this case (5.3.7) is satisfied by any arbitrary c.f. If $\alpha = -1$, then $\beta = -1$, and in this case, condition (5.3.7) is satisfied by any real-valued c.f.

(iii) Let us turn to the only nontrivial case: $\alpha\beta > 0$, and $|\beta| < 1$. Here $\gamma > 0$, and let λ be the unique real solution of the equation $\gamma|\beta|^\lambda = 1$.

(a) If $\lambda \leqq 1$, then $|\alpha| \leqq 1$, and we obtain from (5.3.6) that for $t \in I$, $\phi(t) = \alpha\phi(\beta t) = \cdots = \alpha^n\phi(\beta^n t) \to 0$ as $n \to \infty$, since $\phi(\beta^n t) \to \phi(0) = 0$. Thus, $f(t) \equiv 1$. If $\lambda > 2$, then we have from (5.3.7) for any fixed $t \in I$

$$\frac{\log |f(t)|}{|t|^\lambda} = \frac{\log |f(\beta t)|}{|\beta t|^\lambda} = \cdots = \frac{\log |f(\beta^n t)|}{|\beta^n t|^\lambda} \to 0 \qquad \text{as} \qquad n \to \infty.$$

From part (d) of Lemma 1.1.5, it follows that $X_1 = 0$ with probability one. This proves assertion (iiia).

(b) If $\lambda = 2$, then the sequence $\log |f(\beta^n t)|/(\beta^n t)^2$ is constant-valued, and we conclude from the corollary to Lemma 1.2.1 (on α-decompositions) that f is the c.f. of a normal (possibly degenerate) law. This proves assertion (iiib).

* In fact X_1 follows a semi-stable (c.f.(1.1.15)).

(c) We first show that relation (5.3.7) implies that f is i.d., if only $\lambda > 0$, and that formulas (5.3.3) and (5.3.4) are valid if $0 < \lambda < 2$. We have already proved above with the help of (5.3.6) that, for $\lambda \leqq 1$, the r.v. X_1 is degenerate; this can be regarded as a special case of formulas (5.3.3) and (5.3.4).

We begin by showing that f is nonvanishing. Suppose this is not true, and let $t_0 > 0$ be the zero of f closest to the origin on the positive real axis. Then the relation $f(t) = [f(\beta t)]^\gamma$ is valid in $(0, t_0)$. Taking the absolute values of both sides and letting $t \to t_0 - 0$, we see that, in view of the continuity of f, $f(\beta t_0) = 0$, contrary to assumption. Similarly, or in view of the Hermitian property, $f(t) \neq 0$ on the negative real axis as well. Hence (5.3.7) holds for all real t, since f never vanishes. Then, for all t and all positive integers n,

$$f(t) = [f(\beta t)]^\gamma = \cdots = [f(\beta^n t)]^{\gamma^n}, \qquad \gamma = |\beta|^{-\lambda} > 1$$

if $\lambda > 0$. Hence f^{1/γ^n} is a c.f. for every n and hence f is i.d., provided $\lambda > 0$. Thus, we have the Levy representation $L(\mu, \sigma^2, M, N)$ for $\log f$(cf. (1.1.7)).

$$\log f(t) = i\mu t - \frac{1}{2}\sigma^2 t^2 + \int_{(0,\infty)} \left(e^{itu} - 1 - \frac{itu}{1+u^2}\right) dN(u)$$

$$+ \int_{(-\infty,0)} \left(e^{itu} - 1 - \frac{itu}{1+u^2}\right) dM(u). \tag{5.3.8}$$

Using (5.3.7), we obtain

$$\log f(t) = \gamma \log f(\beta t) = i\mu\gamma\beta t - \frac{1}{2}\sigma^2\gamma\beta^2 t^2$$

$$+ \int_{(0,\infty)} \left(e^{i\beta tu} - 1 - \frac{i\beta tu}{1+u^2}\right) d[\gamma N(u)]$$

$$+ \int_{(-\infty,0)} \left(e^{i\beta tu} - 1 - \frac{i\beta tu}{1+u^2}\right) d[\gamma M(u)]. \tag{5.3.9}$$

Replacing βu by u in (5.3.9) and making a suitable adjustment in the linear term, we have

$$\log f(t) = i\mu' t - \frac{1}{2}\sigma^2\gamma\beta^2 t^2 + \int_{(0,\infty)} \left(e^{itu} - 1 - \frac{itu}{1+u^2}\right) d\left[\gamma N\left(\frac{u}{\beta}\right)\right]$$

$$+ \int_{(-\infty,0)} \left(e^{itu} - 1 - \frac{itu}{1+u^2}\right) d\left[\gamma M\left(\frac{u}{\beta}\right)\right]. \tag{5.3.10}$$

Invoking the uniqueness of the Levy representation, we see from (4.3.8) and (5.3.10) that

$$\mu' = \mu \tag{5.3.11}$$

$$\sigma^2(\gamma\beta^2 - 1) = 0 \tag{5.3.12}$$

and

$$\begin{aligned} N(u) &= \gamma N\left(\frac{u}{\beta}\right), & M(u) &= \gamma M\left(\frac{u}{\beta}\right) & \text{if } \beta > 0 \\ N(u) &= -\gamma M\left(\frac{-u}{\beta}\right), & M(u) &= -\gamma N\left(\frac{-u}{\beta}\right). & \text{if } \beta < 0. \end{aligned} \tag{5.3.13}$$

Since $\gamma\beta^2 \neq 1$ by assumption, we have $\sigma = 0$ from (5.3.12), and the other assertions of (iiic) of the theorem follow from (5.3.13).▌

THEOREM 5.3.2. Let X_1, X_2 be i.i.d.r.v.'s with $EX_1 = 0$. If

$$E(X_1 - \alpha X_2 | X_1 + \beta X_2) = 0, \qquad E(X_1 + \beta X_2 | X_1 - \alpha X_2) = 0 \tag{5.3.14}$$

where $\alpha \neq 0$, $\beta \neq 0$, then the X_i are normal (possibly degenerate) if $\beta\alpha = 1$, and degenerate otherwise.

Proof: Without loss of generality, we may assume that $|\beta| \leqq 1$. Then, in any interval containing the origin in which $f(t) \neq 0$, we have from the first relation in (5.3.14) that

$$\phi(t) = \alpha\phi(\beta t), \qquad \phi = \frac{f'}{f}. \tag{5.3.15a}$$

Again, we have from the second relation in (5.3.14) that in any interval containing the origin in which $f(\alpha t) \neq 0$,

$$\phi(t) = \beta\phi(\alpha t). \tag{5.3.15b}$$

Combining the two relations above, we obtain for t in such an interval.

$$\phi(t) = \alpha\beta\phi(\alpha\beta t) \tag{5.3.16}$$

If $|\alpha\beta| \neq 1$, it follows that $\phi(t) \equiv 0$ in such an interval and hence $f(t) \equiv 1$. If $|\alpha\beta| = 1$, then X_1 is normal (possibly degenerate) if $\alpha\beta = 1$, and degenerate if $\alpha\beta = -1$, by Theorem 5.3.1.▌

We now consider a problem relating to n r.v.'s, which is easily reduced to the case of Theorem 5.3.1. (See Cacoullos, 1967 for a special case)

THEOREM 5.3.3. Let $X_1, \ldots, X_n$ be nondegenerate i.i.d.r.v.'s and $L_i = \sum_j a_{ij} X_j$, $i = 1, \ldots, (n-1)$, be linearly independent linear forms in the X_j, while $L_n = \sum_j a_{nj} X_j$ is a linear form such that the coefficient vector $(a_{n1}, \ldots, a_{nn})$ is not a multiple of any vector with components 0, 1, -1. Then, from the conditions

$$E(L_i | L_n) = 0, \qquad i = 1, \ldots, n-1 \tag{5.3.17}$$

it follows that the X_i are normal, if and only if $\sum_j a_{ij} a_{nj} = 0$, $i = 1, \ldots, (n-1)$. Otherwise, the X_i follow an i.d. law of the type described in part (iiic) of Theorem 5.3.1, with $1 < \lambda < 2$.

***Proof*:** Without loss of generality, we may assume that

$$L_i = X_i + a_{in} X_n, \qquad i = 1, \ldots, n-1. \tag{5.3.18}$$

Then, from the conditions (5.3.17) we have

$$f(a_{ni} t) = [f(a_{nn} t)]^{\gamma_i}, \qquad \gamma_i = -a_{in}/a_{nn}, \qquad i = 1, \ldots, n-1 \tag{5.3.19}$$

where f is the c.f. of the X_i. This equation is the same as we considered in Theorem 5.3.1, and hence the assertion of that theorem is valid here.∎

5.4 CONSTANCY OF REGRESSION OF ONE LINEAR STATISTIC ON ANOTHER (A FINITE NUMBER OF R.V.'S)—I

5.4.1 The Solution of a Functional Equation

In section 5.3, we considered a functional equation of the form

$$f(t) = [f(\beta t)]^{\gamma}, \qquad 0 < |\beta| < 1, \qquad \gamma > 0 \tag{5.4.1}$$

and obtained its complete solution. The generalization of the problem, studied there for two r.v.'s, to the case of an arbitrary, finite number of r.v.'s, is (in some cases) connected with the solution of a functional equation of the form

$$f(t) = \prod_{1}^{p} [f(\beta_j t)]^{\gamma_j} \prod_{p+1}^{n} [f(-\beta_j t)]^{\gamma_j},$$

$$0 < \beta_j < 1, \qquad \gamma_j > 0, \qquad j = 1, \ldots, n. \tag{5.4.2}$$

The equation (5.4.2) is a generalization of a simpler equation

$$f(t) = \prod_{1}^{p} f(\beta_j t) \prod_{p+1}^{n} f(-\beta_j t) \tag{5.4.3}$$

which is obtained in problems of identical distribution of a monomial and a linear form in n independent and identically distributed r.v.'s. The complete solution of (5.4.3) was obtained by Shimizu [165]. Ramachandran and Rao [123], using some basic results of [81], simplified Shimizu's arguments and extended them to the general case of (5.4.2). We shall follow their presentation here.

We introduce some notation. Let A_n be the set of all vectors $\mathbf{B} = (B_1, \ldots, B_n)$ with positive components. For $0 \leqq p \leqq n$, consider the following subsets of A_n:

$A_n(0)$: at least two components of $\mathbf{B}$ are mutually incommensurable;

$A_n(\rho)$: all the components of $\mathbf{B}$ are pairwise commensurable, and $\rho > 0$ is such that $m_j = B_j/\rho$, $j = 1, \ldots, n$, are positive integers with their greatest common factor equal to one;

$B_n^p(\rho)$: the subset of $A_n(\rho)$ such that at least one of $m_1, \ldots, m_p$ is odd, and/or at least one of $m_{p+1}, \ldots, m_n$ is even;

$C_n^p(\rho)$: the subset of $A_n(\rho)$ such that $m_1, \ldots, m_p$ are all even and $m_{p+1}, \ldots, m_n$ are all odd.

We note that $B_n^n(\rho) = A_n(\rho)$, and $C_n^n(\rho)$ is empty, and that any element of A_n belongs either to $A_n(0)$ or, for some $\rho > 0$, to $B_n^p(\rho)$ or to $C_n^p(\rho)$.

THEOREM 5.4.1. Let g and h be nonnegative, nonincreasing, right-continuous functions defined on R_1, with $g(+\infty) = h(+\infty) = 0$, satisfying the relations

$$
\begin{aligned}
g(u) &= \sum_1^p \gamma_j g(u + B_j) + \sum_{p+1}^n \gamma_j h(u + B_j) \\
h(u) &= \sum_1^p \gamma_j h(u + B_j) + \sum_{p+1}^n \gamma_j g(u + B_j)
\end{aligned}
\tag{5.4.4}
$$

for some γ, $\mathbf{B} \in A_n$. Then,

(i) $g \equiv h \equiv 0$ if $\gamma_1 + \cdots + \gamma_n \leqq 1$;

(ii) if $\gamma_1 + \cdots + \gamma_n > 1$, then let $\lambda > 0$ be the unique solution of the equation $\sum_1^n \gamma_j e^{-B_j \lambda} = 1$; then,

(a) if $\mathbf{B} \in A_n(0)$,

$$g(u) = \xi e^{-\lambda u}, \qquad h(u) = \eta e^{-\lambda u} \tag{5.4.5}$$

where ξ and η are nonnegative real constants, with $\xi + \eta > 0$, and further $\xi = \eta$, if $p < n$.

(b) if $\mathbf{B} \in B_n^p(\rho)$, then

$$g(u) = \xi(u) e^{-\lambda u}, \qquad h(u) = \eta(u) e^{-\lambda u} \tag{5.4.6}$$

where ξ and η are nonnegative, right-continuous periodic functions with period ρ; further, $\xi = \eta$ if $p < n$.

(c) if $\mathbf{B} \in C_n^p(\rho)$, so that $p < n$, then

$$g(u) = [\xi(u) + \eta(u)] e^{-\lambda u}, \qquad h(u) = [\xi(u) - \eta(u)] e^{-\lambda u} \tag{5.4.7}$$

where the functions ξ and η are right-continuous, ξ is periodic with period ρ, and $\eta(u + \rho) = -\eta(u)$ for all u.

***Proof*:** Let $B_* = \min B_j$ and $B^* = \max B_j$; and let $k = g + h$, so that $k(u) = \sum_1^n \gamma_j k(u + B_j)$.

(i) If $\sum_1^n \gamma_j \leqq 1$, then $k(u) \leqq k(u + B_*)$ since k is nonincreasing, and the reverse inequality holds for the same reason. Hence $k(u) = k(u + B_*)$ for all u, so that $k(u) = k(+\infty) = 0$. Then, $0 \leqq g, h \leqq k$ implies that $g \equiv h \equiv 0$.

(ii) The basic idea of the proof in this case is simple. We first show that the Laplace transform of g (respectively, h) is analytic in $\operatorname{Re} z > -\lambda$ and agrees there with a function analytic everywhere except possibly for simple poles at a lattice of points lying on the vertical line $\operatorname{Re} z = -\lambda$. Then we employ the method used in [81] and [165], based on the inversion formula for the Laplace transform and the theorem of residues, to obtain the form of the function g.

Let $\gamma_1 + \cdots + \gamma_n = \gamma > 1$, so that

$$k(0) \geqq \gamma k(B^*) \geqq \gamma^2 k(2B^*) \geqq \cdots.$$

Since k is nonincreasing, it follows from the above inequalities that, for all $u \geqq 0$, $k(u) \leqq c_1 \exp(-c_2 u)$, where $c_1 > 0$ and $c_2 = \log \gamma / B^* > 0$. Hence $\int_u^\infty k(t)\,dt$ exists for all u, and the same is true obviously with g or h in place of k. We now set

$$g^*(u) = \int_u^\infty g(t)\,dt, \qquad h^*(u) = \int_u^\infty h(t)\,dt. \tag{5.4.8}$$

g^* and h^* are nonnegative, nonincreasing, with $g^*(+\infty) = h^*(+\infty) = 0$, and they satisfy the relations (5.4.4). In addition to these properties, which g^* and h^* possess in common with g and h, they are also continuous. Obviously, if we show that g^* and h^* have one of the representations (5.4.5)–(5.4.7), then g and h will have the same forms respectively as g^* and h^*.

Thus, it is sufficient to prove Theorem 5.4.1 in the case where g and h are continuous as well, and consequently k is also so. We first establish some lemmas.

LEMMA 5.4.1. $\int_0^\infty e^{xu} k(u)\,du < \infty$ for $x < \lambda$, so that $\chi_g(z) = \int_0^\infty e^{-zu} g(u)\,du$ and $\chi_h(z) = \int_0^\infty e^{-zu} h(u)\,du$ are defined and analytic for $\operatorname{Re} z > -\lambda$.

***Proof*:** Let $k(u) = r(u)e^{-\lambda u}$ (where λ is the unique solution of the equation $\sum_1^n \gamma_j e^{-B_j \lambda} = 1$ (so that $\lambda > 0$). r is a continuous function, and

$$r(u) = \textstyle\sum_1^n p_j r(u + B_j),$$

where $p_j = \gamma_j e^{-B_j \lambda}$, $\sum_1^n p_j = 1$, so that by the intermediate value theorem, $r(u) = r[u + B(u)]$, where $B_* \leqq B(u) \leqq B^*$. Hence there exists a sequence $\{b_m\} \to \infty$ as $m \to \infty$, with $b_0 = 0$, such that $r(b_m) = r(0) = c_3$, or $k(b_m) = c_3 e^{-\lambda b_m}$, where $B_* \leqq b_{m+1} - b_m \leqq B^*$, and consequently $b_m \geqq mB_*$ for all m. Hence, for $0 \leqq x < \lambda$ and for all $m \geqq 0$,

$$\begin{aligned}\int_{b_m}^{b_{m+1}} e^{xu} k(u)\,du &\leqq e^{xb_{m+1}} k(b_m)(b_{m+1} - b_m) \\ &\leqq c_3 B^* \exp(-\lambda b_m + x b_{m+1}) \leqq c_3 B^* \exp(xB^* - (\lambda - x)b_m) \\ &\leqq c_4(x) \exp[-(\lambda - x)mB_*]\end{aligned}$$

where $c_4(x)$ is a constant depending on x but not on m. Hence

$$\int_0^\infty e^{xu} k(u)\,du = \sum_0^\infty \int_{b_m}^{b_{m+1}} < \infty, \qquad \text{if } x < \lambda$$

whence the lemma.▌

LEMMA 5.4.2. There exist entire functions $\sigma(z)$ and $K_g(z)$, given by the formulas (5.4.12) and (5.4.18) below, such that

$$\chi_g(z) = -K_g(z)/\sigma(z) \qquad \text{for Re } z > -\lambda. \tag{5.4.9}$$

***Proof*:** Taking the Laplace transforms of both sides over $(0, \infty)$ in each of the equations (5.4.4), we obtain, in view of Lemma 5.4.1, for Re $z > -\lambda$,

$$\begin{aligned} \chi_g(z)\left(1 - \sum_1^p \gamma_j e^{B_j z}\right) - \chi_h(z)\left(\sum_{p+1}^n \gamma_j e^{B_j z}\right) + E_g(z) = 0 \\ \chi_g(z)\left(-\sum_{p+1}^n \gamma_j e^{B_j z}\right) + \chi_h(z)\left(1 - \sum_1^p \gamma_j e^{B_j z}\right) + E_h(z) = 0 \end{aligned} \tag{5.4.10}$$

where

$$E_g(z) = \sum_1^p \gamma_j e^{B_j z} \int_0^{B_j} e^{-zu} g(u)\, du + \sum_{p+1}^n \gamma_j e^{B_j z} \int_0^{B_j} e^{-zu} h(u)\, du \tag{5.4.11}$$

and $E_h(z)$ is obtained from the right member of (5.4.11) by interchanging g and h. Hence, if we set

$$\begin{aligned} \sigma_1(z) &= 1 - \sum_1^n \gamma_j e^{B_j z} \\ \sigma_2(z) &= 1 - \sum_1^p \gamma_j e^{B_j z} + \sum_{p+1}^n \gamma_j e^{B_j z} \qquad \text{if } p < n \end{aligned} \tag{5.4.12}$$

and

$$\sigma(z) = \begin{cases} \sigma_1(z) & \text{if } p = n \\ \sigma_1(z) \cdot \sigma_2(z) & \text{if } p < n \end{cases}$$

then, eliminating χ_h from the relations (5.4.10), we have for $p < n$ as well as for $p = n$,

$$\sigma(z)\chi_g(z) + K_g(z) = 0 \qquad \text{for Re } z > -\lambda$$

where

$$K_g(z) = \begin{cases} E_g(z) & \text{if } p = n, \\ \left(1 - \sum_1^p \gamma_j e^{B_j z}\right) E_g(z) + \left(\sum_{p+1}^n \gamma_j e^{B_j z}\right) E_h(z) & \text{if } p < n. \end{cases} \tag{5.4.13}$$

Since χ_g is analytic in Re $z > -\lambda$, (5.4.9) follows immediately from the preceding formulas.∎

We now note some facts which are important for what follows.

(A) For $s = 1, 2$, $|\sigma_s(x + iy)| \geqq 1 - \sum_1^n \gamma_j e^{B_j x} > 0$, if $x < -\lambda$, y real, so that $\sigma_s(z) \neq 0$ and hence $\sigma(z) \neq 0$ for Re $z < -\lambda$. Hence the only singularities of the function $K_g(z)/\sigma(z)$ are poles at the zeros of $\sigma(z)$ which lie on the line Re $z = -\lambda$, in view of the fact just noted and relation (5.4.9). Further,

$$|\sigma(z)| > c_5(\gamma) \qquad \text{for Re } z \leqq -\gamma < -\lambda. \tag{5.4.14}$$

(B) For any fixed real c,

$$|K_g(z)| \leqq c_6(c) \qquad \text{for } \operatorname{Re} z \leqq c. \tag{5.4.15}$$

This follows at once from (5.4.11) and (5.4.13).

(C) The function $\sigma(z)$ possesses the following properties:

(i) the number of zeros of $\sigma(z)$ in any closed rectangle $\{a \leqq \operatorname{Re} z \leqq b, y \leqq \operatorname{Im} z \leqq y+1\}$ is bounded by a number $N(a, b)$ which does not depend on y;

(ii) if z_0 is an arbitrary point whose distance from every zero of $\sigma(z)$ exceeds $\delta > 0$, then $|\sigma(z_0)| > c_7(\delta)$, a positive constant which depends on δ but not on z_0.

The properties (C) are consequences of the fact that $\sigma(z)$ is an entire almost periodic function (cf. [76], section 6.2, Lemmas 1 and 2). ($\sigma(z)$ also has the following easily verified property: all its zeros are situated in a strip of the form $-\lambda \leqq \operatorname{Re} z \leqq \mu$, λ being as defined above; but we shall not need this fact.)

Detailed proofs of all these properties of $\sigma(z)$ are to be found in [81], sections 9–10.

Property C(i) implies the existence of a $\delta > 0$, not depending on m, and a sequence $\{T_m\} \to \infty$ as $m \to \infty$, with $m < T_m < m+1$ such that all the zeros of $\sigma(z)$ in $\{-\gamma \leqq \operatorname{Re} z \leqq c,\ m \leqq T_m \leqq m+1\}$ are at a distance of at least δ from the straight line $\operatorname{Im} z = T_m$, so that $|\sigma(z)| > c_8$ for all m if $\operatorname{Im} z = T_m$, in view of C(ii) and (5.4.14). Clearly the same is true for all m if $\operatorname{Im} z = -T_m$. Thus, we finally have

$$|\sigma(z)| > c_8 \qquad \text{for all } m, \qquad \text{if } |\operatorname{Im} z| = T_m. \tag{5.4.16}$$

We now establish a simple lemma concerning the zeros of $\sigma(z)$ on the line $\operatorname{Re} z = -\lambda$.

LEMMA 5.4.3. All the zeros of $\sigma(z)$ on the line $\operatorname{Re} z = -\lambda$ are simple. Further,

(a) If $\mathbf{B} \in A_n(0)$, then $-\lambda$ is the only such zero;

(b) if $\mathbf{B} \in B_n^p(\rho)$, $p \leqq n$, then the set of such zeros for $\sigma_1(z)$ is $\{-\lambda + (2m\pi i/\rho)\colon m \text{ integer}\}$, and $\sigma_2(z)$ has no such zeros; and

(c) if $\mathbf{B} \in C_n^p(\rho)$, so that $p < n$, then the set of such zeros for $\sigma_1(z)$ is $\{-\lambda + (2m\pi i/\rho)\colon m \text{ integer}\}$, and for $\sigma_2(z)$ is $\{-\lambda + (2m+1)\pi i/\rho\colon m \text{ integer}\}$.

***Proof*:** Since $\sigma_1(-\lambda) = 0$, we have $\sigma_1(-\lambda + iy) = 0$ if and only if

$$1 - \cos B_j y = 0 \qquad \text{for } 1 \leqq j \leqq n \tag{5.4.17}$$

and $\sigma_2(-\lambda + iy) = 0$ if and only if

$$\left.\begin{aligned} 1 - \cos B_j y &= 0 \qquad \text{for } 1 \leqq j \leqq p, \qquad \text{and} \\ 1 + \cos B_j y &= 0 \qquad \text{for } p < j \leqq n. \end{aligned}\right\} \tag{5.4.18}$$

Hence σ_1 and σ_2 cannot vanish simultaneously at points of the form $-\lambda + iy$. Also, if $\sigma_s(-\lambda + iy) = 0$ for $s = 1$ or 2, then we have for such y, in view of (5.4.17) and (5.4.18),

$$\sigma_s'(-\lambda + iy) = -\sum \gamma_j B_j e^{-B_j\lambda} < 0.$$

Hence it follows that the zeros of $\sigma(z)$ on the line $\operatorname{Re} z = -\lambda$ are simple. Assertions (a)–(c) then follow from (5.4.17) and (5.4.18).

Let us denote $-K_g(z)/\sigma(z)$ by $G(z)$, so that G is analytic everywhere except possibly for simple poles at the zeros of $\sigma(z)$ located on the line $\operatorname{Re} z = -\lambda$; and $G = \chi_g$ for $\operatorname{Re} z > -\lambda$. By the complex inversion formula for the Laplace transform (cf. [150], p. 73), we have for $t > 0$,

$$\begin{aligned}\int_0^t g(u)\, du &= \lim_{T\to\infty} \frac{1}{2\pi i}\int_{c-iT}^{c+iT} \frac{e^{tz}\chi_g(z)}{z}\, dz \\ &= \lim_{m\to\infty} \frac{1}{2\pi i}\int_{c-iT_m}^{c+iT_m} H(z, t)\, dz, \qquad \text{for any } c > 0 \qquad (5.4.19)\end{aligned}$$

where $H(z, t) = e^{tz}G(z)/z$, and $\{T_m\}$ is a sequence chosen in accordance with (5.4.16). If $S_m(t)$ denotes the sum of the residues of $-H(z, t)$ at the zeros of $\sigma(z)$ lying in the interval $\{\operatorname{Re} z = -\lambda,\ |\operatorname{Im} z| < T_m\}$, then, noting that the residue of $H(z, t)$ at the origin is $\chi_g(0)$, we have for any $R > \lambda$, by the theorem of residues,

$$\begin{aligned}\frac{1}{2\pi i}\int_{c-iT_m}^{c+iT_m} H(z, t)\, dz &= \chi_g(0) - S_m(t) \\ &+ \frac{1}{2\pi i}\left\{\int_{-R+iT_m}^{c+iT_m} - \int_{-R-iT_m}^{c-iT_m} + \int_{-R-iT_m}^{-R+iT_m} H(z, t)\, dz\right\}. \qquad (5.4.20)\end{aligned}$$

It is easily verified that, for fixed m, as $R\to\infty$, $\int_{-R-iT_m}^{-R+iT_m} H(z, t)\, dz \to 0$, in view of (5.4.14) and (5.4.15). Hence the right side of (5.4.20) equals

$$s_m(t) - S_m(t) + \chi_g(0) \qquad (5.4.21)$$

where

$$s_m(t) = \frac{1}{2\pi i}\left[\int_{-\infty+iT_m}^{c+iT_m} - \int_{-\infty-iT_m}^{c-iT_m} H(z, t)\, dz\right].$$

Again, it is easily seen with the help of (5.4.15) and (5.4.16) that $s_m(t)\to 0$ as $m\to\infty$, so that, from (5.4.19)–(5.4.21) and the relation $\chi_g(0) = \int_0^\infty g(t)\, dt$, we have the formula, basic to what follows.

$$\int_t^\infty g(u)\, du = \lim_{m\to\infty} S_m(t). \qquad (5.4.22)$$

In order to prove the theorem, we need only compute the right side of (5.4.22) in the various possible cases.

Suppose we call as "active" a zero of $\sigma(z)$ lying on the line $\operatorname{Re} z = -\lambda$ (this phrase is used here in a different sense from that of Definition 2.2.1).

(a) Let $\mathbf{B} \in A_n(0)$. Then $-\lambda$ is the only active zero and (5.4.22) gives, for $t > 0$,

$$\int_t^\infty g(u)\, du = \xi_1 e^{-\lambda t}, \qquad \xi_1 = \text{constant}$$

or, differentiating with respect to t(g being continuous by assumption),

$$g(t) = \xi e^{-\lambda t}, \qquad \text{and similarly} \qquad h(t) = \eta e^{-\lambda t}. \tag{5.4.23}$$

The validity of (5.4.23) for all t (and not only for $t > 0$) follows from the fact that the study of the functions g and h in the interval $(-A, \infty)$ requires only the consideration of the functions $g_A(t) = g(t - A)$ and $h_A(t) = h(t - A)$ on $(0, \infty)$. Since g_A and h_A satisfy (5.4.4), the above analysis applies to them.

If $p < n$, then, substituting from (5.4.23) in (5.4.4), we see that $\xi = \eta$ in that case.

(b) Let $\mathbf{B} \in B_n^p(\rho)$, $p \leqq n$. The active zeros are $\alpha_k = -\lambda + (2k\pi i/\rho)$, k integer, and from (5.4.22) we have that for $t > 0$

$$\int_t^\infty g(u)\, du = \lim_{m\to\infty} \left\{ \sum_{|\operatorname{Im} \alpha_k| < T_m} c_k e^{2k\pi i t/\rho} \right\} e^{-\lambda t} = \xi_0(t) e^{-\lambda t} \tag{5.4.24}$$

where the function ξ_0 is periodic with period ρ. The left member of (5.4.24) is differentiable, hence ξ_0' exists, and we may differentiate both sides of (5.4.24). As a result, we obtain for $t > 0$

$$g(t) = \xi(t) e^{-\lambda t} \qquad \text{and similarly} \qquad h(t) = \eta(t) e^{-\lambda t} \tag{5.4.25}$$

where the functions ξ and η are periodic with period ρ, in addition to being nonnegative and right-continuous like the functions g and h. The validity of (5.4.25) in the intervals $(-m\rho, \infty)$, where m is a positive integer, is proved as in case (a), so that (5.4.25) is valid for all t. Further if $p < n$, then substituting from (5.4.25) in (5.4.4), we obtain that $\xi(t) = \eta(t)$ in that case.

(c) If $\mathbf{B} \in C_n^p(\rho)$, so that $p < n$, then the active zeros will be $-\lambda + (k\pi i/\rho)$, k integer, and we have for $t > 0$ from (5.4.22) that

$$\int_t^\infty g(u)\, du = \lim_{m\to\infty} [\xi_m(t) + \eta_m(t)] e^{-\lambda t} \tag{5.4.26}$$

where $\xi_m(t + \rho) = \xi_m(t)$ and $\eta_m(t + \rho) = -\eta_m(t)$. Replacing t by $t + \rho$ in (5.4.26), we obtain for $t > 0$

$$\int_{t+\rho}^\infty g(u)\, du = \lim_{m\to\infty} [\xi_m(t) - \eta_m(t)] e^{-\lambda(t+\rho)}. \tag{5.4.27}$$

It follows from the above two relations that, as $m \to \infty$, $\xi_m(t)$, and $\eta_m(t)$ have limits, so that

$$\int_t^{\infty} g(u)\, du = [\xi_0(t) + \eta_0(t)]e^{-\lambda t} \tag{5.4.28}$$

where

$$\xi_0(t+\rho) = \xi_0(t) \qquad \text{and} \qquad \eta_0(t+\rho) = -\eta_0(t),$$

and again

$$\int_{t+\rho}^{\infty} g(u)\, du = [\xi_0(t) - \eta_0(t)]e^{-\lambda(t+\rho)}. \tag{5.4.29}$$

It follows from (5.4.28) and (5.4.29) that each of the functions ξ_0 and η_0 is differentiable, and we immediately obtain for $t > 0$ the representation

$$g(t) = [\xi_1(t) + \eta_1(t)]e^{-\lambda t}, \qquad h(t) = [\xi_2(t) + \eta_2(t)]e^{-\lambda t}. \tag{5.4.30}$$

The validity of (5.4.30) for all t is established as above. Substituting from (5.4.30) in (5.4.4), and recalling that in the present case, $m_1, \ldots, m_p$ are all even, and $m_{p+1}, \ldots, m_n$ are all odd, we find that

$$\xi_1 + \eta_1 = \xi_2 - \eta_2 \qquad \text{and} \qquad \xi_1 - \eta_1 = \xi_2 + \eta_2$$

whence

$$\xi_1 = \xi_2 (= \xi, \text{ say}) \qquad \text{and} \qquad \eta_1 = -\eta_2 (= \eta, \text{ say})$$

which yields the representation (5.4.7).

We recall in conclusion that our argument was based on the assumption that the functions g and h are (also) continuous, but, as noted immediately following Lemma 5.4.1, the general case reduces to this case. Thus, Theorem 5.4.1 stands proved.▌

THEOREM 5.4.2. Let a nonvanishing c.f. f satisfy for all t the relation

$$f(t) = \prod_1^p [f(\beta_j t)]^{\gamma_j} \cdot \prod_{p+1}^n [f(-\beta_j t)]^{\gamma_j} \tag{5.4.31}$$

where $0 \leqq p \leqq n(>1)$, $0 < \beta_j < 1$, $\gamma_j > 0$ for all j. Let λ denote the unique real root of the equation $\sum_1^n \gamma_j \beta_j^{\lambda} = 1$. Then,

(i) f is infinitely divisible;

(ii) f is the c.f. of a degenerate law if $\lambda \leqq 0$ or > 2;

(iii) f is the c.f. of a normal law if $\lambda = 2$; and

(iv) if $0 < \lambda < 2$, then in the Levy representation $L(\mu, \sigma^2, M, N)$ for $\log f$, $\sigma = 0$, and, depending on the nature of the vector $\mathbf{B} = (B_1, \ldots, B_n)$, where $B_j = -\log \beta_j$, the functions M and N have the following forms:

(a) If $\mathbf{B} \in A_n(0)$, then

$$M(u) = \xi |u|^{-\lambda}, \qquad N(u) = -\eta u^{-\lambda} \tag{5.4.32}$$

where ξ and η are nonnegative real numbers with $\xi + \eta > 0$; further, $\xi = \eta$ if $p < n$ or if $\lambda = 1$ (in the latter case irrespective of whether $p < n$ or $p = n$), so that, for some real c, $f(t)e^{ict}$ is the c.f. of a symmetric stable law with exponent λ; if $p = n$ and $\lambda \neq 1$, then f is the c.f. of a stable law with exponent λ.

(b) If $\mathbf{B} \in B_n^p(\rho)$, $p \leqq n$, then

$$M(u) = \frac{\xi(\log |u|)}{|u|^\lambda}, \qquad N(u) = \frac{-\eta(\log u)}{u^\lambda} \tag{5.4.33}$$

where ξ and η are nonnegative, right-continuous functions on the real line, with period ρ; further, if $p < n$, then $\xi = \eta$.

(c) If $\mathbf{B} \in C_n^p(\rho)$, so that $p < n$, then

$$M(u) = \frac{\xi(\log |u|) + \eta(\log |u|)}{|u|^\lambda}, \qquad N(u) = \frac{-\xi(\log u) + \eta(\log u)}{u^\lambda} \tag{5.4.34}$$

where ξ and η are right-continuous functions on R_1, with $\xi(x + \rho) = \xi(x)$ and $\eta(x + \rho) = -\eta(x)$ for all x.

Proof: (i) Let ϕ denote that branch of the logarithm of f which is continuous and vanishes at the origin. Then, for all real t, we have

$$\phi(t) = \gamma_1 \phi(\beta_1 t) + \cdots + \gamma_{p+1}\phi(-\beta_{p+1} t) + \cdots. \tag{5.4.35}$$

Iterating this relation s times, we obtain

$$\begin{aligned} \phi(t) &= \sum \frac{s!}{s_1! \cdots s_n!} \gamma_1^{s_1} \cdots \gamma_n^{s_n} \phi(\beta_1^{s_1} \cdots (-\beta_n)^{s_n} t) \\ &= \sum \gamma_{sj} \phi(\beta_{sj} t) \end{aligned} \tag{5.4.36}$$

where the summation extends over all distinct n-vectors $(s_1, \ldots, s_n)$ with non-negative integer components such that $\sum s_i = s$. If $\beta_0 = \max(\beta_1, \ldots, \beta_n)$, then

$$|\beta_1^{s_1} \cdots (-\beta_{p+1})^{s_{p+1}} \cdots| < \beta_0^s \to 0 \qquad \text{as} \qquad s \to \infty$$

so that $\phi(\beta_{sj} t) \to 0$ or $f(\beta_{sj} t) \to 1$ uniformly in j in any finite t-interval.

Let F_{sj} be the d.f. corresponding to f_{sj}. Then, in precisely the same way as in the solution of the Central Limit Problem, ([91], p. 301), we choose and fix a $\tau > 0$, and set $a_{sj} = \int_{|x|<\tau} x \, dF_{sj}$, $\bar{F}_{sj}(x) = F_{sj}(x + a_{sj})$. Let $\bar{f}_{sj}$ be the c.f. of $\bar{F}_{sj}$, so that $\bar{f}_{sj}(t) = f_{sj}(t) \exp(-ita_{sj})$. Since $f(\beta_{sj} t) \to 1$ as $s \to \infty$ uniformly with respect to j in any finite interval of values of t, it follows (cf. [91], p. 302) that, for any $b > 0$, an $N(b)$ can be found such that

$$|\bar{f}_{sj}(t) - 1| < 1/2 \qquad \text{for } |t| \leqq b \qquad \text{if } s \geqq N(b).$$

From the elementary inequality $|\log z + 1 - z| \leqq |1 - z|^2$ if $|z - 1| < 1/2$, it then follows that for all $s \geqq N(b)$, $|t| \leqq b$, and uniformly in j,

$$|\log \bar{f}_{sj}(t) + 1 - \bar{f}_{sj}(t)| \leqq |\bar{f}_{sj}(t) - 1|^2. \tag{5.4.37}$$

Again (cf. [91], p. 302), for fixed $b > 0$, there exists a constant $c = c(\tau, b) > 0$ such that, for all sufficiently large s, $s \geqq N(\tau, b)$,

$$\max_{|t| \leqq b} |\bar{f}_{sj}(t) - 1| \leq - c \int_0^b \log |f_{sj}(t)| \, dt$$

so that

$$\sum \gamma_{sj} |\bar{f}_{sj}(t) - 1| \leqq - c \int_0^b [\sum \gamma_{sj} \log |f_{sj}(t)|] \, dt$$

$$= - c \int_0^b \log |f(t)| \, dt. \tag{5.4.38}$$

Thus, from (5.4.37) and (5.4.38), we see that, for $|t| \leqq b$ and $s \geqq \max [N(b), N(\tau, b)]$,

$$|\sum \gamma_{sj}\{\log \bar{f}_{sj}(t) + 1 - \bar{f}_{sj}(t)\}| \leqq \sum \gamma_{sj} |\bar{f}_{sj}(t) - 1|^2$$
$$\leqq \max_{j,\, |t| \leqq b} |\bar{f}_{sj}(t) - 1| \{\sum \gamma_{sj} |\bar{f}_{sj}(t) - 1|\}$$

tends to zero as $s \to \infty$, since the first factor in the last expression tends to zero and the second is bounded. Thus we have a sequence of "accompanying" i.d. laws, the logarithms of their c.f.'s being given by the equation

$$\phi_s(t) = it(\sum \gamma_{sj} a_{sj}) + \sum \gamma_{sj} \int (e^{itx} - 1) \, d\bar{F}_{sj} \tag{5.4.39}$$

so that $\phi_s(t) \to \phi(t)$ as $s \to \infty$ for all t.

Thus, f is an i.d.c.f.

(ii) Let $\lambda \leqq 0$, which is equivalent to $\gamma_1 + \cdots + \gamma_n \leqq 1$; then, from (5.4.36), it follows that

$$|\phi(t)| \leqq \sum \gamma_{sj} |\phi(\beta_{sj} t)| \leqq (\gamma_1 + \cdots + \gamma_n)^s \max_j |\phi(\beta_{sj} t)|$$

which tends to zero as $s \to \infty$, so that $\phi(t) \equiv 0$. Hence $f(t) \equiv 1$, and thus corresponds to a degenerate law.

Let now $\lambda > 0$, or, equivalently, $\gamma_1 + \cdots + \gamma_n > 1$. For $t \neq 0$, let $\psi(t) = \log |f(t)|/|t|^\lambda$. Then, (5.4.35) together with $\gamma_j \beta_j^\lambda = p_j$ gives

$$\psi(t) = \sum p_j \psi(\beta_j t) = \psi(\beta(t)) \tag{5.4.40}$$

by the intermediate value theorem, where

$$\beta_* t \leqq \beta(t) \leqq \beta^* t, \qquad \text{with} \qquad \beta_* = \min \beta_j, \; \beta^* = \max \beta_j.$$

Thus we have for any $t > 0$ a sequence of real numbers $\beta^0(t)$, $\beta^1(t)$, ..., tending to zero as $s \to \infty$, such that, for all $s > 0$,

$$\begin{gathered}\beta_* \beta^{(s)}(t) \leqq \beta^{(s+1)}(t) \leqq \beta^* \beta^{(s)}(t) \\ \psi(t) = \psi(\beta^{(s)}(t)).\end{gathered} \tag{5.4.41}$$

In particular, setting $t = 1$, we see that there exist a sequence $\{t_n\}$ of positive numbers converging to zero, such that for all $s \geqq 0$ ($t_0 = 1$)

$$\beta_* t_s \leqq t_{s+1} \leqq \beta^* t_s \qquad \text{and} \qquad \psi(t_s) = \psi(1). \tag{5.4.42}$$

It follows from (5.4.42) and Part (d) of Lemma 1.1.5 that, if $\lambda > 2$, then f is the c.f. of a degenerate law. This proves assertion (ii) of the theorem.

(iii) If $\lambda = 2$, then we set $c = -2\psi(1) \geqq 0$. We then have from (5.4.42) that $|f(t_s)|^2 = \exp(-ct_s^2)$ for all s, so that by Corollary 1 to Lemma 1.2.1, $|f|^2$ is the c.f. of a normal law, but then by Cramer's theorem, f is itself such a c.f. This proves assertion (iii) of the theorem.

(iv) In part (i), we established that f is an i.d.c.f. Hence $\log f$ admits the Levy representation

$$\log f(t) = i\mu t - \frac{1}{2}\sigma^2 t^2 + \int_{(0,\infty)} h(t, u)\, dN(u) + \int_{(-\infty, 0)} h(t, u)\, dM(u),$$

$$h(t, u) = e^{itu} - 1 - \frac{itu}{1 + u^2}. \tag{5.4.43}$$

Invoking the relation (5.4.35), we obtain as in the case of the relations (5.3.8)–(5.3.10) that (for a suitable μ')

$$\mu \sum \gamma_i + \mu' = 0, \tag{5.4.44}$$

$$\sigma^2\left(\sum \gamma_i \beta_i^2 - 1\right) = 0, \qquad \text{so that } \sigma = 0 \quad \text{if } \lambda < 2 \tag{5.4.45}$$

$$\begin{aligned}N(u) &= \sum_1^p \gamma_j N\left(\frac{u}{\beta_j}\right) - \sum_{p+1}^n \gamma_j M\left(\frac{-u}{\beta_j}\right), \qquad u > 0, \\ M(u) &= \sum_1^p \gamma_j M\left(\frac{u}{\beta_j}\right) - \sum_{p+1}^n \gamma_j N\left(\frac{-u}{\beta_j}\right), \qquad u < 0.\end{aligned} \tag{5.4.46}$$

Setting $g(u) = M(-e^u)$ and $h(u) = -N(e^u)$, we obtain the equations

$$\begin{aligned}g(u) &= \sum_1^p \gamma_j g(u + B_j) + \sum_{p+1}^n \gamma_j h(u + B_j) \\ h(u) &= \sum_1^p \gamma_j h(u + B_j) + \sum_{p+1}^n \gamma_j g(u + B_j)\end{aligned} \tag{5.4.47}$$

where $B_j = -\log \beta_j$, and the functions g and h are nonnegative, nonincreasing, and right-continuous, with $g(+\infty) = h(+\infty) = 0$. In this case, on applying Theorem 5.4.1, we obtain all the assertions (iva)–(ivc) of Theorem 5.4.2, with the possible exception of one, namely, that if $\mathbf{B} \in A_n(0)$ and $\lambda = 1$, then f is the c.f. of a Cauchy law, irrespective of whether $p < n$ or $p = n$ (assertion (iva) of the theorem). We note that in that case $M(u) = \xi/|u|$ and $N(u) = -\eta/u$, and, substituting these expressions in (5.4.46) and noting that $\sum \gamma_i |\beta_i| = 1$, we obtain $\xi = \eta$, which yields the Cauchy distribution.∎

Remark. Suppose the same c.f. f satisfies (5.4.31) for two distinct choices of the constants $(n_1, \boldsymbol{\beta}_1, \gamma_1, p_1)$ and $(n_2, \boldsymbol{\beta}_2, \gamma_2, p_2)$. Clearly, both the sets of constants must yield the same λ. If $\mathbf{B}_1 \in A_{n_1}(0)$ or $\mathbf{B}_2 \in A_{n_2}(0)$ or if $\mathbf{B}_1 \in A_{n_1}(\rho_1)$ and $\mathbf{B}_2 \in A_{n_2}(\rho_2)$, where ρ_1/ρ_2 is irrational, then we can show that f is necessarily the c.f. of a stable law. In special cases of such a situation, it may even be possible to go further and show that f is the c.f. of a law differing only by a shift parameter from a symmetric stable law. (We use these considerations in section 10.5.)

5.4.2 Characterization Theorems

We shall now state and prove the basic theorems of section 5.4.

Let $X_1, \ldots, X_n$ be i.i.d.r.v.'s with $EX_1 = 0$. Suppose there exist two linear forms $a_1 X_1 + \cdots + a_n X_n$ and $b_1 X_1 + \cdots + b_n X_n$ such that

$$E(a_1 X_1 + \cdots + a_n X_n | b_1 X_1 + \cdots + b_n X_n) = 0. \tag{5.4.48}$$

We shall assume, without loss of generality, that $|b_n| = \max_i |b_i|$ and that (the corresponding) $a_n \neq 0$. Then, by a suitable change of scale, we can reduce (5.4.48) to the form

$$E(-\alpha_1 X_1 - \cdots - \alpha_{n-1} X_{n-1} + X_n | \beta_1 X_1 + \cdots + \beta_{n-1} X_{n-1} + X_n) = 0 \tag{5.4.49}$$

where $|\beta_j| \leqq 1$ for $j = 1, \ldots, n-1$. Condition (5.4.49) is equivalent to

$$E\{(-\alpha_1 X_1 - \cdots - \alpha_{n-1} X_{n-1} + X_n) \exp it(\beta_1 X_1 + \cdots + X_n)\} = 0$$

which yields the equation

$$\psi(t) = \sum \alpha_i \psi(\beta_i t), \qquad \psi = f'/f, \ |t| < c \tag{5.4.50}$$

where f is the c.f. of the X_i. Integrating the above equation, we obtain for $|t| < c$,

$$f(t) = \prod_1^{n-1} [f(\beta_i t)]^{\delta_i}, \qquad \delta_i = \alpha_i/\beta_i. \tag{5.4.51}$$

Under the conditions of either of the two theorems below, we can show that f never vanishes and that (5.4.51) is indeed valid for all real t.

THEOREM 5.4.3. Let $X_1, \ldots, X_n$ be i.i.d.r.v.'s satisfying (5.4.49). Suppose that exactly k of the numbers β_i are different from one another (we may take them to be $\beta_1, \ldots, \beta_k$) as well as from ± 1. Let, for $j = 1, \ldots, k$,

$$\varepsilon_j = \sum \{\delta_i : \beta_i = \beta_j\}, \qquad \varepsilon_0 = \sum \{\delta_i : \beta_i = 1\}$$

and

$$\varepsilon_0' = \sum \{\delta_i : \beta_i = -1\}.$$

If $\varepsilon_0' = 0$, $\gamma_i = \varepsilon_i/(1 - \varepsilon_0) > 0$ for $i = 1, \ldots, k$, and λ is the unique real solution of the equation $\sum_1^k \gamma_i |\beta_i|^\lambda = 1$, then the following assertions hold:

(i) the X_i are degenerate if $\lambda \leqq 1$ or $\lambda > 2$;

(ii) the X_i are normal (possibly degenerate) if $\lambda = 2$; and

(iii) for $1 < \lambda < 2$, the X_i follow one of the i.d. laws indicated in Parts (iva)–(ivc) of Theorem 5.4.2, depending on to which of the sets $A_k(0)$, $B_k^p(\rho)$ and $C_k^p(\rho)$ the vector $(-\log|\beta_1|, \ldots, -\log|\beta_k|)$ belongs (for suitable p and ρ).

***Proof*:** We start from relation (5.4.51) valid in a neighborhood of the origin in which f does not vanish, this relation following from (5.4.49). Under the conditions of the theorem, we have

$$[f(t)]^{1-\varepsilon_0} = \prod_1^k [f(\beta_i t)]^{\varepsilon_i} \tag{5.4.52}$$

which leads to the relation

$$f(t) = \prod_1^k [f(\beta_i t)]^{\gamma_i}, \qquad \gamma_i > 0, \qquad 0 < |\beta_i| < 1 \tag{5.4.53}$$

(As remarked above, we easily see that (5.4.52) or (5.4.53) is valid for all t.) But then Theorem 5.4.2 applies, and we obtain all the assertions of the present theorem except possibly that the X_i are degenerate if $0 < \lambda \leqq 1$; this assertion follows in fact from the condition $E|X_i| < \infty$ and can be proved as follows.

We take the logarithms of both sides of (5.4.53) and differentiate with respect to t, this being possible since $E|X_1| < \infty$. We obtain

$$\phi(t) = \sum \alpha_i \phi(\beta_i t) \tag{5.4.54}$$

where $\gamma_i \beta_i = \alpha_i$, $\phi = f'/f$. Iterating this relation s times, we obtain

$$\phi(t) = \sum \frac{s!}{s_1! \cdots s_k!} \alpha_1^{s_1} \cdots \alpha_k^{s_k} \phi(\beta_1^{s_1} \cdots \beta_k^{s_k} t)$$

whence

$$|\phi(t)| \leqq \sum \frac{s!}{s_1! \cdots s_k!} |\alpha_1|^{s_1} \cdots |\alpha_k|^{s_k} |\phi(\beta_1^{s_1} \cdots \beta_k^{s_k} t)|. \tag{5.4.55}$$

If $\lambda \leqq 1$, then $|\alpha_1| + \cdots + |\alpha_k| \leqq 1$, and hence the right member above does not exceed

$$\max |\phi(\beta_1^{s_1} \cdots \beta_k^{s_k} t)| \to 0 \qquad \text{as} \qquad s \to \infty \tag{5.4.56}$$

the maximum being taken over all k-tuples $(s_1, \ldots, s_k)$ of nonnegative integers such that their sum is s, in view of the fact that $|\beta_j| < 1$ for all j. Thus, $\phi(t) \equiv 0$ and $f(t) \equiv 1$. Our theorem stands completely proved.▌

THEOREM 5.4.4. Let $X_1, \ldots, X_n$ be i.i.d.r.v.'s satisfying (5.4.51). Suppose exactly k of the $|\beta_i|$ are distinct and different from 1. Let

$$\varepsilon_j = \sum \{\delta_i: |\beta_i| = |\beta_j|\}, \qquad \varepsilon_0 = \sum \{\delta_i : |\beta_i| = 1\}$$

and let $\gamma_i = \varepsilon_i/(1 - \varepsilon_0) > 0$ for $i = 1, \ldots, k$, and λ be the unique solution of the equation $\sum_1^k \gamma_i |\beta_i|^\lambda = 1$. Then

(i) the X_i are degenerate if $\lambda \leqq 1$ or $\lambda > 2$;

(ii) the X_i are normal if $\lambda = 2$; and

(iii) for $1 < \lambda < 2$, the difference $X_i - X_j$, $i \neq j$, follows one of the i.d. laws indicated in Parts (iva)–(ivc) of Theorem 5.4.2, depending on to which of the sets $A_k(0)$ or $B_k^k(\rho)$ for a suitable ρ, the vector $(-\log |\beta_1|, \ldots, -\log |\beta_k|)$ belongs.

Proof: We start with the equation (5.4.51)

$$f(t) = \prod_1^{n-1} [f(\beta_i t)]^{\delta_i} \tag{5.4.57}$$

which follows from (5.4.49). We have from (5.4.57) that

$$w(t) = \prod_1^{n-1} [w(|\beta_i| t)]^{\delta_i} \tag{5.4.58}$$

where $w(t) = f(t) f(-t) = |f(t)|^2$. Under the conditions of the theorem, (5.4.58) takes the form

$$w(t) = \prod_1^{k} [w(|\beta_i| t)]^{\gamma_i} \tag{5.4.59}$$

where $\gamma_i = \varepsilon_i/(1 - \varepsilon_0) > 0$ and $|\beta_i| < 1$ for all i, by assumption.

We note that w is the c.f. of $X_1 - X_2$. Applying Theorem 5.4.2 to the equation (5.4.59), we see that $X_1 - X_2$ is, and consequently all the X_i are, degenerate if $\lambda \leqq 1$ or $\lambda > 2$. Also, if $\lambda = 2$, then $X_1 - X_2$ is normal, and consequently so are the X_i, in the view of Cramer's theorem. Thus, assertions (i) and (ii) of the theorem are proved. (iii) is an immediate consequence of the results (iva)–(ivb) of Theorem 5.4.2.▌

5.5 CONSTANCY OF REGRESSION OF ONE LINEAR STATISTIC ON ANOTHER (A FINITE NUMBER OF R.V.'s)—II

In Theorems 5.4.3 and 5.4.4, we studied the character of the distribution of r.v.'s for which

$$E(a_1 X_1 + \cdots + a_n X_n | b_1 X_1 + \cdots + b_n X_n) = 0 \tag{5.5.1}$$

under certain restrictions on the coefficients a_j, b_j. Now we consider a more general situation, whose analysis is based on the results of Chapter 2. If $X_1, \ldots, X_n$ are i.i.d.r.v.'s with $EX_1 = 0$, and f is the c.f. of the X_i, then (5.5.1) leads to the equation

$$\prod_1^n [f(b_i t)]^{c_i} = 1, \qquad c_i = a_i/b_i \tag{5.5.2}$$

valid for t in some neighborhood of the origin. Let us suppose without loss of generality that $c_1 > 0, \ldots, c_p > 0$, and $c_{p+1} < 0, \ldots, c_n < 0$. Then (5.5.2) can be rewritten in the form

$$\prod_1^p [f(b_i t)]^{c_i} = \prod_{p+1}^n [f(b_i t)]^{|c_i|}. \tag{5.5.3}$$

It is worth noting that equations of the type (5.5.3), considered in Theorems 5.4.3 and 5.4.4. contained only one factor on one side, and in those cases it was shown that f is the c.f. of an i.d. law. In the more general case of equation (5.5.3), a representation for f can be obtained by the method presented in Chapter 2. We consider only the case (describable in terms of roots of the equation $G(\lambda) \equiv \sum c_j |b_j|^{\lambda} = 0$) where the condition (5.5.1) determines the normal law.

THEOREM 5.5.1. Let $X_1, \ldots, X_n$ be i.i.d.r.v.'s with $EX_1 = 0$, satisfying (5.5.1). Suppose that

$$\max(|b_1|, \ldots, |b_p|) \neq \max(|b_{p+1}|, \ldots, |b_n|), \tag{5.5.4}$$

The following conditions constitute a set of necessary and sufficient conditions for the normality of the X_i (which could possibly be degenerate).

(a) All the positive zeros of $G(\lambda)$ which are divisible by 4 must be simple.

(b) All the positive even-integer zeros of $G(\lambda)$, which are not divisible by 4, must have multiplicity at most 2; if such a zero exists with multiplicity 2, then it must be unique and the greatest among the positive zeros of $G(\lambda)$.

(c) If $G(\lambda)$ has a positive zero γ, which is not an even integer, then it must be unique, simple, and the greatest among all the positive zeros of $G(\lambda)$; further, $[\gamma/2]$ must be odd, where $[x]$ as usual denotes the "integral part" of x.

(d) $G(2) = 0$.

THEOREM 5.5.2. Let $X_1, \ldots, X_n$ be i.i.d.r.v.'s with $EX_1 = 0$, satisfying conditions (5.5.1) and (5.5.4). Let γ be the largest of the zeros of $G(\lambda)$; such a γ exists and is necessarily positive. If X_1 has finite moment of order $2m$ where $m = [(\gamma + 2)/2]$, then X_1 is normal (possibly degenerate); in particular, this assertion holds if X_1 has moments of all orders.

The following theorem relates to a case where condition (5.5.4) is not required to be satisfied.

THEOREM 5.5.3. Let $X_1, \ldots, X_n$ be i.i.d.r.v.'s with $EX_1 = 0$, satisfying condition (5.5.1); we do not impose on the coefficients the condition (5.5.4). Suppose $G(\lambda) \not\equiv 0$ and denote by σ the supremum of the real parts of the zeros of $G(\lambda)$. If X_1 has finite moment of order $2m$, where $m = [(\sigma + 2)/2]$, then the X_i are normal (possibly degenerate).

Theorems 5.5.1–5.5.3 follow from the results of Chapter 2. ∎

5.6 CONSTANCY OF REGRESSION OF ONE LINEAR STATISTIC ON ANOTHER (AN INFINITE NUMBER OF R.V.'s)

5.6.1 Functional Equations Involving Infinite Products

In this section, we study solutions of equations of the form

$$f(t) = \prod_{1}^{\infty} [f(\pm \beta_j t)]^{\gamma_j} \tag{5.6.1}$$

where f is a nonvanishing c.f. $\not\equiv 1$, $0 < \beta_j < 1$ and $\gamma_j > 0$ for $j = 1, 2, \ldots$. We note that if all the γ_j are positive integers, then the right member of (5.6.1) may be well-defined even if f has zeros on the real line. If the γ_j are positive integers and relation (5.6.1) is satisfied, then it is possible to *show* that f is indeed nonvanishing, the proof being not quite trivial. We may first note (as in the proof of Theorem 5.6.1) that $\sum \gamma_j \beta_j^2 < \infty$, so that $\beta_j \to 0$ as $j \to \infty$ (since $\gamma_j \geqq 1$ for all j), and then use the fact that the convergence of the sequence of c.f.'s: $\prod_1^n [f(\pm \beta_j t)]^{\gamma_j}$ to the c.f. f is necessarily uniform on every finite interval, to obtain a contradiction to the assumption that f vanishes somewhere on the real line (cf. the argument in section 5.6.2).

We first find conditions (Theorems 5.6.1) on the γ_j and the β_j, which are necessary and sufficient for the only solutions of (5.6.1) to be normal laws; we impose no requirement on the β_j except that $0 < \beta_j < 1$ for all j. Then we prove (Theorem 5.6.2) the infinite divisibility of c.f.'s f satisfying (5.6.1), under the supplementary condition that $\beta_j \to 0$ as $j \to \infty$; this condition is obviously satisfied if the γ_j are bounded away from zero, in particular if they are all positive integers, in view of the relation proved below that $\sum \gamma_j \beta_j^2 < \infty$. This condition is fulfilled in a regression problem studied in section 5.6.2, in which Theorems 5.6.1 and 5.6.2 will be used. Finally we obtain (Theorem 5.6.3) sufficient conditions on the β_j and the γ_j under which assertions similar to those in Theorems 5.4.1 and 5.4.2 will be valid.

THEOREM 5.6.1. Let f be a nontrivial c.f., satisfying, in some neighborhood of the origin where it does not vanish, the equation

$$f(t) = \prod_1^\infty [f(\pm \beta_j t)]^{\gamma_j} \tag{5.6.2}$$

where $0 < \beta_j < 1$ and $\gamma_j > 0$ for all j. Then

(i) $\sum \gamma_j \beta_j^2 \leqq 1$; and

(ii) f is the c.f. of a normal law if and only if $\sum \gamma_j \beta_j^2 = 1$.

Remark. In the interval concerned, $[f(\pm \beta_j t)]^{\gamma_j}$ is taken to be defined by $\exp[\gamma_j \log f(\pm \beta_j t)]$, where $\log f$ denotes the continuous version of the logarithm which vanishes at the origin.

Proof: (*i*) Let $g = |f|^2$ and G be the d.f. corresponding to g. Then we have from (5.6.2) that (I denoting the interval concerned)

$$g(t) = \prod_1^\infty [g(\beta_j t)]^{\gamma_j}, \quad t \in I. \tag{5.6.3}$$

Since F and G are nondegenerate d.f.'s, there exists an $A > 0$ such that $\int_{-A}^{A} x^2 \, dG(x) > 0$. Choose and fix this A so large that $t_0 = \pi/A$ lies in the interval I. Note that

$$g(t) = 1 - 2 \int \sin^2 (tx/2) \, dG(x) \leqq \exp\left[- 2 \int \sin^2 (tx/2) \, dG(x)\right].$$

Then, using the well-known inequality $1 \geqq (\sin \theta)/\theta \geqq 2/\pi$ for $0 \leqq \theta \leqq \pi/2$, we have from (5.6.3) for any positive integer n

$$\begin{aligned} -\log g(t_0) &\geqq 2 \sum_1^n \gamma_j \int \sin^2[(1/2)\beta_j t_0 x)] \, dG(x) \\ &\geqq 2 \sum_1^n \gamma_j \int_{-A}^{A} \sin^2[(1/2)\beta_j t_0 x)] \, dG(x) \\ &\geqq (2t_0^2/\pi^2) \sum_1^n \gamma_j \beta_j^2 \int_{-A}^{A} x^2 \, dG(x) \end{aligned}$$

whence the convergence of the series in question.

We now show that $\sum_1^\infty \gamma_j \beta_j^2$ cannot be greater than one. Suppose to the contrary, and fix n such that

$$\sum_1^n \gamma_j \beta_j^2 = \gamma > 1.$$

Then, setting $\psi(t) = -\log g(t)/t^2$, $t \neq 0$, we have for $t > 0$

$$\psi(t) \geqq \sum_1^n \gamma_j \beta_j^2 \psi(\beta_j t) = \gamma\psi(t\beta(t))$$

by the intermediate value theorem, where

$$\min_{1 \leqq j \leqq n} \beta_j \leqq \beta(t) \leqq \max_{1 \leqq j \leqq n} \beta_j .$$

Thus there exists a sequence $\{b_m\} \to 0$ as $m \to \infty$, depending on the n fixed above, such that $\psi(b_m) \leqq \gamma^{-m}\psi(1)$ for all m, so that $\psi(b_m) \to 0$ as $m \to \infty$. This in turn implies that the d.f.'s G and F are degenerate (Lemma 1.1.5d), contrary to assumption. This proves assertion (i).

For (ii) let now $\sum \gamma_j \beta_j^2 = 1$. We shall show that f is the c.f. of a normal law; the converse is obvious. ψ being defined as above, we have for $0 < |t| < \delta$,

$$\psi(t) = \sum p_j \psi(\beta_j t), \qquad \text{where} \qquad p_j = \gamma_j \beta_j^2, \qquad \sum p_j = 1,$$

which can be rewritten as

$$\sum p_j[\psi(t) - \psi(\beta_j t)] = 0, \qquad 0 < |t| < \delta. \tag{5.6.4}$$

It follows (by contradiction) from (5.6.4) that, for any t, $0 < t < \delta$, there exists at least one β_j, $j = j(t)$, such that $\psi(t) \geqq \psi(\beta_j t)$. Fix t_0 in $(0, \delta)$ and set

$$S(t_0) = \{t : 0 < t \leqq t_0,\ \psi(t) \leqq \psi(t_0)\}.$$

The set $S(t_0)$ is nonempty; let $\tau = \inf S(t_0)$, so that $\tau \geqq 0$. We claim that $\tau = 0$. Suppose not, and let $\tau > 0$. Then, the continuity of ψ implies that $\psi(t_0) \geqq \psi(\tau)$. Also there exists a $k = k(\tau)$ such that $\psi(\tau) \geqq \psi(\beta_k \tau)$, so that $\psi(t_0) \geqq \psi(\beta_k \tau)$, i.e., $\beta_k \tau \in S(t_0)$, which contradicts the definition of τ. Hence we must have $\tau = 0$, and hence there exists a sequence of positive numbers $\{t_n\} \to 0$ such that $\psi(t_n) \leqq \psi(t_0)$ for all n. Then, by Lemma 1.1.5c, G has finite variance, and this in turn implies that lim $\psi(t)$ as $t \to 0$ exists; denoting the limit by $\psi(0)$, we have $\psi(t_0) \geqq \psi(0)$.

Now, from (5.6.4) it also follows that for every $t \in (0, \delta)$, there exists an $l = l(t)$ such that $\psi(t) \leqq \psi(\beta_l t)$. Arguing in a manner dual to the above, we find that $\psi(t_0) \leqq \psi(0)$ for every $t_0 \in (0, \delta)$. Thus, ψ is a constant for $|t| < \delta$, whence it follows that g is a normal c.f. Then, by Cramer's theorem, so is f. ▮

We now turn to study solutions of (5.6.1) in the case where $\sum \gamma_j \beta_j^2 < 1$. Our results here are not as complete as in the finite case, but we single out cases which can be studied by the methods of section 5.4.

We start with an analogue of Theorem 5.4.1, for which we rename the β_j introduced above in the following manner. Let $a_1, a_2, \ldots$ be the sequence of the numbers β_j which appear in the infinite product in (5.6.1) with the positive sign (if such exist), and $b_1, b_2, \ldots$ the sequence of the numbers β_j which appear there with the negative sign (again, if such exist). Denote the exponents γ_j corresponding to the a_j by δ_j and those corresponding to the b_j by ε_j. Let $A_j = -\log a_j$ and $B_j = -\log |b_j|$. Consider the following classification of the pair of infinite vectors $(\mathbf{A}, \mathbf{B})$:

$\mathscr{A}(\rho)$: the set of all $(\mathbf{A}, \mathbf{B})$ such that there exists $\rho > 0$ such that $A_j = k_j/\rho$ and $B_j = l_j/\rho$, where k_j and l_j are positive integers with their greatest common factor equal to one;

$\mathscr{A}(0)$: the set of all $(\mathbf{A}, \mathbf{B})$ for which such a ρ does not exist;

$\mathscr{B}(\rho)$: the subset of $\mathscr{A}(\rho)$ in which at least one of the k_j is odd and or at least one of the l_j is even; and

$\mathscr{C}(\rho)$: the subset of $\mathscr{A}(\rho)$ in which all the k_j are even and all the l_j are odd.

We note that if the b_j are absent, then $\mathscr{B}(\rho) = \mathscr{A}(\rho)$, and $\mathscr{C}(\rho)$ is empty, by the definition of ρ.

THEOREM 5.6.2. Let g and h be nonnegative, nonincreasing, right-continuous functions defined on R^1 with $g(+\infty) = h(+\infty) = 0$, satisfying for all real u the relations

$$g(u) = \sum_1^\infty \delta_j g(u + A_j) + \sum_1^\infty \varepsilon_j h(u + B_j)$$
$$h(u) = \sum_1^\infty \delta_j h(u + A_j) + \sum_1^\infty \varepsilon_j g(u + B_j) \tag{5.6.5}$$

where the δ_j and ε_j are all positive and $A_j \geqq A_0 > 0$ and $B_j \geqq B_0 > 0$ for all j, for some A_0 and B_0. Then,

(i) $g \equiv h \equiv 0$ if $\sum \delta_j + \sum \varepsilon_j \leqq 1$; this assertion is valid even under the weaker conditions: $A_j > 0$, $B_j > 0$ for all j;

(ii) if $1 < \sum \delta_j + \sum \varepsilon_j < \infty$ and λ denotes the unique positive solution of the equation

$$\sum_1^\infty \delta_j e^{-A_j\lambda} + \sum_1^\infty \varepsilon_j e^{-B_j\lambda} = 1$$

then the assertions of Theorem 5.4.1 remain valid if we replace the phrase "if $p < n$" by "if the set of B's is nonempty" and the sets $A_n(0)$, $B_n^p(\rho)$ and $C_n^p(\rho)$ by $\mathscr{A}(0)$, $\mathscr{B}(\rho)$, and $\mathscr{C}(\rho)$ respectively.

(iii) if the series $\sum \delta_j + \sum \varepsilon_j$ diverges, let us further assume that, for some $\nu > 0$,

$$1 < \sum \delta_j e^{-A_j\nu} + \sum \varepsilon_j e^{-B_j\nu} < \infty.$$

Let us denote by λ (obviously $\lambda > \nu$) the unique positive root of the equation $\sum \delta_j e^{-A_j\lambda} + \sum \varepsilon_j e^{-B_j\lambda} = 1$. Then again the assertions of Theorem 5.4.1 remain valid, with the modifications indicated in (ii) above.

Remark. The conditions of (iii) are satisfied for example if $\delta_j = \varepsilon_j = 1$ for all j, $A_j = 2jc$, $B_j = (2j+1)c$ for some $c > 0$.
Proof: (*i*) Let $k = g + h$, so that

$$k(u) = \sum \delta_j k(u + A_j) + \sum \varepsilon_j k(u + B_j). \tag{5.6.6}$$

If $\sum \delta_j + \sum \varepsilon_j \leqq 1$, then (5.6.6) implies that

$$\sum \delta_j[k(u) - k(u + A_j)] + \sum \varepsilon_j[k(u) - k(u + B_j)] \leqq 0$$

but, on the other hand, each summand in each of the series above is nonnegative since k is nonincreasing. Hence each term above must be zero, whence it follows that k is constant and so $= k(+\infty) = 0$, and consequently $g \equiv h \equiv 0$. This proves (i).

For (ii) and (iii): in both these cases, as a consequence of the conditions in (ii), and by assumption in (iii), there exists $\nu > 0$ such that $1 < \sum \gamma_j e^{-C_j \nu} < \infty$, where $C_j = A_j$ if $\gamma_j = \delta_j$ and $C_j = B_j$ if $\gamma_j = \varepsilon_j$, and a unique real number $\lambda > \nu$ exists, satisfying the equation $\sum \gamma_j e^{-C_j \lambda} = 1$. We choose and fix N such that $\sum_1^N \gamma_j = \gamma > 1$; if $C_N^* = \max\{C_j : 1 \leqq j \leqq N\}$, then, as in section 5.4, $k(0) \geqq \gamma^n k(nC_N^*)$ for all positive integers n, whence for all $u \geqq 0$,

$$k(u) \leqq D_1 \exp(-D_2 u); \qquad D_1 > 0,\ D_2 = [\log \gamma]/C_N^* > 0.$$

Consequently $\int_u^\infty k(t)\,dt$ exists for all real u, and the same is obviously true if k is replaced by g or h. As in section 5.4, in view of this circumstance, it suffices to prove Theorem 5.6.2 only for g and h which are also continuous (in addition to satisfying the other conditions of the theorem). Henceforward we shall assume g and h, and hence k also, to be continuous functions.

LEMMA 5.6.1. $\int_0^\infty e^{xu} k(u)\,du < \infty$ for $x < \lambda$, so that the functions

$$\chi_g(z) = \int_0^\infty e^{-zu} g(u)\,du \qquad \text{and} \qquad \chi_h(z) = \int_0^\infty e^{-zu} h(u)\,du$$

are defined and analytic in $\operatorname{Re} z > -\lambda$.
Proof: Setting $k(u) = r(u)e^{-\lambda u}$, so that the function $r(u)$ is also continuous, we have from (5.6.6) that

$$r(u) = \sum_1^\infty \gamma_j e^{-C_j \lambda} r(u + C_j) = \sum_1^\infty p_j r(u + C_j)$$

where $\sum p_j = 1$. For arbitrary but fixed N,

$$r(u) \geqq \sum_1^N p_j r(u + C_j) = \left(\sum_1^N p_j\right) r(u + C(u))$$

where $C \leqq C(u) \leqq C_N^* = \max(C_1, \ldots, C_N)$, by the intermediate value theorem, so that, for a sequence $\{c_m\} \to \infty$ as $m \to \infty$, and depending on N, we have $r(c_m) \leqq r(0) \cdot q_N^{-m}$, where $q_N = \sum_1^N p_j$; it then follows, as in section 5.4, that

$\int_0^\infty e^{xt}k(t)\,dt < \infty$ for all x such that $C(\lambda - x) + \log q_N > 0$. Since N is arbitrary and $q_N \to 1$ as $N \to \infty$, while C is independent of N, our assertion follows.▌

LEMMA 5.6.2. The series $\sum \gamma_j e^{C_j x} \int_0^{C_j} e^{-xu}k(u)\,du$ converges for all real $x < -\nu$, so that

$$\sum \delta_j e^{A_j z} \int_0^{A_j} e^{-zu} g(u)\,du, \qquad \sum \varepsilon_j e^{B_j z} \int_0^{B_j} e^{-zu}\, g(u)\,du$$

as well as the formal sums obtained on replacing g by h, are defined and analytic for $\operatorname{Re} z < -\nu$. Further, all these functions are bounded in every half-plane of the form $\operatorname{Re} z \leqq -\gamma$, where $\gamma > \nu$.

***Proof*:** We fix $x < -\nu$, then choose and fix an α, $0 \leqq \alpha < \lambda$, and θ such that $-x \neq \alpha/(1+\theta) > \nu$. This is possible since $\lambda > \nu$. Then, from Lemma 5.6.1, we have

$$\int_u^{u(1+\theta)} e^{\alpha v} k(v)\,dv \leqq \int_0^\infty e^{\alpha v} k(v)\,dv = D_3(\alpha) < \infty.$$

Since k is nonincreasing, we obtain from the above

$$k(u) \leqq D_4(\alpha, \theta) u^{-1} \exp\left(-\alpha u/(1+\theta)\right) \qquad \text{for all } u > 0.$$

We shall use this estimate for $u > 1$; for $0 < u \leqq 1$, we have in view of the boundedness of k on [0, 1] that

$$k(u) \leqq D_5(\alpha, \theta) \exp\left(-\alpha u/(1+\theta)\right), \qquad 0 \leqq u \leqq 1.$$

Thus, we finally have the estimate

$$k(u) \leqq D_6(\alpha, \theta) \exp\left(-\alpha u/(1+\theta)\right) \qquad \text{for all } u > 0.$$

Recalling that $x + \alpha/(1+\theta) \neq 0$, we obtain

$$\begin{aligned}\sum_1^\infty \gamma_j e^{C_j x} \int_0^{C_j} e^{-xu} k(u)\,du &\leqq D_6(\alpha, \theta) \sum_1^\infty \gamma_j e^{C_j x} \int_0^{C_j} \exp\left[-(x + \alpha/(1+\theta))u\right]\,du \\ &\leqq D_6(\alpha, \theta) \sum_1^\infty \gamma_j [e^{C_j x} - e^{-C_j \alpha/(1+\theta)}]/(x + \alpha/(1+\theta)) \\ &< \infty\end{aligned}$$

since $x < -\nu$ and $\alpha/(1+\theta) > \nu$.

Thus, the basic assertion of the lemma is proved. The others follow immediately from it or from the above estimate.▌

Basing ourselves on (i) the analyticity of χ_g and χ_h for $\operatorname{Re} z > -\lambda$, (ii) that of the functions $1 - \sum \delta_j e^{A_j z} \pm \sum \varepsilon_j e^{B_j z}$ for $\operatorname{Re} z < -\nu$, and (iii) Lemma 5.6.2, we can proceed below as in the proof of Lemma 5.4.2 for $-\lambda < \operatorname{Re} z < -\nu$. In particular, we define for $\operatorname{Re} z < -\nu$, the functions

$$\sigma_1(z) = 1 - \sum_1^\infty \delta_j e^{A_j z} - \sum_1^\infty \varepsilon_j e^{B_j z}$$

$$\sigma_2(z) = 1 - \sum_1^\infty \delta_j e^{A_j z} + \sum_1^\infty \varepsilon_j e^{B_j z}$$

this being defined only if the set of B's is nonempty, and

$$\sigma(z) = \begin{cases} \sigma_1(z) & \text{if the set of } B\text{'s is empty} \\ \sigma_1(z)\sigma_2(z) & \text{if the set of } B\text{'s is nonempty.} \end{cases}$$

Also, let

$$E_g(z) = \sum_1^\infty \delta_j e^{A_j z} \int_0^{A_j} e^{-zu} g(u)\, du + \sum_1^\infty \varepsilon_j e^{B_j z} \int_0^{B_j} e^{-zu} h(u)\, du$$

and $E_h(z)$ the function obtained by interchanging g and h in the above; and

$$K_g(z) = \begin{cases} E_g(z) & \text{if the set of } B\text{'s is empty} \\ \left(1 - \sum_1^\infty \delta_j e^{A_j z}\right) E_g(z) + \left(\sum_1^\infty \varepsilon_j e^{B_j z}\right) E_h(z) & \text{otherwise.} \end{cases}$$

Then the basic relation

$$\chi_g(z) = \frac{-K_g(z)}{\sigma(z)} \qquad \text{for } -\lambda < \operatorname{Re} z < -\nu$$

holds irrespective of whether or not the set of B's is empty. We recall that χ_g is analytic for $\operatorname{Re} z > -\lambda$ and K_g and σ are analytic for $\operatorname{Re} z < -\nu$. Using the complex inversion formula for the Laplace transform, we have for $t > 0$ and any fixed $c > 0$,

$$\begin{aligned} \int_0^t g(u)\, du &= \lim_{T\to\infty} \frac{1}{2\pi i} \int_{c-iT}^{c+iT} \frac{e^{tz}\chi_g(z)}{z}\, dz \\ &= \lim_{T\to\infty} \frac{1}{2\pi i} \int_{a-iT}^{a+iT} \frac{e^{tz}\chi_g(z)}{z}\, dz + \chi_g(0) \end{aligned}$$

where $-\lambda < a < 0$, $\chi_g(0)$ being the residue of the integrand above at the origin. Here we have used the fact that $\chi_g(z)$ is bounded for $\operatorname{Re} z \geqq a$. Thus we have

$$\int_t^\infty g(u)\, du = \lim_{T\to\infty} \frac{1}{2\pi i} \int_{a-iT}^{a+iT} \frac{e^{tz}\chi_g(z)}{z}\, dz, \qquad -\lambda < a < 0.$$

We can now choose and fix a, $-\lambda < a < -\nu$, and proceed as in section 5.4. Then, since $\sigma(z)$ is an analytic almost periodic function in $\operatorname{Re} z < -\nu$, it possesses the properties of the function $\sigma(z)$ in section 5.4. In particular, there exists a sequence $\{T_m\}$, tending to infinity with m, satisfying (5.4.16). Then an obvious analogue of Lemma 5.4.3 holds, and the subsequent argument for the proof of Theorem 5.6.2 is basically the same as in section 5.4. ▌

We now turn to the proof of the infinite divisibility of c.f.'s f satisfying (5.6.1) in the case when $\beta_j \to 0$ as $j \to \infty$. As we have noted earlier, this condition is satisfied if f is a nontrivial c.f. and all the γ_j are positive integers. In this case, the nonvanishing nature of f *follows* from (5.6.1) and need not be formulated as a condition.

THEOREM 5.6.3. Let f be a nontrivial, nonvanishing c.f. satisfying for all real t the relation

$$f(t) = \prod_1^\infty [f(\pm \beta_j t)]^{\gamma_j} \tag{5.6.7}$$

for some fixed sequence of the + and − signs, where $0 < \beta_j < 1$, $\gamma_j > 0$ for all j, and $\beta_j \to 0$ as $j \to \infty$. Then f is an i.d.c.f. Further, if the β_j and γ_j satisfy suitable conditions (in fact, one of the conditions indicated in Theorem 5.6.2), then assertions similar to those in Theorem 5.4.2 are valid for the Levy functions M and N in the Levy representation for $\log f$.

Proof: We shall prove the infinite divisibility of f in the case when all the β_j occur with the positive sign in (5.6.7); a similar argument, with obvious necessary modifications, holds in the general case. We set

$$\phi_n(t) = \prod_{n+1}^\infty [f(\beta_j t)]^{\gamma_j} \tag{5.6.8}$$

so that

$$f(t) = \prod_1^n [f(\beta_j t)]^{\gamma_j} \phi_n(t). \tag{5.6.9}$$

Iterating this relation n times, we obtain

$$f(t) = \xi_n(t)\psi_n(t) \tag{5.6.10}$$

where

$$\xi_n(t) = \prod_{j_1 + \cdots + j_n = n} f(\beta_1^{j_1} \cdots \beta_n^{j_n} t)^{(n; j_1, \ldots, j_n)\gamma_1^{j_1} \cdots \gamma_n^{j_n}} \tag{5.6.11}$$

and

$$\psi_n^{(t)} = \phi_n(t)\{[\phi_n(\beta_1 t)]^{\gamma_1} \cdots [\phi_n(\beta_n t)]^{\gamma_n}\} \times \cdots$$
$$\times \{[\phi_n(\beta_1^{n-1} t)]^{\gamma_1{}^{n-1}} \cdots [\phi_n(\beta_n^{n-1} t)]^{\gamma_n{}^{n-1}}\} \tag{5.6.12}$$

where $(n; j_1, \ldots, j_n) = n!/j_1! \cdots j_n!$. Hence

$$\log \psi_n(t) = \lim_{M \to \infty} \sum_{n+1}^M \gamma_j \Bigg\{ \log f(\beta_j t)$$
$$+ \sum_{r=1}^{n-1} [\textstyle\sum_r^* (r; k_1, \ldots, k_n) \gamma_1^{k_1} \cdots \gamma_n^{k_n} \log f(\beta_j \beta_1^{k_1} \cdots \beta_n^{k_n} t)] \Bigg\}$$

where the summation in $\sum_r^*$ extends over all nonnegative integer n-tuples $(k_1, \ldots, k_n)$ such that $\sum_1^n k_i = r$. Further,

$$\log \psi_n(t) = \lim_{M \to \infty} \sum_{j=n+1}^M \{\textstyle\sum_j \gamma_{j;k_1,\ldots,k_n} \log f(\beta_{j;k_1,\ldots,k_n} t)\} \tag{5.6.13}$$

where

$$\beta_{j;k_1,\ldots,k_n} = \beta_j \beta_1^{k_1} \cdots \beta_n^{k_n},$$
$$\gamma_{j;k_1,\ldots,k_n} = (k_1 + \cdots + k_n; k_1, \ldots, k_n)\gamma_j \gamma_1^{k_1} \cdots \gamma_n^{k_n}$$

and, for fixed j, the summation in $\sum_j$ extends over all nonnegative integer n-tuples $(k_1, \ldots, k_n)$ such that $0 \leqq \sum k_i \leqq n-1$. For the sake of convenience, we omit the rather awkward suffixes; and rewrite (5.6.13) simply as

$$\log \psi_n(t) = \lim_{M\to\infty} \sum_{j=n+1}^{M} [\sum_j \gamma_* \log f_*(t)] \tag{5.6.14}$$

where $f_*(t) = f(\beta_* t)$.

From the condition $\beta_j \to 0$ as $j \to \infty$, it follows that $0 < \max_j \beta_j < 1$, so that $\beta_* \to 0$ uniformly with respect to the index $*$ as $n \to \infty$. Thus, the c.f.'s f_* satisfy the condition of "uniform asymptotic negligibility:" for any fixed $T > 0$,

$$\lim_{n\to\infty} \max_{|t|\leqq T;*} |f_*(t) - 1| = 0. \tag{5.6.15}$$

We fix $\tau > 0$; let F_* denote the d.f. corresponding to f_*. We set

$$\alpha_* = \int_{|x|<\tau} x dF_*(x), \qquad \tilde{F}_*(x) = F(x + \alpha_*), \qquad \tilde{f}_* \text{ is the c.f. of } \tilde{F}_*. \tag{5.6.16}$$

We assert that the functions $\eta_n(t)$ given by

$$\log \eta_n(t) = \sum_{n+1}^{\infty} \sum_j \gamma_* \left\{ i\alpha_* t + \int (e^{itu} - 1)\, d\tilde{F}_*(u) \right\} \tag{5.6.17}$$

are i.d.c.f.'s, "accompanying" the $\psi_n(t)$; in fact, we shall show that $\log \psi_n(t) - \log \eta_n(t) \to 0$ as $n \to \infty$, uniformly in every interval $|t| \leqq T$.

Since, as $n \to \infty$, $\beta_* \to 0$ uniformly with respect to the suffix $*$, we have the usual Central Limit Theorem estimates (cf. the "Central Inequalities" in [91], p. 304): for any given $\varepsilon > 0$, there exists a constant $c(T, \tau)$ such that for all $|t| \leqq T$ and $M > n \geqq N = N(T, \tau)$,

$$\max_{M,*,|t|\leqq T} |\tilde{f}_*(t) - 1| < \varepsilon$$

and

$$\begin{aligned}\sum_{j=n+1}^{M} \sum_j \gamma_* |f_*(t) - 1|^2 &\leqq c(T, \tau) \int_0^T \left\{ \sum_{j=n+1}^{M} \sum_j \gamma_* |\log|f_*(t)|\,| \right\} dt \\ &\leqq c(T, \tau) \int_0^T |\log|\psi_n(t)|\,|\, dt\end{aligned}$$

so that

$$\begin{aligned}\sum_{j=n+1}^{M} \sum_j \gamma_* |\log \tilde{f}_*(t) + 1 - \tilde{f}_*(t)| &\leqq \sum_{j=n+1}^{M} \sum_j \gamma_* |f_*(t) - 1|^2 \\ &\leqq \varepsilon c(T, \tau) \int_0^T \Big|\log|\psi_n(t)|\Big|\, dt.\end{aligned}$$

We now fix $T \geqq 1$ and $n \geqq N(T, \tau)$. Then the series

$$\sum_{j=n+1}^{\infty} \sum_j \gamma_*[\log \tilde{f}_*(t) + 1 - \tilde{f}_*(t)]$$

converges boundedly for all t in $|t| \leqq T$, and since $\log \tilde{f}_*(t) = \log f_*(t) - i\alpha_* t$, it follows from the above and (5.6.14) that the series

$$\sum_{j=n+1}^{\infty} \sum_j \gamma_*[- i\alpha_* t + 1 - \tilde{f}_*(t)]$$

and consequently the series

$$\sum_{j=n+1}^{\infty} \sum_j \gamma_* \int (1 - \cos tx)\, d\tilde{F}_*(x)$$

are boundedly convergent. Let

$$\tilde{G}_*(x) = \int_{(-\infty, x]} \frac{u^2}{1 + u^2}\, d\tilde{F}_*(u)$$

so that the series

$$\sum_{j=n+1}^{\infty} \sum_j \gamma_* \int (1 - \cos tx) \frac{1 + x^2}{x^2}\, d\tilde{G}_*(x)$$

converges for all t. Integrating this series termwise with respect to t over the interval (0, 1), we obtain

$$\sum_{j=n+1}^{\infty} \sum_j \gamma_* \int \left(1 - \frac{\sin x}{x}\right) \frac{1 + x^2}{x^2}\, d\tilde{G}_*(x) < \infty$$

whence, using the fact that, for some $c > 0$ and all x, $(x - \sin x)(1 + x^2)x^3 \geqq c$ we deduce that

$$\sum_{j=n+1}^{\infty} \sum_j \gamma_* \tilde{G}_*(+\infty) < \infty.$$

We set $H_n(x) = \sum_{j=n+1}^{\infty} \sum_j \gamma_* \tilde{G}_*(x)$; then,

$$\sum_{j=n+1}^{\infty} \sum_j \gamma_*\{\log \tilde{f}_*(t) + 1 - \tilde{f}_*(t)\}$$

$$= \sum_{j=n+1}^{\infty} \sum_j \gamma_* \left\{\log f_*(t) - i\alpha_* t + \int (1 - e^{itx})\, d\tilde{F}_*(x)\right\}$$

$$= \sum_{j=n+1}^{\infty} \sum_j \gamma_* \left\{\log f_*(t) - i(\alpha_* + \theta_*)t \right.$$

$$\left. - \int \left(e^{itx} - 1 - \frac{itx}{1 + x^2}\right) \frac{1 + x^2}{x^2}\, d\tilde{G}_*(x)\right\}$$

where $\theta_* = \int [x/(1+x^2)]\, d\tilde{F}_*(x)$, so that it follows from the convergence of the series on the left above that the series $\sum_{j=n+1}^{\infty} \sum_j \gamma_*(\alpha_* + \theta_*)$ converges; denoting its sum by c_n, we then have for all $n \geqq N(T, \tau)$ and $|t| \leqq T$

$$\left| \log \psi_n(t) - ic_n t - \int \left(e^{itx} - 1 - \frac{itx}{1+x^2}\right) \frac{1+x^2}{x^2}\, dH_n(x) \right|$$
$$\leqq \varepsilon c(T, \tau) \int_0^T |\log|\psi_n(t)||\, dt. \quad (5.6.18)$$

Rewriting (5.6.11) in the form $\log \xi_n(t) = \sum \gamma_* \log f_*(t)$—a finite sum depending on n—and using a notation similar to (5.6.16), we obtain for $|t| \leqq T$ and $n \geqq N' = N'(T, \tau)$

$$\left| \log\ \xi_n(t) - it \sum \gamma_* \alpha_* - \sum \gamma_* \int (e^{itu} - 1)\, d\tilde{F}_*(u) \right|$$
$$\leqq \varepsilon c(T, \tau) \int_0^T |\log|\xi_n(t)||\, dt \quad (5.6.19)$$

so that, for $n \geqq \max(N, N')$ and $|t| \leqq T$, the sum of the right-hand members of (5.6.18) and (5.6.19) does not exceed

$$\varepsilon c(T, \tau) \int_0^T |\log|f(t)||\, dt.$$

This shows that there exists a sequence of i.d. laws, accompanying the laws with c.f. $\xi_n(t)\psi_n(t)$. Thus, f is an i.d.c.f. and Theorem 5.6.3 stands proved.∎

If then $L(\mu, \sigma^2, M, N)$ is the Levy representation for $\log f$ and $\{a_n\}$ the subsequence of β_j consisting of the β_j with positive sign (if such exist), and $\{b_n\}$ the subsequence of the β_j with negative sign (if such exist), and δ_n and ε_n the corresponding subsequences of γ_j, then we have

$$\sigma^2(1 - \sum \gamma_j \beta_j^2) = 0,$$

$$M(u) = \sum \delta_j M\left(\frac{u}{a_j}\right) - \sum \varepsilon_j N\left(\frac{-u}{b_j}\right) \qquad \text{for } u < 0,$$

$$N(u) = \sum \delta_j N\left(\frac{u}{a_j}\right) - \sum \varepsilon_j M\left(\frac{-u}{b_j}\right) \qquad \text{for } u > 0,$$

where M and N are assumed (without loss of generality) to be respectively left- and right-continuous on $(-\infty, 0)$ and $(0, \infty)$. Setting $g(u) = -N(e^u)$ and $h(u) = M(-e^u)$, $A_j = -\log a_j$, $B_j = -\log b_j$, we obtain relations of the form (5.6.5). From the fact that $\beta_j \to 0$ as $j \to \infty$, it follows that $\max \beta_j < 1$ so that the conditions $A_j \geqq A_0 > 0$, $B_j \geqq B_0 > 0$ are fulfilled for all j. Hence Theorem 5.6.2 is applicable to our situation and we obtain the further information on the functions M and N, similar to what is contained in Theorem 5.4.2. We shall not pause to formulate this precisely; it is related to Theorem 5.4.2 in the same way as Theorem 5.6.2 is to Theorem 5.4.1.

5.6.2 Applications to Regression Problems

In this section, we shall indicate cases where it is possible, with the help of the results presented above, to obtain solutions of the regression equation

$$E\left(\sum_1^\infty a_j X_j \,\middle|\, \sum_1^\infty b_j X_j\right) = 0 \tag{5.6.20}$$

where the X_j are independent and identically distributed r.v.'s with $EX_1 = 0$.

THEOREM 5.6.4. Let $X_1, X_2, \ldots$ be i.i.d.r.v.'s with $EX_1 = 0$, which are nondegenerate and satisfy (5.6.20). Let $\{a_j\}$ and $\{b_j\}$ be sequences of real numbers such that $\sum |a_j| < \infty$ and the series $\sum b_j X_j$ converges with probability one, and further,

(i) $a_1 \neq 0$, $|b_j| < |b_1|$ for all $j > 1$; and

(ii) $a_j b_1 / a_1 b_j < 0$ for all $j > 1$ such that $a_j b_j \neq 0$.

Then the X_j follow an i.d. law. It is possible to formulate more precise statements on the Levy functions in the Levy representation for the logarithm of the c.f. of the X_j, if $\{a_j\}$ and $\{b_j\}$ satisfy the conditions of one of the Theorems 5.6.1–5.6.3. In particular, under the above conditions, the X_j are normally distributed if and only if $\sum a_j b_j = 0$.

***Proof*:** Since, by assumption, $\sum |a_j| < \infty$ and $E|X_j| < \infty$, $\sum E|a_j X_j| < \infty$ and the series $\sum a_j X_j$ therefore converges with probability one (cf. for instance [14], p. 106). For the same reasons, we have

$$\begin{aligned} 0 &= E(\textstyle\sum a_j X_j \exp(it \sum b_j X_j)) \\ &= \textstyle\sum E(a_j X_j \exp(it \sum b_j X_j)) \\ &= \textstyle\sum a_j [f'(b_j t) \prod_{k \neq j} f(b_k t)] \end{aligned} \tag{5.6.21}$$

where f is the c.f. of the X_j.

In view of the convergence with probability one of the series $\sum b_j X_j$, $\prod_1^\infty f(b_j t)$ represents a c.f., so that $\prod_{n+1}^\infty f(b_k t) \to 1$ as $n \to \infty$, and this convergence is uniform in every finite t-interval, so that $\prod_1^\infty f(b_k t) = 0$ only if $f(b_k t) = 0$ for some k. If now I be the largest (open) interval containing the origin in which f does not vanish, $I = \{t\colon |t| < \delta, f(t) \neq 0\}$, then, dividing both sides of (5.6.21) by $\prod_1^\infty f(b_k t)$, we obtain

$$\sum_1^\infty a_j \frac{f'(b_j t)}{f(b_j t)} = 0, \qquad t \in I. \tag{5.6.22}$$

For every fixed compact subinterval of I there exists a corresponding $\varepsilon > 0$ such that $|f(b_j t)| \geqq \varepsilon$ for all j; also, $|f'(b_j t)| \leqq E|X_j|$ and $\sum |a_j| < \infty$, so that we may integrate the left member of (5.6.22) term by term over $[0, t]$ for $|t| < \delta$. Then we have

$$\sum (a_j/b_j) \log f(b_j t) = \text{constant} = 0, \qquad t \in I$$

the summation being taken over all j for which $a_j b_j \neq 0$. The above relation can be rewritten as

$$f(t) = \prod_1^\infty [f(\beta_j t)]^{\gamma_j}, \qquad t \in I \tag{5.6.23}$$

where $|\beta_j| < 1$ and $\gamma_j > 0$ for all j ($\beta_j = b_{j+1}/b_1$).

We now assert that f is nonvanishing and that (5.6.23) is satisfied for all real t. Suppose it is not; i.e., let $\delta < \infty$. The convergence of $\sum b_j X_j$ and Theorem 5.6.1 imply that $\sum b_j^2 < \infty$, i.e., that $\sum \beta_j^2 < \infty$. In particular, $\beta_j \to 0$ as $j \to \infty$. Therefore, we may speak of $\max_j |\beta_j|$; without loss of generality, let it be $|\beta_1|$, and let $\delta_1 = \delta/|\beta_1|$, so that

$$f(\beta_j t) \neq 0 \qquad \text{for any } j, \qquad \text{if } |t| < \delta_1. \tag{5.6.24}$$

Let now $g = |f|^2$, so that $g(t) = \prod_1^\infty [g(\beta_j t)]^{\gamma_j}$ for $|t| < \delta_1$. We note that g is a real-valued function with $g(-t) = g(t)$, and in particular it satisfies the elementary inequality valid for such c.f.'s

$$1 - g(2t) \leqq 4[1 - g(t)]. \tag{5.6.25}$$

LEMMA 5.6.3. If the infinite product $\prod_1^\infty [g(\beta_j t)]^{\gamma_j}$ converges uniformly for $|t| < \varepsilon$, and $g(\beta_j t) \neq 0$ for all j in $|t| < 2\varepsilon$, then the product converges uniformly in $|t| < 2\varepsilon$.

***Proof*:** Since $\theta \leqq -\log(1 - \theta)$ for $0 \leqq \theta < 1$ and (5.6.25) holds, the uniformity of convergence of the series $\sum \gamma_j \log g(\beta_j t)$ for $|t| < \varepsilon$ implies that of $\sum \gamma_j [1 - g(2\beta_j t)]$ for such t. Since $g(2\beta_j t) \neq 0$ for $|t| < \varepsilon$, $\log g(2\beta_j t)$ is defined for such t; it follows from the fact that $\beta_j \to 0$ as $j \to \infty$ that $g(2\beta_j t) \geqq 1/2$ for all sufficiently large j uniformly for $|t| < \varepsilon$, i.e., $J(\varepsilon)$ exists such that $g(2\beta_j t) \geqq 1/2$ for $|t| < \varepsilon$, if $j \geqq J(\varepsilon)$. For such j and t, we have

$$-\log g(2\beta_j t) \leqq [1 - g(2\beta_j t)]/g(2\beta_j t) \leqq 2[1 - g(2\beta_j t)]$$

so that the series $\sum \gamma_j \log g(2\beta_j t)$ converges uniformly for $|t| < \varepsilon$ and its sum function is therefore continuous. Thus, $\prod_1^\infty [g(\beta_j t)]^{\gamma_j}$ converges uniformly for $|t| < 2\varepsilon$ to a continuous function.∎

Now, from the convergence of $\sum_1^\infty [g(\beta_i t)]^{\gamma_j}$ for $|t| < \delta$ to the continuous function $g(t)$, we conclude that the convergence is uniform in every compact interval $|t| \leqq \varepsilon < \delta$; this follows from the well-known fact that the convergence of a series consisting of nonnegative continuous functions, on a compact interval, to a continuous function, is equivalent to the uniformity of convergence of that series on that interval (cf. for instance [147]). Using the relation (5.6.24) and Lemma 5.6.3, we obtain

$$0 = g(\delta) = \lim_{t \uparrow \delta} g(t) = \lim_{t \uparrow \delta} \prod_1^\infty [g(\beta_j t)]^{\gamma_j} = \prod_1^\infty [g(\beta_j \delta)]^{\gamma_j} \neq 0.$$

This contradiction shows that g and hence f does not vanish anywhere; then as we have seen above, (5.6.23) is valid for all real t, and our theorem stands proved.∎

The particular assertion concerning conditions for the normality of the X_j follows from Theorem 5.6.1.

A well-known theorem due to Marcinkiewicz ([101]; see also the third paragraph of section 2.1 relates the identity of distribution of two linear forms in i.i.d.r.v.'s with the normality of the r.v.'s. Some generalizations thereof permit us to formulate conditions under which the relation (5.6.20) guarantees the normality of the X_j.

THEOREM 5.6.5. Let $X_1, X_2, \ldots$ be i.i.d.r.v.'s, nondegenerate and having moments of all orders, with $EX_1 = 0$. Suppose that the sequences $\{a_j\}$ and $\{b_j\}$ of real constants are such that $\sum |a_j|$ converges, $\sum b_j X_j$ converges with probability one, and (5.6.20) holds. Then the X_j are normal.

Remark. The r.v.'s X_j are not necessarily normally distributed if the condition that moments of all orders exist is not satisfied, even if all the other conditions are. This was proved by Linnik [81] in respect of Marcinkiewicz' original theorem.

***Proof*:** Proceeding as in the proof of Theorem 5.6.4, we arrive at the relation

$$\sum a_j \psi'(b_j t) = 0, \qquad t \in I, \qquad \psi = \log f \tag{5.6.26}$$

from which follows the equation

$$\sum a_j \theta'(\beta_j t) = 0, \qquad t \in I, \qquad \beta_j = |b_j| \tag{5.6.27}$$

where $\theta(t) = \psi(t) + \psi(-t)$. Since the series $\sum b_j X_j$ converges (the X_j are nondegenerate and i.i.d. as well), we have $\sum b_j^2 = \sum \beta_j^2 < \infty$, so that $\beta_j \to 0$ as $j \to \infty$. Further, $\sum |a_j| \beta_j^2 < \infty$ in view of the condition $\sum |a_j| < \infty$. Thus the series $\sum_1 a_j \beta_j^2$ and $\sum_2 a_j \beta_j^2$, where $\sum_1$ and $\sum_2$ denote summations over the positive and the negative a_j respectively, are convergent. Let $\beta^0 = \max_j \beta_j$. Then we may assume without loss of generality that only one term of the form $a^0 \theta'(\beta^0 t)$ occurs in (5.6.27). We then have from (5.6.27) that

$$\begin{aligned} \textstyle\sum_1 a_j \theta'(\beta_j t) &= -\textstyle\sum_2 a_j \theta'(\beta_j t) \\ &= \textstyle\sum_2 a_j^* \theta'(\beta_j^* t) \end{aligned} \tag{5.6.28}$$

where the a_j and a_j^* are all positive, and $\max \beta_j \neq \max \beta_j^*$. We differentiate (5.6.28) termwise $(2s-1)$ times and set $t=0$:

$$\left(\sum_1 a_j \beta_j^{2s} - \sum_2 a_j^* \beta_j^{*2s}\right)\kappa_{2s} = 0 \tag{5.6.29}$$

where κ_{2s} is the semi-invariant of order $2s$ for the c.f. f. We note that the series $\sum_1 a_j \beta_j^{2s}$ and $\sum_2 a_j^* \beta_j^{*2s}$ converge for $s > 1$, since they converge for $s = 1$. The relation $\sum_1 a_j \beta_j^{2s} = \sum_2 a_j^* \beta_j^{*2s}$ can hold only for a finite number of integral values s, since, otherwise, proceeding to the limit as $s \to \infty$ through a sequence of such values, we would have, by Lemma 1.7.2 that $\max \beta_j = \max \beta_j^*$, contrary to assumption. Thus, $\kappa_{2s} = 0$ for all sufficiently large s, and since $\theta(t)$ is a real-valued c.f., $\kappa_{2s+1} = 0$ for all s. Thus, the semi-invariants are zero for all large s, say for $s > s_0$. Thus $\theta(t)$ is a polynomial of degree s_0 at most. Then the usual argument (Marcinkiewicz' theorem) shows that the degree is two (the X_j are nondegenerate) and f is a normal c.f.▮

5.7 CONSTANCY OF REGRESSION AND HOMOSCEDASTICITY OF ONE LINEAR STATISTIC ON ANOTHER

In sections 5.4–5.6, we studied regression problems involving linear statistics in i.i.d.r.v.'s. Now we consider certain problems not involving the condition of identical distribution. The following theorem is due to Lukacs and Laha [95].

THEOREM 5.7.1. Let $X_1, \ldots, X_n$ be independent r.v.'s with finite variance. The linear functions $L = \sum a_j X_j$ and $M = \sum b_j X_j$ with $a_j b_j \neq 0$ for $j = 1, \ldots, n$ satisfy the relations

(i) $E(L \mid M) = \alpha + \beta M$, and
(ii) $\operatorname{Var}(L \mid M) = \sigma_0^2 = \text{constant}$

if and only if the following conditions are satisfied:

(a) the X_j for which $b_j \neq \beta a_j$ are normal, and

(b) $\beta = \left(\sum^* a_j b_j \sigma_j^2\right) / \left(\sum^* a_j^2 \sigma_j^2\right)$, $\sigma_0^2 = \sum^* (b_j - \beta a_j)^2 \sigma_j^2$,

where $\sigma_j^2 = \operatorname{Var} X_j$, and $\sum^*$ denotes that the summation is taken over all j for which $b_j \neq \beta a_j$.

Proof: We first prove the necessity. Let f_j be the c.f. of X_j, and $\theta_j = \log f_j$ (in a neighborhood of the origin where none of the f_j vanishes). By Lemma 1.1.3, we have

$$\sum b_j \theta_j'(a_j t) = i\alpha + \beta \sum a_j \theta_j'(a_j t) \tag{5.7.1}$$

$$\sum b_j^2 \theta_j''(a_j t) = -\sigma_0^2 + \beta^2 \sum a_j^2 \theta_j''(a_j t). \tag{5.7.2}$$

Differentiating (5.7.1) with respect to t, we obtain

$$\sum a_j b_j \theta_j''(a_j t) = \beta \sum a_j^2 \theta_j''(a_j t). \tag{5.7.3}$$

From (5.7.2) and (5.7.3), we derive

$$\sum (b_j - \beta a_j)^2 \theta_j''(a_j t) = \sum{}^* (b_j - \beta a_j)^2 \theta_j''(a_j t) = \sigma_0^2. \tag{5.7.4}$$

Integrating (5.7.4), we obtain

$$\prod{}^* [f_j(a_j t)]^{\gamma_j} = \exp\,[i\mu t - (1/2)\sigma_0^2 t^2] \tag{5.7.5}$$

where $\gamma_j = (b_j - \beta a_j)^2$. Assertion (a) now follows on noting that in Lemma 1.2.1, if (1.2.1) holds with $\phi(t) = \exp\,[i\mu t - (1/2)\sigma^2 t^2]$, then every ϕ_i is a normal c.f. The other results are obtained by setting $t = 0$ in (5.7.3) and (5.7.4).

The sufficiency of the conditions is easily established.▌

We generalize Theorem 5.7.1 to linear functions in a denumerable number of random variables.

THEOREM 5.7.2. Let $X_1, X_2, \ldots$ be a sequence of independent r.v.'s with $\text{Var}\, X_j = \sigma_j^2$. Consider two linear forms $L = \sum a_j X_j$ and $M = \sum b_j X_j$ where $a_j b_j \neq 0$ for any j, $\sum |a_j| \sigma_j < \infty$ and the series $\sum b_j X_j$ converges with probability one. The relations

(i) $E(L|M) = \alpha + \beta M$, and
(ii) $\text{Var}\,(L|M) = \sigma_0^2$

are satisfied if and only if

(a) the X_j for which $b_j \neq \beta a_j$ are normal, and

(b) $\beta = (\sum{}^* a_j b_j \sigma_j^2)/(\sum{}^* a_j^2 \sigma_j^2)$; $\sigma_0^2 = \sum{}^* (b_j - \beta a_j)^2 \sigma_j^2$.

Proof: We proceed according to the same scheme as for the preceding theorem. We note that the condition $\sum |a_j| \sigma_j < \infty$ guarantees the convergence with probability one of the series $\sum a_j X_j$ and the validity of relations (5.7.1)–(5.7.5) for infinite sums.▌

CHAPTER 6

Regression Problems for Nonlinear Statistics

6.1 PRELIMINARY LEMMAS

In Chapter 4, we considered characterizations of distributions through the property of independence of two statistics, at least one of them being nonlinear. Thus, for instance, if $X_1, \ldots, X_n$ is a random sample, then the independence of $\bar{X}$ and $S^2 = (1/n) \sum_1^n (X_i - \bar{X})^2$ (without any supplementary condition on the moments of the X_i) yields the normality of the X_i. We can use the constancy of regression of S^2 on $\bar{X}$, instead of their independence of each other, to establish the same result. On the one hand, this condition is weaker than that of the independence of $\bar{X}$ and S^2, but, on the other, presupposes the existence of the variance of the X_i. Similarly, if X and Y are independent positive r.v.'s, then the independence of $X + Y$ and X/Y turns out to be a characteristic property of the gamma distribution (possibly degenerate). In this chapter, we shall show (*inter alia*) that for i.i.d.r.v.'s X and Y, even the constancy of regression of $X + Y$ on X/Y turns out to be a characteristic property of the gamma distribution (possibly degenerate).

We shall touch on only some of the results available in this area, which will exhibit the nature of the problems and the mathematical apparatus necessary for their solution. The greater part of the results presented below pertain to Khatri and Rao [73, 74].

LEMMA 6.1.1. Let F be the d.f. of the r.v. Y, and g a continuous function on R_1, such that

$$\int g \, dF = \mu \text{ (finite)}. \tag{6.1.1}$$

If, for some real $\rho \neq 0$ and all real t,

$$\int e^{ity} g(y) \, dF(y) = (\mu + i\rho t) \int e^{ity} \, dF(y) \tag{6.1.2}$$

then F is absolutely continuous and has a continuous version f of its density which satisfies the differential equation

$$\rho f'(y) = [\mu - g(y)]f(y) \tag{6.1.3}$$

whose solution is

$$f(y) = \exp\left[-\frac{1}{\rho}\int_a^y [g(y) - \mu]\, dy\right]. \tag{6.1.4}$$

Proof: We set $k(y) = [g(y) - \mu]/\rho$; then (6.1.2) takes the form

$$\int e^{ity}k(y)\, dF(y) = it\int e^{ity}\, dF(y). \tag{6.1.5}$$

Multiplying both sides of (6.1.5) by

$$\frac{1}{2\pi}\frac{1 - e^{-iht}}{it}\frac{e^{-i\alpha t} - e^{-i\beta t}}{it}$$

integrating over the interval $(-T, T)$ with respect to t and then letting $T \to \infty$, we find on using the inversion formula for c.f.'s that the right-hand side of the resulting relation can be given in the form

$$[F(\alpha + h) - F(\alpha)] - [F(\beta + h) - F(\beta)] \tag{6.1.6}$$

if α, $\alpha + h$, β, $\beta + h$ are all continuity points of F. The left-hand side equals

$$\begin{aligned}
&\lim_{T\to\infty}\frac{1}{2\pi}\int_{-T}^{T}\frac{1 - e^{-iht}}{it}\,dt\int_{\alpha}^{\beta} e^{-itu}\,du\int_{-\infty}^{\infty} e^{ity}k(y)\,dF(y)\\
&\qquad = \lim_{T\to\infty}\frac{1}{2\pi}\int_{\alpha}^{\beta} du\left(\int_{-T}^{T}\left[\frac{1 - e^{-iht}}{it}e^{-itu}\int_{-\infty}^{\infty} e^{ity}k(y)\,dF(y)\right]dt\right)\\
&\qquad = \lim_{T\to\infty}\frac{1}{2\pi}\int_{\alpha}^{\beta} du\int_{-\infty}^{\infty} k(y)\,dF(y)\int_{-T}^{T}\frac{1 - e^{-iht}}{it}e^{-it(u-y)}\,dt.
\end{aligned} \tag{6.1.7}$$

Using standard arguments relating to the proof of the inversion theorem, we conclude that

$$\int_{-T}^{T}\frac{1 - e^{-ith}}{it}e^{-it(u-y)}\,dt \tag{6.1.8}$$

is bounded uniformly in T. Hence, passing to the limit in (6.1.7), we obtain

$$\int_{\alpha}^{\beta} f(u, h)\, du \tag{6.1.9}$$

where

$$f(u, h) = \int_{u-0}^{u+h-0} k(y)\, dF(y)$$

(We need not restrict ourselves to continuity points u, $u+h$ of F.) Equating (6.1.6) and (6.1.9), we have

$$\int_{\alpha}^{\beta} f(u, h)\, du = F(\alpha + h) - F(\beta + h) + F(\beta) - F(\alpha). \tag{6.1.10}$$

Letting $h \to -\infty$, we obtain

$$\int_{\alpha}^{\beta} f(u)\, du = F(\beta) - F(\alpha) = \int_{\alpha}^{\beta} dF(y) \tag{6.1.11}$$

at all points of continuity α, β of F, where

$$f(u) = -\int_{-\infty}^{u-0} k(y)\, dF(y). \tag{6.1.12}$$

Equation (6.1.11) shows that F is absolutely continuous with density f. Further,

$$f(u) = -\int_{-\infty}^{u-0} k(y)\, dF(y) = -\int_{-\infty}^{u-0} k(y) f(y)\, dy. \tag{6.1.13}$$

k being continuous, it follows from the above that f is continuous and differentiable. Differentiating both sides of the relation (6.1.13), we obtain

$$f'(u) = -k(u) f(u)$$

or

$$f(u) = \exp\left[-\int_{a}^{u} k(v)\, dv\right] \tag{6.1.14}$$

proving the lemma.▌

Corollary. If, in Lemma 6.1.1, $g(y) = e^y$, then ρ and μ must be positive, and

$$f(y) = \exp\left[-\frac{1}{\rho}(e^y - \mu y - c)\right] \tag{6.1.15}$$

where c is some constant. If Y is an r.v. with density f, then the density of $X = e^Y$ equals

$$\frac{\alpha^{\gamma}}{\Gamma(\gamma)} e^{-\alpha x} x^{\gamma - 1} \tag{6.1.16}$$

where α and γ depend on ρ and μ; thus X has a gamma distribution $G(\alpha, \gamma)$.

LEMMA 6.1.2. Let F be the d.f. of an r.v. Y such that $Ee^Y < \infty$. Suppose

$$\int e^y e^{ity}\, dF(y) = (c + \gamma t) \int e^{ity}\, dF(y) \tag{6.1.17}$$

for $|t| \leqq \varepsilon$. Then (6.1.17) holds for all real t and the functions on either side of the relation are analytic in the strip $-1 < \operatorname{Im} t < 0$.

***Proof*:** Adding to (6.1.17) the relation obtained from it by replacing t by $-t$, we have

$$\int (e^y - c)\cos ty\, dF(y) = i\gamma t \int \sin ty\, dF(y) \tag{6.1.18}$$

whence $\gamma = i\rho$, where ρ is real. Thus, (6.1.17) can be rewritten in the form

$$\int e^y e^{ity}\, dF(y) = (c + i\rho t) \int e^{ity}\, dF(y) \tag{6.1.19}$$

for $|t| \leqq \varepsilon$. Since Ee^Y exists, Ee^{uY} exists for $0 \leqq u \leqq 1$. Hence Ee^{izY} is analytic in the strip $-1 < \operatorname{Im} z < 0$; the function $1 + i\rho z/c$ is also analytic there, and hence

$$g(z) = (1 + i\rho z/c) \int e^{izy}\, dF(y) \tag{6.1.20}$$

is analytic there. Further, $g(t) = \int e^{ity}\, dH(y)$ for $|t| < \varepsilon$, where $dH(y) = (e^y/c)\, dF(y)$. Invoking Theorem 3.2.1 of [121], p. 37, we obtain the assertion of the lemma.▮

LEMMA 6.1.3. Let F be the d.f. of the r.v. X. If $E(1/X)$ exists and the relation

$$\int e^{itx} x^{-1}\, dF(x) = (\mu + i\rho t) \int e^{itx}\, dF(x) \tag{6.1.21}$$

is valid for some $\rho \neq 0$, $\mu \neq 0$, and $|t| < \delta$, then $\rho < 0$, and X has a gamma distribution $G(\alpha, \gamma)$ with $\gamma = 1 - (1/\rho)$ and $\alpha = -\mu/\rho$.
***Proof*:** Since $E(1/X)$ exists, we have that if

$$A(t) = \int (e^{itx}/x)\, dF(x) \tag{6.1.22}$$

then

$$A'(t) = \int ie^{itx}\, dF(x) = i\phi(t). \tag{6.1.23}$$

We have from (6.1.21) that, in some interval around the origin,

$$A'(t)/A(t) = i/(\mu + i\rho t)$$

or

$$A(t) = \mu(1 + i\rho t/\mu)^{1/\rho} \tag{6.1.24}$$

whence

$$\phi(t) = (1 + i\rho t/\mu)^{(1-\rho)/\rho}. \tag{6.1.25}$$

The condition that $|\phi(t)| \leqq 1$ implies that $(1-\rho)/\rho \leqq 0$. Let $\gamma = (\rho - 1)/\rho$ and $\alpha = -\mu/\rho$. Then

$$\phi(t) = (1 - it/\alpha)^{-\gamma} \qquad \text{for } |t| < \varepsilon \tag{6.1.26}$$

for some $\varepsilon > 0$. It follows from the theory of analytic c.f.'s that (6.1.26) must hold for all real t. The boundedness of $A(t)$, equivalently the existence of $E(1/X)$, implies that $\rho < 0\,(\rho \neq 0$ by assumption) and hence $\gamma > 1$. Hence the lemma.▮

6.2 CHARACTERIZATIONS OF THE GAMMA DISTRIBUTION

In this section, we consider some characterizations of the gamma distribution through the property of constancy of regression of one statistic on another. In order to avoid complicated statements of the theorems, we assume that all the r.v.'s concerned are nondegenerate.

THEOREM 6.2.1. Let $X_1, \ldots, X_n$, $n \geqq 3$, be independent, positive, and in general not identically distributed r.v.'s with EX_j finite for all j. If the regression of $\sum_1^n X_j$ on the vector $(X_2/X_1, \ldots, X_n/X_1)$ be constant, then the X_j have gamma distributions; more precisely, $X_j \sim G(\alpha, \gamma_j)$ for $j = 1, \ldots, n$, where the parameter α is the same for all the X_j.

***Proof*:** We set $Y_j = \log X_j$, and rewrite the condition of constancy of regression in the form

$$E(e^{Y_1} + \cdots + e^{Y_n} | Y_2 - Y_1, \ldots, Y_n - Y_1) = m. \tag{6.2.1}$$

By Lemma 1.1.1, a necessary and sufficient condition for (6.2.1) to be satisfied is that

$$E\left[\left(\sum_1^n e^{Y_j}\right) e^{it_1(Y_2 - Y_1) + \cdots + it_{n-1}(Y_n - Y_1)}\right] = mEe^{it_1(Y_2 - Y_1) + \cdots + it_{n-1}(Y_n - Y_1)}. \tag{6.2.2}$$

Let

$$E(e^{Y_j(1+it)}) = \psi_j(t), \qquad Ee^{itY_j} = \phi_j(t), \qquad \xi_j = \psi_j/\phi_j.$$

Then, it follows from (6.2.2) that the relation

$$\xi_1(-t_1 - \cdots - t_{n-1}) + \xi_2(t_1) + \cdots + \xi_n(t_{n-1}) = m \tag{6.2.3}$$

is valid for $|t_j| < \varepsilon_j$, $j = 1, \ldots, n-1$ for some $\varepsilon_j > 0$. Setting $t_3 = \cdots = t_{n-1} = 0$, we obtain from (6.2.3) that

$$\xi_1(-t_1 - t_2) + \xi_2(t_1) + \xi_3(t_2) = m_1. \tag{6.2.4}$$

Using the corollary to Lemma 1.5.1, we have

$$\xi_j(t) = c_j + \gamma t, \qquad j = 1, 2, 3. \tag{6.2.5}$$

Similarly, every ξ_j has the form $\xi_j(t) = c_j + \gamma t$. If F_j is the d.f. of Y_j, then (6.2.5) can be rewritten in the form

$$\int e^y e^{ity}\, dF_j(y) = (c_j + \gamma t) \int e^{ity}\, dF_j(y) \tag{6.2.6}$$

for $|t| < \varepsilon$. By Lemma 6.1.2, we then have that (6.2.6) holds for all real t, and then our result follows from the corollary to Lemma 6.1.1.∎

THEOREM 6.2.2. Let $X_1, \ldots, X_n$, $n \geqq 3$, be independent r.v.'s, not necessarily identically distributed. If $E(1/X_j)$ exists for all j, and is nonzero, and further

$$E(X_1^{-1} + \cdots + X_n^{-1} \mid X_2 - X_1, \ldots, X_n - X_1) = \text{constant} = m$$

then, either $X_j \sim G(\alpha_j, \gamma)$ for all j, or $-X_j \sim G(\alpha_j, \gamma)$ for all j, where $\gamma > 1$.
Proof: As in the proof of Theorem 6.2.1, we have

$$\begin{aligned} &E[(\textstyle\sum X_j^{-1}) \exp\{it_1(X_2 - X_1) + \cdots + it_{n-1}(X_n - X_1)\}] \\ &\qquad = mE \exp\{it_1(X_2 - X_1) + \cdots + it_{n-1}(X_n - X_1)\}. \end{aligned} \tag{6.2.7}$$

Let $\psi_j(t) = EX_j^{-1}e^{itX_j}$ and $\phi_j(t) = Ee^{itX_j}$; if we set $\xi_j(t) = \psi_j(t)/\phi_j(t)$, then (6.2.7) gives

$$\xi_1(-t_1 - \cdots - t_{n-1}) + \xi_2(t_1) + \cdots + \xi_n(t_{n-1}) = m. \tag{6.2.8}$$

Arguing hereafter as in the proof of Theorem 6.2.1, we obtain

$$\int x^{-1}e^{itx}\, dF(x) = (\mu + i\rho t)\int e^{itx}\, dF(x) \tag{6.2.9}$$

for $|t| < \varepsilon$. Applying Lemma 6.1.3, we obtain the desired result.∎

THEOREM 6.2.3. Let $X_1, \ldots, X_n$ be i.i.d. positive r.v.'s such that

$$\begin{aligned} &E(X_1 \log X_1) \text{ is finite and} \\ &E(a_1 X_1 + \cdots + a_n X_n \mid X_1^{b_1} \cdots X_n^{b_n}) = m \end{aligned} \tag{6.2.10}$$

where

$$\begin{aligned} &\textstyle\sum a_j b_j = 0, \qquad |b_n| > \max\{|b_1|, \ldots, |b_{n-1}|\}, \qquad \text{and} \\ &a_j b_j / a_n b_n < 0 \qquad \text{for } j = 1, \ldots, n-1. \end{aligned}$$

Then the X_j have a gamma distribution.
Proof: Let $\psi(t) = Ee^{Y(1+it)}$, $\phi(t) = Ee^{itY}$, and $\xi(t) = \psi(t)/\phi(t)$, where $Y = \log X$. The condition (6.2.10) implies that, for $|t| < \varepsilon$,

$$a_1\xi(b_1 t) + \cdots + a_n \xi(b_n t) = m. \tag{6.2.11}$$

Since $E(Ye^Y) = E(X \log X)$ exists, $\psi(t)$ and $\phi(t)$ have first derivatives continuous at the origin. Hence $\xi(t) - \xi(0) = tu(t)$ where u is continuous at the origin. Then, Lemma 1.5.10 is applicable, and it follows that ξ is a linear function of t. The subsequent argument is the same as for Theorem 6.2.1.▮

Corollary. Let $X_1, \ldots, X_n$ be i.i.d. positive r.v.'s with $E(X_1 \log X_1)$ finite, and

$$E(X_1 + \cdots + X_n | X_1^{n-1} X_2^{-1} \cdots X_n^{-1}) = m. \qquad (6.2.12)$$

Then the X_j have gamma distributions.

THEOREM 6.2.4. Let $X_1, \ldots, X_n$ be i.i.d.r.v.'s. If $E(1/X_1)$ exists and is nonzero, and

$$E\left(\sum_1^n a_j X_j^{-1} \middle| \sum_1^n b_j X_j\right) = m \qquad (6.2.13)$$

where $\sum a_j b_j = 0$, $|b_n| > \max(|b_1|, \ldots, |b_{n-1}|)$, and $a_j b_j / a_n b_n < 0$ for $j = 1, \ldots, n-1$; then, either $X_i \sim G(\alpha, \gamma)$ or $-X_i \sim G(\alpha, \gamma)$ for some $\gamma > 1$.

***Proof*:** The proof is similar to that of Theorem 6.2.3.▮

Corollary. Let $X_1, \ldots, X_n$ be i.i.d.r.v.'s. If $E(1/X_1)$ exists and is nonzero, and

$$E(\sum X_i^{-1} | X_1 - \bar{X}) = m \neq 0 \qquad (6.2.14)$$

then, either $X_i \sim G(\alpha, \gamma)$ or $-X_i \sim G(\alpha, \gamma)$ for some $\gamma > 1$.

In what follows, we shall be using the following notation. $X_1, \ldots, X_n$ are independent random variables. We consider two situations.

(a) If the X_j are positive r.v.'s, we set

$$Y_j = \log X_j; \qquad M_j = a_{j1} Y_1 + \cdots + a_{jn} Y_n; \qquad j = 1, \ldots, p. \qquad (6.2.15)$$

(b) If the X_j are arbitrary r.v.'s, we set

$$Y_j = 1/X_j; \qquad M_j' = a_{j1} Y_1 + \cdots + a_{jn} Y_n; \qquad j = 1, \ldots, p. \qquad (6.2.16)$$

We consider p linearly independent linear functions $M_1, \ldots, M_p$ or $M_1', \ldots, M_p'$; in any case, the M_j or the M_j' can be represented in the canonical form

$$\begin{aligned} M_1 \text{ (or } M_1') &= Y_1 + a_{11} Y_{p+1} + \cdots + a_{1,n-p} Y_n \\ \cdots \quad & \quad \cdots \qquad \cdots \qquad \cdots \\ M_p \text{ (or } M_p') &= Y_p + a_{p1} Y_{p+1} + \cdots + a_{p,n-p} Y_n. \end{aligned} \qquad (6.2.17)$$

A will denote the matrix $((a_{ij}))$.

THEOREM 6.2.5. Let $1 < p \leqq n-1$, and the rank of the matrix **A** be $r = n - p$.

(a) If the X_j are positive, with finite expectation, and the condition

$$E(e^{Y_1} + \cdots + e^{Y_n} | M_1, \ldots, M_p) = m \qquad (6.2.18)$$

is satisfied, then the X_j have gamma distributions.

(b) If $E(1/X_j)$ exists and is nonzero for all j, and the condition

$$E(Y_1 + \cdots + Y_n | M_1', \ldots, M_p') = m \tag{6.2.19}$$

is satisfied, then, either the X_j have gamma distributions, or the $-X_j$ have.

***Proof*:**

(a) We set in this case

$$\xi_j(t) = \int e^{y(1+it)}\, dF_j(y) \Big/ \int e^{ity}\, dF_j(y) \tag{6.2.20}$$

where F_j is the d.f. of X_j.

(b) We set in this case

$$\xi_j(t) = \int y e^{ity}\, dF_j(y) \Big/ \int e^{ity}\, dF_j(y). \tag{6.2.21}$$

Then, respectively, condition (6.2.18) or (6.2.19) is equivalent to

$$\xi_1(t_1) + \cdots + \xi_p(t_p) + \xi_{p+1}(\boldsymbol{\alpha}_1'\mathbf{t}) + \cdots + \xi_n(\boldsymbol{\alpha}_{n-p}'\mathbf{t}) = \text{constant} \tag{6.2.22}$$

for $|t_j| < \delta, j = 1, \ldots, p$, where $\boldsymbol{\alpha}_j$ is the j-th column vector of the matrix $\mathbf{A}$ and $\mathbf{t}' = (t_1, \ldots, t_p)$. Applying Lemma 1.5.4, we find that the ξ_j are linear functions. The rest of the argument is as in the case of Theorems 6.2.1–6.2.4.▌

We may note that, in relations (6.2.18) and (6.2.19), it is possible to replace $e^{Y_1} + \cdots + e^{Y_n}$ by $a_1 e^{Y_1} + \cdots + a_n e^{Y_n}$ in case (a), and $Y_1 + \cdots + Y_n$ by $a_1 Y_1 + \cdots + a_n Y_n$ in case (b), where every a_j is nonzero, to arrive at the same conclusions.

Consider a $q \times n$ matrix $((d_{ij}))$ and an n-dimensional vector $(b_1, \ldots, b_n)$. We say that they satisfy the condition (C) if there exist constants $a_1', \ldots, a_q'$ such that the coefficients

$$a_j = \sum_i a_i' d_{ij}; \qquad j = 1, \ldots, n \tag{6.2.23}$$

satisfy the following conditions: (i) $\sum a_j b_j = 0$, and (ii) if $a_1, \ldots, a_s$ $(s \leq n)$ are the nonzero ones among them (as we may assume without loss of generality), then among $|b_1|, \ldots, |b_s|$, there exists exactly one maximum; further, if, as we may assume without loss of generality, that maximum is $|b_1|$, then $a_2 b_2, \ldots, a_n b_n$ are all of the same sign and $a_1 b_1$ is of opposite sign.

THEOREM 6.2.6. Let $X_1, \ldots, X_n$ be i.i.d. positive r.v.'s and $Y_j = \log X_j$. Consider

$$\begin{aligned} L_i &= \sum_j d_{ij} e^{Y_j}, \qquad i = 1, \ldots, q, \\ M &= b_1 Y_1 + \cdots + b_n Y_n \end{aligned} \tag{6.2.24}$$

where the coefficients d_{ij} and the b_j satisfy the condition (C) enunciated above. If $E(X_j \log X_j)$ is finite and

$$E(L_i | M) = g_i = \text{constant}, \qquad i = 1, \ldots, q$$

then the X_j have gamma distributions.

***Proof*:** We define $\xi(t)$ as in the proof of Theorem 6.2.1; then (6.2.24) reduces to the condition

$$\sum_j d_{ij}\,\xi(b_j t) = g_i, \qquad i = 1, \ldots, q. \tag{6.2.25}$$

Multiplying the successive relations in (6.2.25) by $a_1', \ldots, a_q'$ respectively, and adding, we obtain, in view of condition (C), the relation

$$\sum_1^n a_i\,\xi(b_i t) = g = \text{constant}. \tag{6.2.26}$$

Since $E(X_j \log X_j)$ is finite, we may use the arguments of Theorem 6.2.3, which lead from (6.2.26) to the desired result.▌

THEOREM 6.2.7. Let $X_1, \ldots, X_n$ be i.i.d.r.v.'s with EX_1^{-1} finite and non-zero. Further, let

$$\begin{aligned} L_i &= \sum d_{ij} X_j^{-1} \qquad \text{for } i = 1, \ldots, q, \\ M &= b_1 X_1 + \cdots + b_n X_n \end{aligned} \tag{6.2.27}$$

where the coefficients d_{ij} and the b_j satisfy the condition (C). Then, from the conditions

$$E(L_i | M) = g_i, \qquad i = 1, \ldots, q \tag{6.2.28}$$

it follows that either the X_j have gamma distributions or the $-X_j$ have gamma distributions.

***Proof*:** This is similar to that of Theorem 6.2.6.▌

The following theorem is due to Laha and Lukacs [95].

THEOREM 6.2.8. Let $X_1, \ldots, X_n$ be i.i.d.r.v.'s with $EX_1 = \mu > 0$ and Var $X_1 = \sigma^2$. Further, let

$$Q = \sum a_{jk} X_j X_k + \sum b_j X_j. \tag{6.2.29}$$

Write $B_1 = \sum a_{jj}$, $B_2 = \sum a_{jk}$, and $B_3 = \sum b_j$ and suppose $B_1 \neq 0$, $B_2 \neq 0$, $B_3 = 0$, and $B_1\sigma^2 + B_2\mu^2 = 0$. Then the X_j have a gamma distribution if and only if the regression of Q on $L = X_1 + \cdots + X_n$ is constant.

***Proof*:** The necessity is verified immediately. To prove the sufficiency, we use formula (1.1.17).

$$E(Qe^{itL}) = E(Q) \cdot E(e^{itL}) \tag{6.2.30}$$

Denoting the c.f. of the X_j by f, and noting that

$$f^{(j)}(t) = i^j E(X^j e^{itX}), \qquad j = 1, 2 \tag{6.2.31}$$

we see that (6.2.30) reduces to

$$B_1 f'' f^{n-1} + (B_2 - B_1)(f')^2 f^{n-2} + iB_3 f' f^{n-1} = -Cf^n \tag{6.2.32}$$

where $C = EQ = B_1\sigma^2 + B_2\mu^2 + B_3\mu$. Under the conditions on B_1, B_2, and B_3, we see that (6.2.32) reduces to

$$B_1 f'' f^{n-1} + (B_2 - B_1)(f')^2 f^{n-2} = 0. \tag{6.2.33}$$

Since $f(0) = 1$, and f is continuous, there exists a neighborhood of the origin where $\phi = \log f$ is defined and, in view of (6.2.33), satisfies the relation

$$B_1\phi'' + B_2(\phi')^2 = 0. \tag{6.2.34}$$

Solving this equation and using the condition $B_1\sigma^2 + B_2\mu^2 = 0$, we obtain

$$f(t) = (1 - i\sigma^2 t/\mu)^{-\mu^2/\sigma^2}. \tag{6.2.35}$$

It follows from the theory of analytic c.f.'s that (6.2.35) holds for all real t and hence that the X_j have a gamma distribution.▌

THEOREM 6.2.9. (cf. [95]) Let $X_1, \ldots, X_n$ be i.i.d.r.v.'s with $EX_1 = \mu > 0$ and $\operatorname{Var} X_1 = \sigma^2$. Set $L = X_1 + \cdots + X_n$, $Q = \sum a_{jk} X_j X_k$, $B_1 = \sum a_{jj}$, $B_2 = \sum a_{jk}$ and suppose $B_2 \neq nB_1$. Then the X_j have a gamma distribution if and only if the regression of Q/L^2 on L is constant.
Proof: This is similar to that of Theorem 6.2.8.▌

The following theorem was proved by Linnik, Rukhin and Strelits [90].

THEOREM 6.2.10. Let $X_1, \ldots, X_n$, $(n \geqq 2)$, be positive i.i.d.r.v.'s with probability density function $\phi(x)$ with $\int_0^\infty x^k\phi(x)\,dx < \infty$. Let $P(u_1, \ldots, u_n)$ be a polynomial of degree exactly k such that

$$E\{P(X_1/L, X_2/L, \ldots, X_n/L) \mid L\} = \text{constant} = c. \tag{6.2.36}$$

Then the X_j follow a gamma distribution if the conditions (1)–(3) below are satisfied.

(1) In an arbitrary interval of the form $(0, \varepsilon)$

$$\phi(x) = A_0 x^{p-1} + A_1 x^p + \cdots + (A_s + o(1))x^{p+s-1}$$

for some p and a sufficiently large integer s (the precise condition on s is indicated in the course of the proof).

(2) The equation (6.2.38) below is such that

$$\alpha_0 = \sum_{i_1+\cdots+i_k=k}^{*} A_{i_1,\ldots,i_k} \prod_{\substack{j=1 \\ i_j\neq v}}^{k} \Gamma(p + i_j) \neq 0 \tag{6.2.37}$$

where v is the order of the highest-order derivative in (6.2.38), and the summation in $\sum^*$ is taken over $(i_1, \ldots, i_k)$ such that $\sum i_j = k$ and $i_j = v$ for at least one j.

(3) The function $y(z) = z^{-p} + z^{-p+1}$ is not a solution of (6.2.38), and the equation

$$Q_0(p) = \sum_{i_1+\cdots+i_k=k} A_{i_1,\ldots,i_k}\left\{\sum_{q=1}^{k}\prod_{\substack{j=1\\ j\neq q}}^{k}\frac{\Gamma(p+i_j)}{\Gamma(p)}\frac{\Gamma(p+i_q+\mu)}{\Gamma(p+\mu)}\right\} = 0$$

does not have positive integral roots μ.

***Proof*:** Let $P(u_1, \ldots, u_n) = \sum a_{k_1,\ldots,k_n} u_1^{k_1}\cdots u_n^{k_n}$; then (6.2.36) gives

$$\sum_{s=0}^{k} E\left\{\sum_{k_1+\cdots+k_n=s} a_{k_1,\ldots,k_n} X_1^{k_1}\cdots X_n^{k_n} L^{k-s} \,|\, L\right\} = cL^k.$$

Multiplying both sides of the above relation by e^{-zL}, Re $z > 0$, and then taking the expectations of both sides, we obtain

$$\sum_{s=0}^{k} E\left\{\sum_{k_1+\cdots+k_n=s} a_{k_1,\ldots,k_n} X_1^{k_1}\cdots X_n^{k_n} L^{k-s} e^{-zL}\right\} = cE(L^k e^{-zL}).$$

Setting $y(z) = \int_0^\infty e^{-zx}\phi(x)\,dx$, we can obtain after some simple computations

$$\sum_{i_1+\cdots+i_k=k} A_{i_1,\ldots,i_k} y^{(i_1)}\ldots y^{(i_k)} = 0 \tag{6.2.38}$$

where

$$A_{i_1,\ldots,i_k} = \sum_{s=0}^{k}\sum_{r_1+\cdots+r_n=k-s}\frac{(k-s)!}{r_1!\cdots r_n!}a_{i_1-r_1,\ldots,i_n-r_n} - c\frac{k!}{i_1!\cdots i_k!}.$$

LEMMA 6.2.1. If ϕ satisfies condition (1) of the theorem, and $y(z) = \int_0^\infty e^{-zx}\phi(x)\,dx$ is a solution of (6.2.38), then the following representation is valid in $\mathscr{A} = \{z\colon \text{Re } z > 0, |\arg z| < \pi/2 - \delta_0\}$ for $\delta_0 > 0$:

$$y(z) = \sum_{j=0}^{s} B_j z^{-(p+j)} + R_s(z) \tag{6.2.39}$$

where

$$|R_s^{(j)}(z)| = o(1)|z|^{-(p+s+j)}, \qquad z\in\mathscr{A};\, j = 0, 1, \ldots, v$$

and

$$B_j = A_j\int_0^\infty e^{-u}u^{s+j}\,du \qquad \text{for } j = 0, 1, \ldots, s.$$

***Proof*:** From condition (1) and the form of the function $y(z)$, we find

$$y(z) = \int_0^\varepsilon\left(\sum_{j=0}^{s} A_j u^{p+j-1}e^{-uz}\right)du + \int_0^\varepsilon e^{-uz}w(u)u^{p+s-1}\,du + \int_\varepsilon^\infty e^{-uz}\phi(u)\,du$$

where $w(u) = o(1)$ as $u \to 0$. Hence the required inequality is easily obtained for z in $\mathscr{A}$ if we note that

$$\left|\int_0^{\varepsilon} e^{-uz} w(u) u^{p+s-1}\, du\right| = \frac{o(1)}{x^{p+s}} = \frac{o(1)}{|z|^{p+s}}$$

in view of the fact that $|\text{Im } z| < ax$ in $\mathscr{A}$ and hence

$$|z| = (x^2 + y^2)^{1/2} = x(1 + y^2/x^2)^{1/2} \leqq x(1 + a^2)^{1/2}, \qquad a = \text{constant.} \blacksquare$$

Turning now to the proof of the theorem, we may take $B_0 = 1$ in (6.2.39) without loss of generality. Substituting for $y(z)$ in (6.2.39), we have

$$y^{(j)}(z) = (-1)^j \left\{ \frac{\Gamma(p+j)}{\Gamma(p)} \frac{1}{z^{p+j}} \frac{(B_1 + o(1))\Gamma(p+j+1)}{\Gamma(p+1)} \frac{1}{z^{p+j+1}} \right\};$$

$$\sum_{i_1 + \cdots + i_k = k} A_{i_1, \ldots, i_k} \prod_{j=1}^{k} \left\{ \frac{\Gamma(p+i_j)}{\Gamma(p)} \frac{1}{z^{p+i_j}} + \frac{(B_1 + o(1))\Gamma(p+i_j+1)}{\Gamma(p+1)} \frac{1}{z^{p+i_j+1}} \right\} = 0.$$

Multiplying both sides of the above equality by z^p and letting $z \in \mathscr{A}$ tend to ∞, we see at once that p is a positive root of the equation

$$Q(p) = \sum_{i_1 + \cdots + i_k = k} A_{i_1, \ldots, i_k} \prod_{j=1}^{k} \frac{\Gamma(p+i_j)}{\Gamma(p)} = 0. \tag{6.2.40}$$

We make the substitution $y(z) = w(z) + z^{-p}$ in (6.2.38) and single out the linear terms

$$\left\{(-1)^k \sum_{i_1 + \cdots + i_k = k} A_{i_1, \ldots, i_k} \prod_{q=1}^{k} \frac{\Gamma(p+i_q)}{\Gamma(p)} \right\} \frac{1}{z^{kp+k}}$$
$$+ \sum_{i_1 + \cdots + i_k = k} A_{i_1, \ldots, i_k} \left\{ \sum_{q=1}^{k} \prod_{\substack{j=1 \\ j \neq q}}^{k} (-1)^{i_j} \frac{\Gamma(p+i_j)}{\Gamma(p)} z^{i_q} w^{(i_q)} \right\} \frac{1}{z^{(k-1)p+k}}$$
$$+ R^*(z, w, w^{(1)}, \ldots, w^{(\nu)}) = 0 \tag{6.2.41}$$

where R^* consists of all terms which are nonlinear in w and its derivatives.

We proceed to estimate R^* which is the sum of terms of the form

$$\sigma = A_{i_1, \ldots, i_k} w^{(i_1)} \cdots w^{(i_m)} z^{-[(k-m-j)p + i_{m+1} + \cdots + i_k]}, \qquad m \geqq 2.$$

In view of (6.2.39)

$$|w^{(j)}| = (B_1 + o(1)) \frac{\Gamma(p+j+1)}{\Gamma(p+1)} z^{-(p+j+1)}; \qquad j = 0, 1, \ldots, \nu$$

and consequently

$$\sigma = A_{i_1, \ldots, i_k} \frac{(B_1^{i_1+\cdots+i_m} + o(1))}{z^{kp+k+m}} \prod_{j=1}^{m} \frac{\Gamma(p+i_j+1)}{\Gamma(p+1)}.$$

But $m \geqq 2$ and hence

$$|\sigma| < B^*|z|^{-(kp+k+2)}, \qquad B^* = \text{constant}$$

Summing up the corresponding inequalities for all the terms entering into R^*, we have

$$|R^*| < B|z|^{-(kp+k+2)}, \qquad B = \text{constant}, \qquad z \in \mathscr{A}. \tag{6.2.42}$$

The first term in (6.2.41) vanishes since p is a root of the equation (6.2.40), and (6.2.41) can be written as

$$\sum_{i_1+\cdots+i_k=k} A_{i_1,\ldots,i_k} \sum_{q=1}^{k} \prod_{\substack{j=1\\ j\neq q}}^{k} (-1)^{i_j} \frac{\Gamma(p+i_j)}{\Gamma(p)} z^{i_q} w^{(i_q)} + R^{**}(z)$$
$$= L(w) + R^{**}(z) = 0 \tag{6.2.43}$$

where

$$|R^{**}(z)| < B|z|^{-(p+2)}, \qquad B = \text{constant}. \tag{6.2.44}$$

It is necessary to note that, in view of the conditions of the theorem, the coefficient of $w^{(\nu)}$ does not vanish, so that $L(w)$ contains a term of the form $A^* z^\nu w^{(\nu)}$ with $A^* \neq 0$.

The homogeneous equation $L(w) = 0$ is an equation of Euler's type and has a solution of the form (1.4.6); in particular, the solutions z^μ, where μ is a root of the equation $Q_o(\mu) = 0$. As is easily seen, $\mu = p$ and $\mu = p + 1$ are such solutions: (6.2.40) shows that $Q_o(p) = 0$, and, for $\mu = p + 1$,

$$\sum_{i_1+\cdots+i_k=k} A_{i_1,\ldots,i_k} \sum_{q=1}^{k} \prod_{j=1}^{k} \frac{\Gamma(p+i_j)}{\Gamma(p)} \left(1 + \frac{i_q}{p}\right)$$
$$= k \sum_{i_1+\cdots+i_k=k} A_{i_1,\ldots,i_k} \prod_{j=1}^{k} \frac{\Gamma(p+i_j)}{\Gamma(p)} = 0$$

again in view of (6.2.40).

According to Lemma 1.4.3, there exists a particular solution w_0 of (6.2.43), for which

$$|w_0^{(j)}(z)| < c_0 \frac{(\ln|z|)^{(j+1)s_o}}{|z|^{p+2}}, \qquad z \in \mathscr{A}, \quad c_0 = \text{constant}, \quad j = 0, 1, \ldots, \nu. \tag{6.2.45}$$

Since (6.2.39) must be satisfied, we have

$$w(z) = B_1 z^{-(p+1)} + w_0^*(z)$$

where $w_0^*(z)$ satisfies (6.2.45).

Further, from (6.2.45),

$$y(z) = z^{-p} + B_1 z^{-(p+1)} + w_0^*(z).$$

If we make the substitution

$$y(z) = z^{-p} + B_1 z^{-(p+1)} + w(z)$$

in (6.2.38), then again we obtain an equation of the form (6.2.41) in which the first term does not vanish. In view of (6.2.44) as applied to $w_0^*(z)$, we obtain

$$L(w) = B_2 z^{-(p+2)} + R_2^*(z)$$

where

$$|R_2^*(z)| < c_0^* \frac{(\ln |z|)^{2(s_o+1)}}{|z|^{p+3}}, \qquad c_0^* = \text{constant}. \tag{6.2.46}$$

We now consider the function

$$(z - B_1/p)^{-p} = z^{-p} + B_1 z^{-p-1} + B_2 z^{-p-2} + \cdots \tag{6.2.47}$$

which, as is easily verified, is a solution of (6.2.38).

LEMMA 6.2.2. Let

$$y(z) = z^{-p} + B_1 z^{-p-1} + \cdots + B_N z^{-p-N} + R_N(z), \qquad z \in \mathscr{A} \tag{6.2.48}$$

where $\sum_{j=0}^{N} B_j z^{-p-j}$ is a partial sum of the expansion (6.2.47), satisfy (6.2.38). Then there exists $x_o > 0$ such that, in the set $\mathscr{A}^* = \mathscr{A} \cap \{\text{Re } z > x_o\}$, the following inequality is valid for all $N = 0, 1, \ldots$ and $j = 0, 1, \ldots, v$.

$$|R_N^{(j)}(z)| < \frac{(\ln |z|)^{N(k+1)(s_o+1)+js_o}}{|z|^{p+N+j+1}}. \tag{6.2.49}$$

Here s_o is the same as in Lemma 1.4.3.

***Proof*:** We proceed by the method of induction. Suppose (6.2.49) holds for $N = 1, \ldots, q$. We shall show that it is valid for $N = q + 1$; as we have shown, it holds for $N = 0$.

In (6.2.38), we make the substitution

$$y(z) = \sum_{j=0}^{q} B_j z^{-p-j} + w_q(z) \tag{6.2.50}$$

and single out the linear terms to obtain

$$\sum_{i_1+\cdots+i_k=k} A_{i_1,\ldots,i_k} \prod_{r=1}^{k} \prod_{j=1}^{q} \frac{\Gamma(p+j+i_r)}{\Gamma(p+j)} \frac{B_j}{z^{p+j+i_r}}$$

$$+ \sum_{i_1+\cdots+i_k=k} A_{i_1,\ldots,i_k} \sum_{j=1}^{k} (-1)^{i_j} \prod_{\substack{r=1\\ r\neq j}}^{k} \sum_{i=0}^{q} \frac{\Gamma(p+i+i_r)}{\Gamma(p+i)} \frac{B_i}{z^{p+i+i_r}} w_q^{(i_j)}$$

$$= R_q^*(z, w_q, \ldots, w_q^{(v)}) \quad (6.2.51)$$

where R_q^* is the sum of terms $\sigma_{i_1,\ldots,i_m}$ of the form

$$\sigma_{i_1,\ldots,i_m} = A_{i_1,\ldots,i_k} w_q^{(i_1)} \cdots w_q^{(i_m)} \prod_{r=m+1}^{k} \sum_{j=0}^{q} \frac{\Gamma(p+j+i_r)}{\Gamma(p+j)} \frac{B_j}{z^{p+j+i_r}}$$

$m \geqq 2$. In view of (6.2.49), we have for z in $\mathscr{A}^*$

$$|\sigma_{i_1,\ldots,i_m}| < |A_{i_1,\ldots,i_k}| \frac{(\ln|z|)^{mq(k+1)(s_0+1)+s_0 \sum_{j=1}^{m} i_j}}{|z|^{m(p+q+1)+\sum_{j=1}^{m} i_j}}$$

$$\times \prod_{r=m+1}^{k} \left| \sum_{j=0}^{q} \frac{\Gamma(p+j+i_r)}{\Gamma(p+j)} \frac{B_j}{z^{p+j+i_r}} \right|. \quad (6.2.52)$$

Further, for $|z| > |B_1|/p + 1$,

$$\left| \sum_{j=0}^{q} \frac{\Gamma(p+j+i_r)}{\Gamma(p+j)} \frac{B_j}{z^{p+j+i_r}} \right|$$

$$= \left| \left[\left(z - \frac{B_1}{p} \right)^{-p} - \sum_{j=q+1}^{\infty} B_j z^{-p-j} \right]^{(i_r)} \right|$$

$$= \left| \frac{p(p+1)\cdots(p+i_r-1)}{(z-B_1/p)^{p+i_r}} - \sum_{j=q+1}^{\infty} \frac{(p+j)\cdots(p+j+i_r-1)B_j}{z^{p+j+i_r}} \right|$$

$$\leqq \left| \frac{p(p+1)\cdots(p+i_r-1)}{(|z|-B_1/p)^{p+i_r}} \right| + \left| \frac{p(p+1)\cdots(p+i_r-1)}{(|z|-|B_1|/p)^{p+i_r}} \right|$$

$$< 2\left(1 + \frac{|B_1|}{p}\right)^{p+k} \cdot \frac{p(p+1)\cdots(p+i_r-1)}{|z|^{p+i_r}}. \quad (6.2.53)$$

Returning to (6.2.52), we find

$$|\sigma_{i_1,\ldots,i_m}| < |A_{i_1,\ldots,i_k}| \left(1 + \frac{|B_1|}{p}\right)^{(k-1)(p+k)}$$

$$\times \prod_{i=0}^{k} (p+i)^m \cdot \frac{(\ln|z|)^{mq(k+1)(s_0+1)+s_0 \sum_{j=1}^{m} i_j}}{|z|^{k(p+1)+m(q+1)}}.$$

Setting $(1 + |B_1|/p)^{k(p+k)} \prod_{i=0}^{k} (p+i)^k |A_{i_1, \ldots, i_k}| = H$ for any choice of the indices $i_1, \ldots, i_k$ and

$$\frac{\prod_{i=0}^{k}(p+i)}{(1 + |B_1|/p)^{p+k}} = B$$

where, obviously, H and B do not depend on q, we have

$$|\sigma_{i_1, \ldots, i_m}| \leq H \left\{ \frac{B(\ln|z|)^{2q(k+1)(s_0+1)}}{|z|^{q+1}} \right\}^m \frac{1}{|z|^{kp+k}}.$$

Since $m \geqq 2$, noting that the number of terms in R_q^* is $\leq (k!)^2$,

$$|R_q^*(z, w_q, \ldots, w_q^{(v)})| < \frac{(k!)^2 H}{|z|^{kp+k}} \sum_{m=2}^{\infty} \left[B \frac{(\ln|z|)^{2q(k+1)(s_0+1)}}{|z|^{q+1}} \right]^m.$$

We choose x_0 so large that, for $\operatorname{Re} z > x_0$,

$$\frac{B(\ln|z|)^{3(k+1)(s_0+1)}}{|z|} < \frac{1}{2}.$$

Then

$$|R_q^*(z, w_q, \ldots, w_q^{(v)})| < \frac{2(k!)^2 H B^2}{|z|^{kp+k}} \frac{(\ln|z|)^{2q(k+1)(s_0+1)}}{|z|^{2q+2}}. \tag{6.2.54}$$

In view of (6.2.54), we finally have that, for $z \in \mathscr{A}^* = \mathscr{A} \cap [\operatorname{Re} z > x_0]$,

$$|R_q^*(z, w_q, \ldots, w_q^{(v)})| < |z|^{-(kp+k+q+2)}. \tag{6.2.55}$$

Finally, we leave on the left side of (6.2.51) only the expression

$$L(w_q) = \sum_{i_1 + \cdots + i_k = k} A_{i_1, \ldots, i_k} \sum_{r=1}^{k} (-1)^{i_r} \prod_{\substack{j=1 \\ j \neq r}}^{q} \frac{\Gamma(p + i_j)}{\Gamma(p)} z^{i_r} w_q^{(i_r)} \tag{6.2.56}$$

and shift all the others to the right side; they consist of a sum S with terms of the form

$$T_j = A_{i_1, \ldots, i_k} \left\{ \sum_{r=1}^{q} \frac{\Gamma(p + r + i_j)}{\Gamma(p + r)} \frac{B_r}{z^r} \right\} z^{i_j} w_q^{(i_j)} \tag{6.2.57}$$

and a polynomial in $1/z$.

$$P\left(\frac{1}{z}\right) = -\left\{ \sum_{i_1 + \cdots + i_k = k} A_{i_k, \ldots, i_k} \prod_{j=1}^{k} \sum_{r=1}^{q} \frac{\Gamma(p + j + i_r)}{\Gamma(p + j)} \frac{B_j}{z^{p+j+i_r}} \right\} z^{(k-1)p+k}. \tag{6.2.58}$$

Thus

$$L(w_q) = S + P\left(\frac{1}{z}\right) + R_q^* z^{-(k-1)p-k}. \tag{6.2.59}$$

Using the same arguments as for the derivation of (6.2.53), we can see that for $z \in \mathscr{A}^*$

$$|T_j| < 2|A_{i_1, \ldots, i_k}| \sum_{r=1}^{\infty} \frac{(p+r)\cdots(p+r+i_j-1)|B_j|}{|z|^{p+r+i_j}} \cdot |z|^{2i_j+p}|w_q^{(i_j)}|$$

$$< 2|A_{i_1, \ldots, i_k}| \frac{\prod_{i=1}^{k}(p+i)}{|z|^{i_j}} \left[\left(1 - \frac{|B_1|}{p|z|}\right)^{-p-i_j} - 1\right] |z|^{2i_j}|w_q^{(i_j)}|.$$

Making use of a theorem on finite differences, we obtain

$$\sigma^* = \left(1 - \frac{|B_1|}{p|z|}\right)^{-p-i_j} - 1 < \left(1 - \frac{|B_1|}{p|x_0|}\right)^{-p-i_j-1} \frac{|B_1|}{p|z|}(p+k).$$

But $x_0 > 1 + |B_1|/p$ in $\mathscr{A}^*$ so that

$$\sigma^* < \left(1 + \frac{|B_1|}{p}\right)^{p+k+1} \frac{|B_1|(p+k)}{p|z|}.$$

On the basis of this inequality and (6.2.49), we arrive at the following estimate for $|T_j|$

$$|T_j| < 2(p+k)\left(1 + \frac{|B_1|}{p}\right)^{p+k+1} \frac{|B_1|}{p} \prod_{i=1}^{k}(p+i)$$

$$\times |A_{i_1, \ldots, i_k}| \frac{(\ln|z|)^{q(k+1)(s_0+1)+i_j s_0}}{|z|^{p+q+2}}.$$

Thus

$$|T_j| < H^* \frac{(\ln|z|)^{q(k+1)(s_0+1)+i_j s_0}}{|z|^{p+q+2}}$$

where H^* is some constant, not depending on q. Thus, in the set $\mathscr{A}^*$, we have according to (6.2.54)

$$S \leqq \sum |T_j| < H_0 \frac{(\ln|z|)^{q(k+1)(s_0+1)+ks_0}}{|z|^{p+q+2}} \tag{6.2.60}$$

with $H_0 = k_0 \cdot H^*$, where k_0 is the number of summands in S.

We finally take up the polynomial $P(1/z)$. In this polynomial, all terms containing the powers $z^{-(p+m)}$ with $m \leqq (k-1)p + k + q$ vanish. Indeed, this polynomial is the left member of (6.2.38) multiplied by $z^{(k-1)p+k}$, in which the sum $\tilde{w} = \sum_{j=0}^{q} B_j z^{-p-j}$ has been substituted for $y(z)$. But this sum is a partial sum in the expansion for $(z - B_1/p)^{-p}$, which is a solution of (6.2.38). If the function

$$y(z) = \left(z - \frac{B_1}{p}\right)^{-p} = \tilde{w}(z) + \sum_{j=q+1}^{\infty} \frac{B_j}{z^{p+j}}$$

be substituted in (6.2.38), then the coefficients of any of the powers $z^{-(k-1)p-j}$ must vanish (after similar terms have been collected together). But terms containing the powers $z^{-(k-1)p-j}$, $j \leqq k+q$, are obtained in this expression only as combinations of summands entering into $\tilde{w}$, which proves our assertion. Thus,

$$\sum_{i_1+\cdots+i_k=k} A_{i_1,\ldots,i_k} \prod_{r=1}^{k} \sum_{j=1}^{q} \frac{\Gamma(p+j+i_r)}{\Gamma(p+j)} \frac{B_j}{z^{p+j+i_r}}$$
$$= -(D_{q+1}z^{-p-q-1} + \cdots + D_{q+v}z^{-p-q-v})z^{-(k-1)p-k} \qquad (6.2.61)$$

where $v \geqq 2$. Estimates for the coefficients D_{q+i+1} can be obtained as follows. The polynomial

$$\sum_{r=1}^{q} \frac{\Gamma(p+r+i_j)}{\Gamma(p+r)} \frac{B_r}{z^{p+r+i_j}}$$

is a partial sum of the Newton series of the function

$$p(p+1)\cdots(p+i_j-1)(z-B_1/p)^{-p-i_j}.$$

In other words, the Newton series in powers of z^{-1} of the function above majorizes the polynomial above, and the function

$$[p(p+1)\cdots(p+k)]^k(z-|B_1|/p)^{-kp-k}$$

majorizes each of the derivatives entering into the left member of (6.2.61). Obviously, defining $\mu = k_0 \prod_{i=1}^{k} (p+i)^k H$, where k_0 is the number of coefficients in the equation (6.2.38) and H is the same as in (6.2.54), the series in powers of z^{-1} for the function $\psi(z) = \mu z^{kp+k}(z - |B_1|/p)^{-kp-k}$ majorizes the polynomial $z^p \cdot P(1/z)$. Let $\psi(z) = \sum_1^{\infty} B_j z^{-j}$. This series converges for $|z| \geqq 1/2 + |B_1|/p$. Taking $|z| = 1/2 + |B_1|/p$, we have

$$|\psi(z)| < \mu\left(1 + \frac{2|B_1|}{p}\right)^{kp+k}$$

and, by Cauchy's theorem,

$$|B_j| < \mu\left(1 + \frac{2|B_1|}{p}\right)^{kp+k}\left(\frac{1}{2} + \frac{|B_1|}{p}\right)^j; \qquad j = 0, 1, \ldots.$$

Therefore,

$$|D_{q+i+1}| < \mu\left(1 + \frac{2|B_1|}{p}\right)^{kp+k}\left(\frac{1}{2} + \frac{|B_1|}{p}\right)^{q+i+1}; \qquad i = 0, 1, \ldots$$

and, for $|z| > 1/2 + |B_1|/p$,

$$\sum_{i=1}^{\infty} |D_{q+i+1}||z|^{-p-i-1} < \mu\left(1 + \frac{2|B_1|}{p}\right)^{kp+k} \times \frac{1}{1 - [|B_1|/p + (1/2)](1/|z|)} \left[\frac{(1/2) + |B_1|/p}{|z|}\right]^{q+2}.$$

But $|z| > 1 + |B_1|/p$ in the set $\mathscr{A}^*$. Consequently, a constant H_1, not depending on q, exists such that

$$P\left(\frac{1}{z}\right) = D_{q+1}z^{-p-q-1} + w^*(z)$$

with

$$|w^*(z)| < \frac{H_1[(1/2) + |B_1|/p]^q}{|z|^{p+q+2}}. \tag{6.2.62}$$

Thus, according to (6.2.55), (6.2.59), (6.2.60), and (6.2.62),

$$L(w_q) = D_{q+1}z^{-p-q-1} + w(z) \tag{6.2.63}$$

where

$$|w(z)| < \frac{(\ln|z|)^{q(k+1)(s_0+1)+ks_0}}{|z|^{p+q+2}} \left[1 + H_0 + H_1 \frac{((1/2) + |B_1|/p)^q}{(\ln|z|)^{q(s_0+1)}}\right]$$

where H_0 and H_1 do not depend on q.

For $\operatorname{Re} z > x_0$, where x_0 is sufficiently large and does not depend on q, $((1/2) + |B_1|/p)/(\ln|z|)^{1+s_0} < 1$. In what follows, we shall assume x_0 to be so chosen. Then

$$|w(z)| < (1 + H_0 + H_1)\frac{(\ln|z|)^{q(k+1)(s_0+1)+ks_0}}{|z|^{p+q+2}}. \tag{6.2.64}$$

We now find a particular solution for (6.2.63). For this, we solve two equations:

$$L(w_q^*) = D_{q+1}z^{-p-q-1} \tag{6.2.65}$$

and

$$L(w_q^{**}) = w(z). \tag{6.2.66}$$

Now, by condition (3) of the theorem, the equation $Q_0(p + \mu + 1) = 0$ does not have positive integral roots; hence, (6.2.65) has a solution $w_q^* = B_{q+1}/z^{p+q+1}$ where the constant B_{q+1} is uniquely defined. Also, in accordance with Lemma 1.4.3 and (6.2.64), the equation (6.2.66) has a solution w_q^{**} in the set $\mathscr{A}^* = \mathscr{A} \cap [\operatorname{Re} z > x_0]$, which satisfies the inequality:

$$|[w_q^{**}(z)]^{(j)}| < \frac{A(1 + H_0 + H_1)}{p + q + 2}\frac{(\ln|z|)^{q(k+1)(s_0+1)+(k+j+1)s_0}}{|z|^{p+q+j+2}}$$

$$j = 0, 1, \ldots, \nu - 1. \tag{6.2.67}$$

Since there exist only a finite number of values of q smaller than the sum of the moduli of the roots of the equation $Q_0(p+\mu)=0$, we may take the constant A to be independent of q, in accordance with Lemma 1.4.3.

Thus, $w_0 = w_q^* + w_q^{**}$ will be a particular solution of (6.2.63). The general solution of (6.2.63) is the sum of this solution and a general solution of the homogeneous equation $L(w)=0$, which takes the form of a sum of functions of the type (1.4.6). If $q < s$, where s is the number indicated above, then the solution $w_q(z)$ of (6.2.63) must tend to zero as $z \to \infty$ at least as rapidly as $\tilde{c}|z|^{-p-q-2}$, where $\tilde{c}$ is some constant. Therefore, it is necessary that our solution for (6.2.63) does not contain summands, which are integrals of the equation $L(w)=0$, which decrease less rapidly than $\tilde{c}|z|^{-p-q-2}$, but can possibly contain a summand $c_0 z^{-p-q-2}$. In view of condition (3) of the theorem, the equation $Q_0(p+\mu)=0$ does not have positive integral roots, so that $L(w)=0$ does not have z^{-p-q-2} as a solution. Here it is necessary to note that a solution $c_0 z^{-\lambda-q-1}$ with λ complex is unsuitable, since the desired solution $y(z)$ and hence $w_q(z)$ must take real values for $\operatorname{Re} z > 0$, $\operatorname{Im} z = 0$. If we add a solution w_q of the equation $L(w)=0$, of the form

$$\sum c_j(\ln z)^{s_j^*} z^{-(p_j+q+2)}$$

with $\operatorname{Re} p_j > 0$, to the particular solution w_q^*, and adjoin the resulting function to w_q^{**}, we obtain an integral w_q^{**} for the equation (6.2.63) with an estimate for it given by (6.2.67) with possibly an enhanced value for A being necessary therein.

If, on the other hand, $q \geqq s$, then we cannot add to a particular solution of the equation (6.2.63) a particular solution of the equation $L(w)=0$, without violating (6.2.49) for $N=q$.

Thus, in all cases, our solution has the form

$$w_q(z) = B_{q+1} z^{-p-q-1} + \omega_q(z), \qquad z \in \mathscr{A}^*,$$

where $\omega_q(z)$ satisfies (6.2.67), and the numbers A, H_0, H_1 and s_0 appearing therein do not depend on q. It is important to note that the summand $B_{q+1}z^{-p-q-1}$ is defined uniquely.

We now estimate $\omega_q^{(\nu)}(z)$. $\omega_q(z)$ satisfies (6.2.66) and (6.2.67). By (6.2.37) and (6.2.56), we have

$$|\omega_q^{(\nu)}(z)| \leqq \frac{1}{|\alpha_0|} \sum_{i_1+\cdots+i_k=k} |A_{i_1,\ldots,i_k}| \left\{ \sum_{r=1}^{k} \prod_{j=1}^{q} \frac{\Gamma(p+i_j)}{\Gamma(p)} |z|^{i_r-\nu} |\omega^{(i_r)}| \right\} + |\omega(z)|.$$

Substituting estimates (6.2.64) and (6.2.67) in the preceding relation, we find

$$|\omega_q^{(\nu)}(z)| < H^* A \frac{(1+H_0+H_1)}{p+q+2} \frac{(\ln|z|)^{q(k+1)(s_0+1)+(k+\nu)s_0}}{|z|^{p+q+\nu+2}} + (1+H_0+H_1)\frac{(\ln|z|)^{q(k+1)(s_0+1)+ks_0}}{|z|^{p+q+\nu+2}}$$

where H^* is a constant such that

$$1 + \frac{1}{|\alpha_0|} \sum_{i_1+\cdots+i_k=k} |A_{i_1,\ldots,i_k}| \prod_{j=1}^{k} \frac{\Gamma(p+i_j)}{\Gamma(p)} < H^*$$

which obviously does not depend on q and z. Since Re $z > x_0$, we obtain from (6.2.54)

$$\begin{aligned} |\omega_q^{(v)}(z)| &< \left(1 + \frac{AH^*}{p+q+2}\right)(1 + H_0 + H_1) \frac{1}{\ln|z|} \\ &\quad \times \frac{(\ln|z|)^{(q+1)(k+1)(s_0+1)+vs_0}}{|z|^{p+q+v+2}} \\ &\leq \left\{\left(1 + \frac{AH^*}{p+2}\right)(1 + H_0 + H_1)\frac{1}{\ln x_0}\right\} \frac{(\ln|z|)^{(q+1)(k+1)(s_0+1)+vs_0}}{|z|^{p+q+v+2}}. \end{aligned} \qquad (6.2.68)$$

Since the constants appearing within the curly brackets do not depend on q, we can choose an x_0 so large and independent of q that

$$\left(1 + \frac{AH^*}{p+2}\right) \frac{1 + H_0 + H_1}{\ln x_0} < 1. \qquad (6.2.69)$$

Consequently, according to (6.2.67)–(6.2.69),

$$|\omega_q^{(j)}(z)| < \frac{(\ln|z|)^{(q+1)(k+1)(s_0+1)+js_0}}{|z|^{p+q+j+2}}, \qquad j = 0, 1, \ldots, v.$$

Recalling (6.2.48), (6.2.50), and (6.2.67), we finally obtain

$$y(z) = z^{-p} + B_1 z^{-p-1} + \cdots + B_{q+1} z^{-p-q-1} + R_{q+1}(z)$$

where $R_{q+1}(z) = \omega_q(z)$ satisfies the inequality (6.2.49) with $N = q + 1$. Lemma 6.2.2 stands proved.∎

The proof of Theorem 6.2.10 is now easily concluded in a few words. In fact, as we have seen, the coefficients in the expansion (6.2.48) are uniquely defined, so that they must coincide with the corresponding coefficients in the expansion of the function $(z - B_1/p)^{-p}$ in powers of z^{-1}, which is a solution of (6.2.38). Further, $R_N(z) \to 0$ as $N \to \infty$ in the set $\mathscr{A}^*$, so that $y(z) \equiv (z - B_1/p)^{-p}$.∎

We shall now show that it is possible to choose the polynomial $P(u_1, \ldots, u_n)$, weakening condition (1), in such a way that the relation (6.2.36) will be characteristic not only of r.v.'s X_j having a gamma distribution, but of X_j distributed according to a mixture of gamma distributions.

THEOREM 6.2.11 (cf. [142]). Let $X_1, \ldots, X_n$, $n \geqq 2$, be positive i.i.d.r.v.'s with density function $\phi(x)$ and having moments of all orders; further, for some $p > 0$, in an arbitrary interval $(0, \varepsilon)$, let $\phi(x)$ be of the form

$$\phi(x) = A(1 + o(1))x^{p-1}. \tag{6.2.70}$$

Then, for any choice of distinct positive numbers $p_1, \ldots, p_q$, it is possible to indicate an equation of the form (6.2.38) such that, in the class of functions $y(z) = \int_0^\infty e^{-zx}\phi(x)\,dx$, all the solutions of this equation have the form

$$y(z) = \sum_{i=1}^{q} \frac{B_i}{(z - a_i)^{p_i}} \tag{6.2.71}$$

where the a_i and B_i are arbitrary positive constants. Thus, for the corresponding polynomial $P(u_1, \ldots, u_n)$, the condition (6.2.36) turns out to be characteristic of a mixture of gamma distributions $G(p_i, a_i^{-1})$.

***Proof*:** We shall construct the desired equation of the form (6.2.38) with undetermined coefficients $A_{i_1 \ldots i_k}$. Substituting (6.2.71) in (6.2.38), we obtain an equation which must be satisfied by the coefficients $A_{i_1 \ldots i_k}$. On the left side of this equation stands a linear combination of summands of the form

$$\frac{\prod_{i=1}^{q} B_i^{m_i}}{\prod_{i=1}^{q} (z_i - a_i)^{p_i m_i + n_i}} \tag{6.2.72}$$

where $\sum_1^q m_i = k$, $\sum_1^q n_i = k$. Equating to zero the coefficients of the various distinct terms of the form (6.2.72)—of course, after collecting together similar terms—we obtain a system of linear equations for determining the coefficients $A_{i_1 \ldots i_k}$. We shall further impose on the A's the condition that the equation in μ

$$\sum_{i_1 + \cdots + i_k = k} A_{i_1, \cdots i_k} \left\{ \sum_{r=1}^{k} \prod_{\substack{j=1 \\ j \neq r}}^{k} \frac{\Gamma(p_i + i_r)\Gamma(p_i + \mu + i_r)}{\Gamma(p_i)\Gamma(p_i + \mu)} \right\} = 0 \tag{6.2.73}$$

$$i = 1, \ldots, q$$

should have distinct negative roots. It is not difficult to check that the above conditions can be fulfilled.

We shall now show that in the class of functions $y(z)$ of interest to us, there are no solutions other than (6.2.71). As in Theorem 6.2.10, it can be shown that in the set

$$D = \{\text{Re } z > x_0, |\arg z| < (1/2)\pi - \delta_0\} \text{ where } x_0 > 0,\ \delta_0 > 0,$$

the solution of the equation which we have set up has the form

$$y(z) = \frac{1 + w(z)}{z^p},$$

$$y^{(i)}(z) = \frac{[(-1)^j p(p+1)\cdots(p+j-1) + w_j(z)]}{z^{p+j}}$$

$$j = 1, 2, \ldots$$

where $|w(z)| \leqq w^*(|z|)$, with $w^*(u)$ a nonincreasing function such that $w^*(u) \to 0$ as $u \to \infty$. Replacing $y(z)$ by $z^{-p} + w(z)$ as earlier, we single out the linear terms. An estimate for the remainder R^* in this case has the form

$$|R^*| \leqq A \frac{[w^*(x)]^2}{|z|^{kp}}, \qquad x = \text{Re } z. \tag{6.2.74}$$

This estimate is obtained by the same method as in the proof of Theorem 6.2.10; moreover, the factor $(\log |z|)^{s_0}$ is absent, since (6.2.73) does not have multiple roots. Repeating this argument, we arrive at the estimate

$$|R^*| \leqq A^{2^m - 1}[w^*(x)]^{2^m}|z|^{-kp}.$$

If D^* is the subset of D where $Aw^*(x) < 1$, then, proceeding as in the proof of Lemma 6.2.2, we can show that, in D^*,

$$|R^*| \leqq B|z|^{-kp-v}$$

where v is a number given beforehand, satisfying the condition

$$v < \min_{s,t} |p_s - p_t| < 2v.$$

The subsequent proof is the same as for Theorem 6.2.10.∎

6.3 CHARACTERIZATIONS OF THE NORMAL LAW

In Chapter 4, we considered characterizations of the normal law through the property of independence of two statistics, at least one of which is nonlinear. In this section, we consider such characterizations through regression properties of a nonlinear statistic on the sample mean.

THEOREM 6.3.1. Let $X_1, \ldots, X_n$ bc i.i.d.r.v.'s with Var $X_1 = \sigma^2 < \infty$. Let $L = X_1 + \cdots + X_n$ and $Q = \sum a_{jk} X_j X_k + \sum b_j X_j$. If $B_1 = \sum a_{jj} \neq 0$, $B_2 = \sum a_{jk} = 0$, and $B_3 = \sum b_j = 0$, then the constancy of regression of Q on L is a characteristic property of the normality of the variables X_j.

***Proof*:** By Lemma 1.1.1, we have

$$E(Qe^{itL}) = E(Q) \cdot E(e^{itL}). \tag{6.3.1}$$

Following the scheme of the argument for Theorem 6.2.8, we obtain the differential equation

$$\phi''(t) = -\gamma^2 \qquad \text{for } |t| < \delta \tag{6.3.2}$$

where $\phi = \log f$ and f is the c.f. of the X_j. We find from (6.3.2) that ϕ is a polynomial of the second degree in $|t| < \delta$, and hence that f is a normal c.f.

Conversely, if the X_j are normal, then (6.3.2) holds and hence

$$E(Q|L) = EQ. \blacksquare$$

THEOREM 6.3.2. Let $X_1, \ldots, X_n$ be i.i.d.r.v.'s with $E|X_1|^p$ finite. Let $L = X_1 + \cdots + X_n$ and $P = P(X_1, \ldots, X_n)$ a homogeneous polynomial statistic of degree $p \geqq 2$, which is an unbiased estimator of the p-th semi-invariant (cumulant) κ_p. Then P has constant regression on L if and only if the X_j are normal variables.

***Proof*:** Again, we have the equation

$$E(Pe^{itL}) = E(P) \cdot E(e^{itL})$$

from which, using straightforward calculations, we obtain the differential equation of the p-th order

$$\phi^{(p)}(t) = i^p \kappa_p \qquad \text{for } |t| < \delta \tag{6.3.3}$$

where $\phi = \log f$ and f is the c.f. of the X_j. It follows from (6.3.3) that ϕ is a polynomial of degree at most p, in a neighborhood of the origin. Then, from the theory of analytic c.f.'s and Marcinkiewicz theorem (Lemma 1.4.2), we obtain the conclusion that the X_j are normal.

The converse is easily verified as in the preceding theorem. $\blacksquare$

6.4 CHARACTERIZATIONS OF DISCRETE DISTRIBUTIONS

In this section, we show that some of the well-known discrete distributions are characterized by the property of constancy of regression of suitable (non-homogeneous) polynomial statistics on the sample mean. The results presented here are due to Lukacs [94].

We shall adopt the following notation.

$X_1, \ldots, X_n$ are i.i.d.r.v.'s.

$$L_j = X_1^j + \cdots + X_n^j, \; j = 1, 2, \ldots. \tag{6.4.1}$$

F denotes the d.f. of the X_i, f their c.f., and $\phi(t) = \log f(t)$ in some interval $|t| < \delta$.

The semi-invariants are denoted by the symbols $\kappa_1, \kappa_2, \ldots$ where κ_1 is the mean.

THEOREM 6.4.1. Let

$$T = nL_4 + (n-4)L_3L_1 + (3-2n)L_2^2 + L_2L_1^2 - nL_3 + (n+1)L_2L_1 - L_1^3 \tag{6.4.2}$$

and let the third moment of the X_j exist.

Then the regression of T on L_1 is zero if and only if f has the form given by

$$f(t) = \left(\frac{\kappa_2}{\kappa_1} + \frac{\kappa_1 - \kappa_2}{\kappa_1} e^{it}\right)^{\kappa_1^2/(\kappa_1 - \kappa_2)}. \tag{6.4.3}$$

***Proof*:** We note that T has the explicit representation

$$T = (n-2)\sum X_j^3 X_k - 4(n-2)\sum X_j^2 X_k^2 + 2\sum X_j^2 X_k X_m + (n-2)\sum X_j^2 X_k - 6\sum X_j X_k X_m$$

where the summations are all taken over all distinct indices j, k, m.

To prove the sufficiency, we use the relation $E(Te^{itL_1}) = 0$. Noting that $E(X^k e^{itX}) = i^{-k} f^{(k)}(t)$ for $k = 0, 1, 2, 3$, we hence obtain the differential equation

$$f'''f'f^{n-2} - 2(f'')^2 f^{n-2} + if''f'f^{n-2} + f''(f')^2 f^{n-3} - i(f')^3 f^{n-3} = 0 \tag{6.4.4}$$

which transforms into the equation

$$\phi'''\phi' - 2i(\phi'')^2 - \phi'\phi'' = 0, \ |t| < \delta. \tag{6.4.5}$$

Using the initial conditions $\phi(0) = 0$, $\phi'(0) = i\kappa_1$, etc., we find that the solution of (6.4.5) is, for $|t| < \delta$,

$$\phi(t) = \frac{\kappa_1^2}{\kappa_1 - \kappa_2} \log\left[\frac{\kappa_2}{\kappa_1} + \frac{\kappa_1 - \kappa_2}{\kappa_1} e^{it}\right] \tag{6.4.6}$$

so that (6.4.3) is valid for such t. It follows from the theory of analytic c.f.'s that it is in fact valid for all t. This proves the sufficiency part of the theorem.

The necessity part is established by directly verifying that if f has the form (6.4.3), then (6.4.4) and the relation $E(Te^{itL_1}) = 0$ are valid. ∎

We examine conditions on κ_1 and κ_2 under which $f(t)$ of the form (6.4.3) will be a c.f.

(i) If $\kappa_1 = 0$ or $\kappa_2 = 0$ or $\kappa_1 = \kappa_2$, then f is the c.f. of a degenerate law.

(ii) Let $0 < \kappa_2/\kappa_1 = q < 1$ and $\kappa_1^2/(\kappa_1 - \kappa_2)$ be $= n$, a positive integer; then $f(t) = (q + pe^{it})^n$ is the c.f. of a Binomial distribution.

(iii) Let $0 < \kappa_1/\kappa_2 < 1$, and let $\kappa_1/\kappa_2 = p$, $q = 1 - p$, and $\kappa_1^2/(\kappa_2 - \kappa_1) = k > 0$. Then $f(t) = p^k(1 - qe^{it})^{-k}$ is the c.f. of a negative binomial distribution.

(iv) Let $\kappa_1 < 0$ and let $p = -\kappa_1/(\kappa_2 - \kappa_1)$, $q = 1 - p$, and $m = \kappa_1^2/(\kappa_2 - \kappa_1)$. Then, $f(t) = p^m e^{-itm}(1 - qe^{-it})^{-m}$ is the c.f. of a distribution symmetric to a negative binomial, shifted to the left by m units.

THEOREM 6.4.2. Let

$$T = \frac{n+1}{n-1} L_2 - \frac{2}{n-1} L_1^2 - L_1 = \sum X_j^2 - \frac{4}{n-1} \sum X_j X_k - \sum X_j \quad (6.4.7)$$

and $\int x^2 \, dF(x) < \infty$. Then T has zero regression on L_1 if and only if F has a geometric distribution.

***Proof*:** Following the scheme of the proof of Theorem 6.4.1, we obtain the differential equation

$$\phi''(t) = i\phi'(t) + [\phi'(t)]^2 \quad (6.4.8)$$

with the initial conditions $\phi(0) = 0$, $\phi'(0) = i\kappa_1$. Then

$$f(t) = (1 - q)(1 - qe^{it})^{-1} \quad (6.4.9)$$

where $q = \kappa_1/(1 + \kappa_1)$ so that f is the c.f. of a geometric distribution.

The necessity is proved by tracing the argument backwards.∎

THEOREM 6.4.3. Let K_i be the i-th semi-invariant of the sample, and suppose EK_2 exists. Then $K_2 - K_1$ has constant regression on K_1 if and only if F has a Poisson-type d.f.

***Proof*:** Using the condition

$$E(K_2 - K_1)e^{itK_1} = E(K_2 - K_1) \cdot Ee^{itK_1}$$

we obtain a differential equation which f must satisfy:

$$-f''f^{n-1} + (f')^2 f^{n-2} + if'f^{n-1} = cf^n \quad (6.4.10)$$

so that, for $\phi = \log f$, we obtain the differential equation

$$-\phi'' + i\phi' = c, \qquad |t| < \delta \quad (6.4.11)$$

which has the solution

$$\phi(t) = i\alpha t + \gamma(e^{it} - 1), \qquad |t| < \delta \quad (6.4.12)$$

whence, for such t,

$$f(t) = \exp{[i\alpha t + \gamma(e^{it} - 1)]}, \qquad \gamma > 0. \quad (6.4.13)$$

It then follows from the theory of analytic c.f.'s that (6.4.13) holds for all t, thus defining the c.f. of a Poisson distribution up to a shift parameter. If $E(K_2 - K_1) = 0$, then we obtain a Poisson distribution.

The sufficiency is readily verified.∎

CHAPTER 7

Characterizations of Distributions through the Properties of Admissibility and Optimality of Certain Estimators

It is well-known that the sample mean $\overline{X}$, considered as an estimator of the population mean, possesses a number of optimal properties when the population is normal. The question naturally arises whether the normal distribution is the only one for which $\overline{X}$ possesses one of these properties. In this chapter, we study conditions under which such uniqueness holds. Similar results are obtained for some generalizations of the normal law and also for the gamma distribution.

We establish that, in general, linear estimators of a location parameter are not admissible for non-normal populations. The knowledge of the underlying distribution will always enable us to improve a linear estimator uniformly. Thus a justification of a linear estimator is possible only through normality or in the absence of any knowledge about the parent distribution.

7.1 SOME BASIC THEOREMS FROM THE THEORY OF ESTIMATION

Before proceeding to the main part of this chapter, we formulate some results from estimation theory, which are used for the proofs of the admissibility and optimality of estimators of parametric functions.

We consider a probability space $(\mathscr{X}, \mathscr{A}, P_\theta)$, where $\mathscr{X}$ is the space of observations x, $\mathscr{A}$ is a σ-algebra of events, and $\{P_\theta\}$ is a family of probability measures (distributions) indexed by a parameter $\theta \in \Theta$. Let a statistic $T(x)$ be used as an estimator for a parametric function $t(\theta)$, and $r(T, t)$ be a nonnegative loss function. The mathematical expectation of the loss function,

$$R(T, \theta) = E_\theta r(T, t)$$

is called the *risk* of the estimator T when the true value of the parameter is θ.

As examples of loss functions, we may use:

(a) the quadratic or Gaussian: $r(T, t) = (T - t)^2$;
(b) the Laplacian : $r(T, t) = |T - t|$;
(c) the function : $r(T, t) = \begin{cases} 0 \text{ for } |T - t| \leqq b, \\ 1 \text{ for } |T - t| > b, \end{cases}$

for some $b > 0$, encountered in the problem of confidence estimation (see section 7.9).

DEFINITION 7.1.1. An estimator $T'(x)$, belonging to some class K of estimators of the parametric function $t(\theta)$, will be called *admissible* in this class, under the loss function $r(T, t)$, if there exists no member T of K such that

$$R(T, \theta) \leqq R(T', \theta) \qquad \text{for all } \theta \tag{7.1.1}$$

and further strict inequality holds for at least one θ.

If an estimator T' is admissible in the class of all estimators, then we shall call it *absolutely admissible*.

DEFINITION 7.1.2. An estimator $T^0 \in K$ will be called *optimal* in the class K of estimators of the parametric function $t(\theta)$, under the loss function $r(T, t)$, if for all T in K

$$R(T^0, \theta) \leqq R(T, \theta), \qquad \theta \in \Theta.$$

DEFINITION 7.1.3. An estimator $T_0 \in K$ will be called *optimal at the point* θ_0 (or *locally optimal*) in the class K under the loss function $r(T, t)$, if for all T in K

$$R(T_0, \theta_0) \leqq R(T, \theta_0).$$

We shall now cite without proof some basic theorems from the theory of estimation.

THEOREM 7.1.1. (the Rao-Cramer inequality). Suppose the parameter space Θ is some interval of the real line, all the distributions P_θ are absolutely continuous with respect to some measure μ, with $p(x, \theta) = dP_\theta/d\mu$ as the densities. If T is a statistic such that $E_\theta T = t(\theta) + b(\theta)$, then, under certain "regularity" conditions,

$$E_\theta[T - t(\theta)]^2 \geqq [b(\theta)]^2 + \frac{[t'(\theta) + b'(\theta)]^2}{I(\theta)} \tag{7.1.2}$$

where $I(\theta) = E_\theta(\partial \log p/\partial\theta)^2$ is the Fisherian amount of information.

For the proof, *vide* [131], pp. 323–325.

THEOREM 7.1.2. Suppose that T is an estimator for $t(\theta)$ and that the loss function $r(T, t)$ is convex in T. If the family $\{P_\theta\}$ possesses a sufficient statistic S, then there exists an estimator T_0, depending only on S, such that

$$E_\theta r(T_0, t(\theta)) \leqq E_\theta r(T, t(\theta)) \qquad \text{for } \theta \in \Theta.$$

For the proof, *vide* [7, 69, 124, 131].

It follows from Theorem 7.1.2 that estimators which are not functions of the sufficient statistic turn out to be inadmissible under loss functions which are convex with respect to the first argument. The method of obtaining T_0 from T is known as "Rao-Blackwellization." (see also section 12.1).

If the quadratic loss function is used, then particular attention is given to unbiased estimators of parametric functions; in this connection, *vide* [77], p. 26.

DEFINITION 7.1.4. T will be called an unbiased estimator of $t(\theta)$ if $E_\theta T \equiv t(\theta)$ for all $\theta \in \Theta$.

It can be shown (cf. for instance [131]) that, in Theorem 7.1.2, if T is an unbiased estimator of $t(\theta)$, then T_0 can be chosen to be unbiased.

DEFINITION 7.1.5. Let the family $\{P_\theta\}$, in addition to possessing a sufficient statistic S, have the property that, for a statistic $\psi = \psi(S)$, it follows from the relation $E_\theta \psi \equiv 0$ for all θ that $\psi = 0$ almost everywhere with respect to all the measures P_θ. Such a family will be said to be *complete* with respect to the sufficient statistic S. The sufficient statistic itself will also be called *complete* in such a situation.

For complete families, every parametric function $t(\theta)$ obviously admits at most one unbiased estimator depending on the sufficient statistic (if estimators coinciding almost everywhere (a.e.) with respect to each of the measures P_θ are identified). It follows from Theorem 7.1.2 that, for complete families, every statistic $T_0 = T_0(S)$ is optimal in the class of unbiased estimators of the parametric function $t(\theta) = E_\theta(T_0)$. This circumstance (first observed in [183]) will be used frequently in what follows. (See [131] pp. 320–322 for details).

The following theorem gives simple criteria for the completeness of the family $\{P_\theta\}$.

THEOREM 7.1.3 (Lehmann, [77]). If the distribution of a sufficient vector statistic $S = (S_1, \ldots, S_m)$ is given in R^m by the density function (relative to Lebesgue measure)

$$\pi(s, \theta) = c(\theta)q(s) \exp\left(\sum_1^m c_i(\theta)s_i\right)$$

where $s = (s_1, \ldots, s_m)$ and the vector $(c_1(\theta), \ldots, c_m(\theta))$ ranges over some region in R^m as θ ranges over Θ, then the family of distributions $\{P_\theta\}$ is complete with respect to the sufficient statistic S.

7.2 NECESSARY AND SUFFICIENT CONDITIONS FOR OPTIMALITY IN THE CLASS OF UNBIASED ESTIMATORS

In this section, the concept of optimality of an estimator relates to the quadratic loss function.

Let us denote by L_θ^2 the Hilbert space of measurable functions on $(\mathscr{H}, \mathscr{A})$ which are square-integrable with respect to P_θ with the usual scalar product. Let $L^2 = \bigcap_{\theta \in \Theta} L_\theta^2$.

DEFINITION 7.2.1. A statistic $h = h(x)$ will be called an *unbiased estimator of zero* (u.e.z.) if $E_\theta h \equiv 0$ for all $\theta \in \Theta$.

We shall denote by H_θ the set of all u.e.z. in L_θ^2, and let $H = \bigcap_{\theta \in \Theta} H_\theta$.

In spite of its simplicity, the following lemma, originating in Fisher's work [177], turns out to be useful in many characterization problems.

LEMMA 7.2.1 (cf. [126]). An estimator $T \in L^2$ (respectively, $T \in L_\theta^2$) is optimal (respectively, locally optimal at θ) in L^2 as an unbiased estimator of $t(\theta) = E_\theta T$, if and only if

$$E_\theta(Th) = 0 \qquad \text{for all } h \in H, \qquad \theta \in \Theta \tag{7.2.1}$$

(respectively, if and only if, for the given θ, $E_\theta(Th) = 0$ for all $h \in H_\theta$).

Proof: *Necessity*. Let $T \in L_\theta^2$ be locally optimal at θ as an unbiased estimator of $t(\theta)$, and $h \in H_\theta$. For any constant λ, $T + \lambda h \in L_\theta^2$ and is an unbiased estimator for $t(\theta)$. Further,

$$E_\theta(T + \lambda h - t)^2 = E_\theta(T - t)^2 + 2\lambda E_\theta(Th) + \lambda^2 E_\theta h^2.$$

If $E_\theta(Th) \neq 0$, then by means of a suitable choice of λ, it is always possible to arrange for

$$2\lambda E_\theta(Th) + \lambda^2 E_\theta(h^2) < 0.$$

But this would contradict the optimality of the estimator T at θ. Hence $E_\theta(Th) = 0$ for $h \in H_\theta$.

Sufficiency. Let $T' \in L_\theta^2$ be another unbiased estimator for $t(\theta)$. Then $T' - T = h \in H_\theta$ and we have

$$E_\theta(T' - t)^2 = E_\theta(T - t)^2 + 2E_\theta(Th) + E_\theta(h^2) \geqq E_\theta(T - t)^2$$

since, by the conditions of the lemma, $E_\theta(Th) = 0$.

The assertion of the lemma regarding the optimality of T in L^2 in the class of unbiased estimators of $t(\theta)$ belonging to L^2 is similarly proved.▌

The lemma just proved works only in those cases when the family of distributions in question possesses a fairly abundant supply of u.e.z.'s of a suitable kind. In particular, such turns out to be the case with distributions generated by random samples from a population with parameters of location (shift) and scale. Condition (7.2.1) is necessary not only for optimality but for admissibility of the estimator T in those cases when it turns out that $E_\theta(T - t)^2$ is constant and for the u.e.z's h which are used, $E_\theta(Th)$ and $E_\theta(h^2)$ are constant.

We remark in passing that for loss functions other than quadratic we do not have analogues of Lemma 7.2.1 (and not even a complete idea as to in what classes of estimators it is appropriate to pose the question of conditions for optimality).

We now apply Lemma 7.2.1 to prove that the property of being an optimal unbiased estimator is preserved under functional transformations of a fairly general kind. Suppose that, for the given family $\{P_\theta\}$, the set H' of all u.e.z.'s h with $E_\theta h^4 < \infty$ for all $\theta \in \Theta$ is dense in H in norm in each of the L_θ^2. *We shall refer to this as the condition* (*C*).

If

$$E_\theta(|T|^k) < \infty, \ k = 1, 2, \ldots \tag{7.2.2}$$

then, we shall denote by $L_\theta^2(T)$ the closure in the L_θ^2 norm of the set of all polynomials in T, and by $L^2(T)$ the intersection of all the $L_\theta^2(T)$.

THEOREM 7.2.1 (cf. [126]). Let condition (C) be satisfied by the family of distributions $\{P_\theta\}$. If $T \in L^2$ is the optimal unbiased estimator in L^2 of the function $t(\theta) = E_\theta(T)$ satisfying (7.2.2) for all θ, then every member β of $L^2(T)$ will be optimal in L^2 among the unbiased estimators of its mathematical expectation $b(\theta) = E_\theta \beta$.

***Proof*:** If $T \in L^2$ is optimal in L^2 as an unbiased estimator of $E_\theta T$, then, by Lemma 7.2.1, for every h' in H' and all θ, we will have $E_\theta(Th') = 0$. But, by the Cauchy-Bunyakovski inequality, $E_\theta(T^2h'^2) \leqq (E_\theta T^4)^{1/2} \cdot (E_\theta h'^4)^{1/2}$. We conclude hence that if T satisfies (7.2.2), then $Th' \in H$ for all $h' \in H'$. Then $E_\theta(T^2h') = E_\theta(T \cdot Th') = 0$, for all θ. Since H' is dense in H in the norm of each L_θ^2 by assumption, we obtain from the preceding relation that $E_\theta(T^2h) = 0$ for all θ and all h in H. Similarly, it is shown that, for any $k = 0, 1, \ldots$.

$$E_\theta(T^k h) = 0 \text{ for all } \theta \text{ and } h \text{ in } H. \tag{7.2.3}$$

Since, by the definition of $L^2(T)$, the set of polynomials in T is dense in $L^2(T)$ in the norm of each L_θ^2, we deduce from (7.2.3) that for each element β of $L^2(T)$

$$E_\theta(\beta h) = 0, \text{ for } h \in H \text{ and all } \theta.$$

By Lemma 7.2.1, β is optimal in L^2 as an unbiased estimator of $E_\theta \beta$.▮

We direct our attention to the circumstance that the property of optimal unbiased estimators of being "efficient" in the sense defined by the Rao-Cramer inequality (the right side of (7.1.2) is equal to the variance of such estimators) is not preserved under functional transformations. But, under the conditions of Theorem 7.2.1, functions of efficient unbiased estimators will be optimal unbiased estimators of their respective mathematical expectations.

Remark 1. The condition (C) is perhaps not essential for the validity of Theorem 7.2.1. In any case, it can be omitted if there exists a constant C such that $|T(x)| < C$ with P_θ-probability one for all θ.

Remark 2. The property of an estimator $T \in L_\theta^2$ of being a locally optimal unbiased estimator of $t(\theta)$ is, in general, not preserved under functional transformations.

7.3 SHIFT PARAMETER. PITMAN ESTIMATOR

We assume that the space of observations is R^n and the parameter space Θ is R^1. If the distribution P_θ depends on the parameter θ in the following manner,

$$P_\theta(A) = \int \cdots \int_A dF(x_1 - \theta, \ldots, x_n - \theta) \tag{7.3.1}$$

then we say that θ is a *shift parameter*. Distributions of the form (7.3.1) arise in the considerations of schemes of linear measurements, when the observations have the form

$$X_i = \theta + \varepsilon_i, \qquad i = 1, \ldots, n$$

where the errors ε_i (in general, dependent) have the joint d.f. $F(x_1, \ldots, x_n)$.

For estimating the parameter θ, it is natural to choose a class of estimators $\tilde{\theta} = \tilde{\theta}(X_1, \ldots, X_n)$ satisfying for all real c the condition

$$\tilde{\theta}(x_1 + c, \ldots, x_n + c) = \tilde{\theta}(x_1, \ldots, x_n) + c. \tag{7.3.2}$$

Such estimators were first studied by Pitman [42]. We shall call these estimators *regular*, and the class of regular estimators will be denoted by $\mathscr{T}$.

If the loss function depends only on the difference between the arguments, $r(\tilde{\theta}, \theta) = r(\tilde{\theta} - \theta)$, then it is easily seen that for estimators $\tilde{\theta}$ in $\mathscr{T}$, the risk

$$R(\tilde{\theta}, \theta) = E_\theta r(\tilde{\theta} - \theta) = \text{constant}$$

does not depend on $\theta \in R^1$. Therefore, for such loss functions, an estimator $\tilde{\theta}$ in $\mathscr{T}$ is either optimal in $\mathscr{T}$ or inadmissible in $\mathscr{T}$.

DEFINITION 7.3.1. An estimator $\hat{\theta}$ optimal in the class $\mathscr{T}$ according to $R(\hat{\theta}, \theta) = \min_{\tilde{\theta} \in \mathscr{T}} R(\tilde{\theta}, \theta)$ will be called a Pitman estimator of shift θ, corresponding to the loss function $r(\tilde{\theta} - \theta)$.

If $\min_{\tilde{\theta} \in \mathscr{T}} R(\tilde{\theta}, \theta)$ is attained by a unique statistic $\hat{\theta}$ (we identify statistics coinciding almost everywhere with respect to each of the measures P_θ), then it is clear that a necessary condition for the admissibility of an estimator $\tilde{\theta} \in \mathscr{T}$ is the following,

$$\tilde{\theta} = \hat{\theta} \qquad \text{a.e. } [P_\theta] \text{ for all } \theta \in R^1,$$

which is equivalent to the condition

$$\tilde{\theta} = \hat{\theta} \qquad \text{a.e. } [P_0]. \tag{7.3.3}$$

For certain special loss functions and estimators, condition (7.3.3) turns out to be fairly suitable for characterizing the d.f. $F(\mathbf{x})$ through the property of admissibility of estimator $\tilde{\theta}$.

Suppose that

$$\int x_j^2 \, dF(x_1, \ldots, x_n) < \infty, \qquad j = 1, \ldots, n \tag{7.3.4}$$

and let $L = \sum_1^n c_j X_j$ with $\sum_1^n c_j = 1$ be some linear statistic.

LEMMA 7.3.1.

(i) Under the condition (7.3.4), the estimator given by Rao [126]

$$\begin{aligned} \hat{\theta} &= L - E_0(L \mid X_2 - X_1, \ldots, X_n - X_1) \\ &= L - E_0(L \mid \mathbf{Y}), \qquad \mathbf{Y} = (X_2 - X_1, \ldots, X_n - X_1) \end{aligned} \tag{7.3.5}$$

is the Pitman estimator corresponding to quadratic loss, where, for $\tilde{\theta} \in \mathscr{T}$, $E_\theta(\tilde{\theta} - \theta)^2 > E_\theta(\hat{\theta} - \theta)^2$, $\theta \in R^1$ except when $\tilde{\theta} = \hat{\theta}$ a.e. $[P_0]$ (and consequently $\tilde{\theta} = \hat{\theta}$ a.e. $[P_\theta]$ for all θ).

(ii) If $F(x)$ is absolutely continuous with $f(x)$ as the density function (with respect to Lebesgue measure), then

$$\hat{\theta} = \frac{\int_{-\infty}^{\infty} u f(x_1 - u, \ldots, x_n - u) \, du}{\int_{-\infty}^{\infty} f(x_1 - u, \ldots, x_n - u) \, du}.$$

Proof: (i) Let $\tilde{\theta} \in \mathscr{T}$. Then, by (7.3.2), $\tilde{\theta} = \hat{\theta} + \psi(Y)$. If $E_0 \tilde{\theta}^2 = \infty$, then it is clear that $E_\theta(\hat{\theta} - \theta)^2 < E_\theta(\tilde{\theta} - \theta)^2$, $\theta \in R^1$. If $E_0 \tilde{\theta}^2 < \infty$, then we have

$$E_\theta(\tilde{\theta} - \theta)^2 = E_\theta(\tilde{\theta} - \hat{\theta})^2 + 2E_\theta(\tilde{\theta} - \hat{\theta})(\hat{\theta} - \theta) + E_\theta(\hat{\theta} - \theta)^2.$$

But

$$\begin{aligned} E_\theta(\tilde{\theta} - \hat{\theta})(\hat{\theta} - \theta) &= E_0\, \hat{\theta}(\tilde{\theta} - \hat{\theta}) = E_0\{E_0[\hat{\theta}(\tilde{\theta} - \hat{\theta}) \mid \mathbf{Y}]\} \\ &= E_0[(\tilde{\theta} - \hat{\theta})\, E_0(\hat{\theta} \mid \mathbf{Y})] = 0 \end{aligned}$$

where we have used (7.3.5). Hence, $E_0(\hat{\theta} - \theta)^2 \leqq E_\theta(\tilde{\theta} - \theta)^2$ for all $\tilde{\theta} \in \mathscr{T}$ and assertion (i) is proved. Assertion (2) can be proved in a straighforward manner; it is convenient to take $L = (1/n)\sum_1^n X_i$. (ii) is easily established.▮

We note that assertion (i) of Lemma 7.3.1 is in fact a special case of Lemma 7.2.1: indeed, the optimality of the estimator $\hat{\theta}$ in the class $\mathscr{T}$ implies the orthogonality of $\hat{\theta}$ to unbiased estimators of zero of the form $h(X_2 - X_1, \ldots, X_n - X_1)$. We presented a proof of Lemma 7.3.1 only because it relates to the optimality of the estimator in a special class.

LEMMA 7.3.2. Under the condition that

$$\int |x_j| \, dF(x_1, \ldots, x_n) < \infty, \qquad j = 1, \ldots, n$$

the estimator

$$\bar{\theta} = L - \text{med}_0\,(L \mid \mathbf{Y}) \tag{7.3.6}$$

is a Pitman estimator, under the Laplace loss function

$$r(\tilde{\theta} - \theta) = |\tilde{\theta} - \theta|.$$

Remark. Here, by $\text{med}_0\,(L \mid \mathbf{Y})$ will be meant any version of the median of the conditional distribution under P_0 of the statistic L given $\mathbf{Y}$ such that $\text{med}_0\,(L \mid \mathbf{Y})$ is a statistic.

Proof: Let $\tilde{\theta} \in \mathscr{T}$. Then $\tilde{\theta} = L + \psi(\mathbf{Y})$ and we have

$$\begin{aligned} E_\theta |\tilde{\theta} - \theta| &= E_0 |L + \psi(\mathbf{Y})| = E_0\{E_0[|L + \psi(\mathbf{Y})| \mid \mathbf{Y}]\} \\ &\geqq E_0\{E_0(|L - \text{med}_0\,(L \mid \mathbf{Y})| \mid \mathbf{Y})\} = E_0 |\bar{\theta}| = E_\theta |\bar{\theta} - \theta| \end{aligned}$$

since, for every $\mathbf{y}$, $\min_{\psi(\mathbf{y})} E_0(|L + \psi(\mathbf{Y})| \mid \mathbf{Y} = \mathbf{y})$ is attained for $\psi(\mathbf{y}) = -\,\text{med}_0\,(L \mid \mathbf{Y} = \mathbf{y})$.▮

For general loss functions, there is no convenient representation of the Pitman estimator.

7.4 CHARACTERIZATION OF THE NORMAL LAW THROUGH THE ADMISSIBILITY OF THE OPTIMAL LINEAR ESTIMATOR OF THE SHIFT PARAMETER, UNDER QUADRATIC LOSS

We shall assume here that the r.v.'s $X_1, \ldots, X_n$ are distributed independently with d.f.'s $F_1(x-\theta), \ldots, F_n(x-\theta)$ respectively, depending on a shift parameter $\theta \in R^1$, and further

$$0 < \int x^2 \, dF_j(x) = \sigma_j^2 < \infty, \qquad \text{for all } j. \tag{7.4.1}$$

(The case when $\sigma_j = 0$ for some j is trivial, from the point of view of the estimation of θ.) For simplicity, we assume that

$$\int x \, dF_j(x) = 0 \qquad \text{for all } j \tag{7.4.2}$$

so that θ has the significance of being the population mean for the d.f.'s $F_j(x-\theta)$.

Under quadratic loss, the optimal linear unbiased estimator for θ, given the random sample $X_1, \ldots, X_n$ will be

$$L = \sum c_j^0 X_j \qquad \text{where} \qquad c_j^0 = \frac{1/\sigma_j^2}{\sum_1^n 1/\sigma_i^2}. \tag{7.4.3}$$

(If the condition (7.4.2) is not satisfied, and $\int x \, dF_j(x) = b_j$, then, instead of (7.4.3), it will be necessary to consider the estimator $\hat{L} = \sum_1^n c_j^0(X_j - b_j)$; otherwise, all that follows applies without change.)

The following theorem (for identically distributed X_j) was essentially proved in [56] (also cf. [50, 53]).

THEOREM 7.4.1. Let $n \geqq 3$ and the X_j be independent r.v.'s having the d.f.'s $F_j(x-\theta)$ respectively, $j = 1, \ldots, n$, and let the conditions (7.4.1) and (7.4.2) be satisfied. The optimal linear estimator $\hat{L}$ given by (7.4.3) is admissible, under quadratic loss, in the class of all unbiased estimators of θ if and only if all the d.f.'s F_j are normal.

***Proof*:** *Necessity*. We write the Pitman estimator (7.3.5) in Rao's form

$$\hat{\theta} = \hat{L} - E_0(\hat{L} \mid \mathbf{Y}), \qquad \mathbf{Y} = (X_2 - X_1, \ldots, X_n - X_1) \tag{7.4.4}$$

whence it is seen that $\hat{\theta}$ is an unbiased estimator of θ. According to assertion (i) of Lemma 7.3.1, if $\hat{L}$ is admissible in the class of unbiased estimators, then it is necessary that $\hat{L} = \hat{\theta}$ a.e. $[P_0]$, i.e.,

$$E_0(\hat{L} \mid Y) = 0. \tag{7.4.5}$$

The desired result is now obtained immediately from Corollary 1 to Theorem 5.2.1.

Sufficiency. Let F_j be the d.f. of the normal law $N(0, \sigma_j^2)$. From the expression for the joint density function of the X_j,

$$L(x;\theta) = \prod_1^n (2\pi\sigma_j^2)^{-1/2} \exp\left[-\frac{1}{2}\sum_1^n \frac{x_j^2}{\sigma_j^2} + \theta\sum_1^n \frac{x_j}{\sigma_j^2} - \frac{1}{2}\theta^2 \sum_1^n \frac{1}{\sigma_j^2}\right] \quad (7.4.6)$$

it is seen (cf. Theorem 8.2.1) that the statistic $\sum_1^n (X_j/\sigma_j^2)$ (or the equivalent statistic $S = \sum_1^n c_j^0 X_j$) is a sufficient statistic for the family of densities (7.4.6), for $\theta \in R^1$. The distribution of S is normal with mean θ and variance $\sigma^2 = \sum_1^n c_j^{0^2} \sigma_j^2$, and with density

$$p(s, \theta) = \frac{1}{\sigma\sqrt{2\pi}} \exp\left[-\frac{(s-\theta)^2}{2\sigma^2}\right]$$

whence it is seen (cf. Theorem 7.1.3) that S is a complete sufficient statistic. Hence $\hat{\theta} = \sum_1^n c_j^0 X_j$ is the optimal unbiased estimator for θ, This proves the theorem.∎

Corollary 1. If $F_j = F$ for all j ($n \geqq 3$), and F satisfies (7.4.1) and (7.4.2), then the admissibility of the sample mean $\bar{X}$ in the class of unbiased estimators of the shift parameter θ remains a characteristic property of the normal law.

Corollary 2. Let $X_1, \ldots, X_n$ and $Y_1, \ldots, Y_n$ be independent r.v.'s, $n \geqq 3$, where $X_j \sim F_j(x - \theta - \Delta)$, $Y_j \sim G_j(y - \theta)$, F_j and G_j satisfy (7.4.1) and (7.4.2), with $\theta \in R^1$, $\Delta \in R^1$. In order that an estimator of the form $\sum_1^n a_j(X_j - Y_j)$ be admissible in the class of unbiased estimators of the parameter Δ, under quadratic loss, it is necessary and sufficient that all the F_j and G_j be normal, and further that $\sigma_j^2 = \int x^2\, dF_j(x)$ and $\rho_j^2 = \int y^2\, dG_j(y)$ be connected by the relations $\sigma_j^2 = \lambda\rho_j^2$ for all j.

***Proof*:** First of all, it is easily seen that already the admissibility of $\sum_1^n a_j(X_j - Y_j)$ in the class of estimators of the form $\sum_1^n \alpha_j(X_j - Y_j)$ implies that $a_j \neq 0$, for every j. Consider the r.v. $Z_j = X_j - Y_j$. We have

$$P_{\theta,\Delta}[Z_j < u] = \iint_{x-y<u} dF_j(x - \theta - \Delta)\, dG_j(y - \Delta)$$

$$= \iint_{x-y<u-\Delta} dF_j(x)\, dG_j(y) = H_j(u - \Delta),$$

i.e., the d.f. of Z_j depends only on the shift parameter Δ. From the admissibility of the estimator $\sum_1^n a_j Z_j$ in the class of unbiased estimators for Δ, we conclude, by Theorem 7.4.1, that the Z_j are normally distributed. But then, by Cramer's theorem, the X_j and Y_j are also normal: $X_j \sim N(0, \sigma_j^2)$ and $Y_j \sim N(0, \rho_j^2)$. Further, it is easily verified that for the family of normal d.f.'s

$$F_1(x_1 - \theta - \Delta)\cdots F_n(x_n - \theta - \Delta)G_1(y_1 - \theta)\cdots G_n(y_n - \theta) \quad (7.4.7)$$

$\theta \in R^1$, $\Delta \in R^1$, the pair $\sum_1^n c_j^0 X_j$, $\sum_1^n d_j^0 Y_j$, where

$$c_j^0 = \frac{1}{\sigma_j^2}\left(\sum_1^n \frac{1}{\sigma_i^2}\right)^{-1}, \qquad d_j^0 = \frac{1}{\rho_j^2}\left(\sum_1^n \frac{1}{\rho_i^2}\right)^{-1}$$

serve as complete sufficient statistics. By Theorem 7.1.2, from the admissibility of the estimator $\sum_1^n a_j(X_j - Y_j)$ in the class of unbiased estimators of the parameter Δ, it follows that

$$\sum_1^n a_j(X_j - Y_j) = \phi\left(\sum_1^n c_j^0 X_j, \sum_1^n d_j^0 Y_j\right)$$

so that $c_j^0 = d_j^0/\lambda$ and hence $\sigma_j^2 = \lambda\rho_j^2$, $j = 1, \ldots, n$. The estimator $\sum_1^n a_j(X_j - Y_j)$ is the unique unbiased estimator of Δ based on the sufficient statistics and hence is optimal in the class of unbiased estimators of Δ.∎

Remark 1. The result of Theorem 7.4.1 is valid even in the case $n = \infty$, under the only additional condition that $\sum_1^\infty (1/\sigma_j^2) < \infty$; if this sum diverges, it is easy to see that a linear estimator $\sum_1^\infty c_j X_j$ exists with arbitrarily small variance. If this sum converges, then, under the conditions of Theorem 7.4.1, the estimator $\hat{L} = \sum_1^\infty c_j^0 X_j$ with $c_j^0 = (1/\sigma_j^2)(\sum_1^\infty 1/\sigma_i^2)^{-1}$ is admissible in the class of unbiased estimators if and only if all the F_j are normal d.f.'s. For the proof of the normality of the d.f.'s $F_1, \ldots, F_n$, we note that, from $E_0(\hat{L} \mid X_2 - X_1, \ldots) = 0$, it follows that

$$E_0(\hat{L} \mid X_2 - X_1, \ldots, X_n - X_1) = 0. \tag{7.4.8}$$

Since $\hat{L} = \sum_1^n c_j^0 X_j + \sum_{n+1}^\infty c_j^0 X_j$, and $\sum_{n+1}^\infty c_j^0 X_j$ is independent of $(X_2 - X_1, \ldots, X_n - X_1)$, (7.4.8) reduces to the relation

$$E_0\left(\sum_1^n c_j X_j \mid X_2 - X_1, \ldots, X_n - X_1\right) = 0$$

which is identical with (7.4.5). Conversely, if the F_j are all normal, with $F_j \sim N(0, \sigma_j^2)$, where $\sum_1^\infty 1/\sigma_j^2 < \infty$, then it is easy to check (using Theorem 8.2.1 and the results of [11] or [64]) that $\sum_1^\infty c_j^0 X_j$ is a complete sufficient statistic for the family of distributions $\prod_1^\infty F_j(x - \theta)$, $\theta \in R^1$, in R^∞. Hence follows the optimality of this estimator in the class of unbiased estimators of the parameter θ in this case.

Remark 2. For $n = 2$, the admissibility of $\hat{L}$ in the class of unbiased estimators of θ is not a characteristic property of the normal law. In particular, for $F_1 = F_2 = F$, so that $c_1^0 = c_2^0 = 1/2$, (7.4.5) is satisfied by any symmetric d.f. F, so that the Pitman estimator in this case will be $\bar{X}$. Together with the theorem of Stein [141], this guarantees the (even absolute) admissibility of $\bar{X}$ as an estimator of θ for symmetric d.f.'s F subject to the condition $\int |x|^3 \, dF(x) < \infty$.

THEOREM 7.4.2. Under the conditions of Theorem 7.4.1, absolute admissibility of $\hat{L} = \sum_1^n c_j^0 X_j$ as an estimator of θ is a characteristic property of the normality of the F_j.

***Proof*:** Obviously, we only need to prove that, if the F_j are normal, then $\hat{L}$ is absolutely admissible. We shall follow* Hodges and Lehmann [161]. By Theorem 7.1.2, in considering potential competitors to $\hat{L}$, we may restrict ourselves to estimators which are functions of the sufficient statistic $\hat{L}$. It is easily verified that, for an arbitrary estimator $\tilde{\theta}(\hat{L})$ with $E_\theta \tilde{\theta}^2 < \infty$, $\theta \in R^1$, the regularity conditions required for the derivation of the inequality (7.1.2) are satisfied. Setting $E_\theta \tilde{\theta} = \theta + b(\theta)$, $\sigma^2 = \sum_1^n (c_j^0)^2 \sigma_j^2$, we have

$$E_\theta(\tilde{\theta} - \theta)^2 \geqq [b(\theta)]^2 + (1 + b'(\theta))^2 \sigma^2.$$

If the estimator $\hat{L}$ is inadmissible, then for some $\tilde{\theta}$ we must have

$$[b(\theta)]^2 + (1 + b'(\theta))^2 \sigma^2 \leqq E_\theta(\tilde{\theta} - \theta)^2 \leqq E_\theta(\hat{L} - \theta)^2 = \sigma^2 \qquad (7.4.9)$$

for $\theta \in R^1$. It is seen from (7.4.9) that $b'(\theta) \leqq 0$ and $b(\theta)$ is bounded. Hence $\lim b(\theta)$ exists as $\theta \to \mp \infty$, and it follows in turn that $\overline{\lim}\, b'(\theta)$ exists and $= 0$, as $\theta \to \mp \infty$. But then we see from (7.4.9) that $\lim b(\theta) = 0$ as $\theta \to \mp \infty$. $b(\theta)$ being a monotone function, $b(-\infty) = b(+\infty) = 0$ implies that it must be identically zero. Thus, if (7.4.9) is satisfied, then $\tilde{\theta}(\hat{L})$ must be an unbiased estimator of θ. But, as we have seen in the proof of Theorem 7.4.1, $\hat{L}$ is optimal in the class of unbiased estimators of θ. Therefore, we must have $\tilde{\theta}(\hat{L}) = \hat{L}$ almost surely and Theorem 7.4.2 is proved. ▮

7.5 CONDITIONS FOR THE ADMISSIBILITY OF THE SAMPLE MEAN IN SOME SUBCLASSES OF ESTIMATORS OF A SHIFT PARAMETER

7.5.1 Unbiased Estimators Depending Only on the Sample Mean and Another Linear Statistic

We shall suppose here that the r.v.'s $X_1, \ldots, X_n$ are i.i.d. with d.f. $F(x - \theta)$, $\theta \in R^1$. F will be taken to satisfy the conditions

$$\int x \, dF(x) = 0, \qquad \int x^2 \, dF(x) < \infty. \qquad (7.5.1)$$

Suppose, in addition to the sample mean, the value of one other linear statistic $L = \sum_1^n a_j X_j$ is known to the experimenter. For what d.f.'s F does there exist an unbiased estimator for θ, depending only on $\bar{X}$ and L, better than $\bar{X}$ in the sense of quadratic loss? To answer this question, we consider

* The desired result follows also from the general theorem of Stein [141] on the absolute admissibility of Pitman estimators.

the statistic $V = \sum_1^n b_j X_j$, where $b_j = a_j - \bar{a}$, $\bar{a} = (\sum_1^n a_j)/n$. For the estimator $\tilde{\theta} = \bar{X} - E_0(\bar{X} \mid V)$, the relation below is valid.

$$E_\theta(\tilde{\theta} - \theta)^2 = E_\theta(\bar{X} - \theta)^2 - E_0[E_0(\bar{X} \mid V)]^2.$$

Therefore, a necessary condition for $\bar{X}$ to be an estimator which cannot be improved upon, in the class of unbiased estimators depending only on $\bar{X}$ and L, is that

$$E_0(\bar{X} \mid V) = 0. \tag{7.5.2}$$

This condition has been fully studied in sections 5.4 and 5.5, where in particular it has been shown that (7.5.2) does not guarantee normality of the X_j, in general. However, (7.5.2), together with some additional conditions imposed jointly on the b_j and on F, singles out the normal law. We now present one such result, which follows immediately from Theorem 5.5.2.

We define the auxiliary function

$$G(\lambda) = \sum_{b_i > 0} b_i^{\lambda - 1} - \sum_{b_i < 0} |b_i|^{\lambda - 1}$$

and assume that $G(\lambda) \not\equiv 0$. Let σ be the least upper bound of the real parts of the zeros of $G(\lambda)$.

THEOREM 7.5.1. Let $X_1, \ldots, X_n$ be a random sample from a population with d.f. $F(x - \theta)$, $\theta \in \Theta$, satisfying the condition (7.5.1) and the supplementary condition

$$\int x^{2m} \, dF(x) < \infty, \qquad \text{where} \qquad m = [1 + \sigma/2].$$

Then the admissibility of $\bar{X}$ as an estimator of θ, under quadratic loss, in the class of unbiased estimators of the type $\tilde{\theta}(\bar{X}, \sum_1^n a_j X_j)$ is a characteristic property of the normal law.

7.5.2 The Class of Polynomial Estimators

We shall retain the assumptions of the preceding section, concerning the i.i.d. nature of the r.v.'s $X_1, \ldots, X_n$ and that $X_i \sim F(x - \theta)$, $\theta \in R^1$. We shall further assume that F satisfies (7.5.1) and that, for some integer $k \geqq 1$,

$$\mu_{2k} = \int x^{2k} \, dF(x) < \infty. \tag{7.5.3}$$

In this case, the class of all polynomials $Q(X_1, \ldots, X_n)$ of degree $\leqq k$ constitute a Hilbert space M_k under the usual scalar product, $(Q_1, Q_2) = E_0(Q_1 Q_2)$. M_k contains sufficiently many unbiased estimators of θ (all of them, being elements of M_k, have finite variance); such, for instance, will be estimators of the form $\bar{X} + Q(X_2 - X_1, \ldots, X_n - X_1)$, where the polynomial Q is subject only to the condition $E_0 Q = 0$. In this case, $E_\theta Q = 0$. Let us denote by P_k, $P_k \subset M_k$, the class of all members of M_k which are unbiased estimators of θ.

THEOREM 7.5.2. (cf. [50, 53]) Let $n \geqq 3$ and $X_1, \ldots, X_n$ be a random sample from a population with d.f. $F(x - \theta)$, $\theta \in R^1$, and satisfying conditions (7.5.1) and (7.5.3). Then the admissibility of $\overline{X}$ as an estimator of θ in the class P_k, under quadratic loss, is characteristic of d.f.'s F such that their first $k + 1$ moments coincide with the corresponding moments of some normal law (with zero mean), i.e.,

$$\mu_m = \begin{cases} 0, & \text{if } m \text{ is odd, } 1 \leqq m \leqq k+1 \\ (m-1)!!\,\sigma^m & \text{if } m \text{ is even, } 1 \leqq m \leqq k+1 \end{cases}$$

for some $\sigma \geqq 0$, where $n!! = 1.3.5 \ldots n$ for odd n.

***Proof*:** Denote by Λ_k the subspace of all polynomials of the form $Q(X_2 - X_1, \ldots, X_n - X_1)$ and consider the estimator

$$\hat{\theta}_k = \overline{X} - \hat{E}_0(\overline{X} \mid \Lambda_k) \tag{7.5.4}$$

where $\hat{E}_0(\cdot \mid \Lambda_k)$ is the projection operator on the subspace Λ_k. Estimator (7.5.4) is the analogue of the Pitman estimator (7.3.5). We now establish a few auxiliary results.

LEMMA 7.5.1. For all $\theta \in R^1$,

$$E_\theta \hat{\theta}_k = \theta \tag{7.5.5}$$

and

$$E_\theta(\hat{\theta}_k - \theta)^2 \leqq E_\theta(\overline{X} - \theta)^2 \tag{7.5.6}$$

where the equality or the inequality in (7.5.6) holds uniformly for all $\theta \in R^1$, and equality holds if and only if

$$\hat{E}_0(\overline{X} \mid \Lambda_k) = 0. \tag{7.5.7}$$

***Proof of the Lemma*:** Since $1 \in \Lambda_k$, $E_0(\overline{X} - \hat{E}_0(\overline{X} \mid \Lambda_k)) = 0$, i.e., $E_0 \overline{X} = E_0(\hat{E}_0(\overline{X} \mid \Lambda_k)) = 0$. Hence

$$E_\theta \hat{\theta}_k = E_\theta \overline{X} - E_\theta(\hat{E}_0(\overline{X} \mid \Lambda_k)) = \theta - E_0(\hat{E}_0(\overline{X} \mid \Lambda_k)) = \theta$$

so that $\hat{\theta}_k \in P_k$. Further,

$$\begin{aligned} E_\theta(\overline{X} - \theta)^2 &= E_0(\overline{X}^2) = E_0(\hat{\theta}_k + \hat{E}_0(\overline{X} \mid \Lambda_k))^2 \\ &= E_0(\hat{\theta}_k)^2 + E_0(\hat{E}_0(\overline{X} \mid \Lambda_k))^2 = E_\theta(\hat{\theta}_k - \theta)^2 + E_0(\hat{E}_0(\overline{X} \mid \Lambda_k))^2 \end{aligned}$$

noting that $E_0(\hat{\theta}_k \hat{E}_0(\overline{X} \mid \Lambda_k)) = 0$, in view of the fact that $\hat{\theta}_k \perp \Lambda_k$. ∎

LEMMA 7.5.2. If $n \geqq 3$, then it follows from $\hat{E}_0(\overline{X} \mid \Lambda_k) = 0$ that the first $(k + 1)$ moments of F coincide with the corresponding moments of some normal (possibly degenerate) law.

Proof of the Lemma: We shall use an argument by induction. For $k = 1$, the lemma is trivially true. Since $\Lambda_k \supset \Lambda_{k-1}$, it follows from $\hat{E}_0(\overline{X}|\Lambda_k) = 0$ that $\hat{E}_0(\overline{X}|\Lambda_{k-1}) = 0$. We may therefore assume that the first k moments of F coincide with the corresponding moments of some normal law, and prove only that the moment μ_{k+1} of F coincides with the $(k + 1)$-th moment of the same normal law. The condition (7.5.7) is equivalent to

$$\hat{E}_0(\overline{X}|\Lambda_{k-1}) = 0 \tag{7.5.8}$$

and

$$E_0\{\overline{X}(X_{j_1} - X_1)\cdots(X_{j_k} - X_1)\} = 0 \tag{7.5.9}$$

where (7.5.9) must hold for all choices $j_1, \ldots, j_k$ of integers from among $2, \ldots, n$. The equivalence of (7.5.7) and (7.5.8)–(7.5.9) follows from the fact that Λ_{k-1} and the functions $(X_{j_1} - X_1)\cdots(X_{j_k} - X_1)$ generate Λ_k. Let

$$(X_{j_1} - X_1)\cdots(X_{j_k} - X_1) = (X_{i_1} - X_1)^{\alpha_1}\cdots(X_{i_s} - X_1)^{\alpha_s} \tag{7.5.10}$$

where $i_1, \ldots, i_s$ are all distinct, and $\alpha_1 + \cdots + \alpha_s = k$.
We have [here and below C_k^r stands for the combinatorial symbol $\binom{r}{k}$]

$$(X_{i_1} - X_1)^{\alpha_1}\cdots(X_{i_s} - X_1)^{\alpha_s} = \sum_{p_1=0}^{\alpha_1}\cdots\sum_{p_s=0}^{\alpha_s}(-1)^p C_{\alpha_1}^{p_1}\cdots C_{\alpha_s}^{p_s}$$
$$\times X_{i_1}^{p_1}\cdots X_{i_s}^{p_s} X_1^{k-p}, \qquad p = \sum_1^s p_i \tag{7.5.11}$$

whence

$$E_0\{\overline{X}(X_{i_1} - X_1)^{\alpha_1}\cdots(X_{t_s} - X_1)^{\alpha_s}\}$$
$$= (1/n)\sum_{p_1=0}^{\alpha_1}\cdots\sum_{p_s=0}^{\alpha_s}(-1)^p C_{\alpha_1}^{p_1}\cdots C_{\alpha_s}^{p_s}\mu_{p_1}\cdots\mu_{p_s}\mu_{k+1-p}$$
$$+ (1/n)\sum_{q=1}^{s}\sum_{p_1=0}^{\alpha_1}\cdots\sum_{p_s=0}^{\alpha_s}(-1)^p C_{\alpha_1}^{p_1}\cdots C_{\alpha_s}^{s}\mu_{p_1}\cdots\mu_{p_{q-1}}^{p_{q-1}}\mu_{p_q+1}$$
$$\times \mu_{p_{q+1}}\cdots\mu_{p_s}\mu_{k-p}. \tag{7.5.12}$$

We consider separately the cases of even and odd k.
Case 1. *k Even*. In view of the induction assumption, all the odd order moments up to the $(k - 1)$-th and inclusive of it are zero. Hence we obtain from (7.5.12)

$$nE_0\{\overline{X}(X_{i_1} - X_1)^{\alpha_1}\cdots(X_{i_s} - X_1)^{\alpha_s}\} = \sum{}^*(-1)^p C_{\alpha_1}^{p_1}\cdots C_{\alpha_s}^{p_s}\mu_{p_1}\cdots\mu_{p_s}\mu_{k+1-p}$$
$$+ \sum_{q=1}^{s}\sum{}_q^*(-1)^p C_{\alpha_1}^{p_1}\cdots C_{\alpha_s}^{p_s}\mu_{p_1}\cdots\mu_{p_{q-1}}\mu_{p_q+1}\mu_{p_{q+1}}\cdots\mu_{p_s}\mu_{k-p} \tag{7.5.13}$$

where, in $\sum^*$ the summation extends over even $p_1, \ldots, p_s$ and in $\sum_q^*$ over even $p_1, \ldots, p_{q-1}, p_{q+1}, \ldots, p_s$ and over odd p_q within the limits of summation indicated in (7.5.12). In the sum $\sum^*$, p is always even, hence $k+1-p$ is odd, and if $k+1-p \leqq k-1$, then $\mu_{k+1-p} = 0$ by the induction assumption. In the sum $\sum_q^*$, p is always odd so that $k-p$ is also odd and since $k-p \leqq k-1$, $\mu_{k-p} = 0$. Thus

$$nE_0\{\bar{X}(X_{i_1} - X_1)^{\alpha_1} \cdots (X_{i_s} - X_1)^{\alpha_s}\} = \mu_{k+1}$$

and the condition (7.5.9) is equivalent to

$$\mu_{k+1} = 0. \tag{7.5.14}$$

Case 2. *k Odd.* Again we turn to relation (7.5.13). We divide up the two sums there as follows. Write

$$\sum{}^* = S_0 + S_2 + \cdots + S_{k-1}$$

where S_{2m} is the part of the sum $\sum^*$ corresponding to all $p_1, \ldots, p_s$, $0 \leqq p_s \leqq \alpha_s$, for which $p_1 + \cdots + p_s = 2m$; and

$$\sum_{q=1}^{s} \sum{}_q^* = S_1 + S_3 + \cdots + S_k$$

where S_{2m+1} is the part of the double sum corresponding to those values of $p_1, \ldots, p_s$ for which $p_1 + \cdots + p_s = 2m+1$. We note that, for $n \geqq 3$, polynomials $(X_{i_1} - X_1)^{\alpha_1} \cdots (X_{i_s} - X_1)^{\alpha_s}$ exist, with $s \geqq 2$ and the α's positive.

For what follows, it suffices for us to consider any such polynomial. We shall show that, for $0 < 2m \leqq k-1$, $S_{2m} + S_{2m+1} = 0$. We note that, in view of the conditions $\alpha_1 > 0, \ldots, \alpha_s > 0$, in the sum $\sum_{q=1}^{s} \sum_q^*$, we have always $p_q + 1 \leqq k$ and hence, by the induction assumption,

$$\mu_{p_q+1} = p_q \mu_{p_q-1} \sigma^2 \tag{7.5.15}$$

for some $\sigma^2 \geqq 0$. Then (with the second summation, in the first line below) over $p_1 + \cdots + p_s = 2m+1$),

$$\begin{aligned} S_{2m+1} &= -\sum_{q=1}^{s} \sum{}_q^* C_{\alpha_1}^{p_1} \cdots C_{\alpha_s}^{p_s} \mu_{p_1} \cdots \mu_{p_{q-1}} \mu_{p_{q+1}} \mu_{p_{q+1}} \cdots \mu_{p_s} \mu_{k-1-2m}) \\ &= -\sum_{p_1+\cdots+p_s=2m}^{*} C_{\alpha_1}^{p_1} \cdots C_{\alpha_s}^{p_s} \mu_{p_1} \cdots \mu_{p_s} \mu_{k-1-2m} \\ &\quad \times \sigma^2 \left\{ \frac{C_{\alpha_1}^{p_1+1}}{C_{\alpha_1}^{p_1}} (p_1+1) + \cdots + \frac{C_{\alpha_s}^{p_s+1}}{C_{\alpha_s}^{p_s}} (p_s+1) \right\}. \end{aligned} \tag{7.5.16}$$

But

$$\frac{C_{\alpha_1}^{p_1+1}}{C_{\alpha_1}^{p_1}} (p_1+1) + \cdots + \frac{C_{\alpha_s}^{p_s+1}}{C_{\alpha_s}^{p_s}} (p_s+1) = \sum_{i=1}^{s} (\alpha_i - p_i) = k - 2m.$$

Since

$$\mu_{k-1-2m}\sigma^2(k-2m) = \mu_{k+1-2m}$$

remembering that $m > 0$, we obtain from (7.5.16) that

$$S_{2m} + S_{2m+1} = 0. \tag{7.5.17}$$

In view of (7.5.17), condition (7.5.9) is equivalent to

$$S_0 + S_1 = 0. \tag{7.5.18}$$

But

$$S_0 = \mu_{k+1} \qquad \text{and} \qquad S_1 = -(\alpha_1 + \cdots + \alpha_s)\mu_2\,\mu_{k-1} = -k\sigma^2\mu_{k-1}.$$

Consequently, (7.5.18) is equivalent to

$$\mu_{k+1} = k\sigma^2\mu_{k-1}. \tag{7.5.19}$$

Together with the induction assumption that the first k moments coincide with the corresponding moments of some normal law with mean zero and variance σ^2, (7.5.19) gives

$$\mu_{k+1} = 1 \cdot 3 \cdots k\sigma^{k+1}.$$

Thus Lemma 7.5.2 is proved. ∎

We turn to the proof of Theorem 7.5.3. The normality of the d.f. F is deduced now from the admissibility of $\overline{X}$ in the class P_k immediately. In fact, by Lemma 7.5.1, the estimator $\hat{\theta}_k = \overline{X} - \hat{E}_0(\overline{X}\,|\,\Lambda_k)$ of the parameter θ is unbiased and, for all $\theta \in R_1$, has smaller variance than $\overline{X}$ except in the case $\hat{E}_0(\overline{X}\,|\,\Lambda_k) = 0$. By Lemma 7.5.2, the latter condition guarantees the normality of F.

It remains to show that, if the first $(k + 1)$ moments of F coincide with the corresponding moments of a normal law, then the admissibility of $\overline{X}$ as an estimator of θ in the class P_k follows. Firstly, we note that $\hat{\theta}_k$ is an optimal estimator of θ in the class of polynomials of the form $\overline{X} + Q(X_2 - X_1, \ldots, X_n - X_1)$ of degree at most k. This is proved in precisely the same way as (7.5.6), only $\overline{X}$ is replaced throughout by $\overline{X} + Q$. Further, if the first $(k + 1)$ moments of F coincide with the corresponding moments of some normal law, then for such F, $\hat{\theta}_k = \overline{X}$. In fact, the condition $\hat{E}_0(\overline{X}\,|\,\Lambda_k) = 0$ is either satisfied by all F having identical moments $\mu_1, \ldots, \mu_k$, or by none of them. But, if F is a normal d.f., then $\hat{E}_0(\overline{X}\,|\,\Lambda_k) = 0$, since in this case $\overline{X}$ is independent of the vector $(X_2 - X_1, \ldots, X_n - X_1)$. The desired assertion is immediately obtained from the following lemma.

LEMMA 7.5.3. If μ_{2k} exists, then $\hat{\theta}_k$ is admissible in the class M_k of all polynomials of degree $\leqq k$ (and by the same token in the class P_k).

Proof: Every polynomial of degree at most k can be written in the form

$$Q(\mathbf{X}) = \bar{X}^m q_0(X_2 - X_1, \ldots, X_n - X_1) + \bar{X}^{m-1} q_1(X_2 - X_1, \ldots, X_n - X_1)$$
$$+ \cdots + q_m(X_2 - X_1, \ldots, X_n - X_1)$$

where $m \leqq k$ and we assume that $P_0[q_0 \neq 0] > 0$ (in case $q_0 = 0$ almost surely $[P_0]$, then it will be necessary to take $Q - \bar{X}^m \cdot q_0$ instead of Q). The q_j, $0 \leqq j \leqq m$, are polynomials in their arguments. Suppose that for some Q and for all θ

$$E_\theta(Q - \theta)^2 \leqq E_\theta(\hat{\theta}_k - \theta)^2 = E_0(\hat{\theta}_k)^2. \tag{7.5.20}$$

We have

$$E_\theta(Q - \theta)^2 = E_0[(\bar{X} + \theta)^m q_0(X_2 - X_1, \ldots, X_n - X_1) + \cdots$$
$$+ q_m(X_2 - X_1, \ldots, X_n - X_1) - \theta]^2$$
$$= \theta^{2m} E_0 q_0^2 + \pi_{2m-1}(\theta)$$

where the degree of $\pi_{2m-1}(\theta) \leqq \max(2, 2m - 1)$. Hence it is seen that the inequality (7.5.20) can be satisfied for all θ only if $m \leqq 1$, i.e., only polynomials of the form $Q = \bar{X} q_0 + q_1$ can possibly be better than $\hat{\theta}_k$; for such estimators, we have

$$E_\theta(Q - \theta)^2 = \theta^2 E_0(q_0 - 1)^2 + 2\theta E_0[(q_0 - 1)\bar{X} q_1] + E_0(\bar{X} q_0 + q_1)^2$$

whence it follows that if (7.5.20) is true, then necessarily $q_0(X_2 - X_1, \ldots, X_n - X_1) = 1$ with P_0-probability one (and hence with P_θ-probability one for all θ). But, among polynomials of the form $\bar{X} + q(X_2 - X_1, \ldots, X_n - X_1)$, the estimator $\hat{\theta}_k$ is not only admissible but optimal. Thus (7.5.20) cannot be satisfied for all θ, which proves the admissibility of $\hat{\theta}_k$ in the class M_k. Lemma 7.5.3, and along with it Theorem 7.5.2, stand proved. ∎

Corollary. Let, in addition to the conditions of Theorem 7.5.2, F be assumed to have moments of all orders. Then the admissibility of the sample mean $\bar{X}$ in the class $P_\infty = \bigcup_1^\infty P_k$ is a characteristic property of the normal law.

Proof: In fact, by Theorem 7.5.2, all the moments of F coincide with the corresponding moments of some normal law and hence F must be normal. ∎

We note that though the class P_∞ is in fact smaller than the class of all unbiased estimators of θ, the corollary above is not stronger than Corollary 1 to Theorem 7.4.1, since the latter does not assume the existence of any high-order moments for F.

Remark. A careful inspection of the proof of Theorem 7.5.2 shows that the following result has been proved along with it.

Let (X_1, X_2) be a random sample from a population with d.f. $F(x - \theta)$, satisfying the conditions (7.5.1) and (7.5.3). The admissibility of $(1/2)(X_1 +$

X_2) as an estimator of θ in the class P_k is a characteristic property of d.f.'s F for which all odd moments of orders up to and including $(k+1)$ are zero, i.e.,

$$\mu_1 = \mu_3 = \cdots = \mu_k = 0 \qquad \text{if } k \text{ odd}$$

$$\mu_1 = \mu_3 = \cdots = \mu_{k+1} = 0 \qquad \text{if } k \text{ even.}$$

The admissibility of $(1/2)(X_1 + X_2)$ in the class P_k imposes no restrictions on the even moments.

7.6 CHARACTERIZATION OF THE NORMAL LAW THROUGH THE OPTIMALITY OF FUNCTIONS OF $\hat{L} = \sum c_j^0 X_j$

We return to the assumptions of section 7.4. Let $X_1, \ldots, X_n$ be independent r.v.'s, $X_j \sim F_j(x - \theta)$, $\theta \in R_1$, and F_j satisfy the conditions (7.4.1) and (7.4.2). Under the condition

$$\int x^{2k}\, dF_j(x) < \infty, \qquad j = 1, \ldots, n \tag{7.6.1}$$

the polynomial $q(\hat{L}) = a_0 \hat{L}^k + \cdots + a_k$ is an unbiased estimator of the parameter-polynomial $g(\theta) = E_\theta q = c_0 \theta^k + \cdots + c_k$, having finite variance for all θ. It turns out that the optimality of $q(\hat{L})$ in the class of unbiased estimators of $g(\theta)$, under quadratic loss, characterizes the normal distribution. Indeed we have

THEOREM 7.6.1. Let $X_1, \ldots, X_n$, $n \geqq 3$, be independent r.v.'s, with $X_j \sim F_j(x - \theta)$, where $\theta \in I$, some nondegenerate interval, and F_j satisfy (7.6.1). The polynomial $q(\hat{L}) = a_0 \hat{L}^k + \cdots + a_k$, $a_0 \neq 0$, of degree $k \geqq 1$, is optimal in the class of unbiased estimators in L^2 of the parametric function $g(\theta) = E_\theta(q)$ for all $\theta \in I$ under quadratic loss, if and only if all the F_j are normal.

***Proof*:** *Necessity*. By Lemma 7.2.2, from the optimality of $q(\hat{L})$ in the class of unbiased estimators from L^2 of $g(\theta)$, it follows that for any function

$$h(X_2 - X_1, \ldots, X_n - X_1) = h \in H$$

we have

$$E_\theta[q(\hat{L}) \cdot h] = 0, \qquad \theta \in I. \tag{7.6.2}$$

But

$$E_\theta(qh) = E_0[q(\hat{L} + \theta)h] = E_0\{[a_0(\hat{L} + \theta)^k + \cdots + a_k]\, h\}. \tag{7.6.3}$$

The right-hand side of (7.6.3) is a polynomial in θ; in view of (7.6.2), all its coefficients must be zero. The coefficient of θ^{k-1} is

$$E_0[(a_0 k \hat{L} + a_1)h] = k a_0 E_0(\hat{L} h).$$

Since $a_0 \neq 0$, $E_0(\hat{L}h) = 0$ for all $h \in H$. But this is none other than condition (7.4.5) from which, as we have seen, the normality of the F_j follows.

Sufficiency. If F_j is a normal d.f. (with zero mean), then $\hat{L}$ is a complete sufficient statistic for the family of d.f.'s $\{F_1(x_1 - \theta)\cdots F_n(x_n - \theta)\}$. Hence every function of $\hat{L}$ with finite variance is optimal in L^2 as an unbiased estimator of its expected value. Thus, Theorem 7.6.1 stands proved. ∎

For Borel sets A, let

$$\nu_\theta(A) = \int_{[\hat{L} \in A]} dF_1(x_1 - \theta)\cdots dF_n(x_n - \theta).$$

For a bounded function $\psi(u)$, following the construction in section 7.2 with $dP_\theta = d\nu_\theta(u)$, we construct $L^2_\theta(\psi)$ and $L^2(\psi) = \bigcap_{\theta \in I} L^2_\theta(\psi)$, where I is a nondegenerate interval. We introduce the condition

$$L^2(\psi) \text{ contains a nonconstant polynomial } q. \tag{7.6.4}$$

Then we have the following consequence of Theorem 7.6.1.

Corollary. Let $X_1, \ldots, X_n$, $n \geqq 3$, be independent, with $X_j \sim F_j(x - \theta)$, $\theta \in I$. If the conditions (7.4.1) and (7.6.4) are satisfied, then the optimality in L^2 of the statistic $\psi(\hat{L})$, ψ bounded, in the class of unbiased estimators of $E_\theta \psi(\hat{L})$, is a characteristic property of the normality of the distributions F_j.

Proof: We shall show that from the optimality in L^2 of $\psi(\hat{L})$ follows the normality of the F_j (the converse is proved by reference to the fact that $\hat{L}$ is a complete sufficient statistic). By Theorem 7.2.1 (cf. Remark 1 thereon), the optimality in L^2 of $\psi(\hat{L})$ in the class of unbiased estimators of $E_\theta \psi$ and condition (7.6.4) guarantee the optimality in L^2 of the polynomial $q(\hat{L})$ in the class of unbiased estimators of $E_\theta q$, $\theta \in I$. But then, by Theorem 7.6.1, all the F_j are normal. ∎

We note in conclusion that, if the linear statistic $L^* = \sum_1^n c_j X_j$ is different from $\hat{L}$, then no polynomial $q(L^*)$ of degree $\geqq 1$ and no function $\psi(L^*)$, satisfying the conditions of the Corollary, can be optimal (for $\theta \in I$) as unbiased estimators of their mathematical expectations.

7.7 CHARACTERIZATION OF THE NORMAL LAW THROUGH THE ADMISSIBILITY AND OPTIMALITY OF LEAST SQUARES ESTIMATORS IN THE GAUSS-MARKOV MODEL

Let the random vector $\mathbf{Y}' = (Y_1, \ldots, Y_n)$, made up of independent observations Y_i, have the mean vector $\boldsymbol{\theta}' = (\theta_1, \ldots, \theta_n)$. In the Gauss-Markov model, the mean vector $\boldsymbol{\theta}$ has the form $\boldsymbol{\theta} = \mathbf{C}\boldsymbol{\beta}$, where $\mathbf{C}$ is a known $n \times m$ matrix and the vector parameter $\boldsymbol{\beta}' = (\beta_1, \ldots, \beta_m)$ is unknown and takes values in some region U in R^m. If the rank of $\mathbf{C}$ is r, then, without loss of generality, we may take $\mathbf{C}$ to have the canonical form

$$\mathbf{A} = \begin{pmatrix} \mathbf{I}_r \\ \mathbf{B} \end{pmatrix} \tag{7.7.1}$$

where $\mathbf{I}_r$ is the identity matrix of order $r \times r$, and $\mathbf{B}$ is an $n - r \times r$ matrix such that, if

$$\mathbf{Z} = \begin{pmatrix} -\mathbf{B}' \\ \mathbf{I}_{n-r} \end{pmatrix} \tag{7.7.2}$$

then $\mathbf{A}'\mathbf{Z} = 0$.

We finally assume that the d.f.'s F_j of the errors ε_j in the representations $Y_j = \theta_j + \varepsilon_j$, $j = 1, \ldots, n$, have zero mean and identical (known) variance.

It is well-known (cf. for instance (131]) that the *least squares estimator* (LSE) of every estimable parametric function $\mathbf{p}'\boldsymbol{\beta}$ (i.e., one such that a linear estimator $\mathbf{q}'\mathbf{Y}$ exists with $E_{\boldsymbol{\beta}}(\mathbf{q}'\mathbf{Y}) = \mathbf{p}'\boldsymbol{\beta}$, $\boldsymbol{\beta} \in U$) has the form $\boldsymbol{\lambda}'\mathbf{A}'\mathbf{Y}$ for a suitable choice of the vector $\boldsymbol{\lambda}$, and that every linear statistic of the form $\boldsymbol{\lambda}'\mathbf{A}'\mathbf{Y}$ is the LSE of its expectation. LSE's are optimal in the class of linear unbiased estimators under quadratic loss, whatever be the F_j. It is also well-known (cf. for instance [131]) that, if all the F_j are normal, then the LSE's are optimal among all unbiased estimators of the corresponding parametric functions.

THEOREM 7.7.1. Suppose that every pair of rows of the matrix $\mathbf{B}$ in (7.7.1) have nonzero elements at least in one common column and that no column of $\mathbf{B}$ consists wholly of zeros. Then, if for any vector $\boldsymbol{\lambda}$, $\boldsymbol{\lambda}'\mathbf{A}'\mathbf{Y}$ is admissible in the class of unbiased estimators of the function $E_\beta(\boldsymbol{\lambda}'\mathbf{A}'\mathbf{Y}) = \mathbf{p}'\boldsymbol{\beta}$, then the F_j are normal.

In other words, if (under the conditions on $\mathbf{B}$ above) at least one of the F_j be nonnormal and is completely known, then for some vector $\boldsymbol{\lambda}$, the estimator $\boldsymbol{\lambda}'\mathbf{A}'\mathbf{Y}$ is inadmissible among unbiased estimators of $\mathbf{p}'\boldsymbol{\beta}$, under quadratic loss.

***Proof*:** We consider the following estimator for $\mathbf{p}'\boldsymbol{\beta}$:

$$T = \boldsymbol{\lambda}'[\mathbf{A}'\mathbf{Y} - E_0(\mathbf{A}'\mathbf{Y} \mid \mathbf{Z}'\mathbf{Y})] \tag{7.7.3}$$

which is analogous to Rao's form of the Pitman estimator (7.3.5). It is seen that

$$E_\beta T = E_\beta(\boldsymbol{\lambda}'\mathbf{A}'\mathbf{Y}) = \mathbf{p}'\boldsymbol{\beta},$$

and

$$\begin{aligned} E_\beta(T - \mathbf{p}'\boldsymbol{\beta})^2 &= E_\beta(\boldsymbol{\lambda}'\mathbf{A}'\mathbf{Y} - \mathbf{p}'\boldsymbol{\beta})^2 - E_0[E_0(\boldsymbol{\lambda}'\mathbf{A}'\mathbf{Y} \mid \mathbf{Z}'\mathbf{Y})]^2 \\ &\leqq E_\beta(\boldsymbol{\lambda}'\mathbf{A}'\mathbf{Y} - \mathbf{p}'\boldsymbol{\beta})^2 \end{aligned} \tag{7.7.4}$$

where the equality sign is attained simultaneously for all $\boldsymbol{\beta}$ in U if and only if

$$E_0(\boldsymbol{\lambda}'\mathbf{A}'\mathbf{Y} \mid \mathbf{Z}'\mathbf{Y}) = \boldsymbol{\lambda}'E_0(\mathbf{A}'\mathbf{Y} \mid \mathbf{Z}'\mathbf{Y}) = 0. \tag{7.7.5}$$

If, for any $\boldsymbol{\lambda}$, the estimator $\boldsymbol{\lambda}'\mathbf{A}'\mathbf{Y}$ is admissible in the class of unbiased estimators of $\mathbf{p}'\boldsymbol{\beta}$, then it follows from (7.7.5) that $E(\mathbf{A}'\mathbf{Y}|\mathbf{Z}'\mathbf{Y}) = 0$. Applying now the corollary to Theorem 5.5.2, we conclude that, under the conditions of the theorem on the matrix $\mathbf{A}$ (imposed on it in terms of its component $\mathbf{B}$), all the F_j are normal.▮

We note that the condition of Theorem 7.7.1 on the admissibility of the estimator $\boldsymbol{\lambda}'\mathbf{A}'\mathbf{Y}$ for all $\boldsymbol{\lambda}$ is equivalent to the admissibility of the LSE for some r (linearly) independent parametric functions $\mathbf{p}_1'\boldsymbol{\beta}, \ldots, \mathbf{p}_r'\boldsymbol{\beta}$.

If the F_j are identical, then it suffices to require of the matrix $\mathbf{B}$ that in some one of its columns there be at least two nonzero elements.

We pass on to the study of conditions under which the normality of the F_j can be deduced from the admissibility of the LSE's for some $k < r$ linearly independent parametric functions. We first consider the case of one parametric function.

THEOREM 7.2.2 Let $L = \sum_1^n a_j Y_j$ be the LSE of $\mathbf{p}'\boldsymbol{\beta}$. Suppose that no a_j is zero, and that $\mathbf{B}$ is such that none of its rows is the null vector, or proportional to a row of $\mathbf{I}_r$, or to another row of $\mathbf{B}$. Then the admissibility of L in the class of unbiased estimators of $\mathbf{p}'\boldsymbol{\beta}$ implies the normality of the F_j.

***Proof*:** As in the proof of Theorem 7.7.1, we deduce from the admissibility of L that

$$E_0(L|\mathbf{Z}'\mathbf{Y}) = 0.$$

Applying now Theorem 5.2.1, we obtain the desired result.▮

Let now $\mathbf{D}$ be an $n \times k$ matrix, and $\mathbf{D}'\mathbf{Y}$ be the LSE for the parametric vector function $E_{\boldsymbol{\beta}}(\mathbf{D}'\mathbf{Y}) = \mathbf{D}'\mathbf{A}\boldsymbol{\beta} = \mathbf{G}'\boldsymbol{\beta}$, $\mathbf{G}$ being an $m \times k$ matrix. We shall say that $\mathbf{D}'\mathbf{Y}$ is *admissible* in the class of unbiased estimators of $\mathbf{G}'\boldsymbol{\beta}$, if there exists no unbiased estimator $\mathbf{T} = \mathbf{T}(\mathbf{Y})$ such that

$$E_{\boldsymbol{\beta}}(\mathbf{T} - \mathbf{G}'\boldsymbol{\beta})(\mathbf{T} - \mathbf{G}'\boldsymbol{\beta})' \leqq E_{\boldsymbol{\beta}}(\mathbf{D}'\mathbf{Y} - \mathbf{G}'\boldsymbol{\beta})(\mathbf{D}'\mathbf{Y} - \mathbf{G}'\boldsymbol{\beta})', \qquad \boldsymbol{\beta} \in U \quad (7.7.6)$$

and moreover, for at least one $\boldsymbol{\beta} \in U$, the matrices on the left and on the right are distinct.

As usual, we say that $\mathbf{A}_1 \leqq \mathbf{A}_2$ if $\mathbf{A}_2 - \mathbf{A}_1$ is a nonnegative definite matrix.

THEOREM 7.7.3. Suppose that $\mathbf{D}$ has no row consisting wholly of zeros and that $\mathbf{B}$ satisfies the same conditions as in Theorem 7.7.2. Then, from the admissibility of $\mathbf{D}'\mathbf{Y}$ in the class of unbiased estimators of the parametric vector function $\mathbf{G}'\boldsymbol{\beta} = E_{\boldsymbol{\beta}}(\mathbf{D}'\mathbf{Y})$, it follows that the F_j are normal.

***Proof*:** As in the preceding theorem, from the admissibility of $\mathbf{D}'\mathbf{Y}$ as above, we deduce that $E_0(\mathbf{D}'\mathbf{Y}|\mathbf{Z}'\mathbf{Y}) = 0$. Applying Theorem 5.2.2, we obtain the desired result.▮

In Theorems 7.7.1–7.7.3, we studied the admissibility of LSE's under the assumption that the d.f.'s F_j of the errors ε_j are completely known (do not depend on unknown parameters). Now we shall drop this assumption and obtain conditions in other situations under which the normality of the d.f.'s of the errors can be guaranteed.

As before, we shall consider the Gauss-Markov model; $\mathbf{Y} = \mathbf{C}\boldsymbol{\beta} + \boldsymbol{\varepsilon}$, where $E_{\boldsymbol{\beta}}\mathbf{Y} = \mathbf{C}\boldsymbol{\beta}$ and $\boldsymbol{\varepsilon}$ is the error vector. Let the r.v.'s $Y_j - E_{\boldsymbol{\beta}} Y_j$ have a common distribution F_θ depending on an unknown $\theta \in \Theta$ which may depend on $\boldsymbol{\beta}$. Irrespective of the presence of nuisance parameters, the LSE's are optimal in the class of unbiased estimators linear in $\mathbf{Y}$ of their expectations. Suppose now that LSE's are optimal., i.e., have the smallest variance, for all $\theta \in \Theta$, among the members of some class of unbiased estimators. What can be said in such a case about F_θ? The answer to this question, for special classes of unbiased estimators, is given by Theorems 7.7.4 and 7.7.5, from which it follows, in particular, that the optimality of LSE's in the class of all unbiased estimators, together with certain conditions on F_θ, lead to the normality of F_θ.

We first consider the case when the rank of $\mathbf{C}$ is 1.

THEOREM 7.7.4. Let rank $(\mathbf{C}) = 1$ and $L = b_1 Y_1 + \cdots + b_n Y_n$ be the LSE of the parametric function $E_{\boldsymbol{\beta}}(\mathbf{b}'\mathbf{Y}) = \mathbf{b}'\mathbf{C}\boldsymbol{\beta} = \mathbf{p}'\boldsymbol{\beta}$. Suppose further that $\int x^{2s}\, dF_\theta(x)$ exists for all θ, for some positive integer s, and that the b_j are all nonzero (the last assumption may be made without loss of generality). If L is optimal in the class of polynomials of degree $\leqq s$ which are unbiased estimators of $\mathbf{p}'\boldsymbol{\beta}$, and the vector $\mathbf{b}' = (b_1, \ldots, b_n)$ is not proportional to a vector with components ± 1, then the first $(s+1)$ moments of F_θ coincide with the corresponding moments of some normal law.

***Proof*:** Since rank $(\mathbf{C}) = 1$, there exist $n - 1$ functions

$$Z_j = c_{j1} Y_1 + \cdots + c_{jn} Y_n, \qquad j = 1, \ldots, n-1$$

such that $E_{\boldsymbol{\beta}} Z_j = 0$, $E_{\boldsymbol{\beta},\theta}(Z_i Z_j) = 0$ for $i \neq j$. The statistics Z_j can be regarded as unbiased estimators of zero, further $E_{\boldsymbol{\beta},\theta} Z_j^2 < \infty$. By Lemma 7.2.1 (cf. the conclusion of the proof of Lemma 7.3.1), we conclude that $E_{\boldsymbol{\beta},\theta}(Z_j L) = 0$. But then $Z_j L$ is an unbiased estimator of zero with finite variance if $s \geqq 2$ (the case $s = 1$ is obviously trivial). Repeatedly using Lemma 7.2.1, we obtain

$$E(Z_j L^r) = 0, \qquad r = 1, \ldots, s; \qquad j = 1, \ldots, n-1.$$

We now consider the functions

$$\phi_j(t) = E\{iZ_j \exp it(L - E_{\boldsymbol{\beta}} L)\} = [c_{j1}\psi(b_1 t) + \cdots + c_{jn}\psi(b_n t)] f(b_1 t) \cdots f(b_n t) \tag{7.7.7}$$

where f is the c.f. of F_θ (we shall not indicate the dependence of f on θ), $\psi = f'/f$ and (7.7.7) is valid for $|t| < \delta$ for some $\delta > 0$. Since F_θ has the moment of order $2s$, ψ is differentiable s times. Taking the r-th derivative of ψ for $r \leqq s$, setting $t = 0$ and remembering that $E_{\beta,\theta}(Z_j L^r) = 0$ for $r = 1, \ldots, s$, we obtain

$$\left(\sum_{i=1}^{n} c_{ji} b_i^r\right)\kappa_{r+1}(\theta) = 0, \qquad r \leqq s, \qquad j = 1, \ldots, n-1 \tag{7.7.8}$$

where $\kappa_{r+1}(\theta)$ denotes the $(r+1)$-th semi-invariant of F_θ. We conclude from (7.7.8) that at least one of the two factors on the left-hand side must be zero. Since $E_{\beta,\theta}(Z_j L) = 0$, we have $\sum_{i=1}^n c_{ji} b_i = 0$ for $j = 1, \ldots, n-1$. If now $\kappa_{r+1} \neq 0$, then $\sum_{i=1}^n c_{ji} b_i^r = 0, j = 1, \ldots, n-1$, whence $b_i^r = \lambda b_i$, $i = 1, \ldots, r$, for some λ. For $r > 1$, the last relation contradicts the conditions on **b** and consequently $\kappa_{r+1}(\theta) = 0$ for $r = 2, \ldots, s$, thus proving the theorem.▮

We note that if F_θ has moments of all orders, then, under the conditions of Theorem 7.7.4 on the coefficients in the LSE, $L = b_1 Y_1 + \cdots + b_n Y_n$, the normality of F_θ follows from the optimality of L in the class of all unbiased estimators of $\mathbf{p}'\boldsymbol{\beta}$.

We now consider the general case: rank $\mathbf{C} = r > 1$. Let $L_1, \ldots, L_r$ be the LSE's for r linearly independent parametric functions $\mathbf{p}_1'\boldsymbol{\beta}$, $\mathbf{p}_2'\boldsymbol{\beta}'$, $\ldots$, $\mathbf{p}_r'\boldsymbol{\beta}$. There exist $n - r$ linear functions

$$Z_j = c_{j1} Y_1 + \cdots + c_{jn} Y_n, \qquad j = 1, \ldots, n-r$$

such that $E_\beta Z_j = 0$, $E_{\beta,\theta}(Z_i Z_j) = 0$ for $i \neq j$. Consider the canonical representation for the matrix $\mathbf{C}$ in the form $\begin{pmatrix}\mathbf{I}_r \\ \mathbf{B}\end{pmatrix}$. We shall call the matrix $\mathbf{C}$ *exceptional* if in its canonical representation every row of $\mathbf{B}$ contains at most one nonzero element and this element (if it exists) equals ± 1. We now formulate a theorem in the general case.

THEOREM 7.7.5. Let the rank of $\mathbf{C}$ be r, the matrix $\mathbf{C}$ be nonexceptional, and for some positive integer s, $\int x^{2s}\, dF_\theta < \infty$, for all θ. If the LSE's $L_1, \ldots, L_r$ be optimal in the respective classes of polynomials of degree $\leqq s$ which are unbiased estimators of $\mathbf{p}_1'\boldsymbol{\beta}, \ldots, \mathbf{p}_r'\boldsymbol{\beta}$, then the first $s + 1$ moments of F_θ coincide with the corresponding moments of some normal law.

Proof: The proof is wholly similar to that of Theorem 7.7.4.▮

The results presented in this section are essentially contained in [127] and [130] by Rao (under somewhat different assumptions, Theorem 7.7.3 is proved in [59] by Kagan and Shalaevskii).

7.8 A CASE OF DEPENDENT OBSERVATIONS

Suppose that the process

$$X_j = \xi_j + \theta, \qquad j = 1, \ldots, n \tag{7.8.1}$$

is observed, where the r.v.'s ξ_j(in general, dependent) satisfy the conditions

$E\xi_j = 0$, $E\xi_j^2 < \infty$. The optimal *linear unbiased estimator* (LUE) for the parameter θ, $\theta \in R_1$, is the statistic $\hat{L} = \sum_1^n c_j^0 X_j$, whose coefficients $c_1^0, \ldots, c_n^0$, with $\sum_1^n c_j^0 = 1$, are defined by the covariance function of the process $\{\xi_j\}$. If the process is Gaussian, then the optimal LUE $\hat{L}$ is at the same time optimal in the class of *all* unbiased estimators of θ. In fact, it is easily verified that, for the family of distributions generated by a Gaussian process $\{X_j\}$, $\hat{L}$ is a complete sufficient statistic, whence the desired result is obtained through Theorem 7.1.2. Is the admissibility of $\hat{L}$ in the class of unbiased estimators of θ a characteristic property of Gaussian processes? It is clear that, if no conditions whatever are imposed on the process, this need not be so, the simplest counter example being $\xi_j = \xi$, $j = 1, \ldots, n$, where the distribution of ξ is arbitrary. But, apparently, for "reasonably" structured ξ_j, the admissibility of $\hat{L}$ is a characteristic property of Gaussian processes. We prove this for the simplest case of autoregressive schemes.

THEOREM 7.8.1. Let $n \geqq 3$ and the observations $X_1, \ldots, X_n$ have the structure (7.8.1), where $\{\xi_j\}$ is an autoregressive process of the first order, i.e.,

$$\xi_1 = \varepsilon_1, \qquad \xi_j = \lambda \xi_{j-1} + \varepsilon_j, \qquad j = 2, \ldots, n \tag{7.8.2}$$

where the ε_j are independently but not necessarily identically distributed according to $\varepsilon_j \sim F_j(x)$, with $E\varepsilon_j = 0$, $E\varepsilon_j^2 = \sigma_j^2$, $0 < \sigma_j^2 < \infty$. Then, if $\lambda \neq 1$, the admissibility of the optimal linear unbiased estimator $\hat{L} = \sum_1^n c_j^0 X_j$ for the parameter θ, $\theta \in R_1$, in the class of unbiased estimators of θ, under quadratic loss, is a characteristic property of the normality of the ε_j and consequently of the process $\{X_j\}$.

***Proof*:** We set $\mathbf{Y} = (X_2 - X_1, \ldots, X_n - X_1)$ and construct the Pitman estimator

$$\hat{\theta} = \hat{L} - E_0(\hat{L} \mid \mathbf{Y}). \tag{7.8.3}$$

From Lemma 7.3.1 the next lemma follows.

LEMMA 7.8.1. The estimator (7.8.3) of θ is unbiased; further,

$$E_\theta(\hat{\theta} - \theta)^2 \leqq E_\theta(\hat{L} - \theta)^2$$

and the equality sign in the above is attained, simultaneously for all $\theta \in R_1$, if and only if

$$E_0(\hat{L} \mid \mathbf{Y}) = 0. \tag{7.8.4}$$

Let now

$$f_j(t) = E \exp(it\varepsilon_j) = \int e^{itx}\, dF_j(x); \qquad g_j = f_j'/f_j;$$

$$\Psi(t_1, \ldots, t_n) = E \exp\left(i \sum_1^n t_j \xi_j\right) = E_0 \exp\left(i \sum_1^n t_j X_j\right); \tag{7.8.5}$$

$$\psi(t_1, \ldots, t_n) = \log \Psi(t_1, \ldots, t_n).$$

Here $t, t_1, \ldots, t_n$ are real, the g_j are all defined in some neighborhood of $t = 0$, and $\psi(\mathbf{t})$ in some neighborhood of $\mathbf{t} = \mathbf{0}$.

LEMMA 7.8.2. If $E_0(\hat{L} \mid \mathbf{Y}) = 0$, then, in some neighborhood of $\mathbf{t} = \mathbf{0}$,

$$\sum_1^n c_j^0 \frac{\partial \psi(\tau_1, \ldots, \tau_n)}{\partial \tau_j} = 0 \qquad \text{if only} \sum_1^n \tau_j = 0. \tag{7.8.6}$$

Proof: In fact, we have from (7.8.4) that

$$0 = E_0\left[\hat{L} \exp i \sum_1^n t_j(X_j - X_1)\right] = E_0\left[\left(\sum_1^n c_j^0 X_j\right) \exp i \sum_2^n t_j(X_j - X_1)\right]$$

$$= c_1^0 \frac{\partial \Psi(-\sum_2^n t_j, t_2, \ldots, t_n)}{\partial t_1} + \sum_2^n c_j^0 \frac{\partial \Psi(-\sum_2^n t_j, t_2, \ldots, t_n)}{\partial t_j} \tag{7.8.7}$$

where the notation is obvious. In a sufficiently small neighborhood of $\mathbf{t} = \mathbf{0}$, $\Psi(-\sum_2^n t_j, t_2, \ldots, t_n) \neq 0$. If, in this neighborhood, we divide both sides of (7.8.7) by the above, and set $\tau_1 = -\sum_2^n t_j, \tau_2 = t_2, \ldots, \tau_n = t_n$, then we arrive at (7.8.6).▮

LEMMA 7.8.3. For the autoregressive process (7.8.2)

$$\Psi(t) = f_n(t_n) f_{n-1}(\lambda t_n + t_{n-1}) \cdots f_1(\lambda^{n-1} t_n + \cdots + \lambda t_2 + t_1). \tag{7.8.8}$$

Proof: In fact, we have from (7.8.2) that

$$\xi_1 = \varepsilon_1, \xi_2 = \lambda \varepsilon_1 + \varepsilon_2, \ldots, \xi_n = \lambda^{n-1} \varepsilon_1 + \cdots + \lambda \varepsilon_{n-1} + \varepsilon_n,$$

$$E \exp\left(i \sum_1^n t_j \xi_j\right) = E \exp i[(t_1 + \lambda t_2 + \cdots + \lambda^{n-1} t_n)\varepsilon_1 + \cdots + t_n \varepsilon_n]$$

whence (7.8.8) follows.▮

For the process (7.8.2), the relation (7.8.6) can be written in the form $(g_r = f_r'/f_r)$

$$c_n^0 g_n(t_n) + (\lambda c_n^0 + c_{n-1}^0) g_{n-1}(\lambda t_n + t_{n-1}) + \cdots$$
$$+ (\lambda^{n-1} c_n^0 + \cdots + \lambda c_2^0 + c_1^0) g_1(\lambda^{n-1} t_n + \cdots + t_1) = 0 \tag{7.8.9}$$

if $\sum_1^n t_j = 0$. We introduce new variables $z_1, \ldots, z_n$ according to

$$t_n = z_n, \qquad \lambda t_n + t_{n-1} = z_{n-1}, \ldots, \lambda^{n-1} t_n + \cdots + t_1 = z_1.$$

Then $t_n = z_n, t_{n-1} = z_{n-1} - \lambda z_n, \ldots, t_1 = z_1 - \lambda z_2$ and (7.8.9) can be rewritten as

$$\sum_1^n a_j g_j(z_j) = 0, \qquad \text{if } (1 - \lambda) \sum_1^n z_j + \lambda z_1 = 0 \tag{7.8.10}$$

where

$$a_n = c_n^0, a_{n-1} = \lambda c_n^0 + c_{n-1}^0, \ldots, a_1 = \lambda^{n-1} c_n^0 + \cdots + c_1^0. \tag{7.8.11}$$

LEMMA 7.8.4. The coefficients c_j^0 of the optimal LUE $\hat{L}$ are given by

$$c_1^0 = \frac{c}{\sigma_1^2} - \frac{c\lambda(1-\lambda)}{\sigma_2^2}$$

$$\cdot \quad \cdots \quad \cdots$$

$$c_j^0 = c(1-\lambda)\left(\frac{1}{\sigma_j^2} - \frac{\lambda}{\sigma_{j+1}^2}\right), \qquad 2 \leqq j \leqq n-1 \tag{7.8.12}$$

$$\cdot \quad \cdots \quad \cdots$$

$$c_n^0 = \frac{c(1-\lambda)}{\sigma_n^2}$$

where the constant c is defined by the condition $\sum_1^n c_j^0 = 1$.

***Proof*:** Since the values of the c_j^0 are defined only by the second moments of the process $\{X_j\}$, we are justified in considering it to be Gaussian for the purposes of determining the c_0. In this case, the probability density of $\xi_1, \ldots, \xi_n$ is easily calculated, if we note that, from (7.8.2),

$$\varepsilon_1 = \xi_1, \qquad \varepsilon_2 = \xi_2 - \lambda\varepsilon_1, \ldots, \qquad \varepsilon_n = \xi_n - \lambda\xi_{n-1}.$$

Denoting it by $p(u_1, \ldots, u_n)$ and by $\phi_j(u)$ the density of the normal law $N(0, \sigma_j^2)$, we have

$$p(u_1, \ldots, u_n) = \phi_1(u_1)\phi_2(u_2 - \lambda u_1) \cdots \phi_n(u_n - \lambda u_{n-1}).$$

Hence the density of the X_j (the likelihood function of the observations) is

$$\begin{aligned} L(x_1, \ldots, x_n; \theta) &= p(x_1 - \theta, \ldots, x_n - \theta) \\ &= \phi_1(x_1 - \theta)\phi_2(x_2 - \theta - \lambda(x_1 - \theta)) \cdots \phi_n(x_n - \theta - \lambda(x_{n-1} - \theta)) \\ &= \prod_1^n (2\pi\sigma_j^2)^{-1/2} \exp -\frac{1}{2}\left[\frac{(x_1 - \theta)^2}{\sigma_1^2} + \sum_2^n \frac{(x_j - \theta - \lambda(x_{j-1} - \theta))^2}{\sigma_j^2}\right] \\ &= c(\theta)R(x_1, \ldots, x_n) \exp \theta\left[\frac{x_1}{\sigma_1^2} + (1-\lambda)\sum_2^n \frac{x_j - \lambda x_{j-1}}{\sigma_j^2}\right] \end{aligned} \tag{7.8.13}$$

where the explicit forms of the functions $c(\theta)$ and $R(\mathbf{x})$ are not required for our purposes. It is clear from (7.8.13) that the linear statistic

$$L^* = \sum_1^n (c_j^0/c)X_j \tag{7.8.14}$$

where the coefficients c_j^0/c are defined by (7.8.12) is a sufficient (and incidentally, a complete) statistic for the family of densities (7.8.13). It is clear that the statistic $\hat{L}$ must depend only on L^*, since, otherwise, it cannot be the optimal LUE for θ. Thus we must have the relations (7.8.12), since the corresponding coefficients of L^* and $\hat{L}$ must be proportional. ∎

We now return to (7.8.10) and (7.8.11). We note that $a_j = \lambda a_{j+1} + c_j^0$, $1 \leqq j \leqq n-1$, whence, starting with $a_n = c_n^0 = c(1-\lambda)/\sigma_n^2$, we successively obtain from (7.8.12)

$$a_j = \frac{c(1-\lambda)}{\sigma_j^2} \qquad \text{for } 2 \leqq j \leqq n; \qquad a_1 = \frac{c}{\sigma_1^2}.$$

It is significant for us that, for $\lambda \neq 1$, every a_j is nonzero. Let $z_1, \ldots, z_{n-1}$ be independent variables and

$$z_n = -\frac{\lambda}{1-\lambda} z_1 - \sum_1^{n-1} z_j.$$

Then (7.8.10) gives

$$\sum_1^{n-1} a_j g_j(z_j) = -a_n g_n\left(-\frac{\lambda}{1-\lambda} z_1 - \sum_1^{n-1} z_j\right)$$

whence, on differentiating with respect to z_j (the condition $E_0 X_j^2 < \infty$ permits our doing this), we obtain

$$g_j'(t) = \text{constant} = c_j \ (1 \leqq j \leqq n-1)$$

in some neighborhood of $t = 0$. Remembering that $g_j = f_j'/f_j$, we have $f_j(t) = \exp(-(1/2)c_j t^2)$ for small t, whence follows the normality of the ε_j. If $z_2, \ldots, z_n$ are chosen independent and $z_1 = -(1-\lambda)\sum_2^n z_j$, then, proceeding in the same way as above, we see that ε_n is also normal. To conclude the proof of Theorem 7.8.1, it only remains to verify that, for normally distributed ε_j, $1 \leqq j \leqq n$, the optimal LUE of θ, $\hat{L} = \sum_1^n c_j^0 X_j$ is admissible in the class of all unbiased estimators constructed on the basis of observations of the form (7.8.1), where the ξ_j have the structure (7.8.2). It has already been remarked above that in this case there exists a complete sufficient statistic of the form (7.8.14). Hence the estimator $\hat{L}$, with coefficients defined by (7.8.12), is the only unbiased estimator of θ depending on the sufficient statistic, and by Theorem 7.1.2, in this case, $\hat{L}$ is optimal in the class of unbiased estimators. Theorem 7.8.1 stands completely proved. ▌

Remark. In case $\lambda = 1$, we have $\hat{L} = X_1$ and relation (7.8.10) is satisfied, whatever be the d.f.'s F_j. It can be shown that, in this case, it is in general impossible to obtain any characterization of the laws F_j through the property of admissibility of the estimator $\hat{L}$.

We also note that the result of the present section can be extended to the case when $\{\xi_j\}$ is a general autoregressive process.

7.9 LOSS FUNCTIONS OTHER THAN QUADRATIC

In this section, it will be shown that if $X_1, \ldots, X_n$ be a random sample of size n from a population with d.f. $F(x-\theta)$, $\theta \in R_1$, then the absolute admissibility of $\bar{X}$ as an estimator of θ, under certain loss functions other than quadratic, remains a characteristic property of the normal law. We remark that we study here the phenomenon of absolute admissibility of $\bar{X}$, because restricting ourselves to the class of unbiased estimators, when considering loss functions other than quadratic, is not natural. It is not clear what are "natural" conditions (such as unbiasedness) to be imposed on the classes of estimators when considering loss functions of a general type.

The results presented here are due to Zinger and Kagan (cf. [42, 60]).

7.9.1 Estimation Under Laplacian Loss

THEOREM 7.9.1. Let $X_1, \ldots, X_n$ be a random sample of size $n \geqq 6$ from a population with d.f. $F(x-\theta)$, where F is unimodal and has a continuously differentiable density f (unimodality means then existence of an x_0 such that $f'(x) \geqq 0$ for $x \leqq x_0$ and $f'(x) \leqq 0$ for $x \geqq x_0$). Then, the admissibility of the sample mean $\bar{X}$ as an estimator of θ, under the Laplacian loss function $r(\tilde{\theta}, \theta) = |\tilde{\theta} - \theta|$, is a characteristic property of the normal law.

***Proof*:** From the absolute admissibility of $\bar{X}$ under Laplacian loss, it follows that $\int |x|\, dF(x) < \infty$. But then, by Lemma 7.3.2, the Pitman estimator for θ is the statistic

$$\bar{\theta} = \bar{X} - \text{med}_0\,(\bar{X} \mid \mathbf{Y}), \qquad \mathbf{Y} = (X_1 - \bar{X}, \ldots, X_n - \bar{X}).$$

Therefore, for the admissibility of $\bar{X}$ (even for admissibility in the class of regular estimators) under Laplacian loss, it is necessary that there exist a version of the conditional median of $\bar{X}$ for given $\mathbf{y}$ for which

$$\text{med}_0\,(\bar{X} \mid \mathbf{y}) = 0. \tag{7.9.1}$$

LEMMA 7.9.1. If $\theta = 0$, then the conditional density of $\bar{X}$ at the point u for given $\mathbf{y}$ equals

$$p(u \mid \mathbf{y}) = \frac{\prod_1^n f(u + x_i - \bar{x})}{\int_{-\infty}^{\infty} \prod_1^n f(u + x_i - \bar{x})}\, f(u + x_i - \bar{x})du. \tag{7.9.2}$$

***Proof*:** This assertion is proved in a straightforward manner.∎

Condition (7.9.1) can now be written in the form

$$\int_{-\infty}^{0} \prod_1^n f(u + x_i - \bar{x})\, du = \int_0^{\infty} \prod_1^n f(u + x_i - \bar{x})\, du \tag{7.9.3}$$

for almost all $[P_0]$ $x_1, \ldots, x_n$. We shall take $x_0 = 0$ for simplicity in Theorem 7.9.1. Since f is nonincreasing for $x > 0$, nondecreasing for $x < 0$ and is continuous, we have $f(0) > 0$ and f is bounded: $f(x) < c$ for all x.

LEMMA 7.9.2. The functions $\int_{-\infty}^{0} \prod_1^n f(u + z_i)\, du$, $\int_0^\infty \prod_1^n f(u + z_i)\, du$ are continuous in $(z_1, \ldots, z_n)$.
Proof: In fact,

$$\int_{-\infty}^{-A} \prod_1^n f(u + z_i)\, du \leqq c^{n-1} \int_{-\infty}^{-A} f(u + z_i)\, du = c^{n-1} F(-A + z_i)$$

so that the integrals in (7.9.3) converge uniformly in $(z_1, \ldots, z_n)$ in any bounded set.

If we denote by V a neighborhood of $x = 0$ in which $f(x)$ is positive, then it follows from (7.9.3) and Lemma 7.9.2 by continuity that (7.9.3) holds for all $(x_1, \ldots, x_n)$ in V^n.

For what follows, it is useful to write (7.9.3) as

$$\int_{-\infty}^{0} \prod_1^{n-1} f(u + z_i) f\left(u - \sum_1^{n-1} z_i\right) du = \int_0^\infty \prod_1^{n-1} f(u + z_i) f\left(u - \sum_1^{n-1} z_i\right) du \tag{7.9.4}$$

where $z_1, \ldots, z_{n-1}$ are now independent variables. If we set $V' = (1/(n-1))V$, then (7.9.4) is automatically satisfied for $z_i \in V'$, $i = 1, \ldots, n-1$.

LEMMA 7.9.3. (7.9.4) can be differentiated under the integral sign with respect to each of the variables $z_1, \ldots, z_{n-1}$.
Proof: Since $(d/dz_i)\{\prod_1^{n-1} f(u + z_i) f(u - \sum_1^{n-1} z_i)\}$ is a continuous function of u and $z_1, \ldots, z_{n-1}$, it suffices to verify that the integrals

$$I_1 = \int_{-\infty}^{0} \frac{d}{dz_i}\left\{\prod_1^{n-1} f(u + z_i) f\left(u - \sum_1^{n-1} z_i\right)\right\} du, \quad \text{and}$$

$$I_2 = \int_0^\infty \frac{d}{dz_i}\left\{\prod_1^{n-1} f(u + z_i) f\left(u - \sum_1^{n-1} z_i\right)\right\} du$$

converge uniformly.

$$I_1 = \int_{-\infty}^{0} f'(u + z_i) \prod_{k \neq i} f(u + z_k) f\left(u - \sum_1^{n-1} z_i\right) du$$
$$- \int_{-\infty}^{0} f'\left(u - \sum_1^{n-1} z_i\right) \prod_1^{n-1} f(u + z_i)\, du.$$

Since $f'(x) \geqq 0$ for $x \leqq 0$,

$$\left| \int_{-\infty}^{-A} f'(u + z_i) \prod_{k \neq i} f(u + z_k) f\left(u - \sum_{1}^{n-1} z_i\right) du \right|$$
$$\leqq c^{n-1} \int_{-\infty}^{-A} f'(u + z_i)\, du = c^{n-1} f(-A + z_i). \tag{7.9.5}$$

In view of the monotonicity of $f(x)$ for $x < 0$, by a suitable choice of A, the right side of (7.9.5) can be made as small as desired uniformly with respect to $(z_1, \ldots, z_{n-1})$ in any bounded set. The other integral I_2 is dealt with similarly. ▮

We set

$$w(u) = \begin{cases} -1 & \text{if } u < 0 \\ 1 & \text{if } u > 0 \end{cases}$$

and write (7.9.4) as

$$\int_{-\infty}^{\infty} w(u) \prod_{1}^{n-1} f(u + z_i) f\left(u - \sum_{1}^{n-1} z_i\right) du = 0.$$

We now differentiate the above relation with respect to z_1. Then we have

$$\int_{-\infty}^{\infty} w(u) f'(u + z_1) \prod_{2}^{n-1} f(u + z_i) f\left(u - \sum_{1}^{n-1} z_i\right) du$$
$$- \int_{-\infty}^{\infty} w(u) \prod_{1}^{n-1} f(u + z_i) f'\left(u - \sum_{1}^{n-1} z_i\right) du = 0. \tag{7.9.6}$$

We integrate the second integral by parts. It is

$$= -2 \prod_{1}^{n-1} f(z_i) f\left(-\sum_{1}^{n-1} z_i\right) - \int_{-\infty}^{\infty} w(u) f\left(u - \sum_{1}^{n-1} z_i\right) \frac{d}{du} \left\{ \prod_{1}^{n-1} f(u + z_i) \right\} du$$
$$= -2 \prod_{1}^{n-1} f(z_i) f\left(-\sum_{1}^{n-1} z_i\right) - (n-1) \int_{-\infty}^{\infty} w(u) f'(u + z_1) \prod_{2}^{n-1} f(u + z_i)$$
$$\times f\left(u - \sum_{1}^{n-1} z_i\right) du.$$

Thus, we have from (7.9.6)

$$n \int_{-\infty}^{\infty} w(u) f'(u + z_1) \prod_{2}^{n-1} f(u + z_i) f\left(u - \sum_{1}^{n-1} z_i\right) du$$
$$+ 2 \sum_{1}^{n-1} f(z_i) f\left(-\sum_{1}^{n-1} z_i\right) = 0. \tag{7.9.7}$$

We recall that (7.9.7) holds for at least all $(z_1, \ldots, z_{n-1}) \in (V')^{n-1}$. We write (7.9.7) for the following three sets of values of $z_1, \ldots, z_{n-1}$:

$$0, z_2, z_3, z_4, 0, 0, \ldots, 0;$$

$$0, z_2, z_3, z_2, z_3, 0, \ldots, 0;$$

$$0, z_4, 0, z_4, 0, 0, \ldots, 0;$$

where $z_2 + z_3 + z_4 = 0$ (it is here that we first use the condition $n \geqq 6$). From the relations obtained, it follows that

$$\int_{-\infty}^{\infty} w(u)f'(u)\left[\frac{f(u+z_2)f(u+z_3)}{(z_2)f(z_3)} - \frac{f(u+z_4)f(u)}{f(z_4)f(0)}\right]^2 \times [f(u)]^{n-6}f(u-2z_4)\,du = 0. \quad (7.9.8)$$

Under the conditions of the theorem, $w(u)f'(u) \leqq 0$: we have chosen $x_0 = 0$, the necessary changes in the general case being trivial. It is clear that there exists an interval I such that, for u in I, $w(u)f'(u) < 0$ and $f(u)f(u-2z_4) > 0$. For u in I and sufficiently small z_j, we obtain from (7.9.8)

$$\frac{f(u+z_2)f(u+z_3)}{f(z_2)f(z_3)} - \frac{f(u+z_4)f(u)}{f(z_4)f(0)} = 0,$$

i.e.,

$$\frac{f(u+z_2)f(u+z_3)}{f(z_2)f(z_3)} = \frac{f(u+z_2+z_3)f(u)}{f(z_2+z_3)f(0)}.$$

It is easily seen that for u in I, $f(u+z_2)f(u+z_3)f(u+z_2+z_3) > 0$. Taking logarithms and differentiating with respect to u, we have

$$\frac{f'(u+z_2)}{f(u+z_2)} + \frac{f'(u+z_3)}{f(u+z_3)} = \frac{f'(u+z_2+z_3)}{f(u+z_2+z_3)} + \frac{f'(u)}{f(u)}$$

whence it follows, in view of Cauchy's theorem, that $f'(x)/f(x) = c_1 x + c_2$ and

$$f(x) = \exp P_2(x) \quad (7.9.9)$$

where P_2 is a polynomial of degree 2. Equation (7.9.9) is valid for x in I; in this interval, $f(x) > 0$ and $f'(x)$ does not vanish. By the continuity of f and f', we conclude that these properties hold in some interval containing I. Extending this argument repeatedly, we see that $f(x) = \exp P_2(x)$ for all real x.

Thus, if $\bar{X}$ is an admissible estimator of θ under Laplacian loss, then F is normal; moreover, it is easily seen that $E_0 X_i = 0$. ∎

The converse is also true (even without the assumption that $n \geqq 6$) if θ is the mean value of the Gaussian variables X_i; it follows, for instance, from the general theorem of Fox and Rubin [155] on the admissibility of the Pitman estimator under Laplacian loss or can be proved as in Farrell [157].

7.9.2 Some Generalizations of Laplacian Loss

We consider the following loss function

$$r(\tilde{\theta}, \theta) = \begin{cases} -\alpha(\tilde{\theta} - \theta), & \tilde{\theta} \leqq \theta \\ \beta(\tilde{\theta} - \theta,) & \tilde{\theta} \geqq \theta \end{cases} \tag{7.9.10}$$

where $\alpha > 0$, $\beta > 0$ are some constants. The Laplacian loss function corresponds to $\alpha = \beta = 1$. If we wish to carry over the result of the preceding section to the case of such a loss function as (7.9.10), then we have first of all to bear in mind the fact that for normal variables $X_1, \ldots, X_n$, with $E_\theta X_i = \theta$, the sample mean $\bar{X}$ is not an admissible estimator of the shift parameter $\theta \in R_1$ in the class of regular estimators, under the loss function (7.9.10). It is easily seen that min $E_\theta r(\bar{X} - c, \theta)$ is attained for the value of c given by the condition

$$F(c) = \frac{\beta}{\alpha + \beta} = \gamma. \tag{7.9.11}$$

Hence, for the loss function (7.9.10), it is natural to study conditions for the admissibility of $\bar{X} - c$,where c is defined by (7.9.11).

THEOREM 7.9.2. Under the conditions of Theorem 7.9.1, the admissibility of the estimator $\bar{X} - c$ for $\theta \in R_1$, under the loss (7.9.10), is a characteristic property of the normal law.

***Proof*:** It is established exactly as in the proof of Lemma 7.3.2 that, in the present situation, the Pitman estimator for θ is

$$\theta = \bar{X} - \text{quan}_\gamma\,(\bar{X} \mid \mathbf{Y}), \qquad \mathbf{Y} = (X_1 - \bar{X}, \ldots, X_n - \bar{X})$$

where $\text{quan}_\gamma\,(\bar{X} \mid \mathbf{Y})$ denotes the γ-quantile of the conditional distribution of $\bar{X}$ given $\mathbf{Y}$. From the admissibility of $\bar{X} - c$ (even from its admissibility in the class of regular estimators), it follows that there exists a version of $\text{quan}_\gamma\,(\bar{X} \mid \mathbf{Y})$ for which

$$\text{quan}_\gamma\,(\bar{X} \mid \mathbf{Y}) = c \tag{7.9.12}$$

which may be written as

$$\int_{-\infty}^{c} \prod_{1}^{n} f(u + x_i - \bar{x})\, du = \gamma \int_{-\infty}^{\infty} \prod_{1}^{n} f(u + x_i - \bar{x})\, du.$$

We set $\phi(u) = f(u + c)$, so that the preceding relation can be written as

$$\int_{-\infty}^{0} \prod_{1}^{n} \phi(u + x_i - \bar{x})\, du = \frac{\gamma}{1-\gamma} \int_{0}^{\infty} \prod_{1}^{n} \phi(u + x_i - \bar{x})\, du$$

which is investigated in the same way as (7.9.3), yielding the conclusion that $\phi(u) = \exp P_2(u)$ where P_2 is a polynomial of degree 2. Thus F is a normal d.f. The admissibility of $\overline{X} - c$, as an estimator of $\theta \in R_1$ under the loss (7.9.10), for normal variables $X_1, \ldots, X_n$ is proved in the above-mentioned work [155].▌

7.9.3 Confidence Estimation

To a confidence estimator of a parameter $\theta \in R_1$ corresponds a loss function of the form

$$r(\tilde{\theta}, \theta) = \begin{cases} 0 & \text{if } |\tilde{\theta} - \theta| \leqq b \\ 1 & \text{if } |\tilde{\theta} - \theta| > b \end{cases} \tag{7.9.13}$$

where b is some constant. The corresponding risk is

$$R(\tilde{\theta}, \theta) = E_\theta r(\tilde{\theta}, \theta) = P_\theta[|\tilde{\theta} - \theta| > b] = P_\theta[\theta \notin [\tilde{\theta} - b, \tilde{\theta} + b]]$$

and equals the probability that the random interval $[\tilde{\theta} - b, \tilde{\theta} + b]$ does not contain the unknown value of θ.

Let $X_1, \ldots, X_n$ be a random sample from a population with d.f. $F(x - \theta)$. Then, whatever be F,

$$R(\overline{X}, \theta) = 1 - P_\theta[-b \leqq \overline{X} - \theta \leqq b] = \gamma, \qquad \gamma = \gamma(b)$$

does not depend on θ, and hence $[\overline{X} - b, \overline{X} + b]$ is a random interval having confidence level $1 - \gamma$ (depending on F clearly). We shall show that, for d.f.'s F with continuous and bounded densities, the admissibility of $\overline{X}$ (for $n \geqq 3$) as an estimator of θ, under the loss (7.9.13), for a sequence $\{b_j\} \to 0$, is a characteristic property of the normal law. In other words, for d.f.'s F other than normal and having continuous and bounded densities, for every sufficiently small $b > 0$ it is possible to construct a confidence interval $\Delta = (\underline{\Delta}(X_1, \ldots, X_n), \overline{\Delta}(X_1, \ldots, X_n))$ of constant width $2b$, such that, for all $\theta \in R^1$,

$$P_\theta(\theta \in [\underline{\Delta}, \overline{\Delta}]] > P_\theta[\theta \in [\overline{X} - b, \overline{X} + b]].$$

THEOREM 7.9.3. Let $X_1, \ldots, X_n$ be a random sample of size $n \geqq 3$ from a population with d.f. $F(x - \theta)$, having a continuous and bounded density $f(x - \theta)$. If, for a sequence $\{b_j\} \to 0$, the sample mean $\overline{X}$ is admissible as an estimator of $\theta \in R_1$, under the loss (7.9.13) ,then F is a normal d.f.

***Proof*:** As in similar theorems on quadratic and Laplacian loss functions, we show that, under the conditions of the theorem, the admissibility of $\overline{X}$ in the class of linear estimators $\overline{X} + \psi(\mathbf{Y})$, $\mathbf{Y} = (X_1 - \overline{X}, \ldots, X_n - \overline{X})$ leads to the normality of F. We have

$$P_\theta[\theta \in [\overline{X} + \psi(\mathbf{Y}) - b, \overline{X} + \psi(\mathbf{Y}) + b]]$$
$$= P_0[0 \in [\overline{X} + \psi(\mathbf{Y}) - b, \overline{X} + \psi(\mathbf{Y}) + b]].$$

If, for some $b > 0$, the confidence interval $[\overline{X} - b, \overline{X} + b]$ is admissible, then it is necessary that we have with P_0-probability one that

$$P_0[0 \in [\overline{X} - b, \overline{X} + b] \mid \mathbf{Y}] = \max_\psi P_0[0 \in [\overline{X} + \psi(\mathbf{Y}) - b, \overline{X} + \psi(\mathbf{Y}) + b] \mid \mathbf{Y}]. \tag{7.9.14}$$

In order to use the above condition, we turn to the density (7.9.2) of the conditional distribution of $\overline{X}$ given $\mathbf{Y}$. We note that, in view of the continuity of $f(x)$, an x_0 exists such that $f(x) > 0$ in some neighborhood of x_0. Without loss of generality, we may take $x_0 = 0$, and let x' be the smallest in absolute value of the zeros of $f(x)$; for definiteness, let us suppose that $x' > 0$. Then, as is easily seen, for $|x_i| < x'$, $i = 1, \ldots, n$, $\int_{-\infty}^{\infty} f(u + x_i - \bar{x})\, du$ is positive, and in view of the boundedness of $f(x)$, it is a continuous function of the x_i (cf. Lemma 7.9.2). Hence it follows that the conditional density (7.9.2) is continuous in the x_i for $|x_i| < x'$ for all i.

LEMMA 7.9.4. Let

$$r(u) = \begin{cases} 0 & \text{if } |u| \leqq b \\ 1 & \text{if } |u| > b \end{cases}$$

and ξ be a random variable with continuous density $p(x)$. In order that $\min_a Er(\xi - a)$ be attained at $a = 0$, it is necessary that $p(-b) = p(b)$.
***Proof*:** In fact, $Er(\xi - a) = P[\xi > a + b] + P[\xi < a - b]$ and if the above minimum be attained at $a = 0$, it is necessary that

$$\frac{d}{da} Er(\xi - a)\big|_{a=0} = p(-b) - p(b) = 0. \blacksquare$$

Combining (7.9.14) with Lemma 7.9.4, we arrive at the following necessary condition for the admissibility of $\overline{X}$ for given b under the loss (7.9.13): for almost all $[P_0]$ values of $x_1, \ldots, x_n$ with $|x_i| < x'$ for all i,

$$\prod_1^n f(b + x_i - \bar{x}) = \prod_1^n f(-b + x_i - \bar{x}). \tag{7.9.15}$$

We note first of all that, since $f(x) > 0$ for $|x| < x'$ and is continuous, (7.9.15) must be satisfied for all $x_1, \ldots, x_n$ with $|x_i| < x'$.

We shall now show that $f(x) > 0$ for all x. Suppose not and let x' be as above the smallest in absolute value of the zeros of $f(x)$ and again, as above, assume for definiteness that $x' > 0$. We set $b = b_j$ where $b_j < x'$ and take, in (7.9.15),

$$x_1 = x_2 = (1/2)(b_j - x'); \qquad x_3 = x' - b_j; \qquad x_4 = \cdots = x_n = 0.$$

Then, we have simultaneously

$$\prod_{i=1}^{n} f(b_j + x_i - \bar{x}) = 0$$

and

$$\prod_{i=1}^{n} f(-b_j + x_i - \bar{x}) = [f(-(x' + b_j)/2)]^2 f(x' - 2b_j)[f(-b_j)]^{n-3} > 0$$

and this contradiction to (7.9.15) shows that f cannot vanish anywhere and so is positive everywhere.

Now we set $g(x) = \log f(x)$ and write (7.9.15) in the form

$$\sum_1^n g(b - u_i) = \sum_1^n g(-b - u_i) \tag{7.9.16}$$

where $u_1, \ldots, u_n$ are subject only to the restriction $\sum_1^n u_i = 0$, and b can take any of the values b_j, $j = 1, \ldots$. If we set $v_j = -u_j$ and $G(v) = g(v + b) - g(v - b)$, then we have from (7.9.16)

$$\sum_1^n G(v_i) = 0 \qquad \text{provided only that} \qquad \sum_1^n v_i = 0. \tag{7.9.17}$$

Since $n \geqq 3$, it follows from Cauchy's theorem that

$$g(v + b) - g(v - b) = G(v) = c(b)v, \qquad b = b_j. \tag{7.9.18}$$

But (7.9.18) is a difference equation of the first order, valid for all differences b_j, which $\to 0$ as $j \to \infty$. Hence we conclude that $g(v) = cv^2 + d$ for some c and d, and consequently f is a normal density. Theorem 7.9.3 stands proved. ▌

Remark 1. The condition that $\bar{X}$ be admissible for a sequence $\{b_j\} \to 0$, under the loss (7.9.13), cannot be replaced by the admissibility of $\bar{X}$ for $b \geqq b_0 > 0$. In fact, let F be an arbitrary d.f. with support on $[-b_0/n, b_0/n]$. Then $P_\theta[\theta \in [\bar{X} - b, \bar{X} + b]] = P_0[\bar{X} - b \leqq 0 \leqq \bar{X} + b] = 1$ for $b \geqq b_0$, and consequently for such d.f.'s F, $\bar{X}$ is even optimal under the loss (7.9.13) if $b \geqq b_0$.

Remark 2. If the X_j are distributed normally according to $N(\theta, \sigma^2)$, then $\bar{X}$ is an admissible estimator for θ under the loss (7.9.13) for all $b > 0$, so that the converse of Theorem 7.9.3 is true. It may be found proved, for instance, in [48].

7.10 CHARACTERIZATION OF FAMILIES OF DISTRIBUTIONS FOR WHICH THE PITMAN ESTIMATOR DOES NOT DEPEND ON THE LOSS FUNCTION

Let observations $X_1, \ldots, X_n$ be independent and identically distributed according to the d.f. $F(x - \theta)$. In section 7.2, we saw that the Pitman estimator, corresponding to the quadratic loss function, has the form $\hat{\theta} = \bar{X} - E_0(\bar{X} \mid \mathbf{Y})$, $\mathbf{Y} = (X_1 - \bar{X}, \ldots, X_n - \bar{X})$, and the Pitman estimator under Laplacian loss is $\overline{\theta} = \bar{X} - \text{med}_0\,(\bar{X} \mid Y)$. Thus, generally speaking, for the same F, the Pitman estimator differs from loss function to loss function. However, if F is a normal d.f., then it is easy to see that, for quadratic and Laplacian (and, besides, for many other) loss functions, the Pitman estimator is one and the same, namely, $\bar{X}$.

What are those F for which the Pitman estimator has the property of not being dependent on the choice of the loss function? The posing of this question is due to Rukhin [134]; the results presented in this section are also due to him and relate to cases where the loss function is chosen from among a special class of functions.

For real s, let

$$r_s(\tilde{\theta}, \theta) = |e^{is\tilde{\theta}} - e^{is\theta}|^2 = |e^{is(\tilde{\theta} - \theta)} - 1|^2 = \rho_s(\tilde{\theta} - \theta) \tag{7.10.1}$$

and suppose that the Pitman estimator $T = T(X_1, \ldots, X_n)$ is one and the same, whatever be the loss function (7.10.1); in fact, it is sufficient to assume that the Pitman estimator is independent of s, for all $s \in I$, where I is some neighborhood of the origin. As before, we shall denote by $\mathscr{T}$ the class of regular estimators $\tilde{\theta}(X_1, \ldots, X_n)$, i.e., those for which, for any c,

$$\tilde{\theta}(X_1 + c, \ldots, X_n + c) = \tilde{\theta}(X_1, \ldots, X_n) + c.$$

LEMMA 7.10.1.

(1) If the Pitman estimator T does not depend on the loss function (7.10.1), and $\tilde{\theta}$ is an arbitrary member of $\mathscr{T}$, then for all s

$$e^{isT} = \frac{e^{is\tilde{\theta}} E_0(e^{-is\tilde{\theta}} \mid \mathbf{Y})}{|E_0(e^{-is\tilde{\theta}} \mid \mathbf{Y})|}. \tag{7.10.2}$$

(2) If the distribution F is given by a density f (with respect to Lebesgue measure), then

$$e^{isT} = \frac{\int_{-\infty}^{\infty} e^{isu} \prod_1^n f(x_i - u)\, du}{\left|\int_{-\infty}^{\infty} e^{isu} \prod_1^n f(x_i - u)\, du\right|}. \tag{7.10.3}$$

***Proof*:** We note first of all that the right side of (7.10.2) does not depend on the choice of $\tilde{\theta} \in \mathscr{T}$. In fact, if $\tilde{\theta} \in \mathscr{T}$, then $\tilde{\theta} = \bar{X} + \psi(\mathbf{Y})$. But then

$$\frac{e^{is\tilde{\theta}}E_0(e^{-is\tilde{\theta}}\,|\,\mathbf{Y})}{|E_0(e^{-is\tilde{\theta}}\,|\,\mathbf{Y})|} = \frac{e^{is\bar{X}}E_0(e^{-is\bar{X}}\,|\,\mathbf{Y})}{|E_0(e^{-is\bar{X}}\,|\,\mathbf{Y})|}.$$

It is easily verified that T given by (7.10.2) is a member of $\mathscr{T}$. In fact,

$$\begin{aligned} e^{isT(X_1+c,\,\ldots,\,X_n+c)} &= e^{is\tilde{\theta}(X_1+c,\,\ldots,\,X_n+c)} \frac{E_0(e^{-is\tilde{\theta}}\,|\,\mathbf{Y})}{|E_0(e^{-is\tilde{\theta}}\,|\,\mathbf{Y})|} \\ &= e^{isc}e^{isT(X_1,\,\ldots,\,X_n)}. \end{aligned}$$

Further, $E_\theta\,\rho_s(\tilde{\theta} - \theta) = E_0(2 - 2\,\mathrm{Re}\,e^{is\tilde{\theta}})$, and hence the Pitman estimator T, corresponding to the loss function ρ_s, can be defined by the relation

$$\mathrm{Re}\,E_0\,e^{isT} = \max_{\tilde{\theta} \in \mathscr{T}} \mathrm{Re}\,E_0\,e^{is\tilde{\theta}}.$$

From (7.10.2), we have

$$\begin{aligned} \mathrm{Re}\,E_0\,e^{isT} &= \mathrm{Re}\,E_0\left[\frac{e^{is\tilde{\theta}}E_0(e^{-is\tilde{\theta}}\,|\,\mathbf{Y})}{|E_0(e^{-is\tilde{\theta}}\,|\,\mathbf{Y})|}\right] \\ &= \mathrm{Re}\,E_0\left[\frac{E_0(e^{is\tilde{\theta}}\,|\,\mathbf{Y})E_0(e^{-is\tilde{\theta}}\,|\,\mathbf{Y})}{|E_0(e^{-is\tilde{\theta}}\,|\,\mathbf{Y})|}\right] \\ &= \mathrm{Re}\,E_0\,|E_0(e^{is\tilde{\theta}}\,|\,\mathbf{Y})| \geqq \mathrm{Re}\,E_0[E_0(e^{is\tilde{\theta}}\,|\,\mathbf{Y})] = \mathrm{Re}\,E_0\,e^{is\tilde{\theta}}. \end{aligned}$$

The equality sign in the preceding inequality is attained if and only if $\arg E_0(e^{is\tilde{\theta}}\,|\,\mathbf{Y}) = 0$ with probability $[P_0]$ one. This is equivalent to $\tilde{\theta} = T$ with probability one.

Assertion (2) of the lemma is proved in a straightforward manner. ∎

We now turn to the description of continuous and positive densities $f(x)$, for which the Pitman estimator (7.10.3), continuous in $(x_1, \ldots, x_n)$, does not depend on the loss function if the latter is chosen from among those of the form (7.10.1). Let $N = \{(x_1, \ldots, x_n)\colon T(x_1, \ldots, x_n) = 0\}$. For $(x_1, \ldots, x_n)$ in N, we have from (7.10.3) that

$$\begin{aligned} \int_{-\infty}^{\infty} e^{isu} \prod_1^n f(x_i - u)\,du &= \int_{-\infty}^{\infty} e^{-isu} \prod_1^n f(x_i - u)\,du \\ &= \int_{-\infty}^{\infty} e^{isu} \prod_1^n f(x_i + u)\,du \end{aligned}$$

so that in this case, we have for almost all (in the sense of Lebesgue measure) u in R_1,

$$\prod_1^n f(x_i - u) = \prod_1^n f(x_i + u) \qquad \text{if } (x_1, \ldots, x_n) \in N. \tag{7.10.4}$$

In view of the continuity of f, the relation (7.10.4) is valid for all u. It follows from (7.10.4) that the density f is symmetric with respect to a point c such that $T(c, c, \ldots, c) = 0$. The existence of precisely one such point c follows from the fact that T belongs to $\mathscr{T}$. We shall assume below, without loss of generality, that $c = 0$, so that $f(-x) = f(x)$. We set $g(x) = -\log f(x)$. Then

$$\sum_1^n [g(x_i + z) - g(x_i - z)] = 0 \tag{7.10.5}$$

for all real z, provided only that $(x_1, \ldots, x_n) \in N$. For simplicity of presentation, we take $n = 3$, the general case being treated similarly.

We set $G(u, z) = g(u + z) - g(u - z)$; then $G(u, z) = G(z, u)$ and $G(-u, z) = -G(u, z)$. Let $\phi(u, v)$ be a measurable function such that* $T(u, v, \phi(u, v)) = 0$. We have from (7.10.5) that

$$G(u, z) + G(v, z) = G(\phi(u, v), z). \tag{7.10.6}$$

Besides,

$$\begin{aligned} G(u, z) + G(v, z) + G(w, z) &= G(u, z) + G(\phi(v, w), z) \\ &= G(\phi(u, \phi(v, w)), z) \end{aligned}$$

for all u, v, w, z. On the other hand, the left-most member of the above relations is also

$$= G(\phi(u, v), z) + G(w, z) = G(\phi(\phi(u, v), w), z)$$

so that we have for all z,

$$G(\phi(\phi(u, v), w), z) = G(\phi(u, \phi(v, w)), z).$$

Hence

$$\phi(\phi(u, v), w) = \phi(u, \phi(v, w)).$$

In this case (*vide* [3], p. 176), there exists a monotone function h such that

$$\phi(u, v) = h[h^{-1}(u) + h^{-1}(v)].$$

Returning to (7.10.6), we have

$$G(h(u), z) + G(h(v), z) = G(h(u + v), z)$$

whence, by Cauchy's theorem, $G(h(u), z) = H(z)h(u)$. From the symmetry of G, we have $H(z) = h^{-1}(z)$ and hence

$$G(u, v) = h^{-1}(u)h^{-1}(v).$$

* The existence of such $\phi(u, v)$ can be established, starting from the relation (7.10.4).

Thus we obtain the functional equation

$$g(u+v) - g(u-v) = h^{-1}(u)h^{-1}(v)$$

and the only measurable solutions having probabilistic significance are (cf. [3], p. 131):
either

$$g(u) = \alpha \cosh \beta u, \qquad \alpha, \beta > 0 \tag{7.10.7}$$

or

$$g(u) = \gamma u^2, \qquad \gamma > 0. \tag{7.10.8}$$

The density corresponding to the solution (7.10.7) is

$$f(x) = (2\beta/k(\alpha)) \exp(-\alpha \cosh \beta x), \qquad k(\alpha) = \int_0^\infty e^{-\alpha \cosh x}\, dx,$$

and solution (7.10.8) corresponds to a normal density.

With the help of (7.10.3), we can obtain a Pitman estimator $T(X_1, \ldots, X_n)$ for the density corresponding to (7.10.7). It clearly does not depend on s in R_1 and has the form

$$T(X_1, \ldots, X_n) = (1/2\beta) \log \left[\left(\sum_1^n e^{\beta X_i} \right) \Big/ \left(\sum_1^n e^{-\beta X_i} \right) \right].$$

For the normal density, of course, the Pitman estimator is $T = \bar{X}$.

We formulate the result obtained in the form of a theorem.

THEOREM 7.10.1. Let $(X_1, \ldots, X_n)$ be a random sample of size $n \geqq 3$ from a population with d.f. $F(x-\theta)$, θ real, where F has a positive and continuous density $f(x)$. Then the nondependence of the Pitman estimator $T(X_1, \ldots, X_n)$ on the choice of the loss function from the family (7.10.1) turns out to be a characteristic property of a normal density or of a density $f_{\alpha,\beta}(x) = (2\beta/k(\alpha)) \exp(-\alpha \cosh \beta x)$.

Further as $\beta \to 0$, $f_{\alpha,\beta}(x)$ becomes the density of a normal law, and as $\beta \to \infty$ and $(1/\beta) \log (1/\alpha) \to \delta$, $f_{\alpha,\beta}(x)$ converges to the density of the uniform distribution on $(-\delta, \delta)$ for which also the universality of the Pitman estimator obtains.

7.11 SCALE PARAMETER. PITMAN ESTIMATOR

If the distribution P_σ of the vector of observations $(X_1, \ldots, X_n) \in R^n$ depends on the parameter $\sigma \in R_1^+$ (i.e., $\sigma > 0$) in the following manner:

$$P_\sigma(A) = \int \cdots \int_A dF(x_1/\sigma, \ldots, x_n/\sigma) \tag{7.11.1}$$

then we say that σ is a *scale parameter*. For estimating a scale parameter, it is natural to choose a class of estimators $\tilde{\sigma} = \tilde{\sigma}(X_1, \ldots, X_n)$ satisfying for every $\lambda > 0$ the condition

$$\tilde{\sigma}(\lambda X_1, \ldots, \lambda X_n) = \lambda\tilde{\sigma}(X_1, \ldots, X_n).$$

These were also introduced by Pitman [113]. We shall call such estimators *regular*, again, and denote the class of regular estimators by $\mathscr{T}$.

We suppose that the loss function satisfies the condition

$$r(\tilde{\sigma}, \sigma) = r(\tilde{\sigma} - \sigma); \qquad r(\lambda u) = \lambda^m r(u) \tag{7.11.2}$$

for all $\lambda > 0$, for some m. Then the risk of the estimator $\tilde{\sigma} \in \mathscr{T}$ for the parameter σ is

$$R(\tilde{\sigma}, \sigma) = E_\sigma r(\tilde{\sigma} - \sigma) = \sigma^m E_1 r(\tilde{\sigma} - 1) = \sigma^m R(\tilde{\sigma}, 1)$$

so that, under the loss function (7.11.2), a member of $\mathscr{T}$ is either optimal in that class or inadmissible in that class.

DEFINITION 7.11.1. An estimator $\hat{\sigma} = \hat{\sigma}(X_1, \ldots, X_n)$ optimal in the class, i.e., such that .

$$R(\hat{\sigma}, \sigma) = \min_{\tilde{\sigma} \in \mathscr{T}} R(\tilde{\sigma}, \sigma)$$

will be called the Pitman estimator of the scale parameter σ, corresponding to the loss function (7.11.2).

In what follows, we shall assume that the space of observations is $\mathscr{X} = R^n_+$, i.e., $x_j > 0$ for all j. Then, if we set $L = \sum_1^n c_j X_j$, $c_j > 0$ for all j, then every estimator $\tilde{\sigma}$ in $\mathscr{T}$ can be written in the form

$$\tilde{\sigma} = L \cdot \psi\left(\frac{X_2}{X_1}, \ldots, \frac{X_n}{X_1}\right). \tag{7.11.3}$$

LEMMA 7.11.1. If $EX_j^2 < \infty$ for all j, then

(1) the estimator

$$\hat{\sigma} = L\frac{E_1(L \mid \mathbf{Y})}{E_1(L^2 \mid \mathbf{Y})}, \qquad \mathbf{Y} = \left(\frac{X_2}{X_1}, \ldots, \frac{X_n}{X_1}\right) \tag{7.11.4}$$

is the Pitman estimator for σ, corresponding to quadratic loss; moreover, for any $\tilde{\sigma} \in \mathscr{T}$, $E_\sigma(\tilde{\sigma} - \sigma)^2 > E_\sigma(\hat{\sigma} - \sigma)^2$ for $\sigma \in R^1_+$ except when $\tilde{\sigma} = \hat{\sigma}$ with P_1-probability one (and hence with P_σ-probability one for all $\sigma \in R^1_+$);

(2) if F, the joint d.f. of the X_j, is absolutely continuous, with f as density, then

$$\hat{\sigma} = \frac{\int_0^\infty u^n f(ux_1, \ldots, ux_n)\, du}{\int_0^\infty u^{n+1} f(ux_1, \ldots, ux_n)\, du}. \tag{7.11.5}$$

***Proof*:** First of all, we note that

$$E_\sigma[L\psi(\mathbf{Y}) - \sigma]^2 = \sigma^2 E_1[L\psi(\mathbf{Y}) - 1]^2$$
$$= \sigma^2 E_1\{E_1[(L\psi(\mathbf{Y}) - 1)^2 \mid \mathbf{Y}]\}.$$

For every fixed $\mathbf{y}$, $\min_c E_1[(cL - 1)^2 \mid \mathbf{y}]$ is attained for $c = E_1(L \mid \mathbf{y})/E_1(L^2 \mid \mathbf{y})$. Therefore, $\min_{\psi(\mathbf{Y})} E_\sigma(L\psi(\mathbf{Y}) - \sigma)^2$ is attained simultaneously for all $\sigma \in R^1_+$ by the function $\psi(\mathbf{y}) = E_1(L \mid \mathbf{y})/E_1(L^2 \mid \mathbf{y})$, and the Pitman estimator $\hat{\sigma}(X_1, \ldots, X_n)$ has the form (7.11.4). The second part of assertion (1) is obvious. Assertion (2) is proved in a straightforward manner.▌

We denote by $\mathcal{T}_U$ the class of unbiased regular estimators of σ.

LEMMA 7.11.2. Under the conditions of the above lemma, the optimal estimator of σ in the class $\mathcal{T}_U$ is

$$\hat{\sigma}_U = c_U \tilde{\sigma} \tag{7.11.6}$$

where the constant c_U is defined by the condition $c_U \cdot E_1\hat{\sigma} = 1$. Then, for $\tilde{\sigma}_U \in \mathcal{T}_U$, $E_\sigma(\tilde{\sigma}_U - \sigma)^2 > E_\sigma(\hat{\sigma}_U - \sigma)^2$, $\sigma \in R^1_+$ except in the case when $\tilde{\sigma}_U = \hat{\sigma}_U$ with P_1-probability one.

***Proof*:** Let $\tilde{\sigma}_U \in \mathcal{T}_U$. If $E_1\tilde{\sigma}_U^2 = \infty$, then $E_\sigma\tilde{\sigma}_U^2 = \infty$, and estimator $\tilde{\sigma}_U$ is obviously worse than $\hat{\sigma}_U$. If $E_1\tilde{\sigma}_U^2 < \infty$, then

$$E_\sigma(\tilde{\sigma}_U - \sigma)^2 = \sigma^2 E_1(\tilde{\sigma}_U - 1)^2 = \sigma^2 E_1(\tilde{\sigma}_U - \hat{\sigma}_U + \hat{\sigma}_U - 1)^2$$
$$= \sigma^2\{E_1(\tilde{\sigma}_U - \hat{\sigma}_U)^2 + 2E_1(\tilde{\sigma}_U - \hat{\sigma}_U)(\hat{\sigma}_U - 1) + E_1(\hat{\sigma}_U - 1)^2\}. \tag{7.11.7}$$

But $E_1(\tilde{\sigma}_U - \hat{\sigma}_U) = 0$ by assumption, and from the optimality of the estimator $\hat{\sigma}$ in the class $\mathcal{T}$, it follows that for every unbiased estimator of zero h in $\mathcal{T}$, with $E_1h^2 < \infty$, $E_1(\hat{\sigma}h) = 0$, the proof of this fact being the same as in Lemma 7.2.1. In particular, $E_1\{\hat{\sigma}_U(\hat{\sigma}_U - \tilde{\sigma}_U)\} = 0$, and we have from (7.11.7) that

$$E_\sigma(\tilde{\sigma}_U - \sigma)^2 = E_\sigma(\tilde{\sigma}_U - \hat{\sigma}_U)^2 + E_\sigma(\hat{\sigma}_U - \sigma)^2 \geqq E_\sigma(\hat{\sigma}_U - \sigma)^2$$

where equality holds, simultaneously for all $\sigma \in R^1_+$, if and only if $\tilde{\sigma}_U = \hat{\sigma}_U$ with P_1-probability one.▌

7.12 CHARACTERIZATION OF THE GAMMA DISTRIBUTION THROUGH THE ADMISSIBILITY OF THE OPTIMAL LINEAR ESTIMATOR OF THE SCALE PARAMETER

Suppose that the r.v.'s $X_1, \ldots, X_n$ are nondegenerate, independent, and have d.f.'s $F_1(x/\sigma), \ldots, F_n(x/\sigma)$ respectively, depending on the scale parameter $\sigma \in R^1_+$, where

$$F_j(0) = 0 \quad \text{and} \quad \int x^2\, dF_j(x) < \infty \quad \text{for all } j. \tag{7.12.1}$$

We shall show that it follows, from the absolute admissibility of the optimal linear estimator $\hat{L}=\sum_1^n c_j^0 X_j$ for $\sigma \in R_+^1$ for any two values $n_2 > n_1 \geqq 3$ of n, that the X_j have gamma distributions. Similarly, the property of the optimal linear unbiased estimator $\hat{L}_U = \sum_1^n c_{j,U}^0 X_j$, of being admissible in the class of unbiased estimators of σ for two values $n_2 > n_1 \geqq 3$ of n, also guarantees that the X_j have gamma distributions. The loss function, in both cases, is the quadratic one. The converse result is also true: if the X_j have gamma distributions, then the estimator $\hat{L}$ for σ above is absolutely admissible, and the corresponding estimator $\hat{L}_U$ is admissible (even optimal) in the class of unbiased estimators of σ. The two theorems which follow were established in [57] (and also [53]) for the case of identically distributed X_j having finite moments of all orders. The relaxation of these restrictions is possible, thanks to [73].

THEOREM 7.12.1. Under the conditions (7.12.1), the absolute admissibility of the estimator $\hat{L}=\sum_1^n c_j^0 X_j$ for σ under quadratic loss, for some two values $n_2 > n_1 \geqq 3$ of n is possible if and only if the X_j have gamma distributions

$$F_j(x) = \begin{cases} 0 & \text{for } x \leqq 0 \\ (\alpha_j^{\gamma_j}/\Gamma(\gamma_j)) \int_0^x u^{\gamma_j - 1} e^{-\alpha_j u}\, du & 0 < x < \infty \end{cases}$$

for some positive γ_j and $\alpha_j, j = 1, \ldots, n_1$; we shall write this as $X_j \sim G(\alpha_j, \gamma_j)$. If, conversely, $X_j \sim G(\alpha_j, \gamma_j)$, $j = 1, \ldots, n$, then the estimator $\hat{L}$ based on the X_j is absolutely admissible as an estimator of the parameter σ.

THEOREM 7.12.2. Under condition (7.12.1), the admissibility of the estimator $\hat{L}_U = \sum_1^n c_{j,U}^0 X_j$ in the class of unbiased estimators of the parameter $\sigma \in R_+^1$, under quadratic loss, for some two values of n: $n_2 > n_1 \geqq 3$, is possible only if $X_j \sim G(\alpha_j, \gamma_j)$ for $j = 1, \ldots, n_1$. If $X_j \sim G(\alpha_j, \gamma_j)$ for $j = 1, \ldots, n$, then the estimator $\hat{L}_U$ constructed on the basis of the sample $(X_1, \ldots, X_n)$ is admissible—even optimal—in the class of unbiased estimators of σ.

Remark. Apparently, in Theorem 7.12.1 (respectively, 7.12.2), it is sufficient to assume the absolute admissibility of the estimator $\hat{L}$ (respectively, the admissibility of $\hat{L}_U$ in the class of unbiased estimators of σ) for some one value of $n \geqq 3$.

Proof: First we prove Theorem 7.12.2 and then Theorem 7.12.1. We set

$$\alpha_{1,j} = \int_0^\infty x\, dF_j(x); \qquad \alpha_{2,j} = \int_0^\infty x^2\, dF_j(x); \qquad \sigma_j^2 = \alpha_{2,j} - \alpha_{1,j}^2$$

(if the X_j are nondegenerate, then $\sigma_j^2 > 0$). If $L_U = \sum_1^n c_{j,U} X_j$ be a linear unbiased estimator of σ, then $\sum_1^n c_{j,U} \alpha_{1,j} = 1$. Further, $E_\sigma(L_U - \sigma)^2 = \sigma^2 \sum_1^n c_{j,n}^2 \sigma_j^2$, whence it follows that the coefficients of the optimal linear unbiased estimator $\hat{L}_U$ are

$$c_{j,U}^0 = c\alpha_{1,j}/\sigma_j^2, \qquad \text{where} \qquad c = \left(\sum_1^n \alpha_{1,j}^2/\sigma_j^2\right)^{-1}$$

(We shall not always indicate the dependence of the constants on n.) Let us define a new set of r.v.'s Z_j, where

$$Z_j = \lambda_j X_j, \qquad \lambda_j = \alpha_{1,j}/\sigma_j^2. \tag{7.12.2}$$

The d.f.'s of the Z_j, when $\sigma = 1$, are $G_j(x) = F_j(x/\lambda_j)$. It is easily seen that $\hat{L}_U = \sum_1^n c_{j,U}^0 X_j = c \sum_1^n Z_j$. The condition of the theorem now becomes the admissibility of the estimator $c \sum_1^n Z_j$ in the class of unbiased estimators of $\sigma \in R_+^1$, constructed on the basis of the sample $(Z_1, \ldots, Z_n)$ with $Z_j \sim G_j(x/\sigma)$. By Lemma 7.11.2, it follows from the admissibility of $c \sum_1^n Z_j$ in the class of unbiased estimators of σ that $c \sum_1^n Z_j = \hat{\sigma}_U(Z_1, \ldots, Z_n)$. In view of Lemma 7.11.1, the foregoing condition can be written as

$$E_1\left(\left(\sum_1^n Z_i\right)^2 \middle| \mathbf{Y}_n\right) = c_n E_1\left(\sum_1^n Z_i \,|\, \mathbf{Y}_n\right) \tag{7.12.3}$$

where $\mathbf{Y}_n = (Z_2/Z_1, \ldots, Z_n/Z_1)$, for some constant c_n. This condition is satisfied for two values of n, $n_2 > n_1 \geqq 3$. We consider the relation corresponding to $n = n_2$ and take the conditional mathematical expectation given $\mathbf{Y}_{n_1}$ of both sides of that relation, to obtain

$$E_1\left[\left(\sum_1^{n_2} Z_i\right)^2 \middle| \mathbf{Y}_{n_1}\right] = c_{n_2} E_1\left(\sum_1^{n_2} Z_i \,|\, \mathbf{Y}_{n_1}\right). \tag{7.12.4}$$

We subtract from (7.12.4) the relation (7.12.3), taking $n = n_1$ in the latter, to obtain

$$E_1\left[\left(\sum_{n_1+1}^{n_2} Z_j\right)^2 \middle| \mathbf{Y}_{n_1}\right] + 2E_1\left[\left(\sum_1^{n_2} Z_j\right)\left(\sum_{n_1+1}^{n_2} Z_j\right) \,|\, \mathbf{Y}_{n_1}\right]$$
$$= (c_{n_2} - c_{n_1}) E_1\left(\sum_1^{n_1} Z_j \,|\, \mathbf{Y}_{n_1}\right) + c_{n_2} E_1\left(\sum_{n_1+1}^{n_2} Z_j \,|\, \mathbf{Y}_{n_1}\right). \tag{7.12.5}$$

Considering that, according to (7.12.2), $E_1(Z_j - E_1 Z_j)^2 = E_1 Z_j$, the value of the constant c_n is easily determined from (7.12.3):

$$c_n = 1 + \sum_1^n E_1 Z_j.$$

Since $(Z_1, \ldots, Z_{n_1})$ and $(Z_{n_1+1}, \ldots, Z_{n_2})$ are independent,

$$E_1\left[\left(\sum_{n_1+1}^{n_2} Z_j\right)^2 \Big| \mathbf{Y}_{n_1}\right] = \text{constant}.$$

Further,

$$\begin{aligned}
&E_1\left[\left(\sum_1^{n_1} Z_j\right)\left(\sum_{n_1+1}^{n_2} Z_j\right) \Big| \mathbf{Y}_{n_1}\right] \\
&\quad = E_1\left\{E_1\left[\left(\sum_1^{n_1} Z_j\right)\left(\sum_{n_1+1}^{n_2} Z_j\right) \Big| Z_1, \ldots, Z_{n_1}\right] \Big| \mathbf{Y}_{n_1}\right\} \\
&\quad = E_1\left\{\left(\sum_1^{n_1} Z_j\right) E_1\left[\left(\sum_{n_1+1}^{n_2} Z_j\right) \Big| Z_1, \ldots, Z_{n_1}\right] \Big| \mathbf{Y}_{n_1}\right\} \\
&\quad = E_1\left(\sum_{n_1+1}^{n_2} Z_j\right) E_1\left(\sum_1^{n_1} Z_j \,|\, \mathbf{Y}_{n_1}\right).
\end{aligned}$$

In view of this identity, we can write (7.12.5) in the form

$$E_1\left(\sum_1^{n_1} Z_j \,|\, Z_2/Z_1, \ldots, Z_{n_1}/Z_1\right) = \text{constant}. \tag{7.12.6}$$

As proved in Theorem 6.2.1, it follows from (7.12.6) that, for $n_1 \geqq 3$, $Z_j \sim G(\alpha, \gamma_j)$ for some $\alpha > 0$ and $\gamma_j > 0$ for $j = 1, \ldots, n_1$. Hence it clearly follows that the r.v.'s $X_1, \ldots, X_{n_1}$ have gamma distributions $X_j \sim G(\alpha_j, \gamma_j)$. The first assertion of the theorem is proved.

We now turn to the proof of the second assertion. Let $X_j \sim G(\alpha_j, \gamma_j)$; then it is easily seen that $Z_j \sim G(1, \gamma_j)$, $j = 1, \ldots, n$.

We set up the likelihood of the sample $z_1, \ldots, z_n$ when the value of the parameter is σ.

$$L(z_1, \ldots, z_n; \sigma) = \begin{cases} 0 & \text{if any of the } z_j \leqq 0 \\ \prod_1^n [\sigma^{\gamma_j}\Gamma(\gamma_j)]^{-1} z_j^{\gamma_j - 1} \exp\left(-\dfrac{1}{\sigma}\sum_1^n z_j\right) & \text{otherwise.} \end{cases} \tag{7.12.7}$$

Hence it follows that $U = \sum_1^n z_j$ is a sufficient statistic for the family of distributions (7.12.7) in R_+^n, $\sigma \in R_+^1$. Further, the statistic U has a gamma distribution with density

$$p(u, \sigma) = [\Gamma(N)]^{-1}\sigma^{-N}u^{N-1}e^{-u/\sigma}, \qquad u > 0 \tag{7.12.8}$$

where we have set $N = \sum_1^n \gamma_j$. By Theorem 7.1.3, U is a complete sufficient statistic in this case, and the estimator $c\sum_1^n Z_j$ is optimal as an unbiased estimator of $\sigma \in R_+^1$. Thus, $\hat{L}_U = \sum_1^n c_{j,n}^0 X_j$ is the optimal unbiased estimator of σ given the sample $(X_1, \ldots, X_n)$. Theorem 7.12.2 stands fully proved. ▌

We pass on to prove Theorem 7.12.1. If the estimator $\hat{L}=\sum_1^n c_j^0 X_j$ is absolutely admissible, then, by Lemma 7.11.1, necessarily $\hat{L}=\hat{\sigma}$. But then, by Lemma 7.11.2, the linear unbiased estimator $\hat{L}_U = c\hat{\sigma}$ will be optimal in the class $\mathscr{T}_U$. In this situation, the first part of Theorem 7.12.2 (in fact, for its proof, we used only the class $\mathscr{T}_U$) guarantees that $X_j \sim G(\alpha_j, \gamma_j)$ for $j = 1, \ldots, n_1$.

The second assertion of Theorem 7.12.1 is proved in an entirely different manner from Theorem 7.12.2. We shall follow below essentially [161]. Let $N = \sum_1^n \gamma_j$. Again we set up the variables $Z_1, \ldots, Z_n$ with likelihood function (7.12.7). It is clear from (7.12.7) that in considering possible competitors to $\hat{L}$ as estimators of σ, we may confine ourselves to estimators depending on $U = \sum_1^n Z_j$, in view of Theorem 7.1.2, and (7.12.8) gives the explicit form of $\hat{L}$

$$\hat{L}=\frac{1}{N+1}\sum_1^n Z_j.$$

Let $\tilde{\sigma}(U)$ be an arbitrary estimator with

$$E_\sigma(\tilde{\sigma}-\sigma)=b(\sigma) \qquad \text{and} \qquad E_\sigma(\tilde{\sigma}-\sigma)^2<\infty, \qquad \sigma\in R_+^1.$$

Since $I(\sigma)=\int_{\infty}^0 [(\partial \log p(u, \sigma))/\partial\sigma]^2 p(u, \sigma)\, du = N/\sigma^2$, the inequality (7.1.2) applied to $\tilde{\sigma}(U)$—for the density (7.12.8), all the regularity conditions needed in the derivation of this inequality are satisfied—gives

$$E_\sigma(\tilde{\sigma}-\sigma)^2 \geqq (b(\sigma))^2 + \frac{[1+b'(\sigma)]^2\sigma^2}{N}.$$

Since $E_\sigma(\hat{L}-\sigma)^2 = \sigma^2/(N+1)$, if the estimator $\tilde{\sigma}$ is better than $\hat{L}$, then we must have for $\sigma \in R_+^1$,

$$[b(\sigma)]^2 + \frac{[1+b'(\sigma)]^2\sigma^2}{N} \leq \frac{\sigma^2}{N+1}$$

or, equivalently,

$$\frac{[b(\sigma)]^2}{\sigma^2} + \frac{[1+b'(\sigma)]^2}{N} \leqq \frac{1}{N+1} \qquad \text{for } \sigma\in R_+^1. \tag{7.12.9}$$

It follows from (7.12.9) that $b'(\sigma) < 0$ and $b(0) = 0$, so that $b(\sigma)$ is decreasing on $(0, \infty)$ and $b(\sigma) < 0$ for $\sigma > 0$.

Suppose that for some $\sigma > 0$, $b(\sigma)/\sigma = b'(\sigma)$. Then their common value at such a point must be $-1/(N+1)$ since $u^2 + (1+u)^2/N$ attains its minimum equal to $1/(N+1)$ at $u = -1/(N+1)$.

We shall now show that $b'(\sigma) \leqq b(\sigma)/\sigma$. Indeed, if at some σ_0, $b'(\sigma_0) > b(\sigma_0)/\sigma_0$, then

$$\left[\frac{b(\sigma_0)}{\sigma_0}\right]^2 + \frac{[1+b'(\sigma_0)]^2}{N} > [b'(\sigma_0)]^2 + \frac{[1+b'(\sigma_0)]^2}{N} \geqq \frac{1}{N+1}$$

which contradicts (7.12.9). Further, it then follows by considering the derivative that $b(\sigma)/\sigma$ is nonincreasing.

We next show that there exists a sequence $\{\sigma_j\}$ tending to infinity such that

$$b'(\sigma_j) - \frac{b(\sigma_j)}{\sigma_j} \to 0.$$

In fact, if not, for any $\varepsilon > 0$, a σ_0 can be found such that

$$b'(\sigma) \leqq \frac{b(\sigma)}{\sigma} - \varepsilon, \qquad \sigma \geqq \sigma_0.$$

But then for $\sigma \geqq \sigma_0$, $b(\sigma) \leqq c(\sigma)$, where $c(\sigma_0) = b(\sigma_0)$ and $c'(\sigma) = [c(\sigma)/\sigma] - \varepsilon$ for $\sigma \geqq \sigma_0$. The function $c(\sigma)$ is easily seen to be given by $c(\sigma) = -\varepsilon\sigma \log \sigma + k\sigma$ where the constant k is determined by the condition $c(\sigma_0) = b(\sigma_0)$. But the double inequality

$$-\sigma < b(\sigma) < -\varepsilon\sigma \log \sigma + k\sigma$$

for $\sigma \geqq \sigma_0$ is violated. Hence a sequence $\{\sigma_j\}$ of the kind described above exists.

We shall now show that there exists a sequence $\{\sigma_j\} \to 0$ for which (7.12.10) is satisfied. Otherwise, we would have, for arbitrary $\varepsilon > 0$, $b(\sigma) > -\varepsilon\sigma \log \sigma + k'\sigma$ for $\sigma \leqq \sigma_1$, where σ_1 is sufficiently small. But then $b(\sigma)/\sigma > 0$ for sufficiently small σ, which contradicts the inequality $b(\sigma) < 0$ obtained earlier.

We now write (7.12.9) in the form

$$\begin{aligned}
&\left[\frac{b(\sigma)}{\sigma}\right]^2 + \frac{1}{N}\left[1 + \frac{b(\sigma)}{\sigma} - \frac{b(\sigma)}{\sigma} + b'(\sigma)\right]^2 \\
&\quad = \left[\frac{b(\sigma)}{\sigma}\right]^2 + \frac{1}{N}\left[1 + \frac{b(\sigma)}{\sigma}\right]^2 + \frac{1}{N}\left[b'(\sigma) - \frac{b(\sigma)}{\sigma}\right]^2 \\
&\quad\quad + \frac{2}{N}\left[1 + \frac{b(\sigma)}{\sigma}\right]\left[b'(\sigma) - \frac{b(\sigma)}{\sigma}\right] \leqq \frac{1}{N+1}.
\end{aligned} \tag{7.12.11}$$

We set $\beta(\sigma) = b(\sigma)/\sigma$ and let $\sigma \to \infty$ through a sequence $\{\sigma_j\}$, for which (7.12.10) holds. We obtain from (7.12.11) that $\beta(\infty) = -1/(N+1)$. Similarly, letting $\sigma \to 0$ through a sequence $\{\sigma_j\}$ for which (7.12.10) is satisfied, we see that $\beta(0) = -1/(N+1)$. But we have proved earlier that $\beta(\sigma)$ is nonincreasing, so that it is constant and hence $b(\sigma) = -\sigma/(N+1)$. But, it follows from (7.12.8) that

$$E_\sigma(\hat{L} - \sigma) = E_\sigma\left(\frac{U}{N+1} - \sigma\right) = -\frac{\sigma}{N+1}.$$

Thus,

$$E_\sigma\left[\frac{U}{N+1} - \tilde{\sigma}(U)\right] = 0 \qquad \text{for } \sigma \in R_+^1.$$

In view of U being a complete sufficient statistic, it follows that $\tilde{\sigma}(U) - U/(N+1)$ with P_σ-probability one. Thus the estimator $\hat{L} = (\sum_n^1 Z_j)/(N+1) = \sum_1^n c_j^0 X_j$ for the parameter σ is absolutely admissible. Theorem 7.12.1 stands proved.▌

We note that if the r.v.'s $X_1, \ldots, X_n$ are identically distributed, $X_j \sim F(x/\sigma)$, then the optimal linear estimator has the form $\hat{L} = c_n^0 \overline{X}$ where $c_n^0 = n\alpha_1/(\alpha_2 + (n-1)\alpha_1^2)$, $\alpha_j = \int x^j \, dF(x)$, and the optimal linear unbiased estimator of σ is $\hat{L}_n = \overline{X}/\alpha_1$. From Theorems 7.12.1 and 7.12.2, we obtain:

Corollary. Let $(X_1, \ldots, X_n)$ be a random sample from a population with d.f. $F(x/\sigma)$, $\sigma \in R_+^1$, satisfying (7.12.1). Then the absolute admissibility of the estimator $\hat{L} = c_n^0 \overline{X}$ of σ (or the admissibility of the estimator $\hat{L}_U = \overline{X}/\alpha_1$ in the class of unbiased estimators of σ) for sample sizes n_1 and n_2, where n_1 and n_2 are some two integers such that $n_2 > n_1 \geqq 3$, is a characteristic property of the gamma distribution.

7.13 CHARACTERIZATION OF THE GAMMA DISTRIBUTION THROUGH THE OPTIMALITY OF FUNCTIONS OF THE SAMPLE MEAN

If we assume that the r.v.'s $X_1, \ldots, X_n$ are nondegenerate and i.i.d., then the results of the preceding section can be somewhat extended. The generalization relates to the form of functions $q(\overline{X})$ of $\overline{X}$, the optimality of which in the class of unbiased estimators of the parametric function $E_\sigma q$ is characteristic of the gamma distribution; in Theorem 7.12.2, the case $q(\overline{X}) = a_0 \overline{X} + a_1$ was considered.

We begin with the optimality in L^2 under quadratic loss of the general polynomial $q(\overline{X}) = a_0 \overline{X}^k + \cdots + a_k$ in the class of unbiased estimators of $g(\sigma) = E_\sigma q(\overline{X}) = c_0 \sigma^k + \cdots + c_k$. Naturally, it is assumed that F satisfies the conditions

$$F(0-) = 0, \qquad \alpha_{2k} = \int x^{2k} \, dF(x) < \infty. \tag{7.13.1}$$

The following result is due to Kagan [63].

THEOREM 7.13.1. Let $(X_1, \ldots, X_n)$ be a random sample from a population with d.f. $F(x/\sigma)$, satisfying the condition (7.13.1) and with $\sigma \in I$, some nondegenerate interval. The polynomial $q(\overline{X}) = a_0 \overline{X}^k + \cdots + a_k$, $a_0 \neq 0$, of degree $k \geqq 1$ is optimal in L^2, for $\sigma \in I$, under quadratic loss, in the class of unbiased estimators of $E_\sigma q = g(\sigma)$ based on k samples of sizes $n = m, m+1, \ldots, m+k-1$, for some $m \geqq 3$, if and only if F is a gamma distribution function.

Note: The theorem remains true (and the proof the same *verbatim*) if it is assumed that, for every $n = m, m+1, \ldots, m+k-1$, a polynomial—in general depending on the value of n—$q_n(\overline{X}) = a_{n0}\overline{X}^k + \cdots + a_{nk},\ a_{n0} \neq 0$ is, for $\sigma \in I$, the optimal unbiased estimator of $E_\sigma q_n$.

***Proof*:** *Necessity.* By Lemma 7.2.1, from the optimality in L^2 of the estimator $q(\overline{X})$ in the class of unbiased estimators of $g(\sigma)$ for $\sigma \in I$, it follows that for any $h \in H$

$$E_\sigma[hq(\overline{X})] = 0, \qquad \sigma \in I. \tag{7.13.2}$$

We shall make use of (7.13.2) only for h of the form $h(X_2/X_1, \ldots, X_n/X_1) = h(\mathbf{Y}_n)$. Then the left side of (7.13.2) is a polynomial in σ of degree k with the leading coefficient $a_0 E_1(\overline{X}^k h)$. Then we have from (7.13.2) that $E_1(\overline{X}^k h) = 0$ for $h \in H$; but this is the same as $E_1(\overline{X}^k \mid \mathbf{Y}_n) = \text{constant}$, or

$$E_1\left[\left(\sum_1^n X_i\right)^k \middle| \mathbf{Y}_n\right] = \text{constant}. \tag{7.13.3}$$

In view of the conditions of the theorem, (7.13.3) is satisfied for $n = m, m+1, \ldots, m+k-1$, for some $m \geqq 3$.

LEMMA 7.13.1. Let $Q(z) = b_0 z^k + \cdots + b_k,\ b_0 \neq 0$. If, for $m \geqq 3$ and $n = m, m+1, \ldots, m+k-1$,

$$E_1\left[Q\left(\sum_1^n X_i\right) \middle| \mathbf{Y}_n\right] = \text{constant}, \tag{7.13.4}$$

then $X_i \sim G(\alpha, \gamma)$.

Proof of the Lemma. We shall argue by induction. For $k = 1$, relation (7.13.4) reduces to (7.12.6), which, by Theorem 6.2.1, guarantees that $X_i \sim G(\alpha, \gamma)$. Suppose now that the lemma is true for $k = 1, \ldots, p$; we shall show that it is true for $k = p + 1$. Let then (7.13.4) be satisfied for a polynomial $Q(z) = b_0 z^{p+1} + \cdots + b_{p+1},\ b_0 \neq 0$. We write $Q(\sum_1^n X_i)$ in the form

$$\begin{aligned} Q\left(\sum_1^n X_i\right) &= b_0\left(\sum_1^{n-1} X_i + X_n\right)^{p+1} + \cdots + b_{p+1} \\ &= b_0\left[\left(\sum_1^{n-1} X_i\right)^{p+1} + \binom{p+1}{1}\left(\sum_1^{n-1} X_i\right)^p X_n + \cdots + X_n^{p+1}\right] \\ &\quad + \cdots + b_{p+1}. \end{aligned} \tag{7.13.5}$$

If $k \geqq j$, then

$$\begin{aligned} &E_1\left[\left(\sum_1^{n-1} X_i\right)^j X_n^{k-j} \mid \mathbf{Y}_{n-1}\right] \\ &= E_1\left\{E_1\left[\left(\sum_1^{n-1} X_i\right)^j X_n^{k-j} \mid X_1, \ldots, X_{n-1}\right] \middle| \mathbf{Y}_{n-1}\right\} \\ &= E_1\left\{\left(\sum_1^{n-1} X_i\right)^j E_1(X_n^{k-j} \mid X_1, \ldots, X_{n-1}) \mid \mathbf{Y}_{n-1}\right\} \\ &= E_1 X_n^{k-j} E_1\left[\left(\sum_1^{n-1} X_i\right)^j \mid \mathbf{Y}_{n-1}\right] \end{aligned} \tag{7.13.6}$$

since X_n is independent of $(X_1, \ldots, X_{n-1})$. We now write down also the relations

$$E_1\left[Q\left(\sum_1^{m+p} X_i\right) \middle| \mathbf{Y}_{m+p}\right] = \text{constant} = c_{m+p} \tag{7.13.7}$$

$$E_1\left[Q\left(\sum_1^{m+p-1} X_i\right) \middle| \mathbf{Y}_{m+p-1}\right] = c_{m+p-1}. \tag{7.13.8}$$

It follows from (7.13.7) of course that

$$E_1\left[Q\left(\sum_1^{m+p} X_i\right) \middle| \mathbf{Y}_{m+p-1}\right] = c_{m+p}. \tag{7.13.9}$$

Subtracting (7.13.8) from (7.13.9) and taking the relations (7.13.6) and (7.13.5) into consideration, we obtain

$$E_1\left[Q_1\left(\sum_1^{m+p-1} X_i\right) \middle| \mathbf{Y}_{m+p-1}\right] = \text{constant}$$

where

$$Q_1(z) = b_0\binom{p+1}{1}E_1X_{m+p}z^p + \ldots,$$

the dots standing for a polynomial of degree $\leqq p-1$. It is important to note that the degree of the polynomial $Q_1(z)$ is p, i.e., exactly one less than that of $Q(z)$. If now we repeat the above argument successively for the pairs

$$Q\left(\sum_1^{m+p-1} X_i\right) \quad \text{and} \quad Q\left(\sum_1^{m+p-2} X_i\right); \ldots ; Q\left(\sum_1^{m+1} X_i\right) \quad \text{and} \quad Q\left(\sum_1^{m} X_i\right)$$

then we obtain

$$E_1\left[Q_1\left(\sum_1^{n} X_i\right) \middle| \mathbf{Y}_n\right] = \text{constant} \qquad \text{for } n = m, m+1, \ldots, m+p-1$$

for the same polynomial Q_1 of degree p. Thus, we have passed from a polynomial Q of degree $p+1$, for which (7.13.4) is satisfied for $n = m, m+1, \ldots, m+p$ to a polynomial Q_1 of degree p, for which that relation holds for $n = m, m+1, \ldots, m+p-1$. For $p = 1$, the assertion of the lemma is true, and hence it follows that it is true for $p = k$. ∎

Returning to the proof of Theorem 7.13.1, we note that from the optimality of $q(\overline{X})$ in the class of unbiased estimators of $E_\sigma q(\overline{X})$ for $\sigma \in I$ and for $n = m, m+1, \ldots, m+k-1$, we have obtained the relation (7.13.3) for $n = m, \ldots, m+k-1$. It follows from Lemma 7.13.1 that $X_i \sim G(\alpha, \gamma)$.

Sufficiency. If $X_i \sim G(\alpha, \gamma)$, $i = 1, \ldots, n$, then it follows from (7.12.7) and (7.12.8) that $\overline{X}$ is a complete sufficient statistic for the family of d.f.'s

$$F(x_1/\sigma) \cdots F(x_n/\sigma), \qquad \sigma \in I.$$

We conclude hence that $q(\bar{X})$ is optimal for $\sigma \in I$ (and even for $\sigma \in R_+^1$) as an unbiased estimator of $E_\sigma q(\bar{X})$. Theorem 7.13.1 stands proved.▮

Remark. If we introduce the r.v.'s V_i according to $X_i = \exp(V_i)$, $i = 1, \ldots, n$, and set $G(u) = P[V_i < u;\ \sigma = 1]$, and denote the characteristic function of G by g, then, in terms of the V_i, relation (7.13.3) takes the form

$$E\left\{\left(\sum_1^n e^{V_i}\right)^k \,\middle|\, V_1 - \bar{V}, \ldots, V_n - \bar{V}\right\} = E\left(\sum_1^n e^{V_i}\right)^k$$

where $\bar{V} = (\sum_1^n V_i)/n$, and leads to the following equation involving g:

$$\sum (k; k_1, \ldots, k_n) \prod_1^n g(t_j - ik_j)$$

$$= \sum (k; k_1, \ldots, k_n) \prod_1^n g(-ik_j) \prod_1^n g(t_j) \qquad \text{if } \sum_1^n t_j = 0 \quad (7.13.10)$$

where $(k; k_1, \ldots, k_n)$ is the multinomial coefficient $k!/(k_1! \cdots k_n!)$ and the summation on either side is taken over all $(k_1, \ldots, k_n)$ with $k_j \geqq 0$ for all j and $\sum k_j = k$.

Since $E \exp(2kV_i)$ is finite by assumption, the left and right sides of (7.13.10) are meaningful. The argument used for the proof of Lemma 7.13.1 actually becomes a method of solving (7.13.10) in the case when it is satisfied for some $m \geqq 3$ and for $n = m, m+1, \ldots, m+k-1$. It would be of interest to have a direct method of solving (7.13.10); this would obviously permit the extension of Theorem 7.13.1 to nonidentically distributed r.v.'s $X_1, \ldots, X_n$.

Corollary 1. Under the conditions of Theorem 7.13.1, the optimality of the estimator $\bar{X}/\alpha_1$ in the class of unbiased estimators of the parameter $\sigma \in I$ is a characteristic property of the gamma distribution.

This corollary is not stronger than that of Theorem 7.12.2, which also relates to i.i.d.r.v.'s $X_1, \ldots, X_n$. While Theorem 7.12.2 concerns the optimality of $\hat{L}$ only in the class $\mathscr{T}_U$ (in which optimality and admissibility are equivalent), in the above corollary (to Theorem 7.13.1) the optimality of $\bar{X}/\alpha_1$ in the class of all (not merely regular) unbiased estimators of σ is essentially used.

We now turn to the determination of conditions for the optimality in L^2 of an arbitrary bounded function $\psi(\bar{X})$. For any Borel subset of R^1, we set

$$\nu_\sigma(A) = \int_{\bar{x} \in A} dF(x_1/\sigma) \cdots dF(x_n/\sigma).$$

For a bounded function $\psi(u)$, we construct the spaces $L_\sigma^2(\psi)$ and $L^2(\psi) = \bigcap_{\sigma \in I} L_\sigma^2(\psi)$, I being a nondegenerate interval, following the lines of the construction in section 7.2, with ν_σ in place of P_θ. We assume that, for some $k \geqq 1$,

$$L^2(\psi) \quad \text{contains a nonconstant polynomial} \quad q_n(u) \text{ of degree exactly } k. \tag{7.13.11}$$

The polynomial $q_n(u)$ may depend on n.

Corollary 2. Let $(X_1, \ldots, X_n)$ be a random sample from a population with d.f. $F(x/\sigma)$, $\sigma \in I$. If, for some $m \geqq 3$ and $n = m, m+1, \ldots, m+k-1$, condition (7.13.11) is satisfied, then the optimality of $\psi(\bar{X})$ for $\sigma \in I$ in the class of unbiased estimators of $E_\sigma \psi(\bar{X})$ in L^2 for all such n, is a characteristic property of the gamma distribution.

This corollary also is proved on the whole in the same way as the corollary to Theorem 7.7.1. From the optimality in L^2 of $\psi(\bar{X})$, for given n, and the condition (7.13.11), it follows that $q_n(\bar{X}) = a_{n0}\bar{X}^k + \cdots + a_{nk}$, $a_{n0} \neq 0$, is optimal in L^2 for $\sigma \in I$ as an unbiased estimator of $E_\sigma q_n(\bar{X})$. But then (7.13.3) is satisfied for $n = m, m+1, \ldots, m+k-1$, whence it follows, as we have already seen, that F is the d.f. of some gamma distribution.

CHAPTER 8

Characterization of Families of Distributions Admitting Sufficient Statistics

8.1 INTRODUCTION

Sufficient statistics play an important role in various areas of mathematical statistics. In estimation theory, this is determined by Rao-Blackwellization which fact the reader would have been fairly convinced in the course of the last chapter. Therefore, it is natural to seek characterizations of families of distributions which admit sufficient statistics. The present chapter is basically concerned with this question; firstly, general families of distributions admitting sufficient statistics will be described, and then families generated by random samples from some population. We also deal with some properties characteristic of sufficient statistics.

We start with the definition of sufficient statistics. Let a family of distributions $\{P_\theta;\, \theta \in \Theta\}$ on the space $(\mathscr{X}, \mathscr{A})$, and a statistic S mapping $(\mathscr{X}, \mathscr{A})$ into the space $(\mathscr{S}, \mathscr{B})$ be given.

DEFINITION 8.1.1. The statistic S will be called *sufficient* for the family $\{P_\theta\}$ if for any $A \in \mathscr{A}$ a function $\psi_A = \psi_A[S(x)]$ can be found such that

$$P_\theta(A \mid S) = \psi_A \qquad \text{a.e. } [P_\theta] \tag{8.1.1}$$

where a.e. stands for almost everywhere with respect to the indicated measure. Somewhat freely, a sufficient statistic S is sometimes described thus: "the conditional probability of any event for a fixed value of the statistic S is independent of the parameter."

We note that various statistics S can be sufficient statistics for the same family $\{P_\theta\}$. In this connection, we introduce the following definition.

DEFINITION 8.1.2. The statistic $T: (\mathscr{X}, \mathscr{A}) \to (\mathscr{T}, \mathscr{C})$ will be said to be a *minimal sufficient* statistic for the family $\{P_\theta\}$ if T is a function of every sufficient statistic S. More precisely, T will be minimal sufficient for $\{P_\theta\}$ if whatever be the sufficient statistic $S: (\mathscr{X}, \mathscr{A}) \to (\mathscr{S}, \mathscr{B})$, the following is true: $T^{-1}(\mathscr{C}) \subset S^{-1}(\mathscr{B})$, where $T^{-1}(\mathscr{C})$ and $S^{-1}(\mathscr{B})$ are the inverse images of the algebras $\mathscr{C}$ and $\mathscr{B}$ under the indicated mappings.

We note an obvious fact which will be used in section 8.3: if T is minimal sufficient, and for some sufficient statistic S, it is true that $S^{-1}(\mathscr{B}) \subset T^{-1}(\mathscr{C})$, then $S^{-1}(\mathscr{B}) = T^{-1}(\mathscr{C})$ and S is also a minimal sufficient statistic.

8.2 THE FACTORIZATION THEOREM

The basic theorem of this section gives a description of families of distributions for which a given statistic $S(x)$ is sufficient.

We shall assume that all the distributions $\{P_\theta,\ \theta \in \Theta\}$, are absolutely continuous with respect to some measure μ; in this case, the family $\{P_\theta\}$ will be said to be *dominated* by the measure μ.

THEOREM 8.2.1. In order that the statistic $S: (\mathscr{X}, \mathscr{A}) \to (\mathscr{S}, \mathscr{B})$ be sufficient for the family $\{P_\theta\}$ dominated by μ, it is necessary and sufficient that the density $dP_\theta/d\mu = p(\cdot\,, \theta)$ admit the factorization

$$p(x, \theta) = R[S(x); \theta]\, r(x) \qquad \text{a.e. } [\mu], \qquad \theta \in \Theta \tag{8.2.1}$$

where $R(\cdot\,, \theta)$ is a nonnegative $\mathscr{B}$-measurable function and r is a nonnegative $\mathscr{A}$-measurable function.

In the above form, Theorem 8.2.1 was first established by Halmos and Savage [158]; in a less general form, the factorization theorem was established much earlier by Fisher and Neyman. For the proof, we shall require the following lemma of a metric character (cf. [77], p. 481): we shall omit the proof of the lemma.

LEMMA 8.2.1. If the family $\{P_\theta\}$ is dominated by the measure μ, then a countable subfamily: $\{P_{\theta_1}, P_{\theta_2}, \ldots\}$ exists such that $P_{\theta_i}(A) = 0$ for $i = 1, 2, \ldots$ implies that $P_\theta(A) \equiv 0$ for all θ.

***Proof of Theorem 8.2.1*:** We introduce the probability measure $\lambda = \sum_i c_i P_{\theta_i}$ where every $c_i > 0$ and $\sum c_i = 1$. This measure obviously has the property that $\lambda(A) = 0$ implies that $P_\theta(A) \equiv 0$ for all θ, and (trivially) conversely.

Necessity. Let the statistic S be sufficient for $\{P_\theta\}$. Then, if $P_\theta(A \mid S) = \psi_A$, then $\lambda(A \mid S) = \psi_A$ also. In fact, for any $B \in S^{-1}(\mathscr{B})$, we have

$$\int_B \psi_A \, d\lambda = \int_B \psi_A \, d\Big(\sum_i c_i P_{\theta_i}\Big) = \sum_i c_i \int_B \psi_A \, dP_{\theta_i} = \sum_i c_i \int_{B \cap A} dP_{\theta_i}$$
$$= \sum_i c_i P_{\theta_i}(B \cap A) = \lambda(B \cap A).$$

We set $dP_\theta/d\lambda = f_\theta$. Denoting by I_A the indicator function of the set A, we have for any $A \in \mathscr{A}$,

$$\begin{aligned} E_\lambda(I_A f_\theta) &= P_\theta(A) = E_\theta[P_\theta(A \mid S)] = E_\lambda[f_\theta P_\theta(A \mid S)] \\ &= E_\lambda[f_\theta \lambda(A \mid S)] = E_\lambda[E_\lambda(f_\theta \mid S)\lambda(A \mid S)] \\ &= E_\lambda[E_\lambda\{I_A E_\lambda(f_\theta \mid S) \mid S\}] = E_\lambda[E_\lambda(I_A E_\lambda(f_\theta \mid S))]. \end{aligned}$$

Hence $f_\theta = E_\lambda(f_\theta \mid S)$ a.e. $[\lambda]$, i.e., the density f_θ is $\mathscr{B}$-measurable. Further,

$$p(\cdot\,;\theta) = dP_\theta/d\mu = (dP\ /d\lambda)(d\lambda/d\mu) = f_\theta \cdot r$$

and since $f_\theta = f_\theta\,(S)$, we obtain the required factorization (8.2.1).
Sufficiency. We deduce from (8.2.1) that

$$d\lambda/d\mu = r \sum_i R(S, \theta_i) = r \cdot G(S).$$

We now set

$$\tilde{f}_\theta(x) = \begin{cases} R[S(x), \theta]/G[S(x)] & \text{if } G[S(x)] > 0 \\ 0 & \text{if } G[S(x)] = 0. \end{cases}$$

It is easy to verify that the $\mathscr{B}$-measurable function $\tilde{f}_\theta$ is one version of the derivative $dP_\theta/d\lambda$, and consequently $\tilde{f}_\theta = f_\theta$ a.e. $[\lambda]$. For any $A \in \mathscr{A}$, we then have

$$\begin{aligned} E_\theta[P_\theta(A \mid S)] &= P_\theta(A) = E_\lambda(f_\theta I_A) = E_\lambda[E_\lambda(f_\theta I_A \mid S)] \\ &= E_\lambda[f_\theta E_\lambda(I_A \mid S)] = E_\theta[E_\lambda(I_A \mid S)] = E_\theta[\lambda(A \mid S)]. \end{aligned}$$

Since this relation is valid for any $A \in \mathscr{A}$, it is true for the events $A \cap B$, where $B \in S^{-1}\mathscr{B}$. For such events, it has the form

$$E_\theta[I_B P_\theta(A \mid S)] = E_\theta[I_B \lambda(A \mid S)]$$

whence it is seen that $P_\theta(A \mid S) = \lambda(A \mid S)$ a.e. $[P_\theta]$. This is precisely what is meant by the sufficiency of the statistic S. ▮

Remark. We shall show that if S is a sufficient statistic for a dominated family $\{P_\theta\}$, then, for any real-valued measurable function $\phi(x)$ with $E_\theta|\phi| < \infty$ for all $\theta \in \Theta$, a function $\tilde{\phi}$ exists such that

$$E_\theta(\phi \mid S) = \tilde{\phi} \qquad \text{a.e. } [P_\theta], \qquad \theta \in \Theta. \tag{8.2.2}$$

Proof: We set $\Theta_N = \{\theta \in \Theta : N < E_\theta|\phi| \leqq N + 1\}$; obviously, $\bigcup_{N=0}^{\infty}\Theta_N = \Theta$. For every Θ_N, we construct a measure λ_N in the same way as we constructed λ for Θ, and set $\nu = \sum_{N=0}^{\infty} \lambda_N/2^{N+1}$. The measure ν is different from the measure λ introduced earlier in that $E_\nu|\phi| < \infty$ while $E_\lambda|\phi|$ could be infinite. Proceeding now in exactly the same way as in the proof of the sufficiency part of Theorem 8.2.1, we obtain

$$E_\theta(\phi \mid S) = E_\nu(\phi \mid S) \qquad \text{a.e. } [P_\theta], \qquad \theta \in \Theta. ▮$$

Hence, (8.2.2) can be taken to be the definition of the sufficiency of the statistic $S(x)$. Indeed, the property (8.2.2) of sufficient statistics is used in the theory of estimation.

The following result will be needed in the sequel, and, besides, it is interesting in itself (cf. [158]).

THEOREM 8.2.2. Let the family $\{P_\theta\}$ be a dominated one, and the statistic $S(x)\colon (\mathscr{X}, \mathscr{A}) \to (\mathscr{S}, \mathscr{B})$ be sufficient for each of the families $\{P_{\theta_1}, P_{\theta_2}\}$, $\theta_1 \in \Theta$, $\theta_2 \in \Theta$. Then $S(x)$ is sufficient for $\{P_\theta\}$. In other words, for a dominated family, pairwise sufficiency is equivalent to sufficiency.

***Proof*:** Let $\{P_{\theta_1}, \ldots\}$ be a countable subfamily, chosen as in the proof of Lemma 8.2.1. Fix $\theta \in \Theta$. Using the sufficiency of $S(x)$ for the family $\{P_\theta, P_{\theta_j}\}$ and arguing as in the proof of Theorem 8.2.1, we obtain that the function

$$f_j(\cdot, \theta) = \frac{dP_\theta}{c_j d(P_\theta + P_{\theta_j})}$$

is $\mathscr{B}$-measurable, for any $c_j > 0$. Let $S_j(\theta) = \{x : f_j(x, \theta) = 0\}$. On $\mathscr{X} - \bigcup_j S_j(\theta)$, P_θ is mutually absolutely continuous with respect to $P_\theta + \lambda$ and hence with respect to λ. Further, for $x \in \mathscr{X} - \bigcup S_j(\theta)$,

$$\frac{dP_\theta}{d\lambda} = \frac{1}{d\lambda/dP_\theta} = \frac{1}{\sum (1/f_j) - 1}$$

and, since $S_j(\theta) \in \mathscr{B}$, it follows that $dP_\theta/d\lambda$ is $\mathscr{B}$-measurable. This signifies that the statistic $S(x)$ is sufficient for the family $\{P_\theta\}$. ∎

The structure of the minimal sufficient statistic is easily described with the help of the factorization theorem 8.2.1 if we assume that $p(x, \theta) > 0$ for all $x \in \mathscr{X}$ and all $\theta \in \Theta$. Consider the statistic (θ_0 is an arbitrary member of Θ):

$$T : x \to g_x(\theta) = \log \frac{p(x, \theta)}{p(x, \theta_0)} \tag{8.2.3}$$

(the notation $g_x(\theta)$ is intended to emphasize that we consider it as a function of the parameter θ). The statistic $T(x)$ maps the space $(\mathscr{X}, \mathscr{A})$ into the space $\mathscr{T} = \{g_x(\theta);\ x \in \mathscr{X}\}$ of functions on Θ. In $\mathscr{T}$, we consider the σ-algebra $\mathscr{C}$, generated by cylindrical sets of the form

$$\{\gamma \in \mathscr{T} : (\gamma(\theta_1), \ldots, \gamma(\theta_s)) \in B_s\}$$

where $\theta_1, \ldots, \theta_s$ are points in Θ, B_s is an s-dimensional Borel set. Thus, the statistic (8.2.3) is a measurable mapping from $(\mathscr{X}, \mathscr{A})$ into $(\mathscr{T}, \mathscr{C})$.

LEMMA 8.2.2. If $dP_\theta/d\mu = p(\cdot, \theta) > 0$, then the statistic (8.2.3) is the minimal sufficient statistic for the family $\{P_\theta\}$.

***Proof*:** First, since $p(x, \theta) = \tilde{R}(T(x), \theta)\tilde{r}(x)$, where $\tilde{r}(x) = p(x, \theta_0)$, and the function $\tilde{R}(T(x), \theta) = \exp g_x(\theta)$ is $\mathscr{C}$-measurable for every θ, we conclude from Theorem 8.2.1 that the statistic $T(x)$ is sufficient. Further, if $S(x)$ is any sufficient statistic, then

$$p(x,\theta) = R(S(x),\theta)r(x) \quad \text{and} \quad g_x(\theta) = R(S(x), \theta)/R(S(x), \theta_0),$$

so that T is in fact a function of the sufficient statistic S.∎

8.3 TWO CHARACTERISTIC PROPERTIES OF SUFFICIENT STATISTICS

8.3.1 Invariance of the Fisherian Amount of Information

Let $\{P_\theta, \theta \in I\}$, where I is a real interval, be a family of distributions on the space $(\mathscr{X}, \mathscr{A})$, given by the densities $dP_\theta/d\mu = p(x, \theta)$. We assume that

(i) $p(x, \theta) > 0$ for all x and θ;
(ii) $p(x, \theta)$ is continuously differentiable with respect to θ for all x;
(iii) for any $A \in \mathscr{A}$, $(d/d\theta) \int_A p(x, \theta)\, d\mu = \int_A [\partial p(x, \theta)/\partial\theta]\, d\mu$.

Under the conditions (i)–(iii) above, the Fisherian amount of information for the family $\{P_\theta\}$ is defined as

$$I(\theta) = E_\theta \left[\frac{\partial \log p(x, \theta)}{\partial\theta}\right]^2 = \int \left(\frac{\partial \log p}{\partial\theta}\right)^2 p(x, \theta)\, d\mu.$$

Let $T(x) = T\colon (\mathscr{X}, \mathscr{A}) \to (\mathscr{T}, \mathscr{C})$ be a statistic. On the space $(\mathscr{T}, \mathscr{C})$, it is clear that a family of distributions is generated in a natural manner, namely the family

$$P_\theta^T(C) = P_\theta\{T^{-1}(C)\}, \qquad C \in \mathscr{C}.$$

It is easy to see that P_θ^T is absolutely continuous with respect to μ^T formed by the measure μ under the mapping T; let $q(\cdot, \theta) = dP_\theta^T/d\mu^T$. We assume that the density $q(\cdot, \theta)$ satisfies conditions similar to (i)–(iii). Then the Fisherian amount of information contained in T is defined, being given by

$$\tilde{I}(\theta) = E_\theta \left[\frac{\partial \log q(t, \theta)}{\partial\theta}\right]^2 = \int_{\mathscr{T}} \left(\frac{\partial \log q}{\partial\theta}\right)^2 q(t, \theta)\, d\mu^T.$$

THEOREM 8.3.1. Under the conditions (i)–(iii), and similar conditions for $q(\cdot, \theta)$, we have $\tilde{I}(\theta) \leqq I(\theta)$, $\theta \in I$, where thc sign of equality holds if and only if T is a sufficient statistic for the family $\{P_\theta\}$.

***Proof*:** We set $y = y(\cdot, \theta) = (\partial \log p/\partial\theta)$, $\tilde{y} = \tilde{y}(\cdot, \theta) = (\partial \log q/\partial\theta)$. Then, $I(\theta) = E_\theta y^2$ and $\tilde{I}(\theta) = E_\theta \tilde{y}^2$. It is easily verified that, under condition (iii),

$E_\theta(y \mid T) = \tilde{y}$. Further, we have on applying Jensen's inequality (cf. Lemma 1.7.4) that

$$I(\theta) = E_\theta\, y^2 = E_\theta\, E_\theta(y^2 \mid T) \geqq E_\theta[E_\theta(y \mid T)]^2 = E_\theta\, \tilde{y}^2 = \tilde{I}(\theta),$$

where the equality sign holds (for a given θ) if and only if there exists a function $k(T, \theta)$ such that

$$y(x, \theta) = k[T(x), \theta] \qquad \text{a.e. } [P_\theta]. \tag{8.3.1}$$

We shall denote the exceptional set in the above context, which may depend on θ, by N_θ.

Consider on the product space $I \times \mathscr{X}$ the measure ν which is the product of the Lebesgue measure on I by the measure μ on $\mathscr{X}$: $d\nu = d\theta\, d\mu$. If we denote by N that subset of $I \times \mathscr{X}$ whose θ-section is N_θ, then $\nu(N) = 0$, since the μ-measure of every θ-section is zero, and (8.3.1) is valid for all (θ, x) in the set $(I \times \mathscr{X}) - N$. Hence, for x in $\mathscr{X} - M$, where M is a set of μ-measure zero, the relation (8.3.1) is satisfied for almost all θ. Integrating (8.3.1) with respect to θ for x in $\mathscr{X} - M$, we obtain

$$\log p(x, \theta) - \log p(x, \theta_0) = \int_{\theta_0}^{\theta} y(x, \theta)\, d\theta = \int_{\theta_0}^{\theta} k(T, \theta)\, d\theta = U(T, \theta)$$

whence, for such x, we have

$$p(x, \theta) = \exp\,[U(T, \theta)]p(x, \theta_0) = R(T, \theta)r(x)$$

and, by the factorization theorem 8.2.1, $T(x)$ is a sufficient statistic for the family $\{P_\theta\}$. ▮

The theorem just proved is one of the precise versions of the somewhat vague statement that "sufficient statistics, and only they, contain all the information contained in the sample."

8.3.2 Bahadur's Theorem

We saw in Theorem 7.1.2 that if $T(x)$ is a sufficient statistic for a family of distributions $\{P_\theta, \theta \in \Theta\}$ defined on $(\mathscr{X}, \mathscr{A})$, then, for every estimator $\tau(x)$ of the parametric function $t(\theta)$ we can indicate an estimator $T(x)$ depending only on the sufficient statistic, and not worse than $\tau(x)$ under loss functions $r(\cdot, t)$ which are convex in the first argument; in this sense it is said that the class of estimators dependent on the sufficient statistics is essentially complete.

We shall now show, following Bahadur [4] basically, that for a dominated family $\{P_\theta\}$ and a strictly convex differentiable (all the time with respect to the first argument) loss function $r(\cdot, t)$ satisfying the condition

$$\min_\tau r(\tau, t) \text{ is attained at a unique point } \tau(t). \tag{8.3.2}$$

essential completeness of the class of estimators depending on some statistic $T(x)$ is a characteristic property of sufficient statistics. We remark that condition (8.3.2) is fairly natural in problems of point estimation of parametric functions.

THEOREM 8.3.2. Suppose that the family $\{P_\theta\}$ is dominated and the loss function $r(\tau, t)$ is strictly convex and differentiable with respect to τ and satisfies (8.3.2). If for any function $t(\theta)$ on Θ and for any estimator $\tau(x)$ an estimator $\tilde{\tau}[T(x)]$ can be indicated such that for all $\theta \in \Theta$

$$E_\theta r(\tilde{\tau}, t(\theta)) \leqq E_\theta r(\tau, t(\theta)) \tag{8.3.3}$$

then the statistic $T(x)$ is sufficient for the family $\{P_\theta\}$.

***Proof*:** We take two points θ_1 and θ_2 in Θ and a function $t(\theta)$ such that $t_1 = t(\theta_1) \neq t_2 = t(\theta_2)$. (The case where Θ consists of a single point is trivial; every statistic is sufficient.) We set $r(\tau, t_i) = r_i(\tau)$; $i = 1, 2$. Let $P_{\theta_1} + P_{\theta_2} = \mu$, and $dP_{\theta_1}/d\mu = p(\cdot)$ so that $dP_{\theta_2}/d\mu = 1 - p(\cdot)$. Now consider, for fixed $x \in \mathscr{X}$, the following function of τ.

$$L(\tau) = p(x)r_1(\tau) + (1 - p(x))r_2(\tau).$$

This function is strictly convex and has a unique minimum at the point $\tau^* = \tau^*(x)$, at which

$$p(x)r_1'(\tau^*) + (1 - p(x))r_2'(\tau^*) = 0. \tag{8.3.4}$$

We now consider the estimator $\tau^*(x)$. It is clear that for any $\tilde{\tau}(x)$—including for $\tilde{\tau}[T(x)]$—

$$\int_{\mathscr{X}} [r(\tau^*(x), t_1)p(x) + r(\tau^*(x), t_2)(1 - p(x))]\, d\mu$$
$$\leqq \int_{\mathscr{X}} [r(\tilde{\tau}, t_1)p(x) + r(\tilde{\tau}, t_2)(1 - p(x))]\, d\mu \tag{8.3.5}$$

where equality obtains only if $\tilde{\tau}(x) = \tau^*(x)$ a.e. $[\mu]$. Combining (8.3.5) with (8.3.3), we obtain

$$\tau^*(x) = \tilde{\tau}[T(x)] \qquad \text{a.e. } [\mu]. \tag{8.3.6}$$

If now we take into consideration condition (8.3.2), according to which the functions $r_1'(\tau)$ and $r_2'(\tau)$ do not vanish simultaneously, then we have from (8.3.4) and (8.3.6)

$$p(x) = \tilde{p}[T(x)] \qquad \text{a.e. } [\mu].$$

This means precisely that $T(x)$ is sufficient for the family $\{P_{\theta_1}, P_{\theta_2}\}$. Since θ_1 and θ_2 are arbitrary, and the family $\{P_\theta\}$ is dominated, we conclude from Theorem 8.2.2 that $T(x)$ is sufficient for the family $\{P_\theta\}$. ∎

8.4 ONE-DIMENSIONAL DISTRIBUTIONS, SOME POWERS OF WHICH ADMIT NONTRIVIAL SUFFICIENT STATISTICS

We assume that the distribution P_θ, $\theta \in \Theta$, on the space R^{mn} with a chosen σ-algebra $\mathscr{A}^{mn}$ of Borel sets, has the form

$$P_\theta(A) = \int \cdots \int_A dF(\mathbf{x}_1, \theta) \cdots dF(\mathbf{x}_n, \theta) \tag{8.4.1}$$

where $F(\mathbf{x}, \theta)$ is some d.f. in R^m. In such a case, we say that P_θ is the n-th power of F: $dP_\theta = dF(\mathbf{x}_1, \theta) \cdots dF(\mathbf{x}_n, \theta)$. Under what conditions on $F(\mathbf{x}, \theta)$ will the family $\{P_\theta\}$ have a sufficient statistic, nontrivial in the sense that its dimension is smaller than mn: a precise postulation is given below. This problem will be considered in this section. For simplicity, we confine ourselves to the case $m = 1$, the generalization to the case of an arbitrary m being immediate.

If the d.f. $F(x, \theta)$ is given by a density $f(x, \theta)$ with respect to Lebesgue measure, then the family $\{P_\theta\}$ will be dominated by the Lebesgue measure μ in R^n, and by Theorem 8.2.1, from the sufficiency of $S(\mathbf{x})$ for the family (8.4.1), we derive

$$dP_\theta/d\mu = p(\mathbf{x}, \theta) = \prod_1^n f(x_i, \theta) = R[S(\mathbf{x}), \theta]\, r(\mathbf{x}).$$

Thus, from the sufficiency of $S(\mathbf{x})$ for the family (8..4.1) follows the functional equation

$$\prod_1^n f(x_i, \theta) = R[S(\mathbf{x}), \theta]\, r(\mathbf{x}) \tag{8.4.2}$$

and the problem of identifying the one-dimensional distributions some power of which admits a nontrivial sufficient statistic reduces to that of identifying all $f(x, \theta)$ satisfying (8.4.2) for some R and r.

We now formulate precisely the condition of nontriviality for a (sufficient) statistic $S(\mathbf{x})$.

We shall say that the statistic $S(\mathbf{x})$: $(R^n, \mathscr{A}^n) \to (\mathscr{S}, \mathscr{B})$ is *trivial at the point* $\mathbf{x}$ if in some neighborhood U of that point, the relation $S(\mathbf{x}') = S(\mathbf{x}'')$ implies that $\mathbf{x}' = \mathbf{x}''$.

This definition is obviously equivalent to the following: the statistic $S(\mathbf{x})$ is trivial at $\mathbf{x}$ if there exists a neighborhood U of $\mathbf{x}$ such that

$$S^{-1}(\mathscr{B}) \cap U = \mathscr{A}^n \cap U$$

where the intersection of either σ-algebra with U is taken to mean the σ-algebra formed by the intersections of members of that σ-algebra with U.

A statistic which is not trivial at any point is called *nontrivial.*

We note that a statistic trivial at every point is, in an obvious manner, sufficient for any family of distributions.

We suppose that there exists (a finite or infinite) interval I of R^1 such that

$$\begin{aligned} f(x, \theta) > 0 \quad &\text{for} \quad x \in I, \quad \theta \in \Theta \\ f(x, \theta) = 0 \quad &\text{for} \quad x \notin I, \quad \theta \in \Theta \end{aligned} \tag{8.4.3}$$

and the function $f(x, \theta)$ is continuously differentiable with respect to x, $x \in I$, for $\theta \in \Theta$. If these conditions are satisfied, we shall say that the density $f(x, \theta)$ and the family of densities (8.4.2) are *regular*.

Let us denote by L the minimal linear space of functions on I, containing the constant function 1 and all the functions $g_\theta(x) = \log [f(x, \theta)/f(x, \theta_0)]$, where θ_0 is a fixed element of Θ, and θ ranges over the whole of Θ. Let $\dim L = r + 1$ (the case $r = \infty$ is not excluded). The following lemma is due to Dynkin [26].

LEMMA 8.4.1. We suppose that the family of densities (8.4.2) is regular. Then, for $n \leqq r$, this family does not have a nontrivial sufficient statistic. If $n \geqq r + 1$, and $1, \phi_1, \ldots, \phi_r$ be a basis for the space L, then the system of functions

$$\chi_i(\mathbf{x}) = \sum_{k=1}^{n} \phi_i(x_k), \qquad i = 1, \ldots, r$$

are functionally independent and $S(\mathbf{x}) = \{\chi_1(\mathbf{x}), \ldots, \chi_r(\mathbf{x})\}$ is a minimal sufficient statistic for the family (8.4.2).

***Proof*:** Let $r + 1 \leqq n$, and the functions $1, \phi_1(\mathbf{x}), \ldots, \phi_r(\mathbf{x})$ form a basis for the space L. By the conditions of the lemma, the ϕ_i, being linear combinations of 1 and the $g_\theta(x)$, are continuously differentiable. If the $\chi_j(\mathbf{x})$ were functionally dependent, then for fixed $x_{r+1}, \ldots, x_n$, we would have identically in $x_1, \ldots, x_r$

$$\frac{\partial(\chi_1, \ldots, \chi_r)}{\partial(x_1, \ldots, x_r)} = \begin{vmatrix} \phi_1'(x_1) & \cdots & \phi_1'(x_r) \\ \cdots & \cdots & \cdots \\ \phi_r'(x_1) & \cdots & \phi_r'(x_r) \end{vmatrix} = 0. \tag{8.4.4}$$

We shall show that from (8.4.4) follows the *linear* dependence of the functions $1, \phi_1, \ldots, \phi_r$. This is obvious for $r = 1$. Suppose the assertion is valid for $r - 1$. Then, if

$$\frac{\partial(\chi_1, \ldots, \chi_{r-1})}{\partial(x_1, \ldots, x_{r-1})} \equiv 0$$

the functions $1, \phi_1, \ldots, \phi_{r-1}$ and *a fortiori* the functions $1, \phi_1, \ldots, \phi_{r-1}, \phi_r$ are linearly dependent. Let therefore

$$\frac{\partial(\chi_1, \ldots, \chi_{r-1})}{\partial(x_1^0, \ldots, x_{r-1}^0)} \neq 0.$$

Expanding the determinant (8.4.4) at the point $(x_1^0, \ldots, x_{r-1}^0, x)$ in terms of the elements of the last column, we obtain

$$A_1\phi_1'(x) + \cdots + A_r\phi_r'(x) \equiv 0, \qquad x \in I$$

where $A_r \neq 0$. Hence we conclude that the functions $1, \phi_1, \ldots, \phi_r$ are linearly dependent, on I. Since this contradicts the assumption that $1, \phi_1, \ldots, \phi_r$ constitute a basis for L, the functional independence of the statistics $\chi_1, \ldots, \chi_r$ is proved.

Further,

$$\log \frac{p(\mathbf{x}, \theta)}{p(\mathbf{x}, \theta_0)} = \sum_1^n \log \frac{f(x_i, \theta)}{f(x_i, \theta_0)} = \sum_{i=1}^n g_\theta(x_i)$$

$$= \sum_{i=1}^n \left[\sum_{k=1}^r c_k(\theta)\phi_k(x_i) + c_0(\theta)\right] = \sum_{k=1}^r c_k(\theta)\chi_k(\mathbf{x}) + nc_0(\theta)$$

whence

$$p(\mathbf{x}, \theta) = R[S(\mathbf{x}), \theta]\, r(\mathbf{x})$$

with

$$R(S(\mathbf{x}), \theta) = \exp\left[\sum_{k=1}^n c_k(\theta)\chi_k(\mathbf{x}) + nc_0(\theta)\right],$$

$$r(\mathbf{x}) = p(\mathbf{x}, \theta_0).$$

Thus the statistic $S(\mathbf{x})$ is sufficient for the family (8.4.2).

By the definition of the space L,

$$\phi_k(x) = \sum_s c_{ks}\, g_{\theta_s}(x) + c_{k0}$$

whence

$$\chi_k(\mathbf{x}) = \sum_1^n \phi_k(x_i) = \sum_{i=1}^n \sum_s c_{ks}\, g_{\theta_s}(x_i) + nc_{k0} = \sum_s c_{ks} \sum_{i=1}^n g_{\theta_s}(x_i) + nc_{k0}\,.$$

We know from section 8.2 that the statistic

$$T\colon \mathbf{x} \to \log \frac{p(\mathbf{x}, \theta)}{p(\mathbf{x}, \theta_0)} = \sum_1^n g_\theta(x_i)$$

is minimal sufficient. But we have from the preceding relation that $\chi_k(\mathbf{x}) = \tilde{\chi}_k[T(\mathbf{x})]$ and consequently

$$S(\mathbf{x}) = \{\chi_1(\mathbf{x}), \ldots, \chi_r(\mathbf{x})\}\colon (R^n, \mathscr{A}^n) \to (R^r, \mathscr{A}^r)$$

is a minimal sufficient statistic for the family (8.4.2).

We shall now show that, for $r \geqq n$, the family (8.4.2) does not admit a nontrivial sufficient statistic. In L, we choose the functions $1, \phi_1, \ldots, \phi_n$ such that they are linearly independent. As we have seen in the proof of the first assertion of the lemma, the determinant

$$\frac{\partial(\chi_1, \ldots, \chi_n)}{\partial(x_1, \ldots, x_n)}$$

is nonzero for at least one point $(x_1^0, \ldots, x_n^0)$. By the inverse function theorem, we have

$$x_i = x_i(\chi_1, \ldots, \chi_n), \quad i = 1, \ldots, n$$

and thus

$$\mathscr{A}^n \cap U \subset \chi^{-1}(\mathscr{A}^n) \cap U = T^{-1}(\mathscr{C}) \cap U \tag{8.4.5}$$

where T is the minimal sufficient statistic of section 8.2. Since, by (8.4.5), even the minimal sufficient statistic is trivial at the point $\mathbf{x}^0$, we conclude that the family (8.4.2) does not have nontrivial sufficient statistics.

This concludes the proof of Lemma 8.4.1.▮

Following [26], we can now describe all the regular densities satisfying the relation (8.4.2).

THEOREM 8.4.1. If the regular density $f(x, \theta)$ satisfies (8.4.2) with a nontrivial $S(\mathbf{x})$, then for some $r \leqq n - 1$

$$f(x, \theta) = \exp\left\{\sum_{j=1}^{r} c_j(\theta)\phi_j(x) + c_0(\theta) + \phi_0(x)\right\} \tag{8.4.6}$$

where $\phi_j(x)$ is continuously differentiable and the functions $1, \phi_1, \ldots, \phi_r$ are linearly independent.

Thus, among regular families, only *exponential families of densities* of the form (8.4.6) have the property that some power thereof satisfies (8.4.2) with a nontrivial $S(\mathbf{x})$.

Proof: By Lemma 8.4.1, it follows from the nontriviality of the sufficient statistic $S(\mathbf{x})$ that $\dim L = r + 1 \leqq n$. Let $1, \phi_1, \ldots, \phi_r$ be a basis for the space L. Then,

$$g_\theta(x) = \log\frac{f(x, \theta)}{f(x, \theta_0)} = \sum_1^r c_j(\theta)\phi_j(x) + c_0(\theta). \tag{8.4.7}$$

Setting $\phi_0(x) = f(x, \theta_0)$, we see that (8.4.7) is equivalent to (8.4.6). Conversely, if $f(x, \theta)$ has the form (8.4.6), then

$$\begin{aligned} p(\mathbf{x}, \theta) &= \prod_1^n f(x_i, \theta) \\ &= \exp\left\{\sum_1^r c_j(\theta)\sum_{i=1}^n \phi_j(x_i) + nc_0(\theta) + \sum_{i=1}^n \phi_0(x_i)\right\} \\ &= R(\chi_1, \ldots, \chi_r; \theta)r(\mathbf{x}); \qquad \chi_j = \sum_{i=1}^n \phi_j(x_i) \end{aligned}$$

whence we see that in this case the statistic $S(\mathbf{x}) = \{\chi_1(\mathbf{x}), \ldots, \chi_r(\mathbf{x})\}$ is sufficient for the family (8.4.2). Since $r \leqq n - 1$ and the functions $\phi_1, \ldots, \phi_r$ are continuously differentiable, it follows from considerations of dimension that the sufficient statistic is nontrivial in this case.∎

8.5 EXPONENTIAL FAMILIES WITH SHIFT AND SCALE PARAMETERS

The exponential family (8.4.6) depends on an abstract parameter $\theta \in \Theta$. In the case when θ has a special character, namely, is a shift or scale parameter, it is possible to make even more precise the form of the exponential family. We shall now deal with this problem, following [26] (also cf. [152], where the regularity conditions have been somewhat relaxed).

THEOREM 8.5.1. If the regular density $f(x - \theta)$, depending on a shift parameter $\theta \in R^1$ (there is no change if we assume θ to take values in some interval of R^1, however), has the exponential form (8.4.6), then for some s

$$f(x) = \exp\left[\sum_1^s c_i x^{n_i} e^{\mu_i x}\right], \qquad x \in R^1 \tag{8.5.1}$$

where n_i is a nonnegative integer, and μ_i, c_i are complex numbers.

***Proof*:** Firstly, it follows from the regularity of $f(x - \theta)$ that $f(x - \theta) > 0$ for all real x and θ. Let $\phi(x) = \log f(x)$; then, from the finite dimensionality of the space of functions $\phi(x - \theta)$, $\theta \in R^1$, we have

$$\phi(x - \theta) = \sum_{j=1}^k a_j(\theta)\phi_j(x) \tag{8.5.2}$$

where $\phi_j(x) = \phi(x - \theta_j)$. The argument in the preceding section shows that there exist $x_1, \ldots, x_k$ such that

$$\begin{vmatrix} \phi_1(x_1) & \cdots & \phi_k(x_1) \\ & \cdots & \\ \phi_1(x_k) & \cdots & \phi_k(x_k) \end{vmatrix} \neq 0.$$

Then, in the system

$$\phi(x_i - \theta) = \sum_{j=1}^k a_j(\theta)\phi_j(x_i), \qquad i = 1, \ldots, k$$

the functions $a_j(\theta)$ can be expressed linearly in terms of the $\phi(x_i - \theta)$ and hence are differentiable with respect to θ. Differentiating (8.5.2) with respect to θ and then setting successively $\theta = \theta_1, \ldots, \theta_k$, we obtain

$$\phi_j'(x) = \sum_{i=1}^k A_{ij}\phi_j(x); \qquad j = 1, \ldots, k$$

where $A_{ij} = a_i'(\theta_j)$. Thus the $\phi_j(x)$ satisfy a system of linear differential equations with constant coefficients, the general form of whose solution is well known:

$$\phi_j(x) = \phi(x - \theta_j) = \sum_{i=1}^{s} c_{ij} x^{n_i} e^{\mu_i x}.$$

The assertion of the theorem then follows immediately. ∎

We now turn to the case of a scale parameter $\sigma \in R_+^1$. If $f(x/\sigma)$ is regular, then either $f(x) > 0$ throughout the real axis, or $f(x) > 0$ for $x \in R_+^1$ and $= 0$ for $x \in R_-^1$, or $f(x) > 0$ for $x \in R_-^1$ and $= 0$ for $x \in R_+^1$. All these cases can be treated similarly; we shall consider therefore only the second case.

THEOREM 8.5.2. If the regular density $(1/\sigma)f(x/\sigma)$, where $f(x) > 0$ for $x \in R_+^1$ and $= 0$ for $x \in R_-^1$, depending on the scale parameter $\sigma \in R_+^1$, has the exponential form (8.4.6), then, for some s,

$$f(x) = \exp\left[\sum_1^s c_i (\log x)^{n_i} x^{\mu_i}\right], \qquad x \in R_+^1 \tag{8.5.3}$$

where the n_i are nonnegative integers, and c_i, μ_i are complex numbers.
Proof: Setting $x = e^y$, $\sigma = e^\rho$, we obtain a family with a shift parameter, the new density function being

$$h(y, \rho) = e^{y-\rho} f(e^{y-\rho}) = h(y - \rho).$$

By theorem 8.5.1,

$$h(y) = \exp\left(\sum_1^s c_i\, y^{n_i} e^{\mu_i y}\right)$$

from which the desired representation (8.5.3) for the original family of densities easily follows. ∎

There exists another method of describing one-dimensional distributions, some power of which admits a sufficient statistic. It is applicable in cases when the parameter as well as the sufficient statistic have special forms, but on the other hand we need to impose only almost minimal conditions in such cases; the differentiability of the density is not required, and in fact not even the existence of a density is. We present two results obtained by this method; the first of these is from [53] and the second from [62].

THEOREM 8.5.3. Let $(X_1, \ldots, X_n)$ be a random sample of size $n \geqq 2$ from a population with d.f. $F(x - \theta)$, $\theta \in R^1$, and let $dP_\theta = dF(x_1 - \theta) \cdots dF(x_n - \theta)$. Suppose that $F^{*n}(x)$ is absolutely continuous (with respect to Lebesgue measure). Then, the sufficiency of $\bar{X}$ for the family $\{P_\theta\}$ implies the normality of F.

***Proof*:** Let, for real t and θ

$$E_\theta[\exp it(X_2 - X_1)|\overline{X}] = \phi_t(\overline{X}) \qquad \text{a.e. } [P_\theta] \tag{8.5.4}$$

Then, for real τ,

$$E_\theta \exp[i(\tau\overline{X} + t(X_2 - X_1))] = E_\theta[\phi_t(\overline{X})e^{i\tau\overline{X}}]$$

whence

$$E_0 \exp[i(\tau\overline{X} + t(X_2 - X_1))] = E_0[\phi_t(\overline{X} + \theta)e^{i\tau\overline{X}}].$$

For our purposes, it is important to note that

$$E_0[\phi_t(\overline{X} + \theta)\exp(i\tau\overline{X})] = \psi_t(\tau). \tag{8.5.5}$$

We set $\mu(A) = P_0(\overline{X} \in A)$; by assumption, μ is absolutely continuous with respect to Lebesgue measure. Equation (8.5.5) can be written in the form

$$\int e^{i\tau u}\,\phi_t(u + \theta)\,d\mu(u) = \psi_t(\tau)$$

whence

$$\int e^{i\tau u}[\phi_t(u + \theta) - \phi_t(u)]\,d\mu(u) = 0. \tag{8.5.6}$$

By the uniqueness theorem for Fourier transforms, we obtain from (8.5.6) that

$$\phi_t(u + \theta) - \phi_t(u) = 0 \qquad \text{a.e. } [\mu] \tag{8.5.7}$$

where the exceptional set may depend on θ and t.

We shall now show that, for every t,

$$\phi_t(u) = \text{constant} = c(t) \qquad \text{a.e. } [\mu]. \tag{8.5.8}$$

Again, the exceptional set may depend on t. We fix t and write $\phi(u)$ in place of $\phi_t(u)$. Suppose then that (8.5.8) is not satisfied. Then, for $\phi_1(u) = \operatorname{Re}\phi(u)$ and $\phi_2(u) = \operatorname{Im}\phi(u)$, we will have at least one of the two inequalities

$$\operatorname{essup}\phi_1(u) > \operatorname{essinf}\phi_1(u)$$
$$\operatorname{essup}\phi_2(u) > \operatorname{essinf}\phi_2(u).$$

Suppose the first of these holds. Then there exists a constant k such that, if $A = \{u: \phi_1(u) > k\}$, then $\mu(A) > 0$, $\mu(A') > 0$, where A' is the complement of A. Removing if necessary a subset of μ-measure zero from A', we may always assume that

$$d\mu/du > 0 \qquad \text{for } u \in A'. \tag{8.5.9}$$

Since μ is absolutely continuous with respect to Lebesgue measure, the Lebesgue measures of the sets A and A' are positive: $m(A) > 0$ and $m(A') > 0$. Hence, for some real θ, $m[(A + \theta) \cap A'] > 0$. But then, by condition (8.5.9), we have

$$\mu[(A + \theta) \cap A'] > 0. \tag{8.5.10}$$

Conditions (8.5.7) and (8.5.10) are contradictory; for, on the one hand, by (8.5.7), $\phi_1(u) > k$ for $u \in A + \theta$, and by the choice of the constant k, $\phi_1(u) \leqq k$ for $u \in A'$. Similarly, it can be proved that $\phi_2(u) =$ constant a.e. $[\mu]$.

Thus, $\phi_t(u) =$ constant $= c(t)$ a.e. $[\mu]$ and hence

$$\phi_t(\overline{X}) = c(t) \qquad \text{a.e. } [P_0]. \tag{8.5.11}$$

We have from (8.5.4) and (8.5.11) that

$$E_0[\exp it(X_2 - X_1)|\overline{X}] = c(t). \tag{8.5.12}$$

Writing $f(t) = \int e^{itx}\, dF(x)$, we obtain from (8.5.12) the relation

$$E_0 \exp i\left(\tau \sum_1^n X_i + t(X_2 - X_1)\right) = c(t)\, E_0 \exp\left(i\tau \sum_1^n X_i\right) \tag{8.5.13}$$

where $c(t) = |f(t)|^2$, i.e.,

$$f(\tau + t)f(\tau - t)[f(\tau)]^{n-2} = c(t)[f(\tau)]^n. \tag{8.5.14}$$

We shall now show that the function f never vanishes. It follows from the definition of $c(t)$ that $c(t) \neq 0$ for $|t| < \varepsilon$, where $\varepsilon > 0$ is sufficiently small. Suppose that f vanishes somewhere and let t_0 be the smallest positive zero of f. In (8.5.14) set $\tau = t_0 - \rho$, $t = \rho$, where $0 < \rho < \varepsilon$. Then we will have $c(\rho)[f(t_0 - \rho)]^n = 0$, i.e., $f(t_0 - \rho) = 0$. This contradiction to our definition of t_0 shows that f never vanishes.

Let now $g = \log f$, and $h = \log c$. Then we obtain from (8.5.14) that

$$g(\tau + t) + g(\tau - t) - 2g(\tau) = h(t), \qquad |t| < \varepsilon$$

i.e., the second difference of the function $g(\tau)$ for any increment t is constant. Thus $g(\tau)$ is a polynomial of the second degree, and F is a normal d.f. Theorem 8.5.3 stands proved. ▮

THEOREM 8.5.4. Let $(X_1, \ldots, X_n)$ be a random sample of size $n \geqq 2$ from a population with d.f. $F(x/\sigma)$, $\sigma \in R_+^1$, and let $dP = dF(x_1/\sigma) \cdots dF(x_n/\sigma)$. We assume that

(1) $F(0) = 0$;

(2) F^{*n} is absolutely continuous with respect to Lebesgue measure; and

(3) for some $\delta > 0$, $\int_0^\infty x^\delta\, dF(x) < \infty$.

Then, the sufficiency of the sample mean $\overline{X}$ for the family $\{P_\sigma : \sigma \in R_+^1\}$ implies that F is a gamma distribution function.

***Proof*:** The central position in the proof is occupied by the following lemma.

LEMMA 8.5.1. Under the conditions of Theorem 8.5.4, the d.f. F has finite moments of all orders.

***Proof of the Lemma*:** In view of the sufficiency of $\overline{X}$ and condition (3), we have

$$E_\sigma(X_1^\delta X_2^\delta | \overline{X}) = \psi(\overline{X}) \qquad \text{a.e. } [P_\sigma]. \tag{8.5.15}$$

We hence obtain that for real t,

$$E_\sigma(X_1^\delta X_2^\delta e^{it\overline{X}}) = E_\sigma[\psi(\overline{X})e^{it\overline{X}}]$$

whence we have immediately the relation

$$E_1(\sigma^{2\delta} X_1^\delta X_2^\delta e^{it\sigma\overline{X}}) = E_1[\psi(\sigma\overline{X})e^{it\sigma\overline{X}}]. \tag{8.5.16}$$

We fix σ and set $\tau = t\sigma$; then (8.5.16) takes the form

$$E_1\left[\frac{\psi(\sigma\overline{X})}{\sigma^{2\delta}} e^{i\tau\overline{X}}\right] = E_1(X_1^\delta X_2^\delta e^{i\tau\overline{X}}). \tag{8.5.17}$$

Let $\mu(B) = P_1(\overline{X} \in B)$. Then (8.5.17) can be rewritten as

$$\int_0^\infty \frac{\psi(\sigma u)}{\sigma^{2\delta}} e^{i\tau u}\, d\mu(u) = E_1(X_1^\delta X_2^\delta e^{i\tau\overline{X}}).$$

Hence, taking $\sigma = 1$, we obtain

$$\int_0^\infty \left[\frac{\psi(\sigma u)}{\sigma^{2\delta}} - \psi(u)\right] e^{i\tau u}\, d\mu(u) = 0. \tag{8.5.18}$$

By the uniqueness theorem for Fourier transforms, we see that for any $\sigma \in R_+^1$

$$\frac{\psi(\sigma u)}{\sigma^{2\delta}} = \psi(u) \qquad \text{a.e. } [\mu] \tag{8.5.19}$$

where the exceptional set may depend on σ. Taking into account the fact that $u > 0$, we obtain from (8.5.19)

$$\frac{\psi(\sigma u)}{(\sigma u)^{2\delta}} = \frac{\psi(u)}{u^{2\delta}} \qquad \text{a.e. } [\mu]$$

or

$$h(\sigma u) = h(u) \qquad \text{a.e. } [\mu], \qquad \text{where} \qquad h(u) = \frac{\psi(u)}{u^{2\delta}}. \tag{8.5.20}$$

Suppose that the equality $h(u) = c$ a.e. $[\mu]$ does not hold for any constant $c > 0$. Then,

$$\operatorname{essup} h(u) > \operatorname{essinf} h(u)$$

and a constant k exists such that, if $B = \{u\colon h(u) > k\}$, then $\mu(B) > 0$ and $\mu(B') > 0$. Since, by condition (2), μ is absolutely continuous with respect to

Lebesgue measure, the Lebesgue measures of the sets B and B' are positive: $m(B) > 0$, $m(B') > 0$. Standard arguments based on approximating the sets B and B' by intervals show that it is possible to find a $\sigma > 0$ for which

$$\mu(\sigma B \cap B') > 0 \tag{8.5.21}$$

where, as usual, $\sigma B = \{u : u/\sigma \in B\}$. But (8.5.21) and (8.5.20) are contradictory, since, by the choice of the set B, we have $h(u) > k$ if $u \in \sigma B$, while $h(u) \leqq k$ for $u \in B'$. Thus we have proved that, for some constant c, $h(u) = c$ a.e. $[\mu]$, or $\psi(u) = cu^{2\delta}$ a.e. $[\mu]$. We turn to the relation (8.5.15) we started with. We can now write it in the form

$$E_\sigma(X_1^\delta X_2^\delta \mid \overline{X}) = c\overline{X}^{2\delta}.$$

But in view of condition (3) of the theorem,

$$E_\sigma[E_\sigma(X_1^\delta X_2^\delta \mid \overline{X})] = E_\sigma X_1^\delta \cdot E_\sigma X_2^\delta < \infty.$$

Therefore, $E_\sigma \overline{X}^{2\delta} < \infty$, whence it follows that F has finite moment of order 2δ. Continuing this argument, we see that F has finite moments of all orders. This proves Lemma 8.5.1.▮

Let us denote by f the c.f. of F.

LEMMA 8.5.2. Let

(1) $\int_0^\infty x^2\, dF(x) < \infty$;

(2) F^{*n} be absolutely continuous with respect to Lebesgue measure; and

(3) $E(X_1^2 \mid \overline{X})$ be independent of σ.

Then f coincides with the c.f. of a gamma distribution in some neighborhood of the origin.

***Proof*:** We see from condition (3) that there exists a function ψ such that

$$E_\sigma(X_1^2 \mid \overline{X}) = \psi(\overline{X}) \qquad \text{a.e. } [P_\sigma]. \tag{8.5.22}$$

Using arguments wholly similar to those used for the proof of Lemma 8.5.1, we deduce from (8.5.22) that

$$\psi(u) = cu^2 \qquad \text{a.e. } [\mu]$$

for some constant c. Then we have

$$E_1(X_1^2 \mid \overline{X}) = c\overline{X}^2 \qquad \text{a.e. } [P_1] \tag{8.5.23}$$

whence we obtain

$$E_1\left(X_1^2 \exp\left(it \sum_1^n X_i\right)\right) = cE_1\left(\overline{X}^2 \exp\left(it \sum_1^n X_i\right)\right). \tag{8.5.24}$$

In terms of the c.f. f, the above reduces to the form

$$cf''f^{n-1} = n^{-2}\{nf''f^{n-1} + n(n-1)(f')^2 f^{n-2}\}. \tag{8.5.25}$$

In an interval $|t| < \varepsilon$ in which f does not vanish, (8.5.25) is equivalent to the relation

$$c_1 f'' f - (f')^2 = 0 \tag{8.5.26}$$

which is easily integrated. Since $f(0) = 1$, the solution can be written in the form

$$f(t) = b^{\lambda}(b - t)^{-\lambda}, \ |t| < \varepsilon. \tag{8.5.27}$$

In order that f be the c.f. of a distribution on $(0, \infty)$, it is necessary that $\lambda > 0$ and $b = -ia$, where $a > 0$. But then

$$f(t) = (1 - it/a)^{-\lambda}$$

which is the c.f. of a gamma distribution. This proves Lemma 8.5.2.∎

Theorem 8.5.4 is now obtained very simply. In fact, it follows from Lemma 8.5.1 that $\int_0^\infty x^2 \, dF(x) < \infty$. Then, by Lemma 8.5.2, f coincides with the c.f. of a gamma distribution in $|t| < \varepsilon$. Hence all the moments of F coincide with the corresponding moments of this gamma distribution. But, for a gamma distribution, as easily verified with the help of Lemma 1.1.10, the moment problem has a unique solution. Hence F is the d.f. of a gamma distribution. Thus Theorem 8.5.4 stands proved.∎

Remark. Apparently, condition (3) of the theorem can be omitted, while the remaining conditions are essential for characterizing the gamma distribution through the sufficiency of $\bar{X}$ for the family

$$\{F(x_1/\sigma) \cdots F(x_n/\sigma), \sigma \in R_+^1\}.$$

8.6 PARTIAL SUFFICIENCY

In this section, it will be shown that in the case when θ is a shift or scale parameter, essential information on the distribution of the members of a random sample $(X_1, \ldots, X_n)$ is already contained in the nondependence on θ of the conditional mathematical expectation $E_\theta(\phi|\bar{X})$ of any *one* function $\phi = \phi(X_1, \ldots, X_n)$. Under certain conditions, the (ordinary) sufficiency of $\bar{X}$ for the family of distributions $\{F(x_1, \theta) \cdots F(x_n, \theta)\}$ follows from the partial sufficiency of $\bar{X}$.

THEOREM 8.6.1. (cf. [52]). Let $(X_1, \ldots, X_n)$ be a random sample of size $n \geq 2$ from a population with d.f. $F(x - \theta)$, $\theta \in R^1$. Suppose that F^{*n} is absolutely continuous with respect to Lebesgue measure and $\int x^2 \, dF(x) < \infty$. If for some i and j, $E_\theta(X_i X_j|\bar{X})$ does not depend on $\theta \in R^1$, then F is the d.f. of a normal law and consequently $\bar{X}$ is a sufficient statistic for the family $\{F(x_1 - \theta) \cdots F(x_n - \theta): \theta \in R^1\}$.

Proof: First of all, we note that from the relation

$$\left(\sum_1^n X_i\right)^2 = E_\theta\left[\left(\sum_1^n X_i\right)^2 \middle| \overline{X}\right] = nE_\theta(X_1^2 | \overline{X}) + \sum_{j \neq k} E_\theta(X_j X_k | \overline{X}) \quad (8.6.1)$$

it follows that $E_\theta(X_j X_k | \overline{X})$ is independent of θ for all j and k. Hence there exists a function ψ such that for $S^2 = \sum_1^n (X_i - \overline{X})^2$,

$$E_\theta(S^2 | \overline{X}) = \psi(\overline{X}) \qquad \text{a.e. } [P_\theta],\ \theta \in R^1. \quad (8.6.2)$$

Precisely in the same way as in the proof of Theorem 8.5.3, we can show as a consequence of (8.6.2) that $\psi(\overline{X}) = \text{constant} = c$ a.e. $[P_0]$, i.e.,

$$E_0(S^2 | \overline{X}) = c. \quad (8.6.3)$$

The assertion of the theorem follows then from Theorem 6.3.1 or 6.3.2.∎

We note that Theorem 8.6.1 is not stronger than Theorem 8.5.3, since the latter does not assume that $\int x^2\, dF(x) < \infty$.

In seeking to generalize Theorem 8.6.1, we may consider the nondependence on $\theta \in R^1$ of the conditional mathematical expectation

$$E_\theta\{P(X_1 - \overline{X}, \ldots, X_n - \overline{X}) | \overline{X}\}$$

where P is a polynomial. As in Theorem 8.5.3, we can show that it follows hence that

$$E_0(P(X_1 - \overline{X}, \ldots, X_n - \overline{X}) | \overline{X}) = \text{constant} \quad (8.6.4)$$

which was studied in [39]. It has been shown there that, under certain conditions on P and on the d.f. F, the normality of F can in fact be deduced from (8.6.4). We shall not, however, pause here for the details.

We now turn to the case of a scale parameter.

THEOREM 8.6.2. Let $(X_1, \ldots, X_n)$ be a random sample of size $n \geqq 2$ from a population with d.f. $F(x/\sigma)$, $\sigma \in R_+^1$. Suppose, as earlier, that $F(0) = 0$, F^{*n} is absolutely continuous with respect to Lebesgue measure, and $\int x^2\, dF(x) < \infty$. If, for some i, j, $E_\sigma(X_i X_j | \overline{X})$ does not depend on $\sigma \in R_+^1$, then F is a gamma d.f.

Proof: This theorem is proved in precisely the same way as Lemma 8.5.2.∎

We now take up, following the work of Kagan [62], a similar relation

$$E_\sigma(Q | \overline{X}) = \psi(\overline{X}) \quad (8.6.5)$$

valid for some polynomial $Q = Q(\mathbf{X})$, where of course we naturally assume that $E_1 |Q| < \infty$. First of all, we write Q in the form $Q = q_k + q_{k-1} + \cdots + q_0$, where $q_j = q_j(\mathbf{x})$ is a polynomial which is homogeneous and of degree j. From (8.6.5) we have

$$E_\sigma(e^{it\overline{X}} Q) = E_\sigma[\psi(\overline{X}) e^{it\overline{X}}] \quad (8.6.6)$$

which may also be written in the form

$$E_1\left[\sum_0^k \sigma^i q_j \exp(it\sigma\bar{X})\right] = E_1[\psi(\sigma\bar{X}) \exp(it\sigma\bar{X})]. \tag{8.6.7}$$

We set

$$\tau = t\sigma, \qquad \mu(B) = \int_{\bar{x}\in B} dF(x_1)\cdots dF(x_n). \tag{8.6.8}$$

Since

$$E_1[\psi(\sigma\bar{X}) \exp(i\tau\bar{X})] = \int_0^\infty \psi(\sigma u)e^{i\tau u}\, d\mu(u)$$

and

$$E_1(q_j e^{i\tau\bar{X}}) = E_1[e^{i\tau\bar{X}}E_1(q_j|\bar{X})] = E_1[e^{i\tau\bar{X}}\phi_j(\bar{X})]$$
$$= \int_0^\infty e^{i\tau u}\phi_j(u)\, d\mu(u)$$

where we have set $\phi_j(\bar{X}) = E_1(q_j|\bar{X})$, we have from (8.6.7) that

$$\int_0^\infty e^{i\tau u}\psi(\sigma u)\, d\mu(u) = \int_0^\infty e^{i\tau u}\sum_{j=0}^k \sigma^j\phi_j(u)\, d\mu(u). \tag{8.6.9}$$

By the uniqueness theorem for Fourier transforms, we deduce from (8.6.9) that for every $\sigma \in R_+^1$

$$\psi(\sigma u) = \sum_{j=0}^k \sigma^j\phi_j(u) \qquad \text{a.e. } [\mu] \tag{8.6.10}$$

where the exceptional set may depend on σ.

We now take up the analysis of relation (8.6.10).

LEMMA 8.6.1. If μ is absolutely continuous with respect to Lebesgue measure and (8.6.10) holds, then necessarily

$$\phi_0(u) = \text{constant a.e. } [\mu]. \tag{8.6.11}$$

***Proof*:** Suppose that (8.6.11) is not true, so that

$$\operatorname{essup} \phi_0(u) > \operatorname{essinf} \phi_0(u) \tag{8.6.12}$$

where the essup and essinf relate to the measure μ. It follows from (8.6.12) that there exist constants k_1 and k_2 such that the sets

$$B_1 = \{u: \phi_0(u) > k_1\} \qquad \text{and} \qquad B_2 = \{u: \phi_0(u) < k_2\}$$

have positive μ-measure: $\mu(B_1) > 0$, $\mu(B_2) > 0$. Further, obviously, for any $\varepsilon > 0$, there exists a constant $c = c(\varepsilon)$ such that the set

$$B^* = \{u: |\phi_1(u)| < c, \ldots, |\phi_k(u)| < c\}$$

has the property $\mu(B^*) > 1 - \varepsilon$. Let now $\{\sigma_i\}$ be the set of all positive rationals. Denote by A_i the set of those u for which (8.6.10) is not satisfied for $\sigma = \sigma_i$ and let $A = \bigcup_1^\infty A_i$. Since $\mu(A_i) = 0$ for all i, $\mu(A) = 0$. Now consider the sets $B_1^* = B_1 \cap B^* \cap A'$ and $B_2^* = B_2 \cap B^* \cap A'$. It is clear that for sufficiently small $\varepsilon > 0$, $\mu(B_1^*) > 0$ and $\mu(B_2^*) > 0$. In view of the absolute continuity of μ with respect to Lebesgue measure, we must also have $m(B_1^*) > 0$, $m(B_2^*) > 0$, m being Lebesgue measure. For any preassigned $\delta > 0$, rationals $0 < \sigma_1 < \delta$ and $0 < \sigma_2 < \delta$ exist such that $m(\sigma_1 B_1^* \cap \sigma_2 B_2^*) > 0$ and consequently the set $\sigma_1 B_1^* \cap \sigma_2 B_2^*$ is nonempty. In other words, $u \in B_1^*$ and $v \in B_2^*$ exist such that $\sigma_1 u = \sigma_2 v$. From (8.6.10) we obtain

$$\psi(\sigma_1 u) = \phi_0(u) + \sum_{j=1}^{k} (\sigma_1)^j \phi_j(u),$$

$$\psi(\sigma_2 v) = \phi_0(v) + \sum_{j=1}^{k} (\sigma_2)^j \phi_j(v)$$

whence

$$\phi_0(u) + \sum_{j=1}^{k} (\sigma_1)^j \phi_j(u) = \phi_0(v) + \sum_{j=1}^{k} (\sigma_2)^j \phi_j(v). \tag{8.6.13}$$

But this relation cannot be satisfied for sufficiently small $\delta > 0$, since

$$\sum_{j=1}^{k} (\sigma_1)^j \phi_j(u) \text{ and } \textstyle\sum_{j=1}^{k} (\sigma_2)^j \phi_j(v)$$

can be made arbitrarily small, while $\phi_0(u) > k_1 > k_2 > \phi_0(v)$. This contradiction shows that $\phi_0(u) =$ constant a.e. $[\mu]$, proving the lemma.∎

LEMMA 8.6.2. If μ is absolutely continuous with respect to Lebesgue measure and (8.6.10) is satisfied, then $\psi(u)$ coincides a.e. $[\mu]$ with some polynomial of degree k.

***Proof*:** We shall argue by induction. For $k = 0$, the assertion above is proved in Lemma 8.6.1. Suppose now that it is true for $k = s - 1$ and show its validity for $k = s$. Let then

$$\psi(\sigma u) = \sum_0^s \sigma^j \phi_j(u) \qquad \text{a.e. } [\mu]. \tag{8.6.14}$$

By Lemma 8.6.1, $\phi_0(u) =$ constant $= c$ a.e. $[\mu]$. Then we deduce from (8.6.14)

$$\frac{\psi(\sigma u) - c}{\sigma u} = \sum_{j=1}^{s} \sigma^{j-1} \phi_j^*(u) \tag{8.6.15}$$

where $\phi_j^*(u) = \phi_j(u)/u$. Let $\psi^*(u) = (\psi(u) - c)/u$. Then we have

$$\psi^*(\sigma u) = \sum_0^{s-1} \sigma^j \phi_j^*(u) \qquad \text{a.e. } [\mu]$$

and by the induction assumption we must have

$$\psi^*(u) = \frac{\psi(u) - c}{u} = \sum_{j=0}^{s-1} a_j u^j \qquad \text{a.e. } [\mu]$$

whence the assertion of the lemma follows.▮

Now relation (8.6.9) can be written as

$$\int_0^\infty e^{i\tau u} \sum_{j=0}^{k} a_j(\sigma u)^j \, d\mu(u) = \int_0^\infty e^{i\tau u} \sum_{j=0}^{k} \sigma^j \phi_j(u) \, d\mu(u)$$

whence it follows that

$$\phi_j(u) = a_j u^j \qquad \text{a.e. } [\mu]$$

i.e.,

$$E_1(q_j | \bar{X}) = a_j \bar{X}^j \qquad \text{for } j = 0, 1, \ldots, k.$$

Hence we obtain that

$$E_1\{Q(X_1/\bar{X}, \ldots, X_n/\bar{X}) | \bar{X}\} = \text{constant.} \tag{8.6.16}$$

We have already studied relation (8.6.16) in Theorem 6.2.10; it was shown there that under certain conditions on the polynomial Q, and on the d.f. F of the X_i, (8.6.16) implies that the X_i have a gamma distribution. This enables us to formulate the following result.

THEOREM 8.6.3. Let $(X_1, \ldots, X_n)$ be a random sample of size $n \geqq 2$ from a population with d.f. $F(x/\sigma)$, $\sigma \in R_+^1$. Suppose that F satisfies the conditions of Theorem 6.2.10. If for some polynomial Q satisfying the conditions of the same theorem, we have

$$E_\sigma\{Q(X_1/\bar{X}, \ldots, X_n/\bar{X}) | \bar{X}\} = \psi(\bar{X}) \qquad \text{a.e. } [P_\sigma] \tag{8.6.17}$$

for all $\sigma \in R_+^1$, then F is a gamma d.f.

We also note (cf. [62]) that the validity of (8.6.17) for some polynomial Q of degree k of a "general form" leads to the existence of moments of all orders for F if only $F(0) = 0$, F^{*n} is absolutely continuous, and $\int_0^\infty x^k \, dF(x) < \infty$; it is not necessary to impose any additional conditions on F.

8.7 SUFFICIENT SUBSPACES

Let $(X_1, \ldots, X_n)$ be a random sample from a population with d.f. $F(x, \theta)$, where $x \in R^1$, $\theta \in \Theta$, and let

$$dP_\theta = dF(x_1, \theta) \cdots dF(x_n, \theta).$$

Suppose that for some integer $k \geqq 1$ and all θ

$$\int x^{2k} \, dF(x, \theta) < \infty. \tag{8.7.1}$$

Then the totality of polynomials $Q(X_1, \ldots, X_n)$ of degree $\leqq k$ form a Hilbert space $L_k^{(2)}$ under the scalar product

$$(Q_1, Q_2)_\theta = E_\theta(Q_1 Q_2).$$

Let L be some subspace of $L_k^{(2)}$. By analogy with the concept of sufficient statistics, it is natural to introduce the following definition.

DEFINITION 8.7.1. (cf. [51]). L will be said to form an $L_k^{(2)}$*-sufficient subspace* for the family $\{P_\theta\}$ if, for any Q in $L_k^{(2)}$ there exists a q in L, not depending on θ, such that

$$\hat{E}_\theta(Q \mid L) = q \tag{8.7.2}$$

where $\hat{E}_\theta(\cdot \mid L)$ is the projection operator on L (the conditional mathematical expectation in the broad sense), when the scalar product $(\cdot, \cdot)_\theta$ in $L_k^{(2)}$ corresponds to the measure P_θ.

If L is a sufficient subspace, $1 \in L$, and an element $Q \in L_k^{(2)}$ is used as an unbiased estimator of the parametric function $\gamma(\theta) = E_\theta Q$, then it is easily seen by Rao-Blackwellization that, for the *statistic* $q = \hat{E}_\theta(Q \mid L)$ we have

$$E_\theta q = \gamma(\theta) \quad \text{and} \quad \operatorname{Var}_\theta(q) \leqq \operatorname{Var}_\theta(Q) \quad \text{for all } \theta.$$

Further, for given θ, equality holds if and only if, for that θ, $Q \in L$. Thus, *in the presence of a sufficient subspace* L, *satisfying the condition* $1 \in \mathrm{L}$, *every polynomial* $\mathrm{Q} \in \mathrm{L}_k^{(2)} - \mathrm{L}$ *is inadmissible in the class of unbiased estimators of the parametric function* $\gamma(\theta) = \mathrm{E}_\theta \mathrm{Q}$, *under quadratic loss*. This statement is the analogue of Theorem 7.1.2.

It would be of interest to construct a moment theory of sufficiency, i.e., to describe all $F(x, \theta)$ for which there exists a nontrivial (different from all the $L_k^{(2)}$) sufficient subspace; such a family would be a significant generalization of the exponential families (8.4.6). Clearly, the condition of existence of nontrivial sufficient subspaces is expressible solely in terms of moments

$$\mu_1(\theta) = \int x \, dF(x, \theta), \ldots, \mu_{2k}(\theta) = \int x^{2k} \, dF(x, \theta).$$

We present here results relating to the simplest cases, when θ is a shift or scale parameter and L is a subspace of polynomials in $\bar{X}$.

THEOREM 8.7.1. Let $(X_1, \ldots, X_n)$ be a random sample from a population with d.f. $F(x - \theta)$, $\theta \in R^1$, $dP_\theta = dF(x_1 - \theta) \cdots dF(x_n - \theta)$, and

$$L = \{a_0 \bar{X}^k + \cdots + a_k\}.$$

If the first $2k$ moments of F coincide with the corresponding moments of some normal law, then L is an $L_k^{(2)}$-sufficient subspace for the family $\{P_\theta\}$. If F satisfies the condition $\int x^{2k} \, dF(x) < \infty$, L is an $L_k^{(2)}$-sufficient subspace, and $n \geqq 3$, then the first $2k$ moments of F coincide with the corresponding moments of some normal law.

***Proof*:** Let us first suppose that F is a normal d.f. Every polynomial Q in $L_k^{(2)}$ can be represented as

$$Q = \bar{X}^k Q_0(X_1 - \bar{X}, \ldots, X_n - \bar{X}) + \bar{X}^{k-1} Q_1(X_1 - \bar{X}, \ldots, X_n - \bar{X}) + \cdots + Q_k(X_1 - \bar{X}, \ldots, X_n - \bar{X})$$

where Q_i are polynomials. But, if the X_i are normal, then $\bar{X}$ and the vector $(X_1 - \bar{X}, \ldots, X_n - \bar{X})$ are independent. Hence

$$E_\theta(Q|\bar{X}) = a_0 \bar{X}^k + \cdots + a_k \tag{8.7.3}$$

where

$$a_j = E_\theta(Q_j|\bar{X}) = E_\theta(Q_j) = E_0(Q_j). \tag{8.7.4}$$

Since $E_\theta(Q|\bar{X}) \in L$, $\hat{E}_\theta(Q|\bar{X}) = E_\theta(Q|\bar{X})$ and, by (8.7.3), L is an $L_k^{(2)}$-sufficient subspace, if only F is a normal d.f. But two distributions for which the first $2k$ moments coincide induce the same scalar product in $L_k^{(2)}$. Hence, if the first $2k$ moments of F are the same as those of a normal law, then

$$\hat{E}_\theta(Q|L) = a_0 \bar{X}^k + \cdots + a_k$$

where $a_0, \ldots, a_k$ are the same as in (8.7.4). This proves the first assertion of the theorem.

Let now $\int x^{2k}\, dF(x) < \infty$, $n \geqq 3$ and L be $L_k^{(2)}$-sufficient. We show first of all that if Q in $L_k^{(2)}$ has the form $Q(X_1 - \bar{X}, \ldots, X_n - \bar{X})$, then $\hat{E}_\theta(Q|L) =$ const. $= E_\theta Q = E_0 Q$. In fact, let $\hat{E}_\theta(Q|L) = q(\bar{X}) = a_0 \bar{X}^k + \cdots + a_k$; then $E_\theta Q = E_\theta q$, whence $E_0 Q = E_0 q(\bar{X} + \theta)$. We express $q(\bar{X} + \theta)$ in powers of $\bar{X}$:

$$q(\bar{X} + \theta) = A_0(\theta)\bar{X}^k + \cdots + A_k(\theta)$$

where the $A_j(\theta)$ are all polynomials in θ. If $a_0 \neq 0$, then the degree of $A_k(\theta)$ is k, and of the other $A_j(\theta)$ does not exceed $(k - 1)$. Setting $E_0 Q = c$, $E_0 \bar{X}^j = v_j$, we will have

$$A_0(\theta)v_k + A_1(\theta)v_{k-1} + \cdots + A_k(\theta) = c, \qquad \theta \in R^1.$$

But this identity is possible only if the degree of $A_k(\theta)$ is $\leqq k - 1$. Hence $a_0 = 0$ and in exactly the same way, we can show that $a_1 = \cdots = a_{k-1} = 0$.

Let $\mu_j = \int x^j\, dF(x)$ for $j = 1, 2$. If $Q = Q(X_1 - \bar{X}, \ldots, X_n - \bar{X})$, then as we have shown, $\hat{E}_0(Q|L) = E_0 Q$, which can be written as

$$E_0[Qq] = E_0 Q \cdot E_0 q, \qquad q = q(\bar{X}) \tag{8.7.5}$$

for any $q \in L$. We set $Q = (X_2 - X_1)^2$, $q = \bar{X}$. Then (8.7.5) enables us to express μ_3 in terms of μ_1 and μ_2. Generally, if the first s, $s \leqq 2k - 1$, moments have already been obtained, then to determine μ_{s+1} we have to proceed as follows: let $r = \min(k, s)$; in (8.7.5), set

$$Q = (X_1 - X_2)^{r-1}(X_1 - X_3), \qquad q = \bar{X}^{s+1-r}.$$

Then the resulting relation enables us to express μ_{s+1} in terms of $\mu_1, \ldots, \mu_s$. Thus, if the first two moments of F are fixed, then $\mu_3, \mu_4, \ldots, \mu_{2k}$ are uniquely defined by them. It is clear that they can be none other than the moments of some normal law (for such F, L is automatically an $L_k^{(2)}$-sufficient subspace). Theorem 8.7.1 stands proved.▌

We now turn to the case of scale parameters.

THEOREM 8.7.2 (cf. [62]). Let $(X_1, \ldots, X_n)$ be a random sample from a population with d.f. $F(x/\sigma)$, $\sigma \in R_+^1$, $dP_\sigma = dF(x_1/\sigma) \cdots dF(x_n/\sigma)$, $L = \{a_0 \bar{X}^k + \cdots + a_k\}$. If the first $2k$ moments of F coincide with the corresponding moments of some gamma distribution, then L is an $L_k^{(2)}$-sufficient subspace for the family $\{P_\sigma : \sigma \in R_+^1\}$. If F satisfies the conditions $\mu_{2k} = \int x^{2k}\, dF(x) < \infty$, $F(0) = 0$, and L is $L_k^{(2)}$-sufficient for the family $\{P_\sigma\}$, then either F is a degenerate d.f. or the first $2k$ moments of F coincide with the corresponding moments of some gamma d.f.

***Proof*:** Suppose first that F is a gamma d.f. and consider $E_\sigma(X_1^{j_1} \cdots X_n^{j_n} | \bar{X})$, where $j_1 + \cdots + j_n = j \leq k$. We have

$$E_\sigma(X_1^{j_1} \cdots X_n^{j_n} | \bar{X}) = \bar{X}^j E_\sigma[(X_1/\bar{X})^{j_1} \cdots (X_n/\bar{X})^{j_n} | \bar{X}].$$

Since F is a gamma d.f., $\bar{X}$ and the vector $(X_1/\bar{X}, \ldots, X_n/\bar{X})$ are independent; this follows immediately from the fact that in such a case $\bar{X}$ is a complete sufficient statistic for the family $\{P_\sigma\}$. Hence the right-hand side of the above equality becomes $c\bar{X}^j$, where c is the constant value of

$$E_\sigma[(X_1/\bar{X})^{j_1} \cdots (X_n/\bar{X})^{j_n}].$$

Hence, for any $Q \in L_k^{(2)}$, we have $\hat{E}_\sigma(Q|L) = E_\sigma(Q|\bar{X})$ and L is $L_k^{(2)}$-sufficient. The same holds for d.f.'s F for which the first $2k$ moments are the same as for a gamma distribution.

We now turn to the second assertion of the theorem. Let L be an $L_k^{(2)}$-sufficient subspace; we shall show that then the moments $\mu_3, \ldots, \mu_{2k}$ of F are uniquely defined by the moments μ_1 and μ_2. In fact, let the moments $\mu_3, \ldots, \mu_s$, $s < 2k$, be already defined. If $s \leq k$, then consider the polynomial

$$X_1^s - aX_1^{s-1}X_2,$$

where the constant a is defined by the condition $E_1(X_1^s - aX_1^{s-1}X_2) = 0$, i.e., $a = \mu_s/\mu_{s-1}\mu_1$. We see that a is expressible in terms of moments already known. The condition of $L_k^{(2)}$-sufficiency gives

$$\hat{E}_\sigma(X_1^s - aX_1^{s-1}X_2|L) = \sum_{j=0}^{k} a_j \bar{X}^j.$$

Hence

$$E_\sigma(X_1^s - aX_1^{s-1}X_2) = E_\sigma\left(\sum_{j=0}^{k} a_j \bar{X}^j\right).$$

But the left-hand side above is equal to $\sigma^s E_1(X_1^s - aX_1^{s-1}X_2) = 0$, by our choice of a and

$$E_\sigma\left(\sum_0^k a_j \bar{X}^j\right) = \sum_{j=0}^k a_j \sigma^j E_1 \bar{X}^j.$$

Since $E_1\bar{X}^j > 0$, for $j = 0, 1, \ldots, k$, we must have $a_j = 0$ for all j. Hence

$$\hat{E}_\sigma(X_1^s - aX_1^{s-1}X_2 | L) = 0 \tag{8.7.6}$$

from which we obtain

$$E_1[(X_1^s - aX_1^{s-1}X_2)\bar{X}] = 0$$

whence μ_{s+1} is (uniquely) defined.

If $s > k$, then we consider the polynomial $X_1^k - bX_1^{k-1}X_2$, where b is chosen according to

$$E_1(X_1^k - bX_1^{k-1}X_2) = 0.$$

Arguing similarly, we obtain from the condition of $L_k^{(2)}$-sufficiency:

$$E_1[(X_1^k - bX_1^{k-1}X_2)\bar{X}^{s-k+1}] = 0$$

whence μ_{s+1} is (uniquely) defined.

We now assume that the moments μ_1 and μ_2 are connected by the relation $\mu_2 = \mu_1^2$. It is easily seen that this can happen only for a degenerate d.f. F, for which, of course, L is $L_k^{(2)}$-sufficient. If $\mu_2 > \mu_1^2$, then we can always find a gamma d.f. F^* with μ_1 and μ_2 as its first two moments. Since for a d.f. F, for which the first $2k$ moments coincide with the corresponding moments of some gamma d.f., L is $L_k^{(2)}$-sufficient, the moments $\mu_3, \ldots, \mu_{2k}$, which are uniquely defined by μ_1 and μ_2, must be the same as for some gamma d.f. Thus the theorem stands proved. ▮

CHAPTER 9

Stability in Characterization Problems

9.1 INTRODUCTION

If some property S of a random sample $X_1, \ldots, X_n$ is characteristic of the distribution F, then there arises the natural question whether this characteristic property is "stable." In such discussions, the property S is replaced by a weaker property S_ε, where ε is a "small parameter," and we study what restrictions on the d.f. of the X_i are imposed by the property S_ε of the random sample. If it turns out that in this situation the d.f. F_ε of the X_i differs little from F, then we say that the characterization of F through the property S of the random sample is *stable* under the "perturbations" S_ε. Clearly, various weakenings of the property S are possible, under some of which stability may hold and under some others may not. Of course, not only the fact of stability, but also the degree of closeness of F_ε to F in terms of ε, is of interest.

So far, only isolated results are available concerning the stability of various characteristic properties of probability distributions. Apparently, the first such result is that of Sapogov ([136, 137]) on stability in respect of Cramer's theorem on the normal law. Let X_1 and X_2 be independent and not necessarily identically distributed r.v.'s and $Y = X_1 + X_2$; Cramer's theorem asserts that if Y has a normal distribution (property S in this case), then the X_i are also normal. If now the d.f. G of Y is only ε-normal (property S_ε) in the sense that $\sup|G(x) - \Phi(x; a, \sigma)| < \varepsilon$ for a normal law $\Phi(x; a, \sigma)$ with parameters a and σ^2, then what can be said about the d.f.'s F_1 and F_2 of X_1 and X_2? It was proved in [137] that if $\int_{-N}^{N} x\,dF_1(x) = a_1$, $\int_{-N}^{N} x^2\,dF_1(x) - a_1^2 = \sigma_1^2 > 0$, where $N = (2 \log (1/\varepsilon))^{1/2} + 1$, and $F_1(0) = 1/2$, then

$$\sup_x |F_1(x) - \Phi(x; a_1, \sigma_1)| \leqq C\sigma_1^{-3}(\log (1/\varepsilon))^{-1/2} \qquad (9.1.1)$$

where C is an absolute constant. Later Maloshevskii proved in [97] that the estimate (9.1.1) cannot essentially be improved upon. Shalaevskii [163] studied stability for Raikov's theorem on the Poisson law. In recent times, there has been increased interest in the study of stability for various characteristic properties of distributions (cf. [45, 102, 106]).

We present below results relating to the stability of two characteristic properties of the normal law considered in the preceding chapters: the independence of the sum and the difference of two i.i.d.r.v.'s, and the admissibility of the sample mean $\bar{X}$ as an estimator of the shift parameter on the basis of a random sample.

9.2 ε-INDEPENDENCE OF THE SUM AND THE DIFFERENCE OF TWO I.I.D. RANDOM VARIABLES

A special case of Theorem 3.1.1 is the following result proved by Bernstein [6] even as early as 1941.

If the i.i.d.r.v.'s X_1 and X_2 are such that their sum L_1 and difference $L_2 = X_2 - X_1$ are independent, then the X_i are normally distributed.

We now consider the property of "ε-independence" of L_1 and L_2. Let

$$Q(u, v) = P[L_1 < u, L_2 < v] - P[L_1 < u] \cdot P[L_2 < v]$$
$$= F(u, v) - F_1(u)F_2(v).$$

Q is obviously a function of bounded variation, and the "ε-independence" of L_1 and L_2 signifies the smallness of the function Q in some form or other. We take the following as the definition of ε-independence (cf. [103]).

DEFINITION 9.2.1. The statistics L_1 and L_2 will be said to be ε-independent if for any set of the form

$$A = \{(u, v): a_1 u + b_1 v < c_1; a_2 u + b_2 v < c_2\}$$

we have $\left|\int_A dQ(u, v)\right| < \varepsilon.$ ∎

As we shall see below, such ε-independence of L_1 and L_2, together with certain conditions on the d.f. F of the X_i, guarantees closeness of the order of $\varepsilon^{1/3}$ of F to some normal d.f., so that in the class of distributions considered the characterization of the normal law through the independence of L_1 and L_2 is stable. The result is due to Meshalkin [103], which we shall follow below.

THEOREM 9.2.1. Let X_1 and X_2 be i.i.d.r.v.'s with $EX_1 = 0$ and $EX_1^2 = 1$; let further $E|X_1|^3 = \beta_3 < \infty$. If L_1 and L_2 are ε-independent in the sense of Definition 9.2.1, then the d.f. F of the X_i is close to the d.f. Φ of the standard normal law in the following sense:

$$\sup_x |F(x) - \Phi(x)| \leqq C\varepsilon^{1/3}$$

where the constant C depends only on β_3.

Proof: Clearly, only small $\varepsilon > 0$ are of interest, and we shall take ε to be such. We shall denote by C any constant, not always one and the same, which depends only on β_3. Let f be the c.f. of F. In view of the ε-independence of L_1 and L_2, we have

$$\begin{aligned} E \exp it(L_1 + L_2) = f(2t) &= \iint \exp it(u+v)\, dF(u, v) \\ &= \iint \exp it(u+v)\, dF_1(u)\, dF_2(v) + \iint \exp it(u+v)\, dQ(u, v) \\ &= [f(t)]^3 f(-t) + \psi(t) \qquad (9.2.1) \end{aligned}$$

where

$$\psi(t) = \iint \exp it(u+v)\, dQ(u, v).$$

LEMMA 9.2.1. For all real t, $\psi(t) \leqq C\varepsilon(1 + |t|\varepsilon^{-1/3})$.

Proof: Let $R(x, y) = \iint_\Delta dQ(u, v)$ where Δ is the region $[(u+v) < 2x, (u-v) < 2y]$. We may rewrite $\psi(t)$ as

$$\begin{aligned} \psi(t) &= \iint \exp it(u+v)\, dQ(u, v) = \iint \exp(2itu)\, dR(u, v) \\ &= \int \exp(2itu)\, dR(u, +\infty) \\ &= \int_{[|u| \leqq a]} \exp(2itu)\, d\rho + \int_{[|u| > a]} \exp(2itu)\, d\rho = T_1 + T_2 \end{aligned}$$

where $\rho(u) = R(u, +\infty) - R(u, -\infty) = R(u, +\infty)$. Integrating by parts, we have

$$|T_1| = \left| \rho(a) \exp(2ita) - \rho(-a) \exp(-2ita) - \int_{-a}^{a} 2it \exp(2itu)\rho(u)\, du \right| \leqq 4\varepsilon(1 + 2|t|a).$$

It follows from the conditions of the theorem that $E|L_1|^3$ and $E|L_2|^3$ are both finite, so that $\int |u|^3\, d\rho(u) < \infty$. But then by Chebishev's inequality

$$|T_2| = \left| \int_{|u| > a} \exp(2itu)\, d\rho(u) \right| \leqq C/a^3$$

Choosing $a = \varepsilon^{-1/3}$, we obtain

$$|\psi(t)| \leqq 4\varepsilon(1 + 2|t|\varepsilon^{-1/3}) + C\varepsilon = C_0\varepsilon(1 + |t|\varepsilon^{-1/3}). \qquad (9.2.2)$$

Thus Lemma 9.2.1 stands proved.∎

LEMMA 9.2.2. Let $f(t) = \exp(-(1/2)t^2) + h(t)$. Then, for $T = \varepsilon^{-1/3}$, the estimate below holds.

$$V = \int_{-T}^{T} \left| \frac{h(t)}{t} \right| dt \leqq C\varepsilon^{1/3}. \qquad (9.2.3)$$

Proof: From the conditions $EX_1 = 0$, $EX_1^2 = 1$, $E|X_1|^3 < \infty$, we have

$$|h(t)| < C|t|^3 \tag{9.2.4}$$

for small t. We set $t_i = \varepsilon^{1/3} \cdot 2^i$, $i = 0, 1, \ldots, k$, where k is chosen such that $\varepsilon^{-1/3} \leqq t_k < 2\varepsilon^{-1/3}$. Let

$$\gamma_i = \max\{|h(t)| : t_{i-1} \leqq t \leqq t_i\}; \quad i = 1, \ldots, k.$$

Then

$$\int_{t_{i-1}}^{t_i} \left|\frac{h(t)}{t}\right| dt \leqq \frac{\gamma_i}{t_{i-1}} \int_{t_{i-1}}^{t_i} dt = \gamma_i. \tag{9.2.5}$$

We have from (9.2.4)

$$\int_0^{t_0} \left|\frac{h(t)}{t}\right| dt \leqq Ct_0^3 = C\varepsilon. \tag{9.2.6}$$

Together with the easily verified equality $|h(-t)| = |h(t)|$, the inequalities (9.2.5)–(9.2.6) give, for $T = \varepsilon^{-1/3}$,

$$V = \int_{-T}^{T} \left|\frac{h(t)}{t}\right| dt \leqq C\varepsilon + 2\sum_1^k \gamma_i. \tag{9.2.7}$$

We use (9.2.1) for estimating the right side of (9.2.7). We have

$$\begin{aligned}
|h(2t)| &= |f(2t) - \exp(-2t^2)| = |[f(t)]^3 f(-t) + \psi(t) - \exp(-2t^2)| \\
&= |[h(t) + \exp(-(1/2)t^2)]^3[h(-t) + \exp(-(1/2)t^2)] - \exp(-2t^2) + \psi(t)| \\
&\leqq \sum_{s=0}^{3} \binom{4}{s} \exp(-st^2/2)|h(t)|^{4-s} + |\psi(t)|.
\end{aligned}$$

Taking (9.2.2) into account, we have from the above

$$\gamma_{i+1} \leqq 4a_i\gamma_i(1 + (3/2)\gamma_i + \gamma_i^2) + \gamma_i^4 + C\varepsilon(1 + 2^i) \tag{9.2.8}$$

where we have set $a_i = \exp(-(1/2)t_{i-1}^2)$. We proceed to show by induction that, for the same constant C_0 as in (9.2.2),

$$\gamma_i^4 \leqq C_0\varepsilon(1 + 2^i) \tag{9.2.9}$$

for $i = 1, \ldots, k$ and sufficiently small ε. Evidently (9.2.9) is valid for $i = 1$. Let now (9.2.9) be assumed to be valid for $i \leqq s \leqq k$. Then we have for $i \leqq s$,

$$(3/2)\gamma_i + \gamma_i^2 \leqq C\varepsilon^{1/12} = \delta.$$

Using (9.2.8) s times, we have

$$\gamma_{s+1} \leqq 2C_0\varepsilon(1 + 2^s) + 4a_s\gamma_s(1 + \delta) \leqq \cdots \leqq 2C_0\varepsilon \sum_{j=1}^{s-1} (1 + 2^{s-j})4^j(1 + \delta)^j D_j \tag{9.2.10}$$

where $D_0 = 1$ and

$$D_j = a_s a_{s-1} \cdots a_{s-j+1} = \exp\left(-(1/2)\sum_{m=1}^{j} t_{s-m}^2\right) = \exp[-t_0^2(4^s - 4^{s-j})/6].$$

Inequality (9.2.10) can be rewritten in the form

$$\begin{aligned}\gamma_{s+1} &\leqq 4C_0\,\varepsilon(1+\delta)^s \sum_{j=0}^{s-1} 2^{s+j}D_j + 4^s(1+\delta)^s D_s\gamma_1 \\ &\leqq (4C_0\,\varepsilon + \gamma_1)(1+\delta)^s \sum_{j=0}^{s} 2^{s+j}D_j. \end{aligned} \tag{9.2.11}$$

If now we take into account the facts that

$$\sum_{j=0}^{s} 2^{s+j}D_j \leqq 2^{2s+1}\exp\{-2^{2s-3}t_0^2\} \leqq C\varepsilon^{-2/3}$$

and, according to our choice of k, $\varepsilon^{-2/3} \leqq 2^k < 2\varepsilon^{-2/3}$, so that $(1+\delta)^k \leqq 2$, and further

$$\gamma_1 = \max\{|h(t)|: t_0 \leqq t \leqq t_1\} = \max\{|h(t)|: \varepsilon^{1/3} \leqq t \leqq 2\varepsilon^{1/3}\}$$

whence $\gamma_1 \leqq C\varepsilon$, we have

$$\gamma_{s+1} \leqq 2(4C_0 + C)\varepsilon^{1/3} = C_1\varepsilon^{1/3}.$$

If ε is taken to be sufficiently small: $\varepsilon < C_0^3\, C_1^{-12}$, then we have (9.2.9) for $i = s+1$.

Now we can use (9.2.11) to estimate $\sum_1^k \gamma_i$:

$$\sum_1^k \gamma_i \leqq C\varepsilon \sum_{s=1}^{k} 2^{2s+1}\exp(-2^{2s-3}t_0^2) \leqq \frac{s\varepsilon}{t_0^2} = C\varepsilon^{1/3}.$$

Thus, for $T = \varepsilon^{-1/3}$,

$$V = \int_{-T}^{T}\left|\frac{h(t)}{t}\right| dt \leqq C\varepsilon + 2\sum_1^k \gamma_i \leqq C\varepsilon^{1/3}$$

which proves Lemma 9.2.2.▮

To conclude the proof of Theorem 9.2.1, we use Lemma 1.1.11 with $T = \varepsilon^{-1/3}$. We have

$$\begin{aligned}\sup_x |F(x) - \Phi(x)| &\leqq \int_{-T}^{T}\left|\frac{f(t) - \exp(-(1/2)t^2)}{t}\right| dt + \frac{C}{T} \\ &= \int_{-T}^{T}\left|\frac{h(t)}{t}\right| dt + C\varepsilon^{1/3} \leqq C\varepsilon^{1/3}\end{aligned}$$

as required to prove.▮

As indicated in [103], the method of proof of Theorem 9.2.1 can be used for investigating the stability of the characterization of the normal law due to Polya (also cf. section 2.1), through the identical distribution of the statistics X_1 and $(X_1 + X_2)/2$.

9.3 THE CONDITION OF ε-ADMISSIBILITY OF THE SAMPLE MEAN AS AN ESTIMATOR OF A SHIFT PARAMETER

Let $X_1, \ldots, X_n$ be a random sample of size $n \geqq 3$ from a population with d.f. $F(x, \theta)$, $\theta \in R^1$, where F satisfies the conditions

$$\int x\, dF(x) = 0, \qquad 0 < \int x^2\, dF(x) = \sigma^2 < \infty. \tag{9.3.1}$$

The case $\sigma = 0$ is obviously trivial.

DEFINITION 9.3.1. $\overline{X}$ will be said to be ε-admissible in the class of unbiased estimators of θ under quadratic loss if there exists no estimator $\tilde{\theta} = \tilde{\theta}(X_1, \ldots, X_n)$ with $E_\theta \tilde{\theta} = \theta$ for which

$$E_\theta(\tilde{\theta} - \theta)^2 < E_\theta(\overline{X} - \theta)^2 - \varepsilon \tag{9.3.2}$$

for all θ.

According to the corollary to Theorem 7.4.1, the admissibility of the sample mean $\overline{X}$ in the class of unbiased estimators of θ under quadratic loss is a characteristic property of normal laws. We now take up, following [54], the study of the stability of this characterization if admissibility is replaced by ε-admissibility in the sense of Definition 9.3.1. As will be shown below, stability obtains in this problem (in every case, in the class of *symmetric* d.f.'s) and the closeness of F to the d.f. of a suitable normal law has a very small order in terms of ε; however, estimate (9.3.3) hardly appears to be final. If it is desired to obtain an estimate of the closeness of F to normality, independent of n, then we have to speak of (ε/n)-admissibility; it is convenient to consider $(\sigma^2\varepsilon^2/n)$-admissibility.

THEOREM 9.3.1. Let $(X_1, \ldots, X_n)$ be a random sample of size $n \geqq 3$ from a population with d.f. $F(x - \theta)$, where F is symmetric and has zero mean and finite variance σ^2. Then the $(\sigma^2\varepsilon^2/n)$-admissibility of the sample mean $\overline{X}$ under quadratic loss in the class of unbiased estimators of $\theta \in R^1$ guarantees the following closeness of F to the d.f. Φ_σ of the normal law $N(0, \sigma^2)$:

$$\sup_x |F(x) - \Phi_\sigma(x)| \leqq C(\log(1/\varepsilon))^{-1/2} \tag{9.3.3}$$

where C is an absolute constant.

***Proof*:** We first establish a few auxiliary results which only require that F has zero mean and finite variance σ^2.

LEMMA 9.3.1. If the estimator $\overline{X}$ of θ is ε-admissible, then

$$E_0\{E_0(\overline{X} \mid \mathbf{Y})\}^2 \leqq \varepsilon, \qquad \mathbf{Y} = (X_2 - X_1, \ldots, X_n - X_1). \tag{9.3.4}$$

Proof: The variance of the Pitman estimator $\hat{\theta} = \bar{X} - E_0(\bar{X} \mid \mathbf{Y})$ is given by

$$\begin{aligned}\operatorname{Var}_\theta \hat{\theta} &= E_\theta(\bar{X} - E_0(\bar{X} \mid \mathbf{Y}) - \theta)^2 = E_0 \bar{X}^2 - 2E_0\{\bar{X} E_0(\bar{X} \mid \mathbf{Y})\} + E_0\{E_0(\bar{X} \mid \mathbf{Y})\}^2 \\ &= E_0 \bar{X}^2 - 2E_0\{E_0(\bar{X} E_0(\bar{X} \mid \mathbf{Y}) \mid \mathbf{Y})\} + E_0\{E_0(\bar{X} \mid \mathbf{Y})\}^2 \\ &= E_0 \bar{X}^2 - E_0\{E_0(\bar{X} \mid \mathbf{Y})\}^2.\end{aligned}$$

If $\bar{X}$ is ε-admissible for θ, then we must have

$$E_0 \bar{X}^2 - E_0\{E_0(\bar{X} \mid \mathbf{Y})\}^2 \geqq E_0 \bar{X}^2 - \varepsilon$$

whence (9.3.4) follows. ∎

LEMMA 9.3.2. If $n \geqq 3$ and $\bar{X}$ is ε-admissible, then the c.f. f of F satisfies the relation

$$f''(-t)f(t) + f'(t)f'(-t) + \sigma^2 f(t)f(-t) = R_n(t) \tag{9.3.5}$$

where $|R_n(t)| \leqq C\sigma\sqrt{n\varepsilon}$.

Proof: We set

$$Z = Z(\mathbf{Y}) = \sum_{j=2}^{n-1} (X_j - X_{j-1}) \exp it(X_{j+1} - X_{j-1}),$$

$$h(t) = E_0\{(X_j - X_{j-1}) \exp it(X_{j+1} - X_{j-1})\}, \qquad j = 2, \ldots, n-1.$$

We calculate

$$\begin{aligned}E_0\{\bar{X}(Z - E_0 Z)\} &= E_0(\bar{X}Z) \\ &= \frac{n-2}{n}[f''(-t)f(t) + f'(t)f'(-t) + \sigma^2 f(t)f(-t)].\end{aligned} \tag{9.3.6}$$

On the other hand, since $Z = Z(\mathbf{Y})$, in view of the properties of the conditional expectations and the Cauchy-Schwarz-Bunyakovski inequality,

$$\begin{aligned}|E_0 \bar{X}(Z - E_0 Z)| &= |E_0\{(Z - E_0 Z)E_0(\bar{X} \mid \mathbf{Y})\}| \\ &\leqq (E_0|Z - E_0 Z|^2)^{1/2}(E_0\{E_0(\bar{X} \mid \mathbf{Y})\}^2)^{1/2} \\ &\leqq \sqrt{\varepsilon}(E_0|Z - E_0 Z|^2)^{1/2}\end{aligned} \tag{9.3.7}$$

where we have used Lemma 9.3.1. We now estimate $E_0|Z - E_0 Z|^2$. We fix t and set $U_j = (X_j - X_{j-1}) \exp it(X_{j+1} - X_{j-1})$, for $j = 2, \ldots, n-1$. We have

$$\begin{aligned}E_0|Z - E_0 Z|^2 &= E_0\left\{\sum_2^{n-1} (U_j - h(t)) \sum_2^{n-1} (\bar{U}_j - \overline{h(t)})\right\} \\ &= E_0\left\{\sum_{|j-k|>2} (U_j - h(t))(\bar{U}_k - \overline{h(t)}) + \sum_{|j-k|\leqq 2} (U_j - h(t))(\bar{U}_k - \overline{h(t)})\right\}.\end{aligned} \tag{9.3.8}$$

Further, $|E_0(U_j - h(t))(\overline{U}_k - \overline{h(t)})| \leqq 2\sigma^2$, and for $|j-k| > 2$, the r.v.'s U_j and U_k are independent. The number of summands in the sum corresponding to $|j-k| \leqq 2$ does not exceed $5(n-2)$. In view of all these, we obtain from (9.3.8) the relation

$$E_0|Z - E_0 Z|^2 \leqq 10(n-2)\sigma^2.$$

From (9.3.6), (9.3.7), and the above inequality, we have Lemma 9.3.2. ∎

We pass on to the proof of Theorem 9.3.1. Since F is symmetric, $f(-t) = f(t)$. Then, from the $(\sigma^2\varepsilon^2/n)$-admissibility of $\overline{X}$, we have by Lemma 9.3.2.

$$f''(t)f(t) - [f'(t)]^2 + \sigma^2[f(t)]^2 = R_n(t) \tag{9.3.9}$$

where $|R_n(t)| \leqq C\sigma^2\varepsilon$ (by the symbol C, we shall denote absolute constants). Let

$$\tau(\varepsilon) = \sup\,[\tau\colon |f(t)| \geqq \varepsilon^{1/4} \quad \text{for} \quad |t| \leqq \tau].$$

For $|t| \leqq \tau(\varepsilon)$, relation (9.3.9) can be written as

$$\phi''(t) = -\sigma^2 + \frac{R_n(t)}{[f(t)]^2}, \quad \phi = \log f.$$

Hence, remembering that $f(0) = 1$ and $f'(0) = 0$, we have

$$f(t) = \exp\left\{-\frac{t^2\sigma^2}{2} + \int_0^t\int_0^\tau \frac{R_n(u)}{[f(u)]^2}\,du\,d\tau\right\}, \; |t| \leqq \tau(\varepsilon).$$

Since

$$\left|\operatorname{Re}\int_0^t\int_0^\tau \frac{R_n(u)}{[f(u)]^2}\,du\,d\tau\right| \leqq \left|\int_0^t\int_0^\tau \frac{R_n(u)}{[f(u)]^2}\,du\,d\tau\right| \leqq \frac{1}{2}C\sigma^2 t^2\sqrt{\varepsilon}$$

for $|t| \leqq \tau(\varepsilon)$, we have for such t

$$|f(t)| = \exp\left[-\frac{1}{2}\sigma^2 t^2 + \operatorname{Re}\int_0^t\int_0^\tau \frac{R_n(u)}{[f(u)]^2}\,du\,d\tau\right] \geqq \exp\left[-\frac{1}{2}\sigma^2 t^2(1 + C\sqrt{\varepsilon})\right].$$

Since $|f(t)| \geqq \varepsilon^{1/4}$ for $|t| \leqq \tau(\varepsilon)$, $\tau(\varepsilon)$ is not smaller than the solution of the equation

$$\exp\left[-\frac{1}{2}\sigma^2 t^2(1 + C\sqrt{\varepsilon})\right] = \varepsilon^{1/4}$$

i.e., $\tau(\varepsilon) \geqq (C/\sigma)(\log(1/\varepsilon))^{1/2}$. Further, using the inequality $|e^z - 1| \leqq |z|e^{|z|}$, we have for $|t| \leqq \tau(\varepsilon)$,

$$\left|f(t) - \exp\left(-\frac{1}{2}\sigma^2t^2\right)\right| = \left|\exp\int_0^t\int_0^\tau \frac{R_n(u)}{[f(u)]^2}\,du\,d\tau - 1\right| . \exp\left(-\frac{1}{2}\sigma^2t^2\right)$$

$$\leqq \frac{1}{2}C\sigma^2t^2\sqrt{\varepsilon}\exp\left[-\frac{1}{2}\sigma^2t^2(1 - C\sqrt{\varepsilon})\right] \leqq C\sigma^2t^2\sqrt{\varepsilon}. \quad (9.3.10)$$

By Lemma 1.1.11,

$$\sup_x |F(x) - \Phi_\sigma(x)| \leqq \int_{-T}^{T}\left|\frac{f(t) - \exp(-(1/2)\sigma^2t^2)}{t}\right| dt + \frac{C}{\sigma T} \quad (9.3.11)$$

where $\Phi_\sigma(x)$ is the d.f. of the normal law $N(0, \sigma^2)$. Setting $T = (C/\sigma)(\log(1/\varepsilon))^{1/2}$ in (9.3.11), we obtain, in view of the estimate (9.3.10), that

$$\sup_x |F(x) - \Phi_\sigma(x)| \leqq C\sqrt{\varepsilon}\log(1/\varepsilon) + C(\log(1/\varepsilon))^{-1/2}$$
$$= C(\log(1/\varepsilon))^{-1/2}$$

proving Theorem 9.3.1.∎

CHAPTER 10

Characterization of Random Vectors with Linear Structure

10.1 INTRODUCTION

We say that a p-dimensional vector $\mathbf{X}$ has a linear structure if it can be represented in the form

$$\mathbf{X} = \boldsymbol{\mu} + \mathbf{AY} \tag{10.1.1}$$

where $\boldsymbol{\mu}$ is a vector of constants, $\mathbf{Y}$ is a vector with independent nondegenerate r.v.'s as components, and $\mathbf{A}$ is a matrix of constants no two of whose columns are proportional. If $\mathbf{Y}$ is a q-dimensional vector, then $\mathbf{A}$ is a $p \times q$ matrix. The components of $\mathbf{Y}$ are called *structural* variables.

Let $\mathbf{X} = \mathbf{v} + \mathbf{BZ}$ be another representation of $\mathbf{X}$ of the same type as (10.1.1). Two representations $\boldsymbol{\mu} + \mathbf{AY}$ and $\mathbf{v} + \mathbf{BZ}$ are *equivalent in structure* if every column of $\mathbf{A}$ is proportional to some column of $\mathbf{B}$ and *vice versa*. Otherwise, the two representations are said to be *nonequivalent*. If all structural representations of a random vector are equivalent, then we say that it has an (essentially) unique structure.

It is also possible that $\mathbf{X}$ has a unique structure but two representations of the above forms exist such that the structural variables $\mathbf{Y}$ and $\mathbf{Z}$ have different distributions. We shall say that the *linear model* (10.1.1) *is unique* if $\mathbf{X}$ has a unique structure and, further, the distributions of the structural variables are one and the same to within changes of location and scale.

Let us consider an example,

$$\mathbf{X} = \boldsymbol{\mu} + \mathbf{AY} \tag{10.1.2}$$

where $\mathbf{Y}$ is a vector with independent components each having an $N(0, 1)$ distribution; in this case, the c.f. of $\mathbf{X}$ has the form

$$E \exp(i\mathbf{t}'\mathbf{X}) = e^{i\mathbf{t}'\boldsymbol{\mu}} E \exp(i\mathbf{t}'\mathbf{AY}) = \exp(i\mathbf{t}'\boldsymbol{\mu} - (1/2)\mathbf{t}'\mathbf{AA}'\mathbf{t}) \tag{10.1.3}$$

so that the distribution of $\mathbf{X}$ depends only on $\boldsymbol{\mu}$ and the nonnegative definite matrix $\boldsymbol{\Lambda} = \mathbf{AA}'$. Thus $\mathbf{X}$ has a p-dimensional normal distribution $N_p(\boldsymbol{\mu}, \boldsymbol{\Lambda})$ (cf. for instance [131]). But, for a given matrix $\boldsymbol{\Lambda}$, the factorization $\boldsymbol{\Lambda} = \mathbf{AA}'$ is not unique. If $\boldsymbol{\Lambda} = \mathbf{BB}'$ be another factorization in which $\mathbf{B}$ is a $p \times r$ matrix (the order of $\mathbf{B}$ may be different from that of $\mathbf{A}$, though their ranks are necessarily the same) then

$$\mathbf{X} = \boldsymbol{\mu} + \mathbf{BZ} \tag{10.1.4}$$

where $\mathbf{Z}$ is an r-dimensional vector with independent $N(0, 1)$ components. In particular, the following three representations correspond to one and the same bivariate normal distribution if all the structural variables are independent $N(0, 1)$ variables.

$$\left.\begin{aligned} X_1 &= 2Y_1 \\ X_2 &= 5Y_1 + \sqrt{5}\,Y_2 \end{aligned}\right\} \qquad \left.\begin{aligned} X_1 &= \sqrt{2}\,U_1 + \sqrt{2}\,U_2 \\ X_2 &= \left(\frac{5}{\sqrt{2}} + \frac{\sqrt{5}}{\sqrt{2}}\right)U_1 + \left(\frac{5}{\sqrt{2}} - \frac{\sqrt{5}}{\sqrt{2}}\right)U_2 \end{aligned}\right\}$$

$$\left.\begin{aligned} X_1 &= W_1 + W_2 + W_3 + W_4 \\ X_2 &= W_1 + 2W_2 + 3W_3 + 4W_4 \end{aligned}\right\} \tag{10.1.5}$$

Consequently, the structure of a normal vector is nonunique, both for a given number of structural variables and in relation to the number of such variables.

In this chapter, we study the nature of random vectors admitting non-equivalent structural representations. In particular, we shall show that normal random vectors are completely characterized by nonuniqueness of linear structure. Every random vector having linear structure can be decomposed as the sum of two independent vectors, one of which has a unique structure and hence is nonnormal, and the other is a normal random vector.

The results presented here give a complete solution to the problem of nonidentifiability of parameters in linear structures; this problem was studied some years ago by Reiersol [133]. Most of the results presented here are due to Rao [128].

We also consider linear structural analogues of factor analysis models which are considered by psychometricians.

The problem investigated in this chapter may also be considered as that of identically distributed vector statistics (cf. Chapter 2). When can $\boldsymbol{\mu} + \mathbf{AY}$ and $\mathbf{v} + \mathbf{BZ}$ (or $\boldsymbol{\mu} + \mathbf{AY}$ and $\mathbf{v} + \mathbf{BY}$) have identical distributions?

10.2 AUXILIARY LEMMAS

We present here some lemmas used in the proofs of the basic theorems of this chapter, which are also of independent interest.

LEMMA 10.2.1. Let $\boldsymbol{\alpha}_1, \ldots, \boldsymbol{\alpha}_m$ be given nonzero elements of a vector space in which an inner product is defined. Then there exists a vector $\boldsymbol{\beta}$ which is not orthogonal to any of the above vectors.

***Proof*:** Suppose that a certain vector $\boldsymbol{\beta}_0$ is not orthogonal to the first k vectors, but is orthogonal to $\boldsymbol{\alpha}_{k+1}$. Then consider the vector $\boldsymbol{\beta}_0 + c\boldsymbol{\alpha}_{k+1}$, where c is some constant. The inner product of this vector with $\boldsymbol{\alpha}_i$, $i \leqq k$, is

$$d_i = (\boldsymbol{\beta}_0, \boldsymbol{\alpha}_i) + c(\boldsymbol{\alpha}_{k+1}, \boldsymbol{\alpha}_i)$$

and this is either never zero for any c or vanishes for $c = c_i$. If now we choose a c different from all c_i, $1 \leqq i \leqq k$, then none of the d_i, $1 \leqq i \leqq k$, is zero. Let $\boldsymbol{\beta}_1 = \boldsymbol{\beta}_0 + c\boldsymbol{\alpha}_{k+1}$, where c is chosen as above. Then $\boldsymbol{\beta}_1$ is not orthogonal to any of the vectors $\boldsymbol{\alpha}_i$ for $1 \leqq i \leqq k + 1$. Since we may always take $\boldsymbol{\beta}_0 = \boldsymbol{\alpha}_1$, the required result is obtained by induction.▌

LEMMA 10.2.2. Let **A** and **B** be $p \times k$ and $p \times m$ matrices. Suppose that the first (or some other) column of **A** is not proportional to any other column of **A** or to any column of **B**. Then there exists a $2 \times p$ matrix **H** such that the matrices $\mathbf{C}_1 = \mathbf{HA}$ and $\mathbf{C}_2 = \mathbf{HB}$ of orders $2 \times k$ and $2 \times m$ respectively have the following property; the first column of $\mathbf{C}_1$ is not proportional to any of the other columns of $\mathbf{C}_1$ or to any of the columns of $\mathbf{C}_2$.

***Proof*:** Without loss of generality, we may assume that the first rows of **A** and **B** consist only of zeros and ones as elements and that the element standing at the intersection of the first row and the first column of **A** is nonzero. Denote the columns of **A** whose first elements are 1 by $\boldsymbol{\alpha}_1, \ldots, \boldsymbol{\alpha}_r$, and the similar columns of **B** by $\boldsymbol{\beta}_1, \ldots, \boldsymbol{\beta}_s$.

We take as the first row of the desired matrix **H** the vector $(1, 0, \ldots, 0)$ and as the second the vector $\boldsymbol{\gamma}' = (0, \gamma_2, \ldots, \gamma_p)$ in such a way that

$$(\boldsymbol{\gamma}, \boldsymbol{\alpha}_1 - \boldsymbol{\alpha}_i) \neq 0 \qquad \text{for } i = 2, \ldots, r,$$
$$(\boldsymbol{\gamma}, \boldsymbol{\alpha}_1 - \boldsymbol{\beta}_i) \neq 0 \qquad \text{for } i = 1, \ldots, s.$$

By Lemma 10.2.1, such a vector $\boldsymbol{\gamma}$ exists. Then it is easily seen that the matrix **H**, so chosen, possesses the required property.▌

LEMMA 10.2.3. Let $\mathbf{X} = \boldsymbol{\mu}_1 + \mathbf{AY}$ and $\mathbf{X} = \boldsymbol{\mu}_2 + \mathbf{BZ}$ be two structuarl representations of **X**. Then the linear manifolds generated by the columns of **A** and by the columns of **B** coincide, so that rank **A** = rank **B**. Further, the vector $\boldsymbol{\mu}_1 - \boldsymbol{\mu}_2$ belongs to this (common) linear manifold.

***Proof*:** Let $\boldsymbol{\alpha}$ be a column vector such that $\boldsymbol{\alpha}'\mathbf{A} = \mathbf{0}$. Then

$$\boldsymbol{\alpha}'\mathbf{X} = \boldsymbol{\alpha}'\boldsymbol{\mu}_1 = \boldsymbol{\alpha}'\boldsymbol{\mu}_2 + \boldsymbol{\alpha}'\mathbf{BZ}$$

whence $\boldsymbol{\alpha}'\mathbf{BZ}$ is a degenerate r.v. But this is possible only if $\boldsymbol{\alpha}'\mathbf{B} = \mathbf{0}$. Thus the conditions $\boldsymbol{\alpha}'\mathbf{A} = \mathbf{0}$ and $\boldsymbol{\alpha}'\mathbf{B} = \mathbf{0}$ are equivalent, whence it follows that the linear manifolds generated by the columns of **A** and by those of **B** coincide.

Further, it follows from $\boldsymbol{\alpha}'\mathbf{A} = \mathbf{0}$ that $\boldsymbol{\alpha}'(\boldsymbol{\mu}_1 - \boldsymbol{\mu}_2) = \mathbf{0}$, i.e., $\boldsymbol{\mu}_1 - \boldsymbol{\mu}_2$ belongs to the linear manifold above.∎

It follows from Lemma 10.2.3 that we may take the structural representation of **X** simply as **AY** without loss of generality, omitting the vector of constants.

LEMMA 10.2.4. Let **Y** be a two-dimensional random vector, $\mathbf{Y} = (Y_1, Y_2)$ admitting two representations

$$\left.\begin{aligned} Y_1 &= a_{11}f_1 + \cdots + a_{1k}f_k \\ Y_2 &= a_{21}f_1 + \cdots + a_{2k}f_k \end{aligned}\right\} \qquad \left.\begin{aligned} Y_1 &= b_{11}g_1 + \cdots + b_{1m}g_m \\ Y_2 &= b_{21}g_1 + \cdots + b_{2m}g_m \end{aligned}\right\} \tag{10.2.1}$$

where $f_1, \ldots, f_k$ and $g_1, \ldots, g_m$ are sets of independent r.v.'s. Let the r-th column $(a_{1r}, a_{2r})'$ in the first representation be not proportional to any of the columns $(a_{1j}, a_{2j})'$, $j \neq r$, and to any of the columns $(b_{1j}, b_{2j})'$ of the second representation. Then the r.v. f_r is normal.

***Proof*:** Let ψ_i be the logarithm of the c.f. of f_i and ϕ_j that of the c.f. of g_j (defined as usual in a neighborhood of the origin). Considering the joint c.f. of Y_1 and Y_2 and using the two representations (10.2.1), we obtain for $|u| < \delta_0$, $|v| < \delta_0$ the relations

$$\begin{aligned} \log E \exp(iuY_1 + ivY_2) &= \psi_1(a_{11}u + a_{21}v) + \cdots + \psi_k(a_{1k}u + a_{2k}v) \\ &= \phi_1(b_{11}u + b_{21}v) + \cdots + \phi_m(b_{1m}u + b_{2m}v). \end{aligned} \tag{10.2.2}$$

We may assume without loss of generality that $a_{1r} \neq 0$ and replace nonzero a_{1i} and b_{1j} by unity which would only reflect a change of scale for the r.v.'s f_i and g_j. Then, using the conditions of Lemma 10.2.4, we can reduce (10.2.2) to the form

$$\psi_r(u + a_{2r}v) + \eta_1(u + c_{21}v) + \cdots + \eta_s(u + c_{2s}v) = A(u) + B(v) \tag{10.2.3}$$

if $a_{2r} \neq 0$, and to the form

$$\eta_1(u + c_{21}v) + \cdots + \eta_s(u + c_{2s}v) = \psi_r(u) + B(v) \tag{10.2.4}$$

if $a_{2r} = 0$. In formulas (10.2.3) and (10.2.4), each function η is obtained by adding the functions ψ and subtracting the functions ϕ which have common coefficients in (10.2.2). We note that $a_{2r}, c_{21}, \ldots, c_{2s}$ can be taken to be distinct and that it is impossible to combine ψ_r with the rest.

Applying Lemma 1.5.1, we see that ψ_r is a polynomial of degree $\leq s$ in some neighborhood of the origin. Then ψ_r is necessarily a polynomial of degree 2 if f_r is nondegenerate. Thus f_r is normally distributed.∎

LEMMA 10.2.5. Let the i-th column of the first of the representations (10.2.1) be proportional to the j-th column of the second, but not to any other column of the first or the second. Then, in some neighborhood of the origin, the logarithms of the c.f.'s of f_i and g_j differ by a polynomial.

***Proof*:** Without loss of generality, we may take $i = j$. Noting that one of the functions η in (10.2.3) and (10.2.4) is in fact the difference between the logarithms of the c.f.'s of f_i and g_j, and that by Lemma 1.5.1 each of the functions η is a polynomial, our assertion follows.▮

LEMMA 10.2.6. Let $\mathbf{X}$ be an arbitrary p-dimensional random vector. Then $\mathbf{X}$ admits the decomposition

$$\mathbf{X} = \mathbf{X}_1 + \mathbf{X}_2 \tag{10.2.5}$$

where $\mathbf{X}_1$ and $\mathbf{X}_2$ are independent, and $\mathbf{X}_2$ is a p-dimensional normal vector with maximal dispersion matrix.

Note: We mean by the above that if $\mathbf{X} = \mathbf{Y}_1 + \mathbf{Y}_2$ be another decomposition into independent vectors, where $\mathbf{Y}_2$ is a p-dimensional normal vector, its dispersion matrix is not "greater" than that of $\mathbf{X}_2$ (as usual, we write $\mathbf{A} > \mathbf{B}$ if $\mathbf{A} - \mathbf{B}$ is a nonnegative definite matrix).

***Proof*:** Let $C(\mathbf{t})$ be the c.f. of $\mathbf{X}$ and S the set of all nonnegative definite matrices such that for any $\mathbf{A} \in S$, $C(\mathbf{t}) \exp((1/2)\mathbf{t}'\mathbf{A}\mathbf{t})$ is a c.f. It is easily seen that the set S is bounded above. Let $\mathbf{A} = ((a_{ij}))$ be a member of S. Consider the set $S_1 \subset S$ of matrices with element $a_{11}^{(1)} = \sup_{\mathbf{A} \in S} a_{11}$. From the set S_1 we choose the subset S_2 with element $a_{22}^{(2)} = \sup_{\mathbf{A} \in S_1} a_{22}^{(1)}$, and so on. Then any element of S_p has the property that S has no larger element. Take an arbitrary element $\mathbf{A}^*$ of S_p. Then the factorization

$$C(\mathbf{t}) = C(\mathbf{t})e^{(1/2)\mathbf{t}'\mathbf{A}^*\mathbf{t}} \cdot e^{-(1/2)\mathbf{t}'\mathbf{A}^*\mathbf{t}} \tag{10.2.6}$$

yields the desired decomposition (10.2.5).▮

We have obtained only one maximal decomposition, which is in general not unique. It would be interesting to find conditions under which the maximal decomposition is unique, i.e., there exists a factorization (10.2.6) with $\mathbf{A}^*$ strictly larger than any other element of S.

LEMMA 10.2.7. Let, for every n, the k-dimensional random vector $\mathbf{G}_n$ have independent components. Consider the sequence of p-dimensional random vectors $\mathbf{X}_n = \mathbf{B}\mathbf{G}_n$ where $\mathbf{B}$ is a $p \times k$ matrix. If $\mathbf{X}_n$ converges in law to $\mathbf{X}$: $\mathbf{X}_n \xrightarrow{L} \mathbf{X}$, then $\mathbf{X}$ also has linear structure: $\mathbf{X} = \mathbf{B}\mathbf{G}$, where $\mathbf{G}$ is a vector with independent components.

***Proof*:** Without loss of generality, let $\mathbf{B}$ have no column consisting wholly of zeros as elements. Then $\mathbf{X}_n \xrightarrow{L} \mathbf{X}$ guarantees, using a result of Parthasarathy, Ranga Rao and Varadhan [108], that the sequence $\{\mathbf{G}_n\}$ is shift compact, i.e., there exists a subsequence $\{\mathbf{G}_m\}$ and a corresponding sequence $\{\mathbf{c}_m\}$ of vector constants such that $\mathbf{G}_m - \mathbf{c}_m \xrightarrow{L} \mathbf{G}$. Now consider

$$\mathbf{X}_m = \mathbf{B}(\mathbf{G}_m - \mathbf{c}_m) + \mathbf{B}\mathbf{c}_m.$$

Since $\mathbf{X}_m$ and $\mathbf{G}_m - \mathbf{c}_m$ have limit distributions, it follows that $\mathbf{B}\mathbf{c}_m \to \mathbf{c}$, a constant vector. Then $\mathbf{X} = \mathbf{B}\mathbf{G} + \mathbf{c}$. Let $\mathbf{b}$ be a vector orthogonal to the columns of $\mathbf{B}$. Then,

$$0 = \mathbf{b}'\mathbf{B}\mathbf{c}_m \to \mathbf{b}'\mathbf{c} \quad \text{so that} \quad \mathbf{b}'\mathbf{c} = 0$$

so that the constant vector $\mathbf{c}$ can be included in $\mathbf{G}$ and the structure of $\mathbf{X}$ written simply as $\mathbf{BG}$.▮

10.3 CHARACTERIZATION THEOREMS

We investigate the form of p-dimensional random vectors admitting alternative structural representations. We assume $p \geq 2$.

THEOREM 10.3.1. Let $\mathbf{X} = \mathbf{AF}$ and $\mathbf{X} = \mathbf{BG}$ be two representations of a p-dimensional random vector, where $\mathbf{A}$ and $\mathbf{B}$ are constant matrices of orders $p \times r$ and $p \times s$ respectively, and $\mathbf{F} = (f_1, \ldots, f_r)$ and $\mathbf{G} = (g_1, \ldots, g_s)$ are random vectors with independent components. Then the following assertions hold.

(i) If the i-th column of $\mathbf{A}$ is not proportional to any column of $\mathbf{B}$, then f_i is normal.

(ii) If the i-th column of $\mathbf{A}$ is proportional to the j-th column of $\mathbf{B}$, then the logarithms of the c.f.'s of f_i and g_j differ by a polynomial in a neighborhood of the origin.

***Proof*:** Without loss of generality, we may assume that, under (i), it is the first column of $\mathbf{A}$ that is not proportional to any column of $\mathbf{B}$. Then, by Lemma 10.2.2, a $2 \times p$ matrix $\mathbf{H}$ can be found such that the first column of $\mathbf{C}_1 = \mathbf{HA}$ is not proportional to any of its other columns or to any column of $\mathbf{C}_2 = \mathbf{HB}$. Then

$$\mathbf{HX} = \mathbf{Y} = \mathbf{C}_1\mathbf{F} \qquad \text{and} \qquad \mathbf{HX} = \mathbf{Y} = \mathbf{C}_2\mathbf{G} \tag{10.3.1}$$

and applying Lemma 10.2.4, we see that f_1 is normal.

To prove (ii), we suppose without loss of generality that the first column of $\mathbf{A}$ is the same as the first column of $\mathbf{B}$. By Lemma 10.2.2, there exists a $2 \times p$ matrix $\mathbf{H}$ such that the two matrices $\mathbf{C}_1 = \mathbf{HA}$ and $\mathbf{C}_2 = \mathbf{HB}$ are such that the first column of $\mathbf{C}_1$ is not proportional to either any other column of $\mathbf{C}_1$, or to any column of $\mathbf{C}_2$, other than the first. Then $\mathbf{HX} = \mathbf{Y} = \mathbf{C}_1\mathbf{F}$ and also $= \mathbf{C}_2\mathbf{G}$, and, applying Lemma 10.2.4, we conclude that the logarithms of the c.f.'s of f_1 and g_1 differ by a polynomial in the neighborhood of the origin.▮

THEOREM 10.3.2. (characterization of the multivariate normal law) If the matrix $\mathbf{A}$ in Theorem 10.3.1 is such that none of its columns is proportional to any column of $\mathbf{B}$, then $\mathbf{X}$ is a p-dimensional normal random vector.
***Proof*:** This follows from Theorem 10.3.1, since all the f_i are normal under these conditions.∎

THEOREM 10.3.3. In $\mathbf{X} = \mathbf{AY}$, let $\mathbf{Y}_1$ be the subvector of $\mathbf{Y}$ consisting of the non-normal variables and $\mathbf{Y}_2$ be the subvector of $\mathbf{Y}$ consisting of normal variables. Further, let $\mathbf{A}_1$ and $\mathbf{A}_2$ form the corresponding partition of the matrix $\mathbf{A}$, so that $\mathbf{X} = \mathbf{A}_1\mathbf{Y}_1 + \mathbf{A}_2\mathbf{Y}_2$. Then every structure of $\mathbf{X}$ must have the form

$$\mathbf{X} = \mathbf{A}_1\mathbf{U}_1 + \mathbf{B}_2\mathbf{U}_2 \tag{10.3.2}$$

where $\mathbf{U}_1$ is a vector of nonnormal r.v.'s and $\mathbf{U}_2$ a vector of normal r.v.'s with structure matrix $\mathbf{B}_2$ which is not necessarily equivalent to $\mathbf{A}_2$.

THEOREM 10.3.4. Let all the components of the vector $\mathbf{Y}$ in the representation $\mathbf{X} = \mathbf{AY}$ be nonnormal. Then there does not exist a representation with a smaller number of structural variables.
Theorems 10.3.3 and 10.3.4 are corollaries of Theorem 10.3.1.

THEOREM 10.3.5. Let $\mathbf{X} = \mathbf{AY}$ be a structural representation of $\mathbf{X}$ with a given number s of structural variables (components of $\mathbf{Y}$). If all the structural variables are nonnormal, then $\mathbf{X}$ has essentially unique structure with respect to the given number s of structural variables, i.e., if $\mathbf{X} = \mathbf{BZ}$, where the order of $\mathbf{B}$ is the same as that of $\mathbf{A}$, then $\mathbf{B}$ is equivalent to $\mathbf{A}$.
***Proof*:** Suppose that there are two nonequivalent representations, consisting of the same number s of structural varibles, $\mathbf{X} = \mathbf{AY}$ and $\mathbf{X} = \mathbf{BZ}$. Then, by Theorem 10.3.1, some of the structural variables are normal, which is contrary to assumption.∎

We note that $\mathbf{X}$ can have other structural representations with a larger number of structural variables, some of which will necessarily be normal. It is also possible that $\mathbf{X}$ and $\mathbf{Z}$ in Theorem 10.3.5 have different distributions although $\mathbf{A}$ and $\mathbf{B}$ are equivalent.

The following theorems give conditions for uniqueness of structure, formulated in terms of structural variables and coefficients.

THEOREM 10.3.6. Let $\mathbf{X} = \mathbf{AY}$ be a representation in which none of the structural variables has a normal component. If the columns of $\mathbf{A}$ are linearly independent, then $\mathbf{X}$ has essentially unique structure, and the model $\mathbf{X} = \mathbf{AY}$ is unique.
***Proof*:** We recall that a r.v. is said to have a normal component if it is the convolution of some r.v. with a normal r.v.

If now there exists some other representation, then it must have the form

$$\mathbf{X} = \mathbf{AZ} + \mathbf{BU} \tag{10.3.3}$$

with a larger number of structural variables. By Theorem 10.3.1, the structural variables which are the components of **U** must be normal and $\mathbf{B} = \mathbf{AK}$ for some **K**.

Since the columns of **A** are independent $(\mathbf{A'A})$ is nonsingular in which case

$$(\mathbf{A'A})^{-1}\mathbf{A'X} = \mathbf{Y}, \tag{10.3.4}$$

$$(\mathbf{A'A})^{-1}\mathbf{A'X} = \mathbf{Z} + (\mathbf{A'A})^{-1}\mathbf{A'BU}. \tag{10.3.5}$$

Hence **Y** and $\mathbf{Z} + (\mathbf{A'A})^{-1}\mathbf{A'BU}$ have the same distribution, where the components of $(\mathbf{A'A})^{-1}\mathbf{A'BU}$ are univariate normal. Then at least one element of **Y** has a normal component unless $(\mathbf{A'A})^{-1}\mathbf{A'B} = \mathbf{0}$. Substituting **AK** for **B**, we have $\mathbf{K} = \mathbf{0}$, in which case (10.3.3) reduces to $\mathbf{X} = \mathbf{AZ}$, showing that the structure of **X** unique. Then, from (10.3.4) and (10.3.5), **Y** and **Z** have the same distribution establishing the uniqueness of the model.∎

COROLLARY. The result of Theorem 10.3.6 is true even if one of the structural variables has a normal component provided the others do not.

THEOREM 10.3.7. Let $\mathbf{X} = \mathbf{AY}$ be a structural representation of **X**, in which no linear combination of the structural variables has a normal component. Then **X** has essentially unique structure.

***Proof*:** The proof is similar to that of Theorem 10.3.6. It remains to remark that in Theorem 10.3.7 we drop the condition of linear independence of the columns of **A**, but impose the stronger condition of the absence of normal components for any linear combination of the structural variables.∎

As an example, we consider a two-dimensional vector (X_1, X_2) with the linear structure

$$\left.\begin{aligned} X_1 &= Y_1 + Y_2 + Y_3 + Y_4 \\ X_2 &= Y_1 + 2Y_2 + 3Y_3 + 4Y_4 \end{aligned}\right\} \tag{10.3.6}$$

where the r.v.'s Y_j all have gamma distributions. According to Theorems 10.3.4 and 10.3.5, there is no alternative representation with four or less number of structural variables. Thus, if r.v.'s X_1 and X_2 are constructed as linear combinations of a larger number of nonnormal variables, then a more economical way of representing them, with a smaller number of structural variables, does not exist.

In the above example, the structure is in fact unique, so that there exists no alternative representation even with a larger number of structural variables.

An example of a representation which is partially unique is given by

$$X_1 = Y_1$$
$$X_2 = Y_1 + Y_2 \tag{10.3.7}$$

where the structural variables Y_1 and Y_2 are such that each is itself the convolution of a normal and a gamma variable, i.e., $Y_1 = U_1 + V_1$ and $Y_2 = U_2 + V_2$, where U_1 and U_2 have gamma distributions and V_1 and V_2 have normal distributions, and moreover (U_1, U_2) is independent of (V_1, V_2). Thus, (10.3.7) can be written in the form

$$X_1 = U_1 + V_1$$
$$X_2 = U_1 + U_2 + V_1 + V_2. \tag{10.3.8}$$

Making the transformation

$$V_1 = (Z_1 + Z_2)/\sqrt{2} \qquad \text{and} \qquad V_2 = (Z_1 - Z_2)/\sqrt{2}$$

we can write (10.3.8) as

$$X_1 = U_1 + Z_1/\sqrt{2} + Z_2/\sqrt{2}$$
$$X_2 = U_1 + U_2 + \sqrt{2}\, Z_1. \tag{10.3.9}$$

This structure is not equivalent to (10.3.7).

THEOREM 10.3.8. Let $\mathbf{X} = \mathbf{AY}$ be a representation with a given number of structural variables. If the structural variables are nonnormal and the columns of $\mathbf{A}$ are linearly independent, then the model $\mathbf{X} = \mathbf{AY}$ is unique for the specified number of structural variables.

***Proof*:** Under the conditions of the theorem, $\mathbf{X}$ has unique structure for the given number of structural variables. If now $\mathbf{X} = \mathbf{AY}$ and $\mathbf{X} = \mathbf{AZ}$ be two representations, then we have $\mathbf{Y} = (\mathbf{A}'\mathbf{A})^{-1}\mathbf{A}'\mathbf{X}$ and $\mathbf{Z} = (\mathbf{A}'\mathbf{A})^{-1}\mathbf{A}'\mathbf{X}$, whence it is seen that $\mathbf{Y}$ and $\mathbf{Z}$ have the same distribution.∎

THEOREM 10.3.9. (theorem on uniqueness of decomposition) Let $\mathbf{X} = \mathbf{AY}$ be a structural representation of $\mathbf{X}$ and let the columns of $\mathbf{A}$ be linearly independent. Then $\mathbf{X}$ can be expressed as a sum $\mathbf{X} = \mathbf{X}_1 + \mathbf{X}_2$, where $\mathbf{X}_1$ and $\mathbf{X}_2$ are independent, $\mathbf{X}_1$ has essentially unique structure and $\mathbf{X}_2$ has a multivariate normal distribution with a nonunique structure.

Further, such a decomposition is unique in the sense that if $\mathbf{X} = \mathbf{Z}_1 + \mathbf{Z}_2$ be another decomposition, where $\mathbf{Z}_2$ is a multivariate normal vector and $\mathbf{Z}_1$ has unique structure and $\mathbf{Z}_1$ and $\mathbf{Z}_2$ are independent, then $\mathbf{Z}_1$ and $\mathbf{X}_1$ have the same distribution, and so do $\mathbf{Z}_2$ and $\mathbf{X}_2$. If $\mathbf{X} = \mathbf{U}_1 + \mathbf{U}_2$ is an arbitrary decomposition where $\mathbf{U}_2$ is a multivariate normal vector, then $D(\mathbf{X}_2) - D(\mathbf{U}_2)$ is a nonnegative definite matrix (by $D(\mathbf{Y})$, we mean, of course, the dispersion matrix of $\mathbf{Y}$).

***Proof*:** We note that if an arbitrary one dimensional random variable x is given, then it is always possible to represent it in the form $x_1 + x_2$ where x_1 and x_2 are independent, x_2 is normal and x_1 does not have a normal component. For proving this assertion, let f be the c.f. of x; consider the set of all a for which $f(t) \exp((1/2)at^2)$ is also a c.f. This set has a real supremum a^*. It is easily seen (by appealing to the Levy-Cramer continuity theorem) that $f(t) \exp((1/2)a^*t^2)$ is also a c.f. Let it be the c.f. of x_1 and let $\exp(-(1/2)a^*t^2)$ be the c.f. of x_2; then the decomposition $x = x_1 + x_2$ is the desired one. The variable x_2 is called the maximal normal component of x.

Now consider the representation $\mathbf{X} = \mathbf{AY}$ and let $\mathbf{Y} = \mathbf{Y}_1 + \mathbf{Y}_2$, where $\mathbf{Y}_2$ is the vector of maximal normal components of the elements of $\mathbf{Y}$. Then

$$\mathbf{X} = \mathbf{AY}_1 + \mathbf{AY}_2. \tag{10.3.10}$$

Some elements of $\mathbf{Y}_1$ and $\mathbf{Y}_2$ may be zero; in this case, (10.3.10) can be written in the form

$$\mathbf{X} = \mathbf{A}_1\mathbf{Z}_1 + \mathbf{A}_2\mathbf{Z}_2$$

where only the nondegenerate elements of $\mathbf{Y}_1$ and $\mathbf{Y}_2$ and the columns of $\mathbf{A}$ corresponding to them are left out. By construction, $\mathbf{A}_2\mathbf{Z}_2$ is a multivariate normal vector and $\mathbf{A}_1\mathbf{Z}_1$ has essentially unique structure. The choice $\mathbf{X}_1 = \mathbf{A}_1\mathbf{Z}_1$, $\mathbf{X}_2 = \mathbf{A}_2\mathbf{Z}_2$ shows that our proposition is true. The other properties of such a decomposition mentioned in the statement of the theorem are easily verified. ▌

We note that our proof remains valid if the condition of linear independence of the columns of $\mathbf{A}$ in the theorems is replaced by the condition of linear independence of those columns of $\mathbf{A}$ which are related to the nonnormal components of $\mathbf{Y}$.

Remark. Let $\mathbf{X} = \mathbf{AY}$ be a structural representation such that the columns of $\mathbf{A}$ corresponding to the non-normal components of $\mathbf{Y}$ are linearly independent. Then $\mathbf{X}$ has the decomposition, $\mathbf{X} = \mathbf{BU} + \boldsymbol{\varepsilon}$, where $\boldsymbol{\varepsilon}$ has p-variate normal distribution, $\mathbf{U}$ is a vector of independent variables with no normal components, and $\mathbf{U}$ and $\boldsymbol{\varepsilon}$ are independent. The decomposition is unique, i.e., if $\mathbf{Y} = \mathbf{B}_1\mathbf{U}_1 + \boldsymbol{\varepsilon}_1$ is another decomposition then the matrices $\mathbf{B}$ and $\mathbf{B}_1$ are equivalent, $\mathbf{U}$ and $\mathbf{U}_1$ have the same distribution except for change of scale and location and $\boldsymbol{\varepsilon}$ and $\boldsymbol{\varepsilon}_1$ have the same p-variate normal distribution.

Now we consider the case when the columns of $\mathbf{A}$ corresponding to non-normal components in the representation $\mathbf{X} = \mathbf{AY}$ can be linearly dependent. We formulate a general theorem.

THEOREM 10.3.10. (decomposition theorem) Let $\mathbf{X}$ be a p-dimensional random vector with a linear structure: $\mathbf{X} = \mathbf{AY}$. Then $\mathbf{X}$ admits the decomposition

$$\mathbf{X} = \mathbf{X}_1 + \mathbf{X}_2 \tag{10.3.11}$$

where $\mathbf{X}_1$ and $\mathbf{X}_2$ are independent, $\mathbf{X}_1$ has unique structure and $\mathbf{X}_2$ is a p-dimensional normal vector.

Remark. The distinction between the results of Theorems 10.3.9 and 10.3.10 lies in the fact that in the former we could guarantee not only the uniqueness of structure of $\mathbf{X}_1$, but also the uniqueness of the distribution, whence followed the property of maximality of the dispersion matrix of the normal component $\mathbf{X}_2$. In the general case, we cannot claim either the uniqueness of the distribution of $\mathbf{X}_1$ or the maximality of the dispersion matrix of $\mathbf{X}_2$.

***Proof*:** Consider the representation $\mathbf{X} = \mathbf{AY}$ and the decomposition

$$\mathbf{X} = \mathbf{A}_1\mathbf{Y}_1 + \mathbf{A}_2\mathbf{Y}_2 = \mathbf{U}_1 + \mathbf{U}_2 \tag{10.3.12}$$

where $\mathbf{Y}_1$ is the vector of nonnormal components and $\mathbf{Y}_2$ of the normal ones. $\mathbf{U}_1$ and $\mathbf{U}_2$ are independent and $\mathbf{U}_2$ is a p-dimensional normal vector. But $\mathbf{U}_1 = \mathbf{A}_1\mathbf{Y}_1$ may have nonunique structure. If $\mathbf{U}_1$ has an alternative structure, then, by Theorem 10.3.3,

$$\mathbf{U}_1 = \mathbf{A}_1\mathbf{Y}_{1\alpha} + \mathbf{B}_\alpha\mathbf{Z}_\alpha = \mathbf{X}_{1\alpha} + \mathbf{X}_{2\alpha} \tag{10.3.13}$$

where $\mathbf{Z}_\alpha$ is a vector of standard normal variables. Consider the set S of nonnegative definite matrices $\mathbf{D}_\alpha = \mathbf{B}_\alpha\mathbf{B}'_\alpha$, for which a decomposition of the type (10.3.13) exists. Then, following the proof of Lemma 10.2.6, we find that there exists a matrix $\mathbf{G} = \mathbf{HH}'$, which is the limit of a sequence $\{\mathbf{B}_n\mathbf{B}'_n\}$, such that none of the members of S is "greater than" $\mathbf{G}$. The corresponding sequence of random vectors $\{\mathbf{X}_n\}$ converges in law to a random vector $\mathbf{X}_1$ which, by Lemma 10.2.7, has linear structure, let us say $\mathbf{X}_1 = \mathbf{A}_1\mathbf{V}_1$. Then

$$\mathbf{A}_1\mathbf{Y}_1 = \mathbf{U}_1 = \mathbf{A}_1\mathbf{V}_1 + \mathbf{HV}_2.$$

If we set $\mathbf{X}_1 = \mathbf{A}_1\mathbf{V}_1$, then $\mathbf{X}_1$ will have unique structure. In fact, otherwise, let $\mathbf{X}_1 = \mathbf{A}_1\mathbf{W}_1 + \mathbf{FW}_2$, where $\mathbf{W}_2$ is a vector of standard normal r.v.'s. In that case, $\mathbf{U}_1 = \mathbf{A}_1\mathbf{W}_1 + \mathbf{FW}_2 + \mathbf{HV}_2$, where the dispersion matrix of the normal components $\mathbf{FW}_2 + \mathbf{HV}_2$ is $\mathbf{FF}' + \mathbf{HH}' \geq \mathbf{HH}' = \mathbf{G}$. This contradiction shows the uniqueness of structure of $\mathbf{X}_1$.

We have from (10.3.12) that

$$\begin{aligned}\mathbf{X} = \mathbf{A}_1\mathbf{Y}_1 + \mathbf{A}_2\mathbf{Y}_2 &= (\mathbf{A}_1\mathbf{V}_1 + \mathbf{HV}_2) + \mathbf{A}_2\mathbf{Y}_2\\ &= (\mathbf{A}_1\mathbf{V}_1) + (\mathbf{HV}_2 + \mathbf{A}_2\mathbf{Y}_2) = \mathbf{X}_1 + \mathbf{X}_2\end{aligned}$$

where $\mathbf{X}_1$ and $\mathbf{X}_2$ are independent, $\mathbf{X}_1$ has unique structure, and $\mathbf{X}_2$ is multivariate normal. ∎

We cite an example of two decompositions possessing the properties indicated in Theorem 10.3.10. Let the r.v.'s V_1, V_2 have gamma distributions, and U_1, U_2, U_3, U_4 have normal distributions with zero means and variances

1, 1/2, 1, 1/2 respectively. We set $Y_1 = V_1$, $Y_2 = V_2 + U_1$, $Z_1 = V_1 + U_2$, and $Z_2 = V_2$. Then the representations

$$\left.\begin{aligned} X_1 &= Y_1 + U_3 \\ X_2 &= Y_2 + U_3 \end{aligned}\right\} \qquad \left.\begin{aligned} X_1 &= Z_1 + U_4 \\ X_2 &= Z_2 + 2U_4 \end{aligned}\right\}$$

reduce to the decompositions

$$\mathbf{X} = \begin{pmatrix} Y_1 \\ Y_2 \end{pmatrix} + \begin{pmatrix} U_3 \\ U_3 \end{pmatrix} \qquad \mathbf{X} = \begin{pmatrix} Z_1 \\ Z_2 \end{pmatrix} + \begin{pmatrix} U_4 \\ 2U_4 \end{pmatrix}$$

where, as can be observed, $(Y_1, Y_2)'$ and $(Z_1, Z_2)'$ have unique structures, and $(U_3, U_3)'$ and $(U_4, 2U_4)'$ are bivariate normal vectors whose dispersion matrices are such that neither is greater than the other.

THEOREM 10.3.11. Let $\mathbf{Y}$ be a vector of n independent r.v.'s, and $\mathbf{Z}_1 = \mathbf{AY}$, $\mathbf{Z}_2 = \mathbf{BY}$, $\mathbf{Z}_3 = \mathbf{CY}$ be three sets of linear functions such that the conditional distribution of $\mathbf{Z}_1$ given $\mathbf{Z}_3$ is the same as that of $\mathbf{Z}_2$ given $\mathbf{Z}_3$. Further let $\boldsymbol{\alpha}_i$, $\boldsymbol{\beta}_i$ and $\mathbf{Y}_i$ be the i-th column vectors of $\mathbf{A}$, $\mathbf{B}$ and $\mathbf{C}$ respectively. Then Y_i, the i-th component of $\mathbf{Y}$, is normal if $\begin{pmatrix} \boldsymbol{\alpha}_i \\ \mathbf{v}_i \end{pmatrix}$ is not proportional to any $\begin{pmatrix} \boldsymbol{\beta}_j \\ \mathbf{v}_j \end{pmatrix}$ or $\begin{pmatrix} \boldsymbol{\beta}_i \\ \mathbf{v}_i \end{pmatrix}$ is not proportional to any $\begin{pmatrix} \boldsymbol{\alpha}_j \\ \mathbf{v}_j \end{pmatrix}$.

***Proof*:** The statement on conditional distributions is equivalent to saying that $(\mathbf{Z}_1, \mathbf{Z}_3)$ and $(\mathbf{Z}_2, \mathbf{Z}_3)$ have identical distributions. Then the result follows by equating the c.f.'s and applying Lemma 1.5.2.▌

10.4 FACTOR ANALYSIS MODELS

In this section, we consider models with certain restrictions on the structure coefficients. We shall say that a p-dimensional random vector $\mathbf{X}$ has the structure of a factor analysis model if

$$\mathbf{X} = \mathbf{AF} + \boldsymbol{\varepsilon} \tag{10.4.1}$$

where no column of the $p \times r$ matrix $\mathbf{A}$ is proportional to any other or to any column of the identity matrix $\mathbf{I}_p$ of order p, and the components $f_1, \ldots, f_r$ of $\mathbf{F}$ and $\varepsilon_1, \ldots, \varepsilon_p$ of $\boldsymbol{\varepsilon}$ are independent of one another. The f_j are called "general factors" and the ε_k are called "specific factors."

THEOREM 10.4.1. The factor analysis model (10.4.1) has unique structure if the general factors are nonnormal. If the number of general factors is greater than one, then the condition of "nonnormality of at least $(r - 1)$ general factors" is also necessary for the uniqueness of the structure.

***Proof*:** This theorem follows from the results of Theorem 10.3.1.▌

Uniqueness of structure does not signify that the distribution of the general and the specific factors is also unique. In other words, there may exist two representations

$$\mathbf{X} = \mathbf{A}\mathbf{F}_1 + \boldsymbol{\varepsilon}_1 \quad \text{and} \quad \mathbf{X} = \mathbf{A}\mathbf{F}_2 + \boldsymbol{\varepsilon}_2 \tag{10.4.2}$$

in which the distributions of $\mathbf{F}_1$ and $\mathbf{F}_2$ or of $\boldsymbol{\varepsilon}_1$ and $\boldsymbol{\varepsilon}_2$ are not necessarily identical. We investigate conditions under which the model is unique.

THEOREM 10.4.2. Suppose that the components of $\mathbf{F}$ are nonnormal and that the rank of $\mathbf{A}^\#$ is r (the matrix $\mathbf{A}^\#$ has been defined in Lemma 1.5.4). Then, in some neighborhood of the origin, the c.f.'s of the random vectors $\mathbf{F}$ and $\boldsymbol{\varepsilon}$ are uniquely defined to within an exponential factor.

***Proof*:** Consider two alternative representations

$$\mathbf{X} = \mathbf{A}\mathbf{F}_1 + \boldsymbol{\varepsilon}_1 \text{ and } \mathbf{X} = \mathbf{A}\mathbf{F}_2 + \boldsymbol{\varepsilon}_2 \tag{10.4.3}$$

where $\mathbf{F}_i$ has components $f_{i1}, \ldots, f_{ir}$ and $\boldsymbol{\varepsilon}_i$ has components $\varepsilon_{i1}, \ldots, \varepsilon_{ip}$, $i = 1, 2$. Let ψ_i (respectively, B_i) be the difference of the logarithms of the c.f.'s of f_{i1} and f_{i2} (respectively, ε_{i1} and ε_{i2}). The c.f. of $\mathbf{X}$ can be expressed in two ways corresponding to the two representations (10.4.3). Taking logarithms we obtain the following functional equation for $\mathbf{t}$ in some neighborhood of the origin

$$\sum_1^r \psi_j(\boldsymbol{\alpha}_j'\mathbf{t}) = B_1(t_1) + \cdots + B_p(t_p) \quad \text{for} \quad |t_i| < \delta,\ 1 \leqq i \leqq p \tag{10.4.4}$$

where $\boldsymbol{\alpha}_j$ is the j-th column of $\mathbf{A}$ and $t_1, \ldots, t_p$ are components of $\mathbf{t}$. The condition that rank $\mathbf{A}^\# = r$ together with Lemma 1.5.4 guarantees the linearity of the functions ψ_j and B_k, $1 \leqq j \leqq r$, $1 \leqq k \leqq p$. Hence the c.f.'s of f_{j1} and f_{j2} differ only by a factor of the form $\exp(ic_j t)$; a similar assertion is true for the c.f.'s of ε_{j1} and ε_{j2}. ∎

Remark. If the c.f.'s of the components of $\mathbf{F}$ and $\boldsymbol{\varepsilon}$ are nonvanishing on the real line, then the last statement of Theorem 10.4.2 is valid for the entire real line.

THEOREM 10.4.3. Suppose that all the components of $\mathbf{F}$ are nonnormal and all the components of $\boldsymbol{\varepsilon}$ are normal, and, further, rank $\mathbf{A}$ = rank $\mathbf{A}^\# = r$. Then the distributions of the random vectors $\mathbf{F}$ and $\boldsymbol{\varepsilon}$ are uniquely defined to within a shift parameter.

***Proof*:** Consider, as in (10.4.3), two representations of $\mathbf{X}$ and let ϕ_{j1} and ϕ_{j2} be the c.f.'s of f_{j1} and f_{j2} respectively. Equating the two expressions of the c.f. of $\mathbf{X}$, we have

$$\prod_{j=1}^r \phi_{j1}(\boldsymbol{\alpha}_j'\mathbf{t}) = \prod_{j=1}^r \phi_{j2}(\boldsymbol{\alpha}_j'\mathbf{t}) \exp Q(\mathbf{t}) \tag{10.4.5}$$

where Q is a polynomial of the second degree in $t_1, \ldots, t_p$, since the components of $\boldsymbol{\varepsilon}_1$ and $\boldsymbol{\varepsilon}_2$ are normal. Denoting $\boldsymbol{\alpha}_j'\mathbf{t}$ by τ_j, (10.4.5) reduces to

$$\prod_{j=1}^{r} \phi_{j1}(\tau_j) = \prod_{j=1}^{r} \phi_{j2}(\tau_j) \exp Q(\tau_1, \ldots, \tau_r) \tag{10.4.6}$$

for all $\tau_1, \ldots, \tau_r$, since rank $\mathbf{A} = r$. But, in the neighborhood of the origin, ϕ_{j1} and ϕ_{j2} differ only by a factor of the form $\exp(ic_j t)$, whence it follows that $Q(\tau_1, \ldots, \tau_r)$ contains only linear terms. Now, setting all the τ_j other than τ_1 equal to zero, we obtain

$$\phi_{11}(\tau_1) = \phi_{12}(\tau_1) \exp(\lambda_1 \tau_1)$$

i.e., the distributions of f_{11} and f_{12} differ only by a shift. The same result obtains for all the pairs f_{j1}, f_{j2} and $\varepsilon_{j1}, \varepsilon_{j2}$. Theorem 10.4.3 stands proved.▌

The case when there is a general factor distributed normally requires separate consideration.

THEOREM 10.4.4. Consider the model

$$X_i = a_i f + \varepsilon_i; \qquad i = 1, \ldots, p \tag{10.4.7}$$

where f is normal. Then the structure and the model are unique if at least three of the coefficients $a_1, \ldots, a_p$ are nonzero.

***Proof*:** Suppose that there are two alternative models with coefficient-vectors $(a_1, \ldots, a_p)$ and $(b_1, \ldots, b_p)$, which may be different. Let the corresponding specific factors of the model with coefficients b_j be $\varepsilon_1^*, \ldots, \varepsilon_p^*$. Denoting the c.f.'s of ε_i and ε_i^* by ϕ_i and ϕ_i^*, we have

$$\begin{aligned} &\phi_1(t_1) \cdots \phi_p(t_p) \exp[-(1/2)\sigma_1^2(a_1 t_1 + \cdots + a_p t_p)^2] \\ &\quad = \phi_1^*(t_1) \cdots \phi_p^*(t_p) \exp[-(1/2)\sigma_2^2(b_1 t_1 + \cdots + b_p t_p)^2]. \end{aligned} \tag{10.4.8}$$

Since (10.4.8) is an identity in the t_j, we have

$$\sigma_1^2 a_i a_j = \sigma_2^2 b_i b_j \qquad \text{for all } i \neq j. \tag{10.4.9}$$

But under the condition that at least three of the a_j are nonzero, (10.4.9) is possible only when $\sigma_1 a_i = \pm \sigma_2 b_i$. Then it is easily seen that $\phi_i = \phi_i^*$, yielding the desired result. We may remark that in this theorem no assumptions have been made concerning the distribution of the ε_i.▌

The two results that follow are due to Reiersol [133].

THEOREM 10.4.5. Consider the model

$$X_1 = a_1 f + \varepsilon_1, \qquad X_2 = a_2 f + \varepsilon_2 \tag{10.4.10}$$

where f is a normal r.v. Then a necessary and sufficient condition for the uniqueness of structure (and also of the model) is that ε_1 and ε_2 do not have a normal component.

***Proof*:** This may be based on the analysis of the identity (10.4.8) written for the model (10.4.10).▌

THEOREM 10.4.6. Consider the model

$$X_1 = Y + U, \qquad X_2 = \beta Y + V$$

where β is a constant, (U, V) has a bivariate normal distribution, and Y is a nonnormal r.v. independent of (U, V). Then the structure (X_1, X_2) is unique. The model is unique if and only if the following conditions are satisfied: (i) Y does not have a normal component, and (ii) either $U = 0$ or $V = 0$.

***Proof*:** We need only write the c.f. of (X_1, X_2) in two ways, corresponding to the two alternative representations, and equate the expressions obtained.▌

10.5 REGRESSION PROBLEMS FOR STRUCTURED VARIABLES

Consider the factor analysis model of section 10.4 for a p-dimensional random vector $\mathbf{X}$:

$$\mathbf{X} = \mathbf{AF} + \boldsymbol{\varepsilon} \tag{10.5.1}$$

where $\mathbf{A}$ is a $p \times r$ matrix, every column of which contains at least two nonzero elements and no row of which consists wholly of zeros, $\mathbf{F}$ is an r-dimensional random vector with independent components F_i and $\boldsymbol{\varepsilon}$ is a p-dimensional vector with independent components ε_i, where $\mathbf{F}$ and $\boldsymbol{\varepsilon}$ are further independent themselves. Consider the $(p + 1)$-th r.v. defined by

$$Y = \mathbf{a}'\mathbf{F} + \varepsilon_0 \tag{10.5.2}$$

where $\mathbf{a}$ is a fixed vector, ε_0 is an r.v. independent of $\mathbf{F}$. We investigate conditions under which the regression of Y on $\mathbf{X}$ is linear. We suppose without loss of generality that

$$E(\mathbf{F}) = \mathbf{0}, \qquad E(\boldsymbol{\varepsilon}) = \mathbf{0}, \qquad E(\varepsilon_0) = 0. \tag{10.5.3}$$

Let the regression of Y on $\mathbf{X}$ be linear, i.e., $E(Y|\mathbf{X}) = \boldsymbol{\beta}'\mathbf{X}$. We have from (10.5.1) that

$$\boldsymbol{\beta}'\mathbf{X} = \boldsymbol{\beta}'\mathbf{AF} + \boldsymbol{\beta}'\boldsymbol{\varepsilon}. \tag{10.5.4}$$

THEOREM 10.5.1. Let $p \geqq 2$ and the regression of ε_0 on $\boldsymbol{\varepsilon}$ be linear: $E(\varepsilon_0 | \boldsymbol{\varepsilon}) = \boldsymbol{\delta}'\boldsymbol{\varepsilon}$. Then $E(Y|\mathbf{X}) = \boldsymbol{\beta}'\mathbf{X}$ if and only if the following conditions are satisfied.

(i) Let γ_j be the j-th component of the vector $\mathbf{a} - \mathbf{A}\boldsymbol{\beta}$; if $\gamma_j \neq 0$, then F_j is normal.

(ii) Let ξ_j be the j-th component of the vector $\boldsymbol{\beta} - \boldsymbol{\delta}$; if $\xi_j \neq 0$, then ε_j is a normal r.v.

(iii) If F_j and ε_j are normal r.v.'s and Var $F_j = \sigma_j^2$, Var $\varepsilon_j = \eta_j^2$, then the vector $\boldsymbol{\beta}$ and the numbers σ_j, η_j must be related according to the identity (10.5.7) below, in addition to the conditions on $\boldsymbol{\beta}$ imposed by the vanishing of some components of the vectors $\mathbf{a} - \mathbf{A}\boldsymbol{\beta}$ and $\boldsymbol{\beta} - \boldsymbol{\delta}$.

Proof: Applying Lemma 1.5.1, we obtain

$$E[(Y - \boldsymbol{\beta}'\mathbf{X}) \exp (i\mathbf{t}'\mathbf{X})] = 0.$$

We express Y and $\boldsymbol{\beta}'\mathbf{X}$ by means of the structural variables $\mathbf{F}$, ε_0, and $\boldsymbol{\varepsilon}$. We then have

$$E[(\mathbf{a}'\mathbf{F} + \varepsilon_0 - \boldsymbol{\beta}'\mathbf{A}\mathbf{F} - \boldsymbol{\beta}'\boldsymbol{\varepsilon}) \exp i\mathbf{t}'(\mathbf{A}\mathbf{F} + \boldsymbol{\varepsilon})] = 0$$

or

$$E\left[\left(\sum \gamma_j F_j + \sum \xi_j \varepsilon_j\right) e^{i(\Sigma \boldsymbol{\alpha}_j'\mathbf{t}F_j + \Sigma t_j \varepsilon_j)}\right] = 0 \tag{10.5.5}$$

where $\boldsymbol{\alpha}_j$ denotes the j-th column of $\mathbf{A}$. Denote the c.f. of F_i by f_i and that of ε_i by g_i, and set $\phi_i = f_i'/f_i$; $h_i = g_i'/g_i$ in the neighborhood of the origin. Equation (10.5.5) reduces to

$$\sum_1^r \gamma_j \phi_j(\boldsymbol{\alpha}_j'\mathbf{t}) + \sum_1^p \xi_j h_j(t_j) = 0 \quad \text{for} \quad |\mathbf{t}| < \delta. \tag{10.5.6}$$

If $\gamma_j \neq 0$, then an application of Lemma 1.5.3 shows that ϕ_j is a polynomial. Then, using familiar arguments, we see that the degree of ϕ_j is at most one and that the corresponding r.v. is normal (possibly degenerate). Similarly, if $\xi_j \neq 0$, then ε_j is normal (possibly degenerate).

If in the case when F_j and ε_j are normal, we set Var $F_j = \sigma_j^2$ and Var $\varepsilon_j = \eta_j^2$ and denote by $\sum^*$ summation over all j with $\gamma_j \neq 0$, $\xi_j \neq 0$; then we obtain the identity in $\mathbf{t}$

$$\sum{}^* \gamma_j \sigma_j^2(\boldsymbol{\alpha}_j'\mathbf{t}) + \sum{}^* \xi_j \eta_j^2 t_j = 0. \tag{10.5.7}$$

We note that if $\gamma_j = 0$, then F_j can have an arbitrary distribution, and similarly for ε_j if $\xi_j = 0$.

We have thus proved the necessity part of the theorem. The sufficiency is established by reversing the argument, noting that (10.5.6) holds for all $\mathbf{t}$ if F_j and ε_j are normal.∎

THEOREM 10.5.2. Let $\mathbf{X} = \mathbf{A}\mathbf{F} + \boldsymbol{\varepsilon}$ be a factor analysis model, where rank $\mathbf{A}$ equals the number of columns of $\mathbf{A}$. Consider the r.v. $Y = \mathbf{a}'\mathbf{F} + \varepsilon_0$, and suppose that all the r.v.'s are nondegenerate and have zero means, the r.v. ε_0 is independent of $\mathbf{F}$ and $E(\varepsilon_0 \mid \boldsymbol{\varepsilon}) = \boldsymbol{\delta}'\boldsymbol{\varepsilon}$. In order that the regression of Y on $\mathbf{X}$ be linear for all $\mathbf{a}$, it is necessary and sufficient that all the components of $\mathbf{F}$ and $\boldsymbol{\varepsilon}$ be normal.

***Proof*:** Let $\mathbf{A}$ be a $p \times r$ matrix, rank $\mathbf{A} = r$, $E(Y \mid \mathbf{X}) = \boldsymbol{\beta}'\mathbf{X}$ where $\boldsymbol{\beta}$ depends on $\mathbf{a}$. Equation (10.5.6) can be written as

$$\sum_{j=1}^{r} (a_j - \boldsymbol{\beta}'\boldsymbol{\alpha}_j)\phi_j(\boldsymbol{\alpha}_j'\mathbf{t}) + \sum_{k=1}^{p} (\delta_k - \beta_k)h_k(t_k) = 0 \tag{10.5.8}$$

where $\boldsymbol{\alpha}_1, \ldots, \boldsymbol{\alpha}_r$ are the columns of $\mathbf{A}$; a_k, δ_k, β_k are the k-th components of the vectors $\mathbf{a}$, $\boldsymbol{\delta}$, $\boldsymbol{\beta}$ respectively.

Suppose that for all $\mathbf{a}$

$$a_j = \boldsymbol{\beta}'\boldsymbol{\alpha}_j \qquad \text{for } j = 1, \ldots, s < r \tag{10.5.9}$$

$$\delta_k = \beta_k \qquad \text{for } k = 1, \ldots, m < p. \tag{10.5.10}$$

We may assume that $s \neq r$ and $m \neq p$.

Then, it follows easily from Lemma 1.5.3 and standard arguments that the r.v.'s $F_{s+1}, \ldots, F_r$ and $\varepsilon_{m+1}, \ldots, \varepsilon_p$ are normal. We set Var $F_i = \sigma_i^2$ and Var $\varepsilon_j = 1$ (without loss of generality) in case F_i and ε_j have normal distributions. Noting that in such a case $\phi_i(u) = -\sigma_i^2 u$ and $h_j(u) = -u$, we can reduce (10.5.8), with help of conditions (10.5.9) and (10.5.10), to the following.

$$\sum_{j=s+1}^{r} \sigma_j^2(a_j - \boldsymbol{\beta}'\boldsymbol{\alpha}_j)(\boldsymbol{\alpha}_j'\mathbf{t}) + \sum_{k=m+1}^{p} (\delta_k - \beta_k)t_k = 0 \tag{10.5.11}$$

We denote by $\mathbf{C}$ the matrix with the columns $\boldsymbol{\alpha}_{s+1}, \ldots, \boldsymbol{\alpha}_r$; by $\mathbf{D}$ the diagonal matrix with diagonal elements $\sigma_{s+1}^2, \ldots, \sigma_r^2$; and by $\boldsymbol{\xi}$ the vector with components $a_{s+1}, \ldots, a_r$. In matrix notation, (10.5.11) can be written as

$$\boldsymbol{\beta}'(\mathbf{CDC}' + \mathbf{I})\mathbf{t} = \boldsymbol{\xi}'\mathbf{DC}'\mathbf{t} + \boldsymbol{\delta}'\mathbf{t}$$

whence

$$(\mathbf{CDC}' + \mathbf{I})\boldsymbol{\beta} = \mathbf{CD}\boldsymbol{\xi} + \boldsymbol{\delta}. \tag{10.5.12}$$

Since the rank of $(\mathbf{CDC}' + \mathbf{I})$ is p, for given $\boldsymbol{\xi}$, $\boldsymbol{\beta}$ is uniquely defined by (10.5.12). But, in addition, $\boldsymbol{\beta}$ must satisfy (10.5.9) for arbitrary a_j, $j = 1, \ldots, s$. This is obviously impossible and hence, necessarily, $s = 0$, Thus, in (10.5.8), the coefficient of $\phi_j(\boldsymbol{\alpha}_j'\mathbf{t})$ is nonzero for at least one value of $\mathbf{a}$. Hence it follows that all the components of $\mathbf{F}$ are normal.

If now we remove the condition (10.5.9), then (10.5.12) can written as

$$(\mathbf{ADA}' + \mathbf{I})\boldsymbol{\beta} = \mathbf{AD}\boldsymbol{\xi} + \boldsymbol{\delta} \tag{10.5.13}$$

where the diagonal matrix $\mathbf{D}$ is formed by the variances of all the r components of $\mathbf{F}$ and $\boldsymbol{\xi} = \mathbf{a}$. From (10.5.13),

$$\boldsymbol{\beta} = \boldsymbol{\delta} + \mathbf{A}\boldsymbol{\gamma} \tag{10.5.14}$$

for a suitable $\boldsymbol{\gamma}$, so that (10.5.13) takes the form

$$(\mathbf{ADA}' + \mathbf{I})\mathbf{A}\boldsymbol{\gamma} = \mathbf{AD}(\boldsymbol{\xi} - \mathbf{A}'\boldsymbol{\delta}). \tag{10.5.15}$$

As $\boldsymbol{\xi}$ ranges over p-dimensional space, so does $\boldsymbol{\gamma}$. For every $\boldsymbol{\gamma}$, we obtain $\boldsymbol{\beta}$ in accordance with (10.5.14). If now $\beta_i = \delta_i$ for all $\boldsymbol{\gamma}$, then the i-th row of $\mathbf{A}$ is necessarily the null vector. Hence, under the conditions of Theorem 10.5.2, relation (10.5.10) cannot be satisfied for all $\mathbf{a}$. Hence the coefficient of $h_j(t_j)$ in (10.5.8) is nonzero for any j, for at least one value of $\mathbf{a}$. Hence all the ε_j are normal. ▮

THEOREM 10.5.3. Let $r = 1$, so that $\mathbf{F}$ has only one component F_1 and let $F_1, \varepsilon_1, \ldots, \varepsilon_p$, and ε_0 be independent and nondegenerate (ε_0 may, however, be degenerate) r.v.'s. In order for the regression of Y on $\mathbf{X}$ to be linear (and not constant)

$$E(Y \mid \mathbf{X}) = \alpha + \boldsymbol{\beta}'\mathbf{X} = \alpha + \beta_1 X_1 + \cdots + \beta_p X_p$$

it is necessary and sufficient that F_1 and those ε_i corresponding to nonzero β_i be normal.

***Proof*:** The equation (10.5.6) reduces in this case to

$$(a_1 - \gamma_1)\phi_1(\boldsymbol{\alpha}_1'\mathbf{t}) - \beta_1 h_1(t_1) - \cdots - \beta_p h_p(t_p) = 0 \qquad (10.5.16)$$

where $\beta_1, \ldots, \beta_p$ are the regression coefficients, not all zero. Then, $a_1 - \gamma_1 \neq 0$, for, otherwise, it would follow that some of the ε_i are degenerate, contrary to assumption. Then, by Lemma 1.5.3, ϕ_1 is linear and hence F_1 is normal. But then h_i is a linear function if $\beta_i \neq 0$, whence it follows that ε_i is also normal. ▮

In Theorems 10.5.1–10.5.3, we have dealt with various generalizations of the following problem, posed by Ragnar Frisch before the Oxford Conference of the Econometric Society in 1936.

"Let X_1 and X_2 be r.v.'s of the form $X_1 = au + v$, $X_2 = bu + w$, where u, v, w are independent r.v.'s. Under what conditions will the regression of X_1 on X_2 be linear for all a and b?"

In such a form, the problem is a special case of Theorem 10.5.1 corresponding to $p = 1$. Unfortunately, the proof of that theorem holds only for $p \geqq 2$ and the investigations of the problem for $p = 1$ requires special methods. Its solution is also of interest because it shows that it is not necessary that the r.v.'s u, v, w be normal in order for the regression of X_1 on X_2 to be linear.

We determine below conditions under which the regression of X_1 on X_2 will be linear, not for all a and b, but only for some values of these constants. We shall require the following lemma.

LEMMA 10.5.1. Let f be the c.f. of some nondegenerate r.v., satisfying the condition

$$f(t) = [f(\beta_1 t)]^{\gamma_1} = [f(\beta_2 t)]^{\gamma_2}, \qquad t \in R^1 \qquad (10.5.17)$$

where $0 < |\beta_i| < 1$ and $\gamma_i > 0$ for $i = 1, 2$. Let $B_i = -\log |\beta_i|$. Then we have:

(i) there exists a real λ such that $\gamma_1 |\beta_1|^\lambda = \gamma_2 |\beta_2|^\lambda$;

(ii) if $\lambda = 2$, then f is the c.f. of a normal law;

(iii) let $0 < \lambda < 2$; if B_1/B_2 is irrational, then f is the c.f. of a stable law (*vide* formula (1.1.11)) if β_1 and β_2 are positive and $\lambda \neq 1$, and a symmetric stable law (*vide* formula (1.1.14)) if $\lambda = 1$.

(iv) let $0 < \lambda < 2$; if B_1/B_2 is rational, then f is the c.f. of a semistable law given by (1.1.15).

***Proof*:** It has been shown in Theorem 5.3.1 that, if a c.f. f satisfies the equation

$$f(t) = [f(\beta t)]^\gamma, \qquad 0 < |\beta| < 1, \qquad \gamma > 0 \tag{10.5.18}$$

then f is infinitely divisible and if λ be the unique real solution of the equation $\gamma |\beta|^\lambda = 1$, then the Levy representation $L(\mu, \sigma^2, M, N)$ for $\log f$ has the following properties:

$$\text{(i) } \sigma \neq 0,\ M = 0,\ N = 0 \text{ if } \lambda = 2; \tag{10.5.19}$$

$$\text{(ii) } \sigma = 0,\ \left.\begin{aligned} M(u) &= \xi(\log |u|)/|u|^\lambda \\ N(u) &= -\eta(\log u)/u^\lambda \end{aligned}\right\} \text{ if } 0 < \lambda < 2,\ \beta > 0, \tag{10.5.20}$$

where ξ and η are nonnegative right-continuous functions defined on R^1 and periodic with period $B = -\log \beta$;

$$\text{(iii) } \sigma = 0,\ \left.\begin{aligned} M(u) &= \xi(\log |u|)/|u|^\lambda \\ N(u) &= -\xi(\log u + B)/u^\lambda \end{aligned}\right\} \text{ if } 0 < \lambda < 2,\ \beta < 0, \tag{10.5.21}$$

where ξ is a nonnegative right-continuous function on R^1 with period $2B = -2 \log |\beta|$.

To prove assertion (i) of the lemma, we note that a d.f. whose c.f. satisfies (10.5.18) has absolute moments of all orders less than λ (and does not have moment of order λ), where λ is defined by $\gamma |\beta|^\lambda = 1$ (cf. Lemmas 1.1.5–1.1.7). Since f satisfies (10.5.17), which is the same as (10.5.18) written out for two pairs of values for (β, γ), the just-noted property of d.f.'s with c.f. f satisfying (10.5.18) shows that we must have

$$\gamma_1 |\beta_1|^\lambda = \gamma_2 |\beta_2|^\lambda = 1.$$

Assertion (ii) of the lemma follows from the Levy representation (10.5.19).

To prove assertion (iii), we consider the representation (10.5.20) and note that the function ξ is periodic with two periods B_1 and B_2. Since B_1/B_2 is irrational, the set of points

$$\{mB_1 + nB_2 : m \text{ and } n \text{ integers}\}$$

is everywhere dense on the real line, and $\xi(mB_1 + nB_2) = \xi(0)$. The right continuity of ξ implies that ξ is constant. Similarly η is constant also. Then we find from (1.1.12) that f is the c.f. of a stable law and its logarithm is given by formula (1.1.11).

$$\log f(t) = i\alpha t - c|t|^{\lambda}\left\{1 + i\delta \frac{t}{|t|} \tan\left(\frac{1}{2}\pi\lambda\right)\right\} \qquad \text{if } \lambda \neq 1 \quad (10.5.22)$$

$$\log f(t) = i\alpha t - c|t|\left\{1 + i\frac{2\delta}{\pi}\frac{t}{|t|}\log|t|\right\} \qquad \text{if } \lambda = 1. \quad (10.5.23)$$

Substituting from (10.5.23) in (10.5.17) for one pair of values of (β, γ), say (β_1, γ_1), we see that $\delta = 0$ if $\lambda = 1$; in this case the stable law is symmetric.

If, however, even one of β_1, β_2 is negative, then the Levy representation for $\log f$ is given by (10.5.21); then using arguments similar to those above, we can show that ξ and η are constant and further are equal. But then, by (1.1.14), f is the c.f. of a symmetric stable law.

To prove assertion (iv) of the lemma, we note that $\log f$ has a Levy representation of either the form (10.5.20) or the form (10.5.21), depending on whether both β_1 and β_2 are positive or at least one of them is negative. No further simplification of the functions ξ and η is possible when B_1/B_2 is rational: we can only claim that they are periodic with some period ρ which depends on B_1 and B_2. ∎

We now formulate and prove a characterization theorem which follows from Lemma 10.5.1.

THEOREM 10.5.4. Let X, Y, and Z be nondegenerate r.v.'s with finite expectations which we take to be all zero (without loss of generality). Let further X be independent of (Y, Z) and $E(Y|Z) = 0$. Then the following hold.

(i) For the regression of $X + Y$ on $aX + Z$ to be linear for at least two values of a: $a = a_1$ and $a = a_2$, such that $|a_1/a_2| \neq 1$, it is necessary and sufficient that X and Z have semistable distributions.

(ii) Let a_1, a_2, and a_3 be three values of a such that $|a_1| > |a_2| > |a_3| > 0$; let $\beta_1 = a_2/a_1$ and $\beta_2 = a_3/a_1$, $B_1 = -\log|\beta_1|$ and $B_2 = -\log|\beta_2|$. For the regression of $X + Y$ on $aX + Z$ to be linear for the three values of a: $a = a_1, a_2, a_3$ indicated above, it is necessary and sufficient that

(1) X and Z have stable distributions if $\beta_1 > 0$, $\beta_2 > 0$, and B_1/B_2 is irrational;

(2) X and Z have a symmetric stable distribution if $\beta_1 \leqq 0$, $\beta_2 \leqq 0$, and B_1/B_2 is irrational; and

(3) X and Z have semistable distributions if B_1/B_2 is rational.

In all cases, the exponent λ corresponding to the distributions satisfies the conditions: $1 < \lambda \leq 2$.

Proof: We prove the necessity of the conditions; the sufficiency part is easily verified.

Let

$$E(X + Y | aX + Z) = c(aX + Z).$$

This implies the existence of a constant c such that

$$E[(1-ac)X+Y-cZ \mid aX+Z]=0.$$

Using standard arguments already repeatedly used, we obtain

$$(1-ac)\log f_1(at)=ac\log f_2(t) \qquad \text{for } |t|<\delta \tag{10.5.24}$$

where f_1 and f_2 are the c.f.'s of X and Z respectively. We note that $c \neq 0$ since, otherwise, X would be degenerate. We have from (10.5.24) that

$$f_2(t)=[f_1(at)]^{\gamma^*}, \qquad |t|<\delta \tag{10.5.25}$$

for a suitable value of γ^*. Setting $a=a_1$ and $a=a_2$ in (10.5.25), we obtain the functional equation for f_1

$$f_1(t)=[f_1(\beta t)]^{\gamma}, \qquad 0<|\beta|<1,\ |t|<\delta. \tag{10.5.26}$$

Hence we see that $\gamma>0$ in the above and that (10.5.26) is satisfied for all real t. Hence f_1 is the c.f. of a semistable law and similarly for f_2. Thus assertion (i) is proved.

To prove (ii), we substitute a_1, a_2, a_3 for a in (10.5.25); we obtain the equation

$$f_1(t)=[f_1(\beta_1 t)]^{\gamma_1}=[f_1(\beta_2 t)]^{\gamma_2} \tag{10.5.27}$$

from which all the assertions of (ii) follow in view of Lemma 10.5.1.

Since the first moments of the r.v.'s under consideration exist, the exponent λ of the distributions satisfies the inequality $1<\lambda\leqq 2$. For $\lambda>2$, the distribution is degenerate.▌

Consider now two independent r.v.'s X and Y. We determine conditions under which the regression of $X+Y$ on $aX+bY$ is linear for two or three pairs of values of (a, b).

THEOREM 10.5.5. Let X and Y be independent r.v.'s with EX and EY finite, which we take to be zero without loss of generality. Then the following hold.

(i) For the regression of $X+Y$ on $aX+bY$ to be linear for two pairs of values of (a, b): (a_1, b_1) and (a_2, b_2) with the condition $|a_1b_2/a_2b_1|\neq 1$, it is necessary and sufficient that X and Y have semistable distributions.

(ii) Let (a_1, b_1), (a_2, b_2), and (a_3, b_3) be three pairs of values of (a, b) such that $\prod_1^3 a_i b_i \neq 0$, $|a_1/b_1|<|a_2/b_2|<|a_3/b_3|$; $\beta_1=b_2a_1/b_1a_2$ and $\beta_2=b_3a_1/b_1a_3$ and $B_j=-\log|\beta_j|$. For the regression of $X+Y$ on $aX+bY$ to be linear for the three pairs of values of (a, b) above, it is necessary and sufficient that

(1) X and Y have stable distributions if $\beta_1>0$, $\beta_2>0$, and B_1/B_2 is irrational;

(2) X and Y have symmetric stable distributions if $\beta_1 \leqq 0$, $\beta_2 \leqq 0$, and B_1/B_2 is irrational; and
(3) X and Y have semistable distributions if B_1/B_2 is rational.

This theorem is proved similarly to Theorem 10.5.4. ∎

The results of section 10.5 constitute a generalization of the previous work work by Rao [125], Ferguson [181] and Laha [182].

CHAPTER 11

Polynomial Statistics of a Normal Sample

11.1 NORMAL RANDOM SAMPLES

Let $(X_1, \ldots, X_n) = \mathbf{X}$ be a nondegenerate normal vector of a general type. It is well known that by means of a nonsingular linear transformation it can be reduced to a vector with independent standard normal components. Thus many questions concerning statistics of a normal vector can be reduced to a study of a normal random sample which is considerably more simply distributed. It is well-known that a standard normal random sample vector is transformed into another such by any orthogonal transformation. Less well-known are nonlinear transformations of such a vector into another such, among them being fairly simple algebraic and rational transformations. Following the note [89], we shall indicate some such transformations here.

Let $(X_1, \ldots, X_n)$ be a standard normal random sample vector. We split up the n variables into two groups, say $X_1, \ldots, X_{n_1}$ and $X_{n_1+1}, \ldots, X_n$. Let $\mathbf{F}$ be an orthogonal matrix made up of rational fractions of the form

$$r_{ij} = r_{ij}(X_{n_1+1}, \ldots, X_n); \qquad i, j = 1, \ldots, n_1. \tag{11.1.1}$$

Such a matrix can be constructed, for instance, with the help of the Cayley parametric representation: if I is the identity matrix and $\mathbf{S}$ is a skew-symmetric matrix ($\mathbf{S}' = -\mathbf{S}$) whose eigenvalues λ_i are all less than one in absolute value, then the matrices $\mathbf{I} - \mathbf{S}$ and $\mathbf{I} + \mathbf{S}$ are both nonsingular, and the matrix $\mathbf{F} = (\mathbf{I} - \mathbf{S})(\mathbf{I} + \mathbf{S})^{-1}$ is an orthogonal one. In fact, $\mathbf{F}' = (\mathbf{I} - \mathbf{S})^{-1}(\mathbf{I} + \mathbf{S})$ and $\mathbf{FF}' = \mathbf{I}$ in view of the commutativity of the factors. If $\mathbf{S}$ is taken to be a skew-symmetric matrix of the type indicated above, with rational fractions of $X_{n_1+1}, \ldots, X_n$ as elements, then $\mathbf{F}$ will be a matrix of the kind we need. Now set up the matrix

$$\mathbf{F}_1 = \left(\begin{array}{c|c} \mathbf{F} & \mathbf{0} \\ \hline \mathbf{0} & \mathbf{I} \end{array}\right)$$

where the matrices occupying the principal diagonal are respectively the $n_1 \times n_1$ orthogonal matrix F and the $(n - n_1) \times (n - n_1)$ identity matrix. Then $\mathbf{F}_1$ will be an orthogonal $n \times n$ matrix with elements of the form (11.1.1). If $\mathbf{X}$ is a standard normal random sample vector of size n, then $\mathbf{Y} = \mathbf{F}_1\mathbf{X}$ will also be a vector of the same kind; in fact, for any fixed $X_{n_1+1}, \ldots, X_n$, we obtain one and the same normal random sample* $(Y_1, \ldots, Y_{n_1})$ with every $Y_j \sim N(0, 1)$; considering the distribution of $X_1, \ldots, X_{n_1}$ under such conditional distributions, we see that our assertion about $\mathbf{Y}$ is true.

Another type of rational transformation which preserves the normality of the sample has been indicated by Zinger. Consider

$$Y_1 + iY_2 = (X_1 + iX_2)^m/(X_1^2 + X_2^2)^{(m-1)/2}; \qquad m = 2, 3, \ldots. \qquad (11.1.2)$$

Separating the real and imaginary parts, we obtain a transformation of the desired kind. For instance, for $m = 2$, we have

$$Y_1 = (X_1^2 - X_2^2)/(X_1^2 + X_2^2)^{1/2}; \qquad Y_2 = 2X_1X_2/(X_1^2 + X_2^2)^{1/2}.$$

For odd m, (11.1.2) gives rational transformations. It is natural to attempt to generalize (11.1.2) to normal samples of sizes 4 and 8, using quaternions and Cayley numbers, but this fails because the analogue of (11.1.2) for these cases does not preserve the normality of the measure.

Nonlinear transformations preserving normality can be set up in the form of entire functions; for instance, the matrix F_1 described above could have entire functions as elements. Another form of transformation is given by the formula

$$\begin{aligned} Y_1 &= \quad X_1 \cos[\phi(X_1^2 + X_2^2)] + X_2 \sin[\phi(X_1^2 + X_2^2)] \\ Y_2 &= -X_1 \sin[\phi(X_1^2 + X_2^2)] + X_2 \cos[\phi(X_1^2 + X_2^2)] \end{aligned}$$

where ϕ is some entire function. The geometric interpretation is that any circle with center at the origin is rotated through an angle which depends on the radius of the circle. Similar formulas can be obtained for samples of any size.

Compositions of the transformations indicated above again lead to transformations preserving the normality of the sample. However, the general problem of describing all rational, all algebraic, or all entire transformations which preserve normality is apparently very difficult. Only some isolated facts are known in this area (cf. the note [89]). Among these we note a result due to Eidlin, communicated by him to the authors.†

* That is, the corresponding measure in R^n is preserved.

† Announced by Eidlin in December, 1968 at the Statistical Seminar in Leningrad University.

THEOREM 11.1.1. Let $\sigma_0 > 0$ be a given number. Consider a random sample $(X_1, \ldots, X_n)$ with every $X_j \sim N(0, \sigma_0^2)$. Every algebraic transformation preserving the normality of such a sample also preserves spheres $x_1^2 + \cdots + x_n^2 = r^2$ and also Lebesgue measure.

In all the examples above of nonlinear transformations preserving normality, we assumed that $EX_i = 0$. If $EX_i = \mu$, where μ is a given number, then all the nonlinear transformations constructed above have to be modified in an obvious manner.

However, if nonlinear transformations are sought which simultaneously preserve the normality of two samples with different means, then according to a conjecture due to Basu (cf. [20]), such do not exist even for samples of size one. We formulate a theorem in this direction, proved by Ghosh [20].

THEOREM 11.1.2. Let $X_i \sim N(\mu_i, \sigma_i^2)$, $i = 1, 2$, be two normal r.v.'s, where $\mu_1 \neq \mu_2$, and T a single-valued transformation which preserves normality and is such that $E\mu_1 T(X_1) \neq E\mu_2 T(X_2)$. Then T is a linear transformation.

***Proof*:** We refer the reader to [20].∎

In modern statistics, complex normal variables are often met with (cf. [17, 72, 12]); they have the form $Z = X + iY$ where X and Y are independent standard normal variables. We may consider a random sample $(Z_1, \ldots, Z_n)$ and transformations preserving the corresponding measure, or, as we shall say, preserving the normality of the sample. Here the situation is different from what obtains in the usual random normal samples. In the note [89] by Linnik and Eidlin the following result was obtained.

THEOREM 11.1.3. There does not exist a nonlinear meromorphic transformation from C^n to C^1 which maps a complex normal vector into a complex normal scalar.

We shall not present the proof of the theorem here.

In this chapter, we shall study polynomial statistics of a random normal sample; they are involved in the construction of almost all tests in multivariate analysis. It is clear that such statistics have moments of all orders; if $P(\mathbf{X})$ be the polynomial statistic, we use the notation

$$\alpha_r = E[P(\mathbf{X})]^r; \qquad r = 1, 2, \ldots. \tag{11.1.3}$$

The following theorem holds.

THEOREM 11.1.4. There exists an integer constant $R_0 = R_0(m)$ such that for polynomials of given degree m, $R_0(m)$ relations of the form (11.1.3) themselves determine all the remaining relations for $r > R_0$.

***Proof*:** We shall denote by $P(\mathbf{X}) = P_{\mathbf{a}}(\mathbf{X})$ a polynomial statistic of degree m with coefficient $a_{k_1 \dots k_n}$ for the monomial $X_1^{k_1} \cdots X_n^{k_n}$, where $k_1 + \cdots + k_n \leqq m$. Let, for a polynomial $P_{\mathbf{a}'}(\mathbf{X})$ of degree m,

$$\alpha_r = E[P_{\mathbf{a}'}(\mathbf{X})]^r, \qquad r = 1, 2, \dots .$$

Then the relations (11.1.3) have the form

$$E[P_{\mathbf{a}}(\mathbf{X})]^r - E[P_{\mathbf{a}'}(\mathbf{X})]^r = 0; \qquad r = 1, 2, \dots . \tag{11.1.4}$$

These relations have polynomials in **a** and **a′** on the left, and by the Hilbert basis theorem (Lemma 1.3.3) they are equivalent to a finite number of those relations, say for $r \leqq R_0$. ▮

The question arises as to how far the countable number of relations (11.1.3), (equivalently, as we have seen, a finite number of those, for $r \leqq R_0$) determine the distribution of the polynomial $P(\mathbf{X})$. We are thus interested in the question of to what extent the distribution of $P(\mathbf{X})$ is determined by its moments.

For polynomials of the first and second degree, the distribution is determined by their moments α_r. In fact, by Holder's inequality, we have for polynomials of the second degree (which may degenerate into linear forms), for $r = 1, 2, \dots$

$$\begin{aligned} |P(\mathbf{X})|^2 &= \left|\sum a_{jk} X_j X_k + \sum b_j X_j\right|^2 \\ &\leqq M^2 C^2 \left(\sum |X_j X_k|^2 + \sum |X_j|^2\right) \end{aligned}$$

where $M = \max_{j,k} (|a_{jk}| + |b_j|)$ and $C > 0$ are constants.

Further, for any j and k ($j, k = 1, 2, \dots, n$),

$$E(|X_j|^r |X_k|^r) \leqq E X_j^{2r}$$

so that

$$\beta_r = E|P(\mathbf{X})|^r \leqq 2M^r C^r E X_1^{2r} = 2M^r C^r (2r)!!$$

Hence $\sum \beta_r^{-1/r}$ diverges, so that, by the Hamburger moment theorem (Lemma 1.1.10), the distribution of polynomials of the first and second degrees in a normal sample is determined by their moments. However, even for polynomials of the third degree, the above estimates for the rate of growth of the moments do not hold.

For polynomial statistics of a normal sample, $P(\mathbf{X})$, of a general type, it is not yet known whether their distribution is determined by their moments. For homogeneous polynomials ("forms") of any degree, this question is solved without any difficulty.

THEOREM 11.1.5. Let $P(\mathbf{X})$ be a homogeneous polynomial ("form") of degree $m \geqq 1$. The distribution of the polynomial $P(\mathbf{X})$, where $\mathbf{X}$ is a normal sample vector, is determined if its first $R_0 = R_0(m)$ moments

$$\alpha_r = E[P(\mathbf{X})]^r, \qquad r = 1, \ldots, R_0 \tag{11.1.5}$$

are given.

***Proof*:** By Theorem 11.1.4, the first R_0 relations of the form (11.1.5) are equivalent to all the moments of P being given. We turn to spherical coordinates, setting

$$\begin{aligned} x_1 &= \rho \sin \phi_1 \\ x_2 &= \rho \cos \phi_1 \sin \phi_2 \\ &\cdots \\ x_i &= \rho \cos \phi_1 \cos \phi_2 \cdots \cos \phi_{i-1} \sin \phi_i \\ &\cdots \\ x_n &= \rho \cos \phi_1 \cos \phi_2 \cdots \cos \phi_{n-1} \end{aligned}$$

where $\phi_i \in [-(1/2)\pi, (1/2)\pi]$, $\phi_{n-1} \in [0, 2\pi]$, and

$$J = \frac{\partial(x_1, \ldots, x_n)}{\partial(\rho, \phi_1, \ldots, \phi_{n-1})} = (-1)^{n-1}\rho^{n-1} \cos^{n-2}\phi_1 \ldots \cos \phi_{n-2}.$$

We have now

$$E[P(\mathbf{X})]^r = (2\pi)^{-n/2} \int_{(\rho, \boldsymbol{\phi})} [P(\mathbf{x})]^r \exp(-(1/2)\rho^2)|J| \, d\rho \, d\boldsymbol{\phi}.$$

Since P is a homogeneous polynomial of degree m, we have

$$P(\mathbf{x}) = \rho^m \pi(\boldsymbol{\phi}) \tag{11.1.6}$$

where $\pi(\boldsymbol{\phi}) = \pi(\phi_1, \ldots, \phi_{n-1})$ is a trigonometric polynomial. Hence we have

$$E[P(\mathbf{X})]^r = \frac{(-1)^{n-1}}{(2\pi)^{n/2}} \int_0^\infty \rho^{mr+n-1} e^{-1/2\rho^2} \, d\rho$$

$$\times \int_{(\boldsymbol{\phi})} (\pi(\boldsymbol{\phi}))^r \cos^{n-2}\phi_1 \ldots \cos \phi_{n-2} \, d\boldsymbol{\phi} \tag{11.1.7}$$

where $(\boldsymbol{\phi})$ is the torus defined by the inequalities for the ϕ_j. The first integral on the right is a function of $(mr + n - 1)$, and is easily expressed as a gamma integral. Let us denote this integral together with the factor in front of it by $\sigma(mr + n - 1)$. Taking (11.1.5) into account, we set

$$\alpha_r / \sigma(mr + n - 1) = \alpha_r'.$$

Then we obtain from (11.1.5) and (11.1.6)

$$\int_{(\boldsymbol{\phi})} (\pi(\boldsymbol{\phi}))^r \cos^{n-2}\phi_1 \cdots \cos \phi_{n-2}\, d\boldsymbol{\phi} = \alpha_r' \qquad (11.1.8)$$

$$\text{for } r = 1, 2, \ldots .$$

Here $\cos \phi_i \geqq 0$ for all $i \leqq n-2$ on the torus $(\boldsymbol{\phi})$. Equation (11.1.8) represents the moment problem on the unit sphere for trigonometric polynomial $\pi(\boldsymbol{\phi})$. The solution of such problems is unique (cf. [167]). Thus the distribution of $\pi(\boldsymbol{\phi})$ on $(\boldsymbol{\phi})$ is determined, and that of ρ^m (cf. formula (11.1.6)) does not depend on it and corresponds to the distribution of $(X_1^2 + \cdots + X_n^2)^{m/2}$, where $X_i \sim N(0, 1)$. This proves the theorem.∎

We now turn to the problem of distribution of polynomial statistics of normal samples.

11.2 THE DISTRIBUTION OF POLYNOMIAL STATISTICS OF NORMAL SAMPLES ("PROPER STATISTICS")

Very many tests of multivariate analysis are based on polynomial and rational statistics of normal samples. A systematic study of such statistics is in itself desirable. Besides, the problem of unlinking of polynomial and rational statistics, presented below, is connected with such studies and is of interest in connection with characterizations of families of the form $H(X)$ where $X \sim N(a, \sigma^2)$. In this section, we shall consider the distribution of polynomial statistics.

In what follows, **X** will be a random sample vector.

As is well known, the c.f.'s of linear and quadratic statistics based on a normal sample are solutions of differential equations of a very special form (cf. for instance [32]). For example, if f is the c.f. of a real quadratic form with eigenvalues $\lambda_1, \ldots, \lambda_n$, then it satisfies the equation

$$2\prod_1^n (1 - 2i\lambda_j t) f'(t) + \left(\prod_1^n (1 - 2i\lambda_j t)\right)' f(t) = 0.$$

This property turns out to be very useful for investigating various problems connected with the distribution of quadratic forms, and in particular for the problem of classifying them.

In connection with the problem of classification of independent polynomials based on a normal sample, there arises the question of the possibility of generalizing properties similar to the above to the distribution of arbitrary polynomials and applying them to the classification problem.

In this direction, certain results (cf. [32, 33]) were obtained in 1961–62 by Zinger, reducing basically to this: the c.f. of the distribution of a polynomial statistic P of a normal sample is the solution of a linear homogeneous differential equation with polynomial coefficients and the moment EP^z, as a function of z, satisfies a finite-difference equation. Our presentation below will follow Zinger's dissertation [37]. His results on polynomials of degree greater than two will be given; for quadratic and linear statistics, similar properties are immediately verified.

In addition to normal samples, which are of the most interest to us, we shall also touch upon some samples similar to them, from populations having so-called contiguity to the normal distributions.

We shall study polynomial statistics with real coefficients. Some of the results obtained here can be immediately extended to polynomials with complex coefficients, but we shall not stop to prove them. We shall first investigate "proper" polynomials, which permit us then to pass on to the general case (cf. section 11.3).

The properties considered below of distributions of polynomial statistics of a normal sample are closely connected with the invariance properties of such samples under groups of orthogonal transformations and with some aspects of the theory of polynomial ideals. In particular, we shall make use of the wellknown Hilbert basis theorem (cf. Lemma 1.3.3). The following simple algebraic lemma will be of use to us.

LEMMA 11.2.1. If, for some integer $n > 1$, $P_1(\mathbf{x}), \ldots, P_n(\mathbf{x})$ be forms of degree $m > 1$, where $\mathbf{x} = (x_1, \ldots, x_n)$, and have no zeros in common in C^n other than the trivial zero: $(0, \ldots, 0)$, then an integer $k_0 \geqq m$, depending on m and n but not on the values of the coefficients of the forms, exists such that every form $G(\mathbf{x})$ of degree $M \geqq k_0$ can be expressed in the form

$$G = b_1 P_1 + \cdots + b_n P_n \tag{11.2.1}$$

where $b_1(\mathbf{x}), \ldots, b_n(\mathbf{x})$ are forms of degree $M - m$, whose coefficients are rational functions of the coefficients of the original forms $P_j(\mathbf{x})$.

***Proof*:** Under the conditions of the lemma, in view of Hilbert's basis theorem and properties of the resultant of n forms in n variables (cf. [8], sections 79, 82), we can write

$$x_1^{\alpha_1} \cdots x_n^{\alpha_n} R = a_1 P_1 + \cdots + a_n P_n \tag{11.2.2}$$

where R is the resultant of the forms $P_j(\mathbf{x})$, the α_j are nonnegative integers such that $\alpha_1 + \cdots + \alpha_n = M \geqq k_0$, k_0 being the upper bound of the powers, defined by m and n, and the $a_j(\mathbf{x})$ are forms of degree $M - m$, whose coefficients are polynomials in the coefficients of the original forms $P_j(\mathbf{x})$.

Since an arbitrary form is a linear combination of monomials of the same degree, dividing both sides of (11.2.2) by $R \neq 0$, we may pass from (11.2.2) to (11.2.1), as required. The lemma stands proved.∎

We introduce the basic definitions which will be required in the course of this and the following sections. Let $n \geqq 1$ be some integer. Consider $\mathscr{A} = \{\alpha\}$, where $\boldsymbol{\alpha} = (\alpha_1, \ldots, \alpha_n)$ is a multiple index with α_j nonnegative integers. We shall enumerate the elements of $\mathscr{A}$ in some manner, but subject to the following rule: if $\alpha', \alpha'' \in \mathscr{A}$ and m', m'' are the numbers corresponding to them, then we must have $m' < m''$ if $\|\alpha'\| < \|\alpha''\|$, where

$$\|\alpha\| = \sum_1^n \alpha_j. \tag{11.2.3}$$

We shall associate with every set of complex numbers

$$\mathscr{F} = \{q_\alpha : K \leqq \|\alpha\| \leqq M\},$$

where $K \leqq M$ are nonnegative integers, a vector $\mathbf{Q}_{KM}$ whose components are elements of $\mathscr{F}$, ordered in accordance with the chosen principle of enumeration. We denote by $\mathbf{X}_{KM}$ a vector with components from the set

$$\mathscr{X}_{KM} = \{x_1^{\alpha_1} \cdots x_n^{\alpha_n}; \alpha \in \mathscr{A}, K \leqq \|\alpha\| \leqq M\}.$$

We shall denote by $\bar{k}(M)$* the number of components of the vector $\mathbf{X}_{0M}$. With the notations adopted, the polynomial

$$Q_{KM}(\mathbf{x}) = \sum_{K \leqq \|\alpha\| \leqq M} q_{\alpha_1 \cdots \alpha_n} x_1^{\alpha_1} \cdots x_n^{\alpha_n}$$

can be written in the form

$$Q_{KM}(\mathbf{x}) = \mathbf{Q}'_{KM} \mathbf{X}_{KM}$$

where $\mathbf{Q}_{KM}$ is the vector corresponding to the set of coefficients of $Q_{KM}(\mathbf{x})$.

We shall consider a standard normal random sample $\mathbf{X} = (X_1, \ldots, X_n)$. We take some real polynomial statistic $P(\mathbf{X})$ and let

$$f(\tau, Q) = E[Q(\mathbf{X}) \exp(\tau P(\mathbf{X}))], \qquad \tau = it,\ t \text{ real} \tag{11.2.4}$$

where Q is a polynomial.

If $Q(\mathbf{X}) = X_1^{\alpha_1} \cdots X_n^{\alpha_n}$, then we denote the corresponding $f(\tau, Q)$ by $f_\alpha(\tau)$, so that

$$f_\alpha(\tau) = E\{X_1^{\alpha_1} \cdots X_n^{\alpha_n} \exp[\tau P(\mathbf{X})]\}. \tag{11.2.5}$$

We shall then call $\|\alpha\| = \sum_1^n \alpha_j$ the *order* of $f_\alpha(\tau)$. We shall denote the vector corresponding to the set $\{f_\alpha(\tau), K \leqq \|\alpha\| \leqq M\}$ by $\mathbf{f}_{KM}(\tau)$.†

With these notations, we may write

$$\mathbf{f}_{KM}(\tau) = E[\mathbf{X}_{KM} \exp(\tau P(\mathbf{X}))]$$
$$f(\tau, Q_{KM}) = \mathbf{Q}'_{KM} \mathbf{f}_{KM}(\tau). \tag{11.2.6}$$

* As well-known (cf., for instance, [105], p. 61), $\bar{k}(M) = \binom{n+M}{n}$: also cf. Lemma 1.3.1.

† Such a notion is close to the modern concept of a "string" (cf. [98]).

Similarly we define

$$\mathbf{W}_{KM}(z) = E\{\mathbf{X}_{KM}[P(\mathbf{X})]^z\}, \ \mathrm{Re}\, z > 0 \tag{11.2.7}$$

$$W_\alpha(z) = E\{X_1^{\alpha_1} \cdots X_n^{\alpha_n}[P(\mathbf{X})]^z\} \tag{11.2.8}$$

$$W(z, Q_{KM}) = \mathbf{Q}'_{KM}\mathbf{W}_{KM}(z) \tag{11.2.9}$$

where we take

$$[P(\mathbf{x})]^z = \exp(z \ln P(\mathbf{x})) \qquad \text{if } P(\mathbf{x}) \neq 0.$$

It is easily seen from the definitions that (11.2.5) represents a function having bounded, uniformly continuous derivatives of all orders for all real t, and (11.2.8) represents a function analytic in the halfplane $\mathrm{Re}\, z > 0$.

The constructions introduced above, in (11.2.4) and the following relations, can be defined in more general cases, for example, when, $\mathbf{X}$ is a random vector in R^n, whose components have finite moments of all positive orders.

We introduce yet another concept, that of a "proper" polynomial statistic (cf. [33]).

DEFINITION 11.2.1. The form $P(\mathbf{x})$, $n > 1$, of degree $m + 1$ will be called *proper* if the n forms $\partial P/\partial x_1, \ldots, \partial P/\partial x_n$ do not have common zeros in C^n other than the trivial $(0, \ldots, 0)$, i.e., the gradient of P vanishes only at the origin. A polynomial $P(\mathbf{x})$ of degree $m + 1$ will be called *proper* if it can be represented in the form

$$P(\mathbf{x}) = P_0(\mathbf{x}) + P_1(\mathbf{x}) \tag{11.2.10}$$

where $P_0(\mathbf{x})$ is a proper form and $P_1(\mathbf{x})$ is a polynomial of degree at most m.

For $n = 1$, all polynomials of nonzero degree may be considered proper.

For proper polynomials, the following simple lemma holds, being an immediate consequence of Lemma 11.2.1.

LEMMA 11.2.2. If $P(\mathbf{x})$ be a proper polynomial of degree $m + 1$, then a positive integer $K > m$, depending on m and n, exists such that for any polynomial $Q(\mathbf{x})$ of degree $M \geqq K > m$ the representation

$$Q(\mathbf{x}) = \sum_1^n q_j(\mathbf{x})(\partial P/\partial x_j) + Q_1(\mathbf{x}) \tag{11.2.11}$$

holds, where q_j are forms of degree $M - m$, and Q_1 is a polynomial of degree at most $M - 1$, whose coefficients are rational functions of the coefficients of P and Q.

***Proof*:** Let $P_0(\mathbf{x})$ be a proper form of degree $m + 1$ in the representation (11.2.10) for $P(\mathbf{x})$. By definition, its partial derivatives satisfy the conditions of Lemma 11.2.1, so that there exists an integer $K > m$ such that any form $q(\mathbf{x})$ of degree $M \geqq K$ can be expressed as

$$q(\mathbf{x}) = \sum_1^n q_j(\mathbf{x})(\partial P_0/\partial x_j) \tag{11.2.12}$$

where the q_j are forms of degree $M - m$. If now an arbitrary polynomial Q of degree $M \geqq K$ is taken, and the form of the highest degree contained in Q is expressed in the form (11.2.12), then, substituting there the expression for $P_0(\mathbf{x})$ from (11.2.10), we obtain the desired result. The lemma is proved.▌

An application of this lemma permits us to establish the following simple fact, useful for what follows.

LEMMA 11.2.3. Let $\mathbf{X}$ be a standard normal random sample vector, and $P(\mathbf{X})$ a real proper polynomial statistic of degree $m + 1 \geqq 3$. Then,

(1) there exist positive integers $k_1(m, n)$ and $s_1(m, n)$, not depending on the values of the coefficients of $P(\mathbf{x})$, such that for any $s_1 = s_1(m, n)$ polynomials $Q_{11}, \ldots, Q_{1s_1}$ of degree not less than $k_1(m, n)$ the relation below holds:

$$\sum_{j=1}^{s_1} U_{1j}(\tau) f(\tau, Q_{1j}) = 0, \qquad \tau = it, t \text{ real} \tag{11.2.13}$$

where the U_{1j} are polynomials, not all of them identically zero;

(2) there exists a positive integer $k_2(m, n)$, not depending on the values of the coefficients of $P(\mathbf{x})$, such that for any integers $k'' \geqq k' > k_2(m, n)$, an integer $s_2 = s_2(m, n, k', k'') \leqq \bar{k}(k'') - \bar{k}(k' - 1)$, also not depending on the coefficients of $P(\mathbf{x})$, can be found such that for any s_2 polynomials $Q_{21}, \ldots, Q_{2s_2}$ whose degrees lie between k' and k'', the following relation holds:

$$\sum_{1}^{s_2} U_{2j}(z) W(z, Q_{2j}) = 0 \qquad \text{for } \operatorname{Re} z > 0 \tag{11.2.14}$$

where the U_{2j} are polynomials, not all of which are identically zero.

The coefficients of the polynomials $U_{k1}, \ldots, U_{ks_k} (k = 1, 2)$ are polynomials in the coefficients of P, $Q_{k1}, \ldots, Q_{ks_k} (k = 1, 2)$ and their degrees are bounded by some number depending only on m and n and the maximum of the degrees of $Q_{k1}, \ldots, Q_{ks_k} (k = 1, 2)$.

***Proof*:** We take k_0 corresponding to the proper polynomial statistics of degree $m + 1 > 2$, k_0 being determined on the basis of Lemma 11.2.2, and consider an arbitrary polynomial $Q(\mathbf{x})$ of degree $M > k_0$.

Using (11.2.11) for this polynomial, we can write

$$\tau f(\tau, Q) = \tau (2\pi)^{-n/2} \sum_{j=1}^{n} \int_{-\infty}^{\infty} \cdots \int_{-\infty}^{\infty} q_j (\partial P / \partial x_j) \exp(\tau P - (1/2)|\mathbf{x}|^2)\, d\mathbf{x}$$
$$+ \tau f(\tau, Q_1), \qquad \tau = it, t \text{ real} \tag{11.2.15}$$

where $|\mathbf{x}|^2 = x_1^2 + \cdots + x_n^2$, $d\mathbf{x} = dx_1 \cdots dx_n$. Integrating by parts in the above, with respect to each of the variables, and keeping in view the decreasing nature of the integrand at infinity, we have

$$\tau f(\tau, Q) = \tau f(\tau, Q_1) - f(\tau, Q_{\mathrm{I}} - Q_{\mathrm{II}}) \tag{11.2.16}$$

where

$$Q_{\mathrm{I}} = \sum_1^n (\partial q_j/\partial x_j) \qquad \text{and} \qquad Q_{\mathrm{II}} = \sum_1^n x_j q_j. \tag{11.2.17}$$

It is easily seen from (11.2.12) that Q_{I} is of degree $M - m - 1$ and Q_{II} of degree $M - m + 1$. If on the right side of (11.2.16) Q_{I} or $Q_{\mathrm{I}} - Q_{\mathrm{II}}$ has degree greater than k_0, then this process of reducing the degree can be extended. As a result, in a finite number of steps we arrive at a polynomial of degree at most k_0 and define the vector $\mathbf{Q}_{0k_0}(\tau)$ whose components are polynomials in τ whose degree is bounded by some number depending on M, and (^{T} standing for transpose)

$$\tau^{s'} f(\tau, Q) = \mathbf{Q}_{0k_0}^T(\tau)\mathbf{f}_{0k_0}(\tau) \tag{11.2.18}$$

where $s' < M/(m - 1)$.

If we take s polynomials, where $s = \bar{k}(k_0) + 1$, whose degrees are greater than k_0, then we will obviously have (11.2.13) in view of (11.2.18).

We turn to the derivation of (11.2.14). We take integers $k_0 < k_2 \leqq k' \leqq k''$, where the value of k_2 will be fixed later. Consider the polynomial $Q(\mathbf{x})$ of degree M such that $k' \leqq M \leqq k''$. Proceeding in the same way as for the derivation of (11.2.16), we can see that

$$(z + 1)W(z, Q) = (z + 1)W(z, Q_{\mathrm{I}}) - W(z + 1, Q_{\mathrm{I}} - Q_{\mathrm{II}}). \tag{11.2.19}$$

Proceeding thereafter in the same way as for deriving (11.2.18), we define vectors $\mathbf{Q}_{0k_0}^1(z), \ldots, \mathbf{Q}_{0k_0}^{s''}(z)$, where $s'' \leqq 1 + k''/(m - 1)$, the components of which are polynomials in z, the degrees of which are bounded by some number depending on k'', and (^{T}standing for transpose)

$$(z + s'') \cdots (z + 1)W(z, Q) = \sum_{j=0}^{s''} (\mathbf{Q}_{0k_0}^j(z))^T \mathbf{W}_{0k_0}(z + j). \tag{11.2.20}$$

If now $k_2(m, n)$ is chosen such that, for $k'' > k' > k_2(m, n)$,

$$\bar{k}(k'') - \bar{k}(k' - 1) > [(k''/(m - 1)) - 1]\bar{k}(k_0) \tag{11.2.21}$$

then we easily establish the required (11.2.14) in this case, using (11.2.20).

It is obvious from the course of the argument that the coefficients in (11.2.13) and (11.2.14) possess the required properties. The lemma stands proved.▌

THEOREM 11.2.1. Let $Q(\mathbf{x})$ be an arbitrary polynomial. Under the conditions of Lemma 11.2.3, $f(\tau, Q)$ and $W(z, Q)$ respectively satisfy equations of the form

$$\sum_{j=0}^{s_1} U_j(\tau) f^{(j)}(\tau, Q) = 0 \tag{11.2.22}$$

$$\sum_{j=0}^{s_2} V_j(z) W(z + j, Q) = 0 \tag{11.2.23}$$

where $U_j(\tau), j = 0, 1, \ldots, s_1$ are polynomials, not all of them identically zero, and $V_j(z)$ are also such polynomials.

Then the orders s_1 and s_2 of the equations and the degrees of the polynomials U_j and V_j are bounded by some number, depending only on n, and the degrees of P and Q, but not on the values of their coefficients; also, the coefficients of the polynomials U_j and V_j are polynomials in the coefficients of P and Q.

***Proof*:** Some formulas which are immediate consequences of the definitions —cf. relations (11.2.5), (11.2.9)—can be used in the proof; these are

$$f'(\tau, Q) = f(\tau, QP) \tag{11.2.24}$$

$$W(z + 1, Q) = W(z, QP). \tag{11.2.25}$$

To derive (11.2.22), we can apply the first part of Lemma 11.2.3 to choose polynomials of the form

$$Q_s(\mathbf{x}) = Q(\mathbf{x})[P(\mathbf{x})]^s, \qquad s = 1, \ldots, N \tag{11.2.26}$$

where $Q(\mathbf{x})$ is an arbitrary polynomial and N is chosen suitably.

To prove (11.2.23), it is more convenient to apply (11.2.20) to polynomials of the form (11.2.26), first choosing $Q(\mathbf{x})$ of degree at most k_0 and then turning to polynomials $Q(\mathbf{x})$ of arbitrary degree.∎

We further consider a case when $P(\mathbf{x})$ is a form (cf. [33]).

THEOREM 11.2.2. If, under the conditions of Lemma 11.2.3, $P(\mathbf{x})$ is a form of degree $m + 1$ ($m \geqq 2$), then $k_1 > 0$, depending only on m and n, exists such that for any integer $M \geqq k_1$

(1) the vector $\mathbf{W}_{MM}(z)$ corresponding to $P(\mathbf{x})$ satisfies for Re $z > 0$ a finite-difference relation in vector form:

$$\prod_{j=0}^{m-1} \{(m+1)(z+1) + M + j\}(z+1)\mathbf{W}_{MM}(z) = \begin{cases} (\mathbf{D}_{11}z + \mathbf{D}_{12})\mathbf{W}_{MM}(z+1) & \text{for } m \text{ odd} \\ (\mathbf{D}_{21}z + \mathbf{D}_{22})\mathbf{W}_{M-1,\,M-1}(z+1) & \text{for } m \text{ even} \end{cases} \tag{11.2.27}$$

where the $\mathbf{D}_{jk}$ are constant matrices;

(2) if $m > 2$, then the vector $\mathbf{f}_{M,\,M+m}(\tau)$ corresponding to $P(\mathbf{x})$ is a solution of a differential equation in vector form:

$$d\mathbf{f}/d\tau = [(\mathbf{R}_1 + \mathbf{R}_2\tau)/\tau^2]\mathbf{f}, \qquad \tau = it, \qquad t \neq 0 \text{ real} \tag{11.2.28}$$

where R_1 and R_2 are constant matrices.

Remark. If in the conditions of (2), $m = 2$, then the equation for $\mathbf{f}_{M,M+m}(\tau)$ takes the form

$$d\mathbf{f}/d\tau = (\mathbf{R}_1/\tau + \mathbf{R}_2/\tau^2 + \mathbf{R}_3/\tau^3)\mathbf{f} \tag{11.2.29}$$

where $\mathbf{R}_1$, $\mathbf{R}_2$, and $\mathbf{R}_3$ are constant matrices.

***Proof*:** We begin with the proof of (11.2.28). Let $P(\mathbf{x})$ be a proper form of degree $m+1 \geqq 3$. By (11.2.16), an integer $k_0 > m$ exists such that for any monomial of degree $M_1 \geqq k_0$

$$x_1^{\alpha_1} \cdots x_n^{\alpha_n} = \sum_{j=1}^{n} q_j(\partial P/\partial x_j) \tag{11.2.30}$$

where the q_j are forms of degree $M_1 - m$.

We take $f_{\boldsymbol{\alpha}}(\tau)$ and use (11.2.30). Then we obtain (cf. (11.2.16)),

$$\tau f_{\boldsymbol{\alpha}}(\tau) = -f(\tau, q_{\mathrm{I}} - q_{\mathrm{II}}), \qquad \tau = it, t \text{ real} \tag{11.2.31}$$

where, for brevity, we have used the notation

$$q_{\mathrm{I}} = \sum_{j=1}^{n} (\partial q_j/\partial x_j), \qquad q_{\mathrm{II}} = \sum_{j=1}^{n} x_j q_j. \tag{11.2.32}$$

It is easy to see (cf. (11.2.17)) that q_{I} is a form of degree $M - m - 1$ and q_{II} a form of degree $M - m + 1$.

We consider next

$$f'_{\boldsymbol{\alpha}}(\tau) = E[X_1^{\alpha_1} \cdots X_n^{\alpha_n} P(\mathbf{X}) \exp(\tau P(\mathbf{X}))]. \tag{11.2.33}$$

The right member can be transformed in the same way as (11.2.15), using, instead of (11.2.30), only Euler's identity for homogeneous functions as applied to $P(\mathbf{x})$. As a result, we have, analogously to (11.2.16),

$$f'_{\boldsymbol{\alpha}}(\tau) = -[(M_1 + n)/(m+1)\tau] f_{\boldsymbol{\alpha}}(\tau) + \tau^{-1} \sum_{j=1}^{n} f_{\alpha_1, \alpha_2, \ldots, \alpha_{j-1}, \alpha_j + 2, \alpha_{j+1}, \ldots, \alpha_n}(\tau). \tag{11.2.34}$$

Relations (11.2.31) and (11.2.34) enable us to obtain the required relation for the vector $\mathbf{f}_{M_1 - m, M_1}(\tau)$. For $\|\boldsymbol{\alpha}\| \leqq M_1 - 2$, according to (11.2.34), the components of the vector $\mathbf{f}'_{M_1 - m, M_1}$ can be linearly expressed in terms of the components of $\mathbf{f}_{M_1 - m, M_1}$. For the other components of the vector $\mathbf{f}'_{M_1 - m, M_1}$, those corresponding to $\|\boldsymbol{\alpha}\| = M_1 - 1, M_1$, an application of (11.2.34) leads to linear combinations involving also those components of $\mathbf{f}_{M_1 - m, M_1}$ for which $\|\boldsymbol{\alpha}\| = M_1 + 1, M_1 + 2$, but in this case we may use (11.2.31) and express $f_{\boldsymbol{\alpha}}(\tau)$ in terms of those components of the vector $\mathbf{f}_{M_1 - m, M_1}$ for which $\|\boldsymbol{\alpha}\| = M_1 - m, M_1 - m + 1, M_1 - m + 2, M_1 - m + 3$. The result of successively applying (11.2.31) and (11.2.34) to the vector $\mathbf{f}'_{M_1 - m, M_1}$ can be written in the form (11.2.28). According to the procedure indicated, the elements of the matrices $\mathbf{R}_1$ and $\mathbf{R}_2$ can be expressed in terms of the coefficients of the right hand members of the transformations (11.2.31) and (11.2.34). This proves (11.2.28).

We note that for $m = 2$, $f_{\boldsymbol{\alpha}}(\tau)$ of order $M_1 + 2$ can be expressed, using (11.2.31), in terms of elements of orders $M_1 - 1$ and $M_1 + 1$, so that yet another application of (11.2.31) is required for the latter. This fact makes for a change in the equation in this case and leads to (11.2.29).

We turn to (1). Consider the proper form $P(\mathbf{x})$ of even degree $m + 1 = 2R$. Choose $k > 0$ such that for

$$q(\mathbf{x}) = x_1^{\alpha_1} \cdots x_n^{\alpha_n}(x_1^2 + \cdots + x_n^2)^{R-1}, \qquad \|\boldsymbol{\alpha}\| = k \tag{11.2.35}$$

(11.2.12) is satisfied. We rewrite it in that case as

$$\sum_1^n q_j(\partial P/\partial x_j). \tag{11.2.36}$$

Here the forms $q_j(\mathbf{x})$ have degree $(k - 1)$. We apply (11.2.20) to

$$\begin{aligned} &E\{X_1^{\alpha_1} \cdots X_n^{\alpha_n}(X_1^2 + \cdots + X_n^2)^R(P(\mathbf{X}))^z\} \\ &\quad = -(z+1)^{-1}E\Bigg[\bigg(|\mathbf{X}|^2 \sum_1^n X_j q_j - 2\sum_1^n X_j q_j \\ &\qquad + |\mathbf{X}|^2 \sum_1^n (\partial q_j/\partial X_j)\bigg)(P(\mathbf{X}))^{z+1}\Bigg]. \end{aligned} \tag{11.2.37}$$

Then we can use the simple relation

$$E[|\mathbf{X}|^2 U(\mathbf{X})(P(\mathbf{X}))^z] = ((m+1)(z+1) + N)E[U(\mathbf{X})(P(\mathbf{X}))^z] \tag{11.2.38}$$

where $U(\mathbf{x})$ is a form of degree N. This relation is the analogue here of (11.2.34) and is derived in the same manner, using, for example, Euler's identity for homogeneous functions.

Successively applying (11.2.27) and (11.2.38), we find

$$\prod_{j=0}^{m-1} [(m+1)(z+1) + R + j](z+1)W_{\boldsymbol{\alpha}}(z) = zW(z+1, q_1^*) + W(z+1, q_2^*),$$

where

$$\begin{aligned} q_1^* = -(m+1)\sum_1^n x_j q_j; \qquad q_2^* &= -(2m+R)\sum_1^n x_j q_j \\ &= |\mathbf{x}|^2 \sum_1^n (\partial q_j/\partial x_j). \end{aligned} \tag{11.2.39}$$

It follows from (11.2.36) that the forms q_1^* and q_2^* have degree k. Considering (11.2.39) for all multiple indices $\boldsymbol{\alpha}$, for which $\|\boldsymbol{\alpha}\| = k$, and denoting by $\mathbf{D}_{11}$ and $\mathbf{D}_{12}$ the matrices whose rows are made up of the coefficients of q_1^* and q_2^* respectively, we obtain (11.2.27) for even $(m + 1)$ if $k \geqq k_0 - m + 1$.

We now turn to the case of odd $m + 1$. Let $m + 1 = 2R + 1$, where $R \geqq 1$. For this R and $k \geqq k_0 - m + 2$, the forms $q_j(\mathbf{x})$, $1 \leqq j \leqq n$, in the identity (11.2.36) have degree $k - 2$. Proceeding hereafter as in the case of even

$m+1$, we obtain (11.2.39), where the forms q_1^* and q_2^* have degree $k-1$. Concluding the argument as in the preceding case, we establish (11.2.27) for odd $m+1$.

For $k \geqq k_1 = k_0 - m + 2$, all the relations obtained will be valid. The theorem stands proved.▮

The equations obtained may be used for investigating properties of the distributions of polynomial statistics in normal samples.

Considering (11.2.27) for $z = 1, 2, \ldots$, these may be used as recurrence relations for computing high order moments of forms based on a normal sample.

11.3 THE DISTRIBUTION OF POLYNOMIAL STATISTICS OF NORMAL AND RELATED SAMPLES (THE GENERAL CASE)

In this section, we consider the principal results of section 11.2 for an arbitrary polynomial statistic. We shall make use of the following simple lemma for the purpose.

LEMMA 11.3.1. If

$$P(\mathbf{x}) = \sum_{0 \leqq |\alpha| \leqq m+1} p_{\alpha_1, \ldots, \alpha_n} x_1^{\alpha_1} \cdots x_n^{\alpha_n}$$

be a polynomial statistic of degree $m+1$, then a sequence of *proper* polynomial statistics of the same degree

$$P_N(\mathbf{x}) = \sum_{0 \leqq \|\boldsymbol{\alpha}\| \leqq m+1} p_{N; \alpha_1, \ldots, \alpha_n} x_1^{\alpha_1} \cdots x_n^{\alpha_n}$$

exists such that for all $\boldsymbol{\alpha}$ with $0 \leqq \|\boldsymbol{\alpha}\| \leqq m+1$

$$\lim_{N \to \infty} p_{N, \boldsymbol{\alpha}} = p_{\boldsymbol{\alpha}}. \tag{11.3.1}$$

If P is real, then so are the P_N.

***Proof*:** Consider the form of highest degree contained in $P(\mathbf{x})$.

$$P^0(\mathbf{x}) = \sum_{\|\alpha\| = m+1} p_{\alpha_1 \cdots \alpha_n} x_1^{\alpha_1} \cdots x_n^{\alpha_n}.$$

Let D be the resultant of the n components of the gradient

$$\partial P^0/\partial x_1, \ldots, \partial P^0/\partial x_n. \tag{11.3.2}$$

If $D \neq 0$, then obviously $P^0(\mathbf{x})$ is a proper form and the lemma is trivially true. Let then $D = 0$; then, according to the properties of the resultant (cf. [8], p. 23), it is possible to choose numbers as close as we please to the values of the coefficients of $P^0(\mathbf{x})$ (as real numbers if P_α is real), such that the resultant of the n components of the gradient of the form obtained by choosing these numbers as coefficients is nonzero; thus a sequence $\{P_N^0(\mathbf{x})\}$ of proper forms of degree $m+1$ can be constructed, whose coefficients converge to the corresponding coefficients of $P^0(\mathbf{x})$.

Setting

$$P_N(\mathbf{x}) = P_N^0(\mathbf{x}) + \sum_{0 \leq \|\alpha\| \leq m} p_{\alpha_1, \ldots, \alpha_n} x_1^{\alpha_1} \cdots x_n^{\alpha_n}$$

we obtain the desired result, and the lemma stands proved.∎

Consider an arbitrary real polynomial statistic $P(\mathbf{X})$ and construct a sequence of real proper polynomial statistics $\{P_N(\mathbf{X})\}$, for which (11.3.1) is satisfied. We take the functions (cf. (11.2.6))

$$\begin{aligned} f_N(\tau) &= E[Q(\mathbf{X}) \exp\{\tau P_N(\mathbf{X})\}], \\ f(\tau, Q) &= E[Q(\mathbf{X}) \exp\{\tau P(\mathbf{X})\}], \qquad \tau = it, \ t \text{ real} \end{aligned} \tag{11.3.3}$$

where Q is some polynomial. By construction, for every real t, $f_N(\tau)$ and its derivatives of any order obviously converge to $f(\tau, Q)$ and its derivatives respectively, as $N \to \infty$.

Since, by construction, $P_N(\mathbf{x})$ is a proper polynomial, by Theorem 11.2.1 $f_N(\tau)$ is the solution of a linear homogeneous differential equation with polynomial coefficients. We take this equation

$$U_{0N}(\tau) f_N^{(s)}(\tau) + \cdots + U_{sN}(\tau) f_N(\tau) = 0. \tag{11.3.4}$$

In view of Theorem 11.2.1 again, we can take one and the same order s in (11.3.4) for all N as the degree of the polynomial coefficients of (11.3.4) bounded by some number dependent on m, n, and the degree of Q but not on N. Let us denote by a_N the largest in absolute value of the coefficients of the polynomials $U_{jN}(\tau)$, $j = 0, 1, \ldots, s$. Dividing out both sides of (11.3.4) by a_N, we obtain the same equation as (11.3.4):

$$\tilde{U}_{0N}(\tau) f_N^{(s)}(\tau) + \cdots + \tilde{U}_{sN}(\tau) f_N(\tau) = 0 \tag{11.3.5}$$

but now the coefficients of the polynomials $\tilde{U}_{jN}(\tau)$ are uniformly bounded in N. For every N, at least one of the coefficients is equal to ± 1. Thus we can construct a sequence of positive integers N_k: $k = 1, 2, \ldots$ such that the coefficients of all the polynomials $U_{jN_k}(\tau)$, $j = 1, \ldots, s$, converge to finite limits as $k \to \infty$, and one particular coefficient takes the value $+1$ or -1 for all k.

Letting $N \to \infty$ through the above sequence N_k, we find

$$\tilde{U}_0(\tau) f^{(s)}(\tau, Q) + \cdots + \tilde{U}_s(\tau) f(\tau, Q) = 0$$

where $\tilde{U}_j(\tau)$ for $j = 1, \ldots, s$ are polynomials, not all of them identically zero, in view of our choice of the sequence $\{N_k\}$. Precisely the same construction can be made for obtaining a finite-difference equation which is satisfied by $W(z, Q)$ for arbitrary $P(\mathbf{x})$.

Thus we can establish the following proposition as an extension of Theorem 11.2.1.

THEOREM 11.3.1. If under the conditions of Theorem 11.2.1 an arbitrary real polynomial statistic $P(\mathbf{X})$ is taken, then this theorem remains valid.

We shall dwell on yet another question which is of importance for the study of the properties of distributions of polynomial statistics based on normal samples and their applications: the analytic continuation of $W(z, Q)$ to the half plane $\operatorname{Re} z \leqq 0$.

For this purpose, we may use the finite-difference equation introduced in the preceding section in the same way as the basic functional relation for the gamma function is used for its analytic continuation.

We take a proper form $P(\mathbf{x})$ of even degree and consider the vector $\mathbf{W}_{MM}(z)$ corresponding to it, defined by (11.2.7) for $\operatorname{Re} z > 0$. By Theorem 11.2.7, for $M \geqq k_1 - m$ and $\operatorname{Re} z > 0$, this vector satisfies (11.2.27). We set

$$\mathbf{W}_{MM}(z) = (m+1)^{mz}\Gamma(z+1)\prod_{j=0}^{m-1}\Gamma\left(z+1+\frac{M+j}{m+1}\right)\tilde{\mathbf{W}}(z), \qquad \operatorname{Re} z > 0. \tag{11.3.6}$$

It follows from the definition and (11.2.27) that the vector $\tilde{\mathbf{W}}(z)$ satisfies the equation

$$\tilde{\mathbf{W}}(z) = (\mathbf{D}_{11}z + \mathbf{D}_{12})\tilde{\mathbf{W}}(z+1). \tag{11.3.7}$$

This equation permits us to make an analytic continuation of $\tilde{\mathbf{W}}(z)$ defined by (11.3.6) for $\operatorname{Re} z > 0$ to the halfplane $\operatorname{Re} z \leqq 0$ in such a way that all the components of $\tilde{\mathbf{W}}(z)$ will be entire functions and $\tilde{\mathbf{W}}(z)$ will satisfy (11.3.7) for all complex z. Since $\Gamma(z)$ is analytic everywhere in the complex plane except at the points $z = 0, -1, -2, \ldots$, at which it has simple poles, we can use (11.3.6) to analytically extend $\mathbf{W}_{MM}(z)$ to the halfplane $\operatorname{Re} z \leqq 0$. Thus we define all the components of $\mathbf{W}_{MM}(z)$ as functions analytic everywhere except possibly at the points $z = -k/(m+1)$, $k = 1, 2, \ldots$, at which they may have poles, whose order is bounded by some number independent of k. The vector $\mathbf{W}_{MM}(z)$ satisfies (11.2.27) everywhere except at the poles of its components.

In the case when $P(\mathbf{x})$ is a proper form of odd degree, we can proceed in the same way, basing ourselves on (11.2.27), or on the equation which can be derived from it by applying it twice, to obtain $\mathbf{W}_{MM}(z)$ on the right-hand side.

Thus, if we also take (11.2.9) and (11.2.38) into consideration, then the following fact may be taken to have been established.

THEOREM 11.3.2. If under the conditions of Lemma 11.2.3, $P(\mathbf{x})$ is a form, then for any polynomial $Q(\mathbf{x})$, $W(z, Q)$ can be extended as an analytic function to the halfplane $\operatorname{Re} z \leqq 0$, with the possible exception only of the points $z = -k/(m+1)$, $k = 1, 2, \ldots$, at which it may have poles, the orders of which are bounded.

To consider the case of arbitrary $P(\mathbf{x})$, we can set up for $W(z, Q)$, in $\operatorname{Re} z > 0$, a suitable finite-difference equation (see Theorem 11.3.1) of the form (11.2.23), and using this equation, similarly make an analytic continuation thereof to the halfplane $\operatorname{Re} z \leqq 0$.

The analytic continuation of $W(z, Q)$ to the halfplane $\operatorname{Re} z \leqq 0$ can also be effected by starting directly from the integral representation (11.2.9).*

It is easily seen from this expression that the singularities which arise in passing to the halfplane $\operatorname{Re} z \leqq 0$ must be connected with that part of the integral which is taken over the set $\{\mathbf{x} : |P(\mathbf{x})| < \varepsilon\}$ for arbitrarily small $\varepsilon > 0$. In this case, for the analytic extension, it is necessary to effect a deformation of this set into a suitable complex space.

Equations similar to (11.2.22) can be set up for distribution functions of polynomial statistics. Evidently some changes will be necessary, since these functions may not be differentiable the necessary number of times.

Consider for example the following modification. Let, as before,

$$f(\tau) = E \exp [\tau P(\mathbf{X})], \qquad \tau = it, \ t \text{ real} \tag{11.3.8}$$

where $P(\mathbf{X})$ is a polynomial statistic.

By Theorem 11.3.1, $f(\tau)$ satisfies a differential equation of the form (11.2.22). We write

$$g(t) = f(it) \exp \left(-\frac{1}{2} \sigma^2 t^2 \right) \tag{11.3.9}$$

where $\sigma > 0$. Obviously, g also satisfies an equation of the form (11.2.22). If we denote by $F(x)$ the d.f. of $P(\mathbf{X})$, then, clearly,

$$F_\sigma(y) = \frac{1}{\sigma\sqrt{2\pi}} \int_{-\infty}^{\infty} \exp \left[\frac{-(y-u)^2}{2\sigma^2} \right] dF(u) \tag{11.3.10}$$

will be the d.f. having the c.f. $g(t)$. Using elementary properties of the Fourier transform, and in view of (11.3.10), we find that for any nonnegative integers s and k,

$$(x^s F_\sigma(x))^{(k)} = \frac{(-i)^{s+k}}{2\pi} \int_{-\infty}^{\infty} e^{-ixt} g^{(s)}(t) t^k \, dt. \tag{11.3.11}$$

It follows hence that $F_\sigma(y)$ also is a solution of an equation of the form (11.2.22) for any $\sigma > 0$.

We may also use other transformations of $F(y)$ similar to (11.3.10).

* When this book was being written, the authors came to know that similar integrals of a general character were the subject of study by I. M. Gelfand and his collaborators and that they find application in the theory of group representations (*vide* the program of the November 30, 1968 meeting of the Moscow Mathematical Society, Communication from I. M. Gelfand, and also [5]).

The equations constructed above for various purposes connected with the distribution of polynomial statistics of normal samples can be considered also for samples from populations having distributions of a more general character. Such results for distributions similar to the normal are to be found in [33].

DEFINITION 11.3.1. A distribution function will be said to be *similar to the normal* if it has a density f such that, for some positive integer s, the s-th derivative of $f(x)$ is defined for all real x, and f satisfies a linear differential equation of the form

$$f^{(s)}(y) + \sum_{j=1}^{s} P_j(y) f^{(s-j)}(y) = 0 \qquad \text{for all real } y \tag{11.3.12}$$

where the P_j are polynomials, and further, as $|y| \to \infty$,

$$f(y) = O(|y|^{-M})$$

for any $M > 0$.

We shall denote by μ the maximum of the degrees of the coefficient polynomials P_j in (11.3.12).

It is easy to see that the normal distributions satisfy the above conditions for $s = \mu = 1$.

The distributions of polynomial statistics of samples from a population similar to the normal possess properties fully similar to those in the normal case. We can develop by similar methods a linear theory similar to that considered above, and construct corresponding equations of the same type. As an example, we indicate the extension of equation (11.2.23) to this case and briefly indicate the derivation (cf. [33]).

THEOREM 11.3.3. Let $\mathbf{X} = (X_1, \ldots, X_n)$ be a random vector in R^n, all the components of which are mutually independent and have the same distribution which is similar to the normal for some s and μ. If $P(\mathbf{X})$ is a real polynomial statistic of degree $m + 1 > \mu$, then

$$f(\tau, Q) = E[Q(\mathbf{X}) \exp(\tau P(\mathbf{X}))], \qquad \tau = it, \ t \text{ real},$$

where $Q(\mathbf{x})$ is an arbitrary polynomial, is a solution of a linear homogeneous differential equation with polynomial coefficients

***Proof*:** We may consider, similarly to $f_{\boldsymbol{\alpha}}(\tau)$ the function (cf. (11.2.5))

$$f_{\boldsymbol{\alpha},\boldsymbol{\beta}}(\tau) = \int x_1^{\alpha_1} \cdots x_n^{\alpha_n} \exp(\tau P(\mathbf{x})) f^{(\beta_1)}(x_1) \cdots f^{(\beta_n)}(x_n)\, d\mathbf{x} \tag{11.3.13}$$

where $\boldsymbol{\alpha} = (\alpha_1, \ldots, \alpha_n)$ and $\boldsymbol{\beta} = (\beta_1, \ldots, \beta_n)$ have nonnegative integers as components. The existence of the derivatives of f, the density function, and of the moments concerned obviously follows from the definition.

From the functions (11.3.13) for a proper $P(\mathbf{x})$, a corresponding linear basis is constructed, constituting the basis of our considerations as in the case of normal samples. We then derive and use relations similar to (11.2.18). An important role is played here by the condition $m + 1 > \mu$ which permits the use of (11.3.12) to diminish the orders of the derivatives in (11.3.13) by using (11.2.11) without increasing the degrees of the polynomials concerned.∎

11.4 THE *U*-CONJECTURE ON "UNLINKING." SOME EXAMPLES

We shall consider here the problem of characterizing simplest types of pairs of independent statistics of standard normal random samples $(X_1, \ldots, X_n)$. We shall for the most part be interested only in polynomial statistics but some results will also be valid for sufficiently smooth statistics.

Such investigations are useful in the theory of tests of multivariate analysis and also for characterizations of distributions through properties of statistics. For constructing multivariate statistical tests, often pairs of independent statistics are used, which are usually conjectured to be such; no classification of such pairs, as.far as known to the authors, exists. In sections 11.4 and 11.5, an attempt at taking the first steps in this direction is made.

The U-conjecture (cf. [82]) is that two independent polynomial statistics in our normal sample can be "unlinked" by means of a suitable orthogonal transformation, i.e., that there exists an orthogonal transformation which reduces our polynomials to functions of distinct sets of variables.

As is well-known, the converse assertion is trivial; an unlinked pair of arbitrary statistics of our normal sample remains a pair of independent statistics under any orthogonal transformation.

Until now, the U-conjecture has been neither proved nor disproved; in recent notes, Zinger and Linnik [41] and Linnik [87] outlined a proof of this conjecture, roughly speaking, for "almost all" pairs of independent polynomials; a detailed proof obtained by the methods of algebraic geometry will be presented below.

Now we consider some examples showing that the U-conjecture is false if we consider independent entire functions instead of polynomials, and other examples relating to pairs of identically distributed polynomial statistics (as is well-known, the property of identical distribution is closely connected with that of independence of statistics). In connection with these examples, see section 11.1 and the note [89].

Example 1. Let the r.v.'s X_1, X_2, X_3 be i.i.d. standard normal variables. Consider the pairs

$$Y_1 = X_1 \cos X_3 + X_2 \sin X_3$$
$$Y_2 = -X_1 \sin X_3 + X_2 \cos X_3 .$$

We shall show that these statistics are independent, but nevertheless cannot be unlinked by means of an orthogonal transformation of (X_1, X_2, X_3).

We first verify the independence of Y_1 and Y_2.

$$\begin{aligned} &E \exp [i(t_1 Y_1 + t_2 Y_2)] \\ &\quad = E_{X_3}\{E[\exp i(t_1 Y_1 + t_2 Y_2) \mid X_3]\} \\ &\quad = E_{X_3} \exp \{ -(1/2)[(t_1 \cos X_3 + t_2 \sin X_3)^2 + (-t_1 \sin X_3 + t_2 \cos X_3)^2]\} \\ &\quad = \exp [-(1/2)(t_1^2 + t_2^2)] \end{aligned} \tag{11.4.1}$$

On the other hand, let us take any orthogonal transformation

$$X_j = a_{j1} Z_1 + a_{j2} Z_2 + a_{j3} Z_3; \qquad j = 1, 2, 3. \tag{11.4.2}$$

If we compute for instance $\partial Y_1/\partial Z_1$,

$$\partial Y_1/\partial Z_1 = a_{11} \cos X_3 + a_{21} \sin X_3 + a_{31}(-X_1 \sin X_3 + X_2 \cos X_3). \tag{11.4.3}$$

We easily see from (11.4.3) that the condition $\partial Y_1/\partial Z_1 \equiv 0$ is equivalent to $a_{11} = a_{21} = a_{31} = 0$ which is impossible since the transformation (11.4.2) was chosen to be orthogonal. The other $\partial Y_j/\partial Z_k$ are considered in precisely the same way.

Example 2. Let the X_j, $j = 1, 2, 3, 4$, be i.i.d. standard normal variables. Consider two polynomial statistics given by

$$A(\mathbf{X}) = X_1(X_3^2 - X_4^2) + 2X_2 X_3 X_4; \; B(\mathbf{X}) = X_1(X_3^2 + X_4^2). \tag{11.4.4}$$

We shall show that the statistics A and B are identically distributed. We consider their characteristic functions for this purpose.

$$\begin{aligned} E \exp (itA) &= E\{E[\exp (itA) \mid X_3, X_4]\} \\ &= E \exp [-(1/2)t^2(X_3^2 + X_4^2)^2] = E \exp (itB). \end{aligned} \tag{11.4.5}$$

We shall show that $A(\mathbf{X})$ cannot be reduced to a function of a smaller number of variables by means of orthogonal transformations of $(X_1, \ldots, X_4)$. Consider such a transformation

$$X_j = \sum_{k=1}^{4} a_{jk} Y_k, \qquad j = 1, \ldots, 4. \tag{11.4.6}$$

We compute

$$\begin{aligned} \partial A/\partial Y_1 = a_{11}(X_3^2 - X_4^2) + 2a_{21} X_3 X_4 + 2a_{31}(X_1 X_3 + X_2 X_3) \\ + 2a_{41}(-X_1 X_4 + X_2 X_3). \end{aligned} \tag{11.4.7}$$

It follows from (11.4.7) as earlier that $\partial A/\partial Y_1 \equiv 0$ is equivalent to $a_{j1} = 0$ for all j, which is impossible. Similarly for the other $\partial A/\partial Y_j$, $j = 2, 3, 4$.

Since B depends on three variables, it follows hence, in particular, that the statistics A and B cannot be reduced to each other by means of any orthogonal transformation.

The first example shows that even in the class of entire functions, the U-conjecture cannot be expected to hold. As we shall see below, this is related to the nature of the growth of the statistics in various directions.

The second example shows that, when we come to consider the property of identical distribution of statistics, the situation is even more complex. Even in the class of polynomial statistics, it is possible for polynomial statistics depending on different numbers of variables to be identically distributed, whence it follows in particular that it is not true that only those statistics based on a normal random sample can be identically distributed which can be transformed into each other by orthogonal transformations.

The proof of Theorem 11.1.4 shows that the investigation of identical distribution:

$$A \cong B \tag{11.4.8}$$

where A and B are polynomial statistics based on $\mathbf{X}$, a random vector in R^n, can be related to the study of algebraic sets generated by the relations

$$EA^s - EB^s = 0, \qquad s = 1, 2, \ldots. \tag{11.4.9}$$

which must hold under (11.4.8) if all the components of $\mathbf{X}$ have finite moments of all positive orders. The set of left members of (11.4.9)—polynomials in the coefficients of the statistics A and B as independent variables—form a polynomial ideal, which possesses, in view of the well-known theorem due to Hilbert (Lemma 1.3.3) a finite basis. Thus, as in the proof of Theorem 11.1.4, we see that the fulfilment of the countable set of relations (11.4.9) is equivalent to that of a finite number of them.

If then restrictions on the growth of the moments are imposed in such a way as to guarantee the solvability of the moment problem for (11.4.9)*, then (11.4.8) turns out to be equivalent to a finite number of relations of the form (11.4.9).

A similar situation obtains when we consider the independence of two polynomial statistics A and B. In this case, an algebraic set can be constructed, starting from the so-called uncorrelatedness relations

$$E(A^r B^s) - E(A^r) \cdot E(B^s) = 0 \tag{11.4.10}$$

for $r, s = 1, 2, \ldots$, which follow from the independence of A and B under the condition that the corresponding moments exist.

* This can be realized, in particular, if we assume

$$EX_j^s = O[\exp(o(s \ln s))] \quad \text{as} \quad s \to \infty, \quad j = 1, \ldots, n$$

(cf. e.g., [167], p. 19 or [2], p. 110, and also Lemma 1.6.1).

It is necessary to note that though the equivalence of the relations (11.4.9) or (11.4.10) to a finite number of the corresponding relations is a fundamental fact, it can hardly be used to study (11.4.9) or (11.4.10) in a concrete way in view of the fact that it merely asserts the existence of a bound for the number of relations (11.4.9) or (11.4.10) which are respectively equivalent to all of them and does not indicate the possible values of such a number. The basic value of these considerations lies in the fact that they help establish a connection between the independence and identity of distribution of polynomial statistics on the one hand and the theory of polynomial ideals on the other.

The statement of the problem of reducing the condition of identical distribution (or independence) to a finite number of relations was substantially widened in scope in [86]. Instead of relations (11.4.9) or (11.4.10), similar relations of a very general type are considered there, called Φ-moments, applicable to statistics depending on a finite number of parameters. Some very general restrictions on the Φ-moments enable us to apply the theory of ideals in the ring of functions of several complex variables analytic on a compact simply-connected polycylinder, and to establish that a countable set of relations in terms of Φ-moments is finitely generated (cf. [86], Chapter IV, Section 6).

We shall now consider the U-conjecture as applied to various special types of pairs of statistics of a normal sample $\mathbf{X}$ with standard normal components $X_1, \dots, X_n$.

We consider polynomials

$$Q(\mathbf{x}) = \sum_{a_1 + \cdots + a_n \leq m} q_{a_1 \cdots a_n} x_1^{a_1} \cdots x_n^{a_n} \tag{11.4.11}$$

where the a_j are nonnegative integers, m is a nonnegative integer, and the coefficients $q_{\mathbf{a}}$ are in general complex numbers.

If among the coefficients $q_{a_1 \dots a_n}$ for which $a_1 + \cdots + a_n = m$, there is at least one which is nonzero, then, as usual, m is the degree of Q. As before, homogeneous polynomials will be called forms.

In addition to polynomial statistics, we shall consider statistics, which we call 0-statistics for brevity, and are given by the following

DEFINITION 11.4.1. The statistic $S(X_1, \dots, X_n)$ will be called an 0-statistic if

(1) $S(\mathbf{X})$ is homogeneous and of positive degree of homogeneity γ; and

(2) at every point of the unit sphere in R^n, the second order derivatives of all orders $\partial^2 S(\mathbf{x})/\partial x_j \, \partial x_k$; $j, k = 1, \dots, n$, exist and are continuous.

We note that it follows immediately from the definition that for 0-statistics of normal samples, absolute moments of any positive order exist.

11.5 UNLINKING OF CERTAIN PAIRS OF UNCORRELATED PAIRS OF STATISTICS OF NORMAL SAMPLES. SYMMETRIC STATISTICS

In this section, we shall study some classes of uncorrelated pairs of real polynomial statistics and 0-statistics of normal samples. The results obtained will be applied to the study of symmetric statistics. Some information on them is to be found in [65]. In the presentation below, we shall follow the dissertation [37] of Zinger's and the work of Linnik and Zinger [41].

We shall be interested in conditions under which a pair of uncorrelated statistics $A(\mathbf{X})$ and $B(\mathbf{X})$ of a normal vector $\mathbf{X}$ in R^n can be reduced to functions involving distinct sets of coordinates by means of orthogonal transformations (can be "unlinked"). For this we shall use the property mentioned earlier of the invariance of the distribution of a random sample from a standard normal population under the group of orthogonal transformations of R^n.

We establish a lemma which will be basic to our constructions.

LEMMA 11.5.1. Let

(1) $\mathbf{X} = (X_1, \ldots, X_n)$ be a random vector in R^n, with i.i.d. components, each of them having a standard normal distribution, and let two statistics of this vector: a polynomial statistic A and an 0-statistic B be uncorrelated*, i.e.,

$$E(A^r B^s) = E(A^r) \cdot E(B^s); \qquad r, s = 1, 2, \ldots; \tag{11.5.1}$$

(2) integers n_1, n_2, and $n_3 \geqq 0$ with $n_1 + n_2 + n_3 = n$ exist such that $A(\mathbf{x}) = A(x_1, \ldots, x_{n_1+n_2})$ and, under any orthogonal transformation: $(x_1, \ldots, x_n) \to (x'_1, \ldots, x'_n)$, has positive degree in each of the components $x'_{n_1+1}, \ldots, x'_{n_1+n_2}$;† and let $B(\mathbf{x}) = B(x_{n_1+1}, \ldots, x_n)$ and attain its maximum absolute value on the sphere $x^2_{n_1+1} + \cdots + x^2_n = 1$ in the subspace $\{(x_{n_1+1}, \ldots, x_n)\}$ at a point $(x^0_{n_1+1}, \ldots, x^0_n)$ such that

$$(x^0_{n_1+1})^2 + \cdots + (x^0_{n_1+n_2})^2 \neq 0. \tag{11.5.2}$$

Then conditions (1) and (2) are incompatible.

Proof: Suppose that (1) and (2) hold simultaneously.

We may assume that the given statistics are nonnegative, without loss of generality, since we may always consider A^2 and B^2—or the condition (11.5.1) for even values of r and s. We may also assume without loss of generality

* The existence of the moments in (11.5.1) follows from the nature of the sample and the statistics.

† That is, in the expansion of A in powers of each of the components, say x'_{n_1+1}, the leading coefficient is not identically zero (A depends on x'_{n_1+1}).

that the largest value of B on the unit sphere $\sum_{n_1+1}^{n} x_j^2 = 1$ in $\{(x_{n_1+1}, \ldots, x_n)\}$ is unity and is attained at a point

$$\mathbf{x}_0 = (\underbrace{\xi_0, 0, \ldots, 0}_{n_2}, \underbrace{\eta_0, 0, \ldots, 0}_{n_3})$$

where ξ_0 is nonzero. If we make the orthogonal transformation

$$\begin{aligned} x_{n_1+1} &= y_{n_1+1} \cos\alpha - y_{n_1+n_2+1} \sin\alpha \\ x_{n_1+n_2+1} &= y_{n_1+1} \sin\alpha + y_{n_1+n_2+1} \cos\alpha \\ x_j = y_j \qquad &\text{for } j = 1, 2, \ldots, n; \qquad j \neq n_1+1 \quad \text{or} \quad n_1+n_2+1 \\ \xi_0 &= \cos\alpha; \qquad \eta_0 = \sin\alpha \end{aligned} \tag{11.5.3}$$

then $\mathbf{x}_0$ transforms into $(\underbrace{1, 0, \ldots, 0}_{n_2+n_3})$.

For sufficiently small $\delta > 0$, if $(y_1, \ldots, y_n)$ satisfies the conditions

$$y_1^2 + \cdots + y_n^2 = 1; \qquad |y_{n_1+1} - 1| < \delta \tag{11.5.4}$$

then, in view of the properties of the statistic $B(\mathbf{X})$, the following estimate holds.

$$B \geqq 1 - C_1(y_1^2 + \cdots + y_{n_1}^2 + y_{n_1+2}^2 + \cdots + y_n^2). \tag{11.5.5}$$

We fix $\delta > 0$ such that the above is satisfied.

We then choose an arbitrary $0 < \varepsilon < 1$ and C_2, and construct the set

$$\mathcal{N}_{\varepsilon s} = \{(y_1, \ldots, y_n) \colon y_1^2 + \cdots + y_{n_1}^2 + y_{n_1+2}^2 + \cdots + y_n^2 \leqq C_2;\ y_{n_1+1} \geqq \varepsilon\sqrt{\gamma s}\} \tag{11.5.6}$$

for $s > 0$, where γ is the degree of homogeneity of B.

For sufficiently large s, all the points in the projection of $\mathcal{N}_{\varepsilon s}$ on the unit sphere $y_1^2 + \cdots + y_n^2 = 1$ will satisfy (11.5.4). Thus we can find a C_3 such that, for $s > C_4$,

$$B(\mathbf{y}) \geqq y_{n_1}^{\gamma} (1 - C_3/s) \qquad \text{for } \mathbf{y} \text{ in } \mathcal{N}_{\varepsilon s}. \tag{11.5.7}$$

Further, we also have the obvious upper bound for $B(\mathbf{y})$

$$B(\mathbf{y}) \leqq (y_{n_1+1}^2 + \cdots + y_n^2)^{\gamma/2}. \tag{11.5.8}$$

We turn to the statistic $A(\mathbf{X})$. In view of (11.5.3), we can write it in the form (cf. (2))

$$A(\mathbf{y}) = a_0\, y_{n_1+1}^{2p} + a_1 y_{n_1+1}^{2p-1} + \cdots + a_{2p}, \tag{11.5.9}$$

where $a_0, a_1, \ldots, a_{2p}$ are polynomials in the remaining coordinates. In view of the properties of A,

$$p > 0; \qquad a_0 \not\equiv 0, \qquad \text{and} \qquad a_0 \geqq 0. \tag{11.5.10}$$

In view of (11.5.10), we can choose positive ε_1 and ε_2, and indicate a set

$$\mathcal{N} \subset \{(y_1, \ldots, y_{n_1}, y_{n_1+2}, \ldots, y_n): y_1^2 + \cdots + y_{n_1}^2 + y_{n_1+2}^2 + \cdots + y_n^2 \leqq C_5\}$$

of Lebesgue measure at least ε_1 such that

$$a_0 \geqq \varepsilon_2 > 0 \qquad \text{for } (y_1, \ldots, y_{n_1}, y_{n_1+2}, \ldots, y_n) \in \mathcal{N}. \tag{11.5.11}$$

If we let

$$\tilde{\mathcal{N}}_{\varepsilon s} = \{(y_1, \ldots, y_n): (y_1, \ldots, y_{n_1}, y_{n_1+2}, \ldots, y_n) \in \mathcal{N}, y_{n_1+1} \geqq \varepsilon\sqrt{\gamma s}\}, \tag{11.5.12}$$

then, obviously by construction, $\tilde{\mathcal{N}}_{\varepsilon s} \subset \mathcal{N}_{\varepsilon s}$ and a C_6, sufficiently large, can be indicated such that for $s > C_6$,

$$A \geqq C_7 y_{n_1+1}^{2p} \qquad \text{in the set } \tilde{\mathcal{N}}_{\varepsilon s}. \tag{11.5.13}$$

We choose and fix r subject to the restriction

$$2pr > n_2 + n_3. \tag{11.5.14}$$

We shall now estimate the right side of (11.5.1) as $s \to \infty$. Using (11.5.8), we easily see that

$$E(B^s) \leq C_8 2^{\gamma s/2} \Gamma\left(\frac{\gamma s + n_2 + n_3}{2}\right) \tag{11.5.15}$$

where Γ as usual denotes the gamma function.

Applying the asymptotic formula for the gamma function (cf. [105], p. 951) we find from (11.5.15) as $s \to \infty$ that

$$E(B^s) = O[(\gamma s)^{(1/2)(\gamma s + n_2 + n_3)} e^{-(1/2)\gamma s}]. \tag{11.5.16}$$

On the other hand,

$$E(A^r B^s) \geqq C_9 \int_{\tilde{\mathcal{N}}_{\varepsilon s}} A^r B^s e^{-(1/2)|\mathbf{y}|^2} \, d\mathbf{y}. \tag{11.5.17}$$

Hence, keeping (11.5.7) and (11.5.13) in mind and using the elementary inequality

$$\left(1 - \frac{C_3}{s}\right)^s > C_{10} \qquad \text{for } s > C_{11}$$

we see that for sufficiently large $s > C_{12}$

$$E(A^r B^s) \geqq C_{13} \int_{\varepsilon\sqrt{\gamma s}}^{\infty} y^{\gamma s + 2pr} e^{-(1/2)y^2}\, dy. \tag{11.5.18}$$

Using the same arguments as for the right side of (11.5.15), we easily conclude that as $s \to \infty$

$$E(A^r B^s) \geqq C_{14}(\gamma s)^{(1/2)(\gamma s + 2pr)} e^{-(1/2)\gamma s}[1 + O(s)^{-1/2}] \tag{11.5.19}$$

In view of the choice of r, (11.5.16) and (11.5.19) are in contradiction as $s \to \infty$ in view of the uncorrelatedness relation, showing that our original assumption that conditions (1) and (2) hold simultaneously is false. The lemma stands proved.▌

From the above lemma, we can immediately deduce a useful fact concerning the partial unlinking of uncorrelated statistics.

THEOREM 11.5.1. If condition (1) in the statement of Lemma 11.5.1. is satisfied, then an orthogonal transformation $\mathbf{x} \to \mathbf{y}$ exists such that the statistic A is a function solely of the coordinates $y_2, \ldots, y_n$.

Then, if the statistics A and B are forms of positive degree, the above orthogonal transformation can be chosen such that, further, the statistic B is a function only of the coordinates $y_1, y_2, \ldots, y_{n-1}$.

Corollary. The U-conjecture is valid for pairs of forms (A, B) depending only on two variables.

***Proof*:** In this case we have in fact $A = A(y_2)$ and $B = B(y_1)$.▌

***Proof of the Theorem*:** Suppose the first assertion of the theorem is false; then we can apply Lemma 11.5.1. with $n_2 = n$ and arrive at a contradiction. To prove the second assertion, it is sufficient to note that we can apply Lemma 11.5.1, interchanging A and B. The theorem stands proved.▌

We now establish sufficient conditions for the complete unlinking of an uncorrelated pair of statistics, one of them a polynomial and the other an 0-statistic.

THEOREM 11.5.2. If condition (1) of Lemma 11.5.1 is satisfied, and an integer $1 \leqq m \leqq n$ can be indicated such that $B(\mathbf{x}) = B(x_1, \ldots, x_m)$ and $B(\mathbf{x})$ attains its maximum absolute value on the unit sphere $x_1^2 + \cdots + x_m^2 = 1$ in the subspace $\{(x_1, \ldots, x_m)\}$ at m points $(x_{1j}, \ldots, x_{mj})$; $j = 1, 2, \ldots, m$, such that

$$\det((x_{ij})) \neq 0 \tag{11.5.20}$$

then there exists an orthogonal transformation $(x_1, \ldots, x_n) \to (x'_1, \ldots, x'_n)$ with $x_j = x'_j$ for $m < j \leqq n$, under which A reduces to a function of $x'_{m+1}, \ldots, x'_n$ alone.

Proof: Suppose the theorem is false. Then there exists an integer s, $1 \leqq s \leqq m$, and an orthogonal transformation $(x_1, \ldots, x_m) \to (y_1, \ldots, y_m)$ such that A depends on $(y_s, \ldots, y_m)$ and no orthogonal transformation $(y_s, \ldots, y_m) \to (y'_s, \ldots, y'_m)$ can decrease the number of components of $(y'_s, \ldots, y'_m)$ on which A depends. Let under the above transformation

$$(x_{1j}, \ldots, x_{mj}) \to (y_{1j}, \ldots, y_{mj}) \qquad \text{for } j = 1, 2, \ldots, m.$$

By the conditions of the theorem, for at least one j, say $j = j_0$

$$\sum_{k=s}^{m} y_{kj_0}^2 \neq 0 \tag{11.5.21}$$

but, in view of Lemma 11.5.1, the above relation contradicts the conditions of the theorem. Thus the theorem must be true.∎

As an immediate corollary from Theorem 11.5.2, we can derive the following result.

THEOREM 11.5.3. If condition (1) of Lemma 11.5.1 is satisfied and an integer $1 \leqq m \leqq n$ can be indicated such that $B(\mathbf{x}) = B_1(x_1^2, \ldots, x_m^2)$ and $B(\mathbf{x})$ attains its maximum modulus on the unit sphere $x_1^2 + \cdots + x_m^2 = 1$ at a point $(x_{11}, x_{21}, \ldots, x_{m1})$ such that $x_{j1} \neq 0$ for any j, then an orthogonal transformation $(x_1, \ldots, x_n) \to (x'_1, \ldots, x'_n)$ exists with $x_j = x'_j$ for $m < j \leqq n$, under which A reduces to a function of $x'_{m+1}, \ldots, x'_n$ alone.

Proof: We shall show that in such a case the conditions of Theorem 11.5.2 are satisfied. Indeed, $B(\mathbf{x})$ attains its maximum modulus on the unit sphere above at $(m-1)$ other points $(x_{1k}, \ldots, x_{mk})$; $k = 2, \ldots, m$, as well as at $(x_{11}, \ldots, x_{m1})$ which are given by the relations

$$x_{jk} = \begin{cases} x_{j1} & \text{if } k \neq j \\ -x_{j1} & \text{if } k = j. \end{cases}$$

For these points, it is easy to verify that the determinant det $((x_{jk}))$ is non zero being equal to

$$\begin{vmatrix} x_{11} & x_{11} & x_{11} & \cdots & x_{11} \\ x_{21} & -x_{21} & x_{21} & \cdots & x_{21} \\ x_{31} & x_{31} & -x_{31} & \cdots & x_{31} \\ \cdots & \cdots & \cdots & \cdots & \\ x_{m1} & x_{m1} & x_{m1} & \cdots & -x_{m1} \end{vmatrix} = x_{11}x_{21}\cdots x_{m1}(-2)^{m-1}.∎$$

The theorems established by us show that, in the study of the unlinking of uncorrelated (independent) statistics $A(\mathbf{X})$ and $B(\mathbf{X})$ based on normal samples $\mathbf{X}$, a very important role in the conditions imposed by us is played by the points on an appropriate unit sphere at which one of the statistics, say B, attains its maximum absolute value. If this statistic is sufficiently smooth, then the coordinates of such points must obviously satisfy the following system of equations.

$$x_j \, \partial B/\partial x_k - x_k \, \partial B/\partial x_j = 0; \qquad j, k = 1, \ldots, n, \tag{11.5.22}$$
$$x_1^2 + \cdots + x_n^2 = 1.$$

In the case when B is a polynomial (under our conditions, a form), which is in fact the case of basic interest, the system (11.5.22) turns out to be algebraic. The set of real solutions of (11.5.22), among which are to be found the points of interest to us, can, in general, have very small linear dimension, which is important for Theorems 11.5.1–11.5.2. The complexity of our problem is in a significant degree due to the difficulty of establishing a connection between (11.5.1) and, for instance, (11.5.22). For the proof of Lemma 11.5.1, we used rough asymptotic expressions for the members of (11.5.1) as $s \to \infty$. In principle, we can construct here asymptotic expansions containing information relating to the mutual character of A and B. These expansions are essentially connected with the character of the location of the solutions of (11.5.22). So far such expansions have been obtained in a "visible" form only for the case of isolated points.

A consideration of the proof of Lemma 11.5.1 shows that in fact we have used there, not the polynomial nature but only the nature of the growth of A as a polynomial. Quasi-polynomials, studied in connection with the characterization problems in Chapter 4, possess a similar property.

In particular, Lemma 11.5.1 remains valid if the statistic A satisfies the relation $0 \leqq A_1 \leqq A \leqq A_2$, where A_1 and A_2 are polynomials, A satisfies (11.5.1) and the lower polynomial A_1 satisfies condition (2) of the lemma. Similarly, in the conditions of Lemma 11.5.1, we may take as the statistic A satisfying condition (1) a rational function whose denominator does not have real zeros and whose numerator satisfies condition (2), and which under any orthogonal transformation of $(x_{n_1+1}, \ldots, x_{n_1+n_2})$, has the degree of each component larger than that of the denominator.

Such a method does not, however, enable us to consider rational functions which are homogeneous functions of degree zero. Such statistics are very important since they enable us to indicate nontrivial examples of statistics which are independent, for instance, of polynomial statistics. The property of the normal distribution is well-known, whereby a homogeneous statistic of degree zero in a random sample from a standard normal population is independent of the statistic $X_1^2 + \cdots + X_n^2$ (regarding the statistical character of such a phenomenon of independence, *vide* [86]). This property has wide statistical applications, for instance in the study of procedures eliminating a scale parameter.

We may note that in the proof of Lemma 11.5.1, (11.5.1) was in fact used for a fixed value of r subject only to the condition (11.5.14). It follows that the assertion of the lemma remains valid if (11.5.1) is replaced by the requirement of constancy of regression of the statistic A^{2r} on the statistic B if a restriction of the type (11.5.14) is imposed on r.

We shall now consider the case of symmetric statistics, i.e., statistics which are invariant under any permutation of $X_1, \ldots, X_n$, which are of the greatest importance for studying random samples. For instance, the sample mean and the higher order central moments, which are widely used in statistical practice, are examples of such statistics.

The symmetry property enables us to find a sufficient number of points on the unit sphere at which a statistic attains its maximum absolute value and consequently permits the application of Theorem 11.5.2. For locating these points, we use the following simple lemma.

LEMMA 11.5.2. Let $\mathbf{y} = (y_1, \ldots, y_n)$ be a nonnull vector in R^n. Consider all the vectors obtained by permuting the components of $\mathbf{y}$ in all possible ways, and let ρ denote the number of linearly independent vectors among them. Then

$$\rho = \begin{cases} 1 & \text{if } y_1 = y_2 = \cdots = y_n, \\ n-1 & \text{if } y_1 + y_2 + \cdots + y_n = 0, \\ n & \text{in all other cases.} \end{cases}$$

***Proof*:** The assertion of the lemma follows immediately from the construction of a basis for the vectors concerned. If all the components of $\mathbf{y}$ are equal, then $\rho = 1$. If at least two of the components are distinct, then the vectors can be constructed from $\mathbf{y}$ permuting pairs of adjacent components in a suitable manner. We demonstrate this by an example; the general case is dealt with similarly.

Let $n = 7$, and $\mathbf{y} = (y_1, \ldots, y_7)$ with $y_5 = y_6 = y_7 = a$, with none of the other y_j being $= a$. The procedure, with $\mathbf{y}$ as a column vector, is illustrated below.

$$\begin{array}{ccccccc} y_1 & y_1 & y_1 & y_1 & y_1 & y_1 & a \\ y_2 & y_2 & y_2 & y_2 & y_2 & a & y_1 \\ y_3 & y_3 & y_3 & y_3 & a & y_2 & y_2 \\ y_4 & a & a & a & y_3 & y_3 & y_3 \\ a & y_4 & a & a & a & a & a \\ a & a & y_4 & a & a & a & a \\ a & a & a & y_4 & y_4 & y_4 & y_4 \end{array} \quad \blacksquare$$

The application of this lemma enables us to obtain a complete unlinking of uncorrelated statistics, on the basis of Theorem 11.5.2, in certain cases where the statistics involved are either themselves symmetric or are reducible to such by means of orthogonal transformations.

We have the following result (cf. [41]).

THEOREM 11.5.4. If condition (1) of Lemma 11.5.1 is satisfied, and if, for some integer $1 \leqq m \leqq n$, $B(\mathbf{x}) = B(x_1, \ldots, x_m)$ and is symmetric in $x_1, \ldots, x_m$ and attains its maximum modulus on the sphere $x_1^2 + \cdots + x_m^2 = 1$ at the point $(x_{10}, x_{20}, \ldots, x_{m0})$ which does not lie on the plane $x_1 + x_2 + \cdots + x_m = 0$ or on the line $x_1 = x_2 = \cdots = x_m$, then an orthogonal transformation $(x_1, \ldots, x_n) \to (x_1', \ldots, x_n')$ with $x_j = x_j'$ for $m < j \leqq n$ exists which reduces A to a function only of $x_{m+1}', \ldots, x_n'$.

***Proof*:** Since B is symmetric, it takes its maximum absolute value on the above sphere, not only at $(x_{10}, \ldots, x_{m0})$ but also at all points which are obtained from it by permuting the coordinates. In view of the conditions of the theorem, among the coordinates of the above point, at least two are distinct, and the sum of the coordinates is not zero, so that we may apply Theorem 11.5.2 on the basis of Lemma 11.5.2. This concludes the argument and proves the theorem.▮

The case where the condition $\sum_{j=1}^{m} x_{j0} \neq 0$ is not satisfied is of independent interest, since symmetric statistics of the form $B(X_1 - \bar{X}, \ldots, X_n - \bar{X})$ are widely used in statistical practice, for instance in the construction of tests eliminating a shift parameter. The sample central moments, for instance, are of this kind.

The following theorem relates to this case (cf. [41]).

THEOREM 11.5.5. Let condition (1) of Lemma 11.5.1 be satisfied. If $B = B(x_1 - \bar{x}, \ldots, x_n - \bar{x})$ and symmetric in such a case, then A is a function of $\bar{x}$ alone.

***Proof*:** $B(y_1, \ldots, y_n)$ takes its maximum absolute value on the unit sphere $\{\mathbf{y}: y_1 + \cdots + y_n = 0;\ y_1^2 + \cdots + y_n^2 = 1\}$ at least at one point $y_0 = (y_{10}, \ldots, y_{n0})$ such that a pair of its coordinates are distinct. Hence the maximum absolute value is attained also at those points which can be obtained from y_0 by permuting the coordinates. On the basis of Lemma 11.5.1, we can now apply Theorem 11.5.1, according to which, an orthogonal transformation $(x_1, \ldots, x_n) \to (x_1', \ldots, x_n')$ with $x_n' = \sqrt{n}\,\bar{x}$ transforms A into a function of x_n' alone, which is what is required to prove. The theorem stands proved.▮

The results obtained can be used to prove yet another theorem relating to the case when both the statistics A and B are symmetric.

THEOREM 11.5.6. Let condition (1) of Lemma 11.5.1 hold. If then A and B are symmetric forms, then one of them is a function of $\bar{x}$ alone, and the other of $x_i - x_j$, $i, j = 1, \ldots, n$ alone.

***Proof*:** It is easy to establish on the basis of Lemma 11.5.1 that the forms A and B cannot attain their maximum moduli on the unit sphere at one and the same point. Thus, at least one form, say B, attains its maximum modulus on this sphere at some point $(x_{10}, \ldots, x_{n0})$ which does not lie on the line $x_1 = x_2 = \cdots = x_n$.

Then we have the alternative possibilities:

(I) $x_{10} + \cdots + x_{n0} \neq 0$;
(II) $x_{10} + \cdots + x_{n0} = 0$.

In case (I), using Theorem 11.5.4, we see immediately that A is necessarily a constant, and in case (II), successively applying Theorem 11.5.1 and Lemma 11.5.2, we establish that B is a function solely of $x_i - x_j$, $i, j = 1, \ldots, n$. Further, $A(\mathbf{x}) = c\bar{x}^k$. The theorem stands proved.∎

11.6 UNLINKING OF "ALMOST ALL" PAIRS OF INDEPENDENT PAIRS OF STATISTICS OF NORMAL SAMPLES

In the preceding section, special cases of the U-conjecture were considered, where it was found possible to confirm it; the most important of these were apparently the cases when at least one of the two independent polynomial statistics was symmetric. Further, in view of the corollary to Theorem 11.5.1, the U-conjecture is true for independent forms in two variables. In the general case, the U-conjecture has not been as yet established.

In this section, we show, following the notes [41] and [87], that, roughly speaking, almost all pairs of independent homogeneous polynomial statistics of a normal random sample $(X_1, \ldots, X_n)$ are unlinkable.

For the formulation as well as the proof of the theorems here, we shall use the tools of algebraic geometry, which are apparently natural in this area.

We shall consider the affine space $\tilde{\mathscr{M}} = \tilde{\mathscr{M}}_{pq}$ of all possible pairs of forms P and polynomials Q of degrees p and q respectively. We write $P(\mathbf{x}) = \sum_{\|\boldsymbol{\alpha}\|=p} p_{\boldsymbol{\alpha}} x_1^{\alpha_1} \cdots x_n^{\alpha_n}$ and $Q(\mathbf{x}) = \sum_{\|\boldsymbol{\beta}\| \leqq q} q_{\boldsymbol{\beta}} x_1^{\beta_1} \cdots x_n^{\beta_n}$, where $\|\boldsymbol{\alpha}\| = \alpha_1 + \cdots + \alpha_n$ and $\|\boldsymbol{\beta}\| = \beta_1 + \cdots + \beta_n$; $\deg(P) = p$, $\deg(Q) = q$.

We associate with this space another space whose elements are the coefficients of the polynomials which figure in the definition of $\tilde{\mathscr{M}}_{pq}$, the coefficients being ordered in some manner

$$\mathscr{M} = \mathscr{M}_{pq} = \{(p_{\boldsymbol{\alpha}}, q_{\boldsymbol{\beta}}) : \|\boldsymbol{\alpha}\| = p,\ \|\boldsymbol{\beta}\| = q\}. \tag{11.6.1}$$

Let $\mathscr{N} \subset \mathscr{M}$ be the subset of points of $\mathscr{M}$ corresponding to pairs of independent P and Q, and $\mathscr{V} \subset \mathscr{M}$ be the subset corresponding to uncorrelated forms (the definition was given in the formulation of Lemma 11.5.1; cf. equation (11.5.1)). The set $\mathscr{V}$ consists of points for which

$$E[P^r Q^s] = EP^r \cdot EQ^s; \qquad r, s = 0, 1, \ldots. \tag{11.6.2}$$

Obviously, $\mathscr{N} \subset \mathscr{V}$. We further define $\mathscr{L} \subset \mathscr{M}$ to be the subset corresponding to unlinkable pairs (P, Q). Then we have the inclusion relations $\mathscr{L} \subset \mathscr{N} \subset \mathscr{V}$.

The U-conjecture reduces to $\mathscr{L} = \mathscr{N}$, but we have not succeeded in proving this. We shall establish here a weaker assertion to the effect, roughly speaking, that under certain conditions the topological dimension of $\mathscr{N} \backslash \mathscr{L}$ is much lower than that of $\mathscr{N}$.

LEMMA 11.6.1. The denumerable set of relations (11.6.2) is equivalent to a finite number of them. In fact, there exists an integer constant $m_0 = m_0(n, p, q)$ such that if (11.6.2) is true for $0 \leqq r \leqq m_0$, $0 \leqq s \leqq m_0$, then it is true for any positive integers r and s.

***Proof*:** This follows immediately from the Hilbert basis theorem for polynomial ideals (Lemma 1.3.3). In fact, the equality $E(P^r Q^s) - (EP^r)(EQ^s) = 0$ is a polynomial relation between the coefficients p_α and q_β of the polynomials P and Q. The left members of these relations constitute the basis of a polynomial ideal. By Lemma 1.3.3, we can choose a finite basis for it with the corresponding values of r and s not exceeding $m_0(n, p, q)$.∎

It follows hence that $\mathscr{V}$ forms a real algebraic set over the real field (cf. the definition in section 1.3)—in fact, an algebraic correspondence between p_α and q_β over the real field.

We note that the sets $\mathscr{V}$, $\mathscr{N}$, and $\mathscr{L}$ are obviously invariant under orthogonal transformations of $\mathbf{x}$. Hence, in particular, $\mathscr{V}$ can be given by algebraic relations between orthogonal invariants, and not by all the p_α, q_β but we shall not deal with this aspect.

Under orthogonal transformations of $(x_1, \ldots, x_n)$, the cogredient tensor of coefficients (p_α, q_β) undergoes a corresponding transformation: p'_α and q'_β turn out to be polynomials in the coefficients f_{ij} of the matrix $\mathbf{F} = ((f_{ij}))$ of the orthogonal transformation and of p_α, q_β in which latter they are linear functions.

Thus, in the relations (11.6.2), we may assume that $p'_\alpha = p'_\alpha(f_{ij}, p_\alpha)$; $q'_\beta = q'_\beta(f_{ij}, q_\beta)$ are polynomials in the f_{ij} and linear in p_α, q_β, and with the relations (11.6.2) for $r, s \leqq m_0$ are adjoined the algebraic relations of orthogonality:

$$\sum_{j=1}^{n} f_{ij}^2 = 1; \qquad \sum_{j=1}^{n} f_{ij} f_{kj} = 0, \qquad i \neq k. \tag{11.6.3}$$

We turn to the relations (11.6.2). By Theorem 11.5.1, for every pair (P, Q) belonging to $\mathscr{N} \subset \mathscr{V}$, an orthogonal transformation F, depending on P and Q exists such that the polynomial Q depends only on the variables $(x'_1, \ldots, x'_{n-1})$. This means that for $(P, Q) \in \mathscr{N}$, we can add the relation below to the relations (11.6.2) and (11.6.3):

$$q'_\beta = 0 \qquad \text{for} \quad \boldsymbol{\beta} = (\beta_1, \ldots, \beta_n) \quad \text{such that} \quad \beta_n > 0.$$

The totality of these relations, polynomials in f_{ij} and linear in q_α, will be denoted by Φ. We shall consider the real algebraic set $\mathscr{V}_1 \subset \mathscr{V}$ given by the relations (11.6.2), (11.6.3), and Φ. Since $\mathscr{L} \subset \mathscr{N} \subset \mathscr{V}$, $\mathscr{V}_1$ is not empty.

We now estimate the topological dimension of $\mathscr{N}$ in the space of variables $p_\alpha, q_\beta, f_{ij}$, The set $\mathscr{L}$ is the set of unlinkable pairs of polynomials of degrees p and q. It is a linear space, formed by the coefficients p'_α, q'_β of polynomials $P(x_{\alpha_1}, \ldots, x_{\alpha_s})$, $Q(x_{\beta_1}, \ldots, x_{\beta_t})$ where no α_i is a β_j and conversely (unlinked pairs). It is known from Lemma 1.3.1 that the number of distinct monomials possible in a form of degree d in m variables is equal to

$$m(m+1)\cdots(m+d-1)/d! \tag{11.6.4}$$

In what follows, we shall take the degree p and the sample size n to be fixed numbers, and the degree $q > n$ to be large. Then, for $d = q$, it will be convenient for us to rewrite (11.6.4) in the form

$$\frac{(q+1)(q+2)\cdots(q+m-1)}{(m-1)!}. \tag{11.6.5}$$

Therefore, the topological dimension of $\mathscr{N}$ over the space $\{(p'_\alpha, q'_\beta)\}$ will not exceed the maximum of the quantities

$$\frac{s(s+1)\cdots(s+p-1)}{p!} + \sum_{d=0}^{q} \frac{t(t+1)\cdots(t+d-1)^*}{d!}$$

for $s > 0$, $t > 0$, $s + t \leqq n$. For sufficiently large q, say $q \geqq q_0$, the maximum of these quantities will be attained for $s = 1$, $t = n - 1$, and will be equal (for $n \geqq 3$) to

$$D_0(n) = \sum_{d=0}^{q} \frac{(n-1)n\cdots(d+n-2)}{d!} + 1. \tag{11.6.6}$$

If we consider the same dimension over the space $\{(p_\alpha, q_\beta, f_{ij})\}$, then we obtain the obvious relation

$$D_0(n) \leqq \dim \mathscr{N} \leqq \dim \mathscr{V}_1.$$

Then we note that the topological dimension of $\{f_{ij}\}$, the set of coefficients of the orthogonal matrix, is, as is well-known, $n(n+1)/2$.

We can now formulate the basic result.

THEOREM 11.6.1. For $n \geqq 3$ and

$$q \geqq 30\left[\frac{n(n+1)\cdots(n+p-1)}{p!} + \frac{n(n-1)}{2} + 1\right], \tag{11.6.7}$$

* For $d = 0$, this summand will be taken as 1.

the topological dimension of the set $\mathcal{N}\backslash\mathcal{Z}$ does not exceed

$$\delta_0(n) = D_0(n) - \frac{q}{6} \leqq \dim \mathcal{N} - \frac{q}{6}. \tag{11.6.8}$$

Thus, if the inequality (11.6.7) is satisfied, the set of all those elements of $\mathcal{N}$ which do not belong to $\mathcal{Z}$ (the set of unlinkable pairs of independent polynomials) has topological dimension which is significantly smaller than that of the leading components of $\mathcal{Z}$, which is not smaller than $D_0(n)$. Thus, "almost all" pairs of independent P and Q are unlinkable, in the indicated strong sense.

***Proof*:** The algebraic set $\mathcal{V}_1$ contains an algebraic subset corresponding to unlinkable pairs of the special form $P(\mathbf{x}) = a_0 x_n^p$ and $Q(\mathbf{x}) = Q(x_1, \ldots, x_{n-1})$. It is obtained by adjoining to the relations (11.6.2), (11.6.3) and Φ, the relations

$$p'_{\boldsymbol{\alpha}} = 0, \qquad \boldsymbol{\alpha} = (\alpha_1, \ldots, \alpha_n), \qquad \alpha_1 + \cdots + \alpha_{n-1} > 0. \tag{11.6.9}$$

We denote the algebraic set so obtained by $\mathcal{V}_2$: $\mathcal{V}_2 \subset \mathcal{V}_1$. By Lemma 1.3.4, the sets $\mathcal{V}_1$ and $\mathcal{V}_2$ can be decomposed into sums of a finite number of real algebraic varieties over the real field

$$\mathcal{V}_1 = V_1 \cup V_2 \cup \cdots \cup V_N; \qquad \mathcal{V}_2 = W_1 \cup W_2 \cup \cdots \cup W_M.$$

We may take these varieties to be arranged in order of decreasing topological dimension. Let s_0 be the last of the indices $s \leqq N$ for which

$$\dim V_s \geqq \delta_0(n). \tag{11.6.10}$$

Consider some V_s with $s \leqq s_0$, and the intersections $V_s \cap W_m$ for $m = 1, \ldots, M$. $V_s \cap W_m$ is an algebraic variety according to Lemma 1.3.5. Considering the various possible cases, with reference to the dimension of $V_s \cap W_m$, we have

$$\dim V_s \geqq \dim (V_s \cap W_m).$$

If $\dim V_s = \dim (V_s \cap W_m)$, then, by Lemma 1.3.10, in view of the fact that $V_s \cap W_m$ is a variety, $V_s \cap W_m = V_s$, so that $V_s \subset W_m$. Thus, all elements (P, Q) of V_s correspond to unlinkable pairs. If for every V_s with $s \leqq s_0$ a W_m can be found such that the present case obtains, then all the V_s correspond to unlinkable pairs, and the theorem is true. If this be not so, then a V_s $(s \leqq s_0)$ exists such that $\dim V_s > \dim (V_s \cap W_m)$ for all m, $m = 1, \ldots, M$. Consider the correspondence between $p'_{\boldsymbol{\alpha}}$, $q'_{\boldsymbol{\beta}}$ under the condition (11.6.3) defined by V_s. This correspondence pertains to certain uncorrelated pairs (P, Q); also the pairs belonging to $V_s - (V_s \cap \mathcal{V}_2)$ will have topological dimension greater than the pairs belonging to $V_s \cap \mathcal{V}_2$. It follows from

Lemma 1.3.8 that, for the set $V_s - (V_s \cap \mathscr{V}_2)$ we can fix $\varepsilon > 0$ and real parameters $\tau_1, \ldots, \tau_k$ such that $k = \dim V_s$, and

$$p_{\boldsymbol{\alpha}} = p_{\boldsymbol{\alpha}}(\tau_1, \ldots, \tau_k), \quad p'_{\boldsymbol{\alpha}} \equiv 0 \text{ for } \boldsymbol{\alpha} = (\alpha_1, \ldots, \alpha_n),\ \alpha_1 > 0;$$
$$q_{\boldsymbol{\beta}} = q_{\boldsymbol{\beta}}(\tau_1, \ldots, \tau_k), \quad q'_{\boldsymbol{\beta}} \equiv 0 \text{ for } \boldsymbol{\beta} = (\beta_1, \ldots, \beta_n),\ \beta_n > 0;$$
$$f_{ij} = f_{ij}(\tau_1, \ldots, \tau_k)$$

where $(\tau_1^0, \ldots, \tau_k^0)$ is some parameter point, and

$$(p_{\boldsymbol{\alpha}}, q_{\boldsymbol{\beta}}, f_{ij}) \in V_s - (V_s \cap \mathscr{V}_2)$$

and then

$$|\tau_i - \tau_i^0| \leqq \varepsilon \qquad \text{for } i = 1, \ldots, k. \tag{11.6.11}$$

The rank of the matrix $((\partial\pi/\partial\tau_r))$, where π ranges over the values $p_{\boldsymbol{\alpha}}, q_{\boldsymbol{\beta}}, f_{ij}$; $r = 1, \ldots, k$ equals $k = \dim V_s$ under the condition (11.6.11); an index $\boldsymbol{\alpha}_0 = (\alpha_1^0, \ldots, \alpha_n^0)$ exists where

$$\alpha_1^0 + \cdots + \alpha_{n-1}^0 > 0 \qquad \text{and} \qquad p'_{\boldsymbol{\alpha}_0} \neq 0$$

subject to the condition (11.6.11). This is because the form P_0 of maximal degree in P does not reduce to a monomial $a_0 x_n^p$.

In view of these circumstances, according to the well-known implicit function theorem, we can fix the quantities

$$p'_{\boldsymbol{\alpha}} = p'_{\boldsymbol{\alpha}}(\tau_1^0, \ldots, \tau_k^0), \qquad f_{ij} = f_{ij}(\tau_1^0, \ldots, \tau_k^0)$$

such that their values will remain unaltered under changes of $\tau_1, \ldots, \tau_k$ satisfying (11.6.11), and also

$$q'_{\boldsymbol{\beta}} = 0 \qquad \text{for } \boldsymbol{\beta} = (\beta_1, \ldots, \beta_n), \qquad \beta_n > 0 \tag{11.6.12}$$

and the points $(p_{\boldsymbol{\alpha}}, q_{\boldsymbol{\beta}}, f_{ij}) \in V_s - (V_s \cap \mathscr{V}_2)$ under the condition (11.6.11). Then the topological dimension of the set obtained will be not less than

$$\dim V_s - \frac{n(n+1)\cdots(n+p-1)}{p!} - \frac{n(n-1)}{2}$$
$$\geqq \delta_0(n) - \frac{n(n+1)\cdots(n+p-1)}{p!} - \frac{n(n-1)}{2} \tag{11.6.13}$$

in view of (11.6.4) and (11.6.10).

All the points constructed by us will correspond to pairs (P_0, Q) where $P_0 = P_0(x_1, \ldots, x_n)$ is a fixed form not depending on x_n alone, and $Q = Q(x_1, \ldots, x_{n-1})$. Then we have the uncorrelatedness relations

$$E(P_0^r\, Q^s) = E(P_0^r)E(Q^s). \tag{11.6.14}$$

These relations in the coordinates $q'_{\boldsymbol\beta}$, $\boldsymbol\beta = (\beta_1, \ldots, \beta_{n-1}, 0)$ under consideration define a real algebraic set in which we can select a variety $\mathscr{L}$ of maximal dimension. Then, in view of (11.6.13), we will have

$$\dim(\mathscr{L}) \geqslant \delta_0(n) - \frac{n(n+1)\cdots(n+p-1)}{p!} - \frac{n(n-1)}{2}. \qquad (11.6.15)$$

The real variety $\mathscr{L}$ is given by the coordinates $q'_{\boldsymbol\beta}$ with $\beta_n = 0$. According to formula (11.6.5), the number of variables $q'_{\boldsymbol\beta}$ is

$$D = D_0(n) - 1 = \sum_{d=0}^{q} \frac{(n-1)n\cdots(d+n-2)}{d!}. \qquad (11.6.16)$$

We have fixed the f_{ij} and the numbers $q'_{\boldsymbol\beta}$ are linear functions of the $q_{\boldsymbol\beta}$. For the chosen values of f_{ij}, consider the linear space of all $q'_{\boldsymbol\beta}$ such that

$$q'_{\boldsymbol\beta} = 0; \qquad \boldsymbol\beta = (\beta_1, \ldots, \beta_n) \quad \text{with} \quad \beta_n > 0. \qquad (11.6.17)$$

By Lemma 1.3.2, the dimension of this linear space will be equal to D. The form $P_0(x_2, \ldots, x_n)$ depends not only on x_n; without loss of generality, we may assume that it depends on x_1.

We now take the following important step. In accordance with Lemma 1.3.9, we complexify the variety $\mathscr{L}$ in such a way that its complex dimension $\dim_C(\mathscr{L})$ is equal to the real (topological) dimension of $\mathscr{L}$. Then the relation (11.6.14) will hold for given real P_0, depending not on x_n alone, and suitable polynomials $Q = Q(x_1, \ldots, x_{n-1})$ with complex coefficients lying in a complex linear space $\{q'_{\boldsymbol\beta}\}$ of complex dimension D.

Then, in view of (11.6.7), (11.6.15), and the definition of $\delta_0(n)$, the complex dimension of $\mathscr{L}$ will have the following estimate below.

$$\begin{aligned}\dim_C \mathscr{L} &\geqq \delta_0(n) - \frac{n(n+1)\cdots(n+p-1)}{p!} - \frac{n(n-1)}{2} \\ &\geqq D - \frac{q}{6} - \frac{q}{30} = D - \frac{q}{5}.\end{aligned} \qquad (11.6.18)$$

We have thus arrived at the following situation. There exists a real form $A_0 = A_0(x_1, \ldots, x_n)$ essentially dependent on x_1 and a complex algebraic variety W of complex dimension

$$\dim_C W \geqq D - \frac{q}{5} \qquad (11.6.19)$$

formed by the coefficients of polynomials $B = B(x_1, \ldots, x_{n-1})$ of degree q, all of them uncorrelated with A_0. The coefficients of the polynomials B belong to a complex linear space C with dimension D.

We now take a linear complex subspace $L \subset C$, obtained by equating to zero all the coefficients in the polynomials B for those monomials which are not of the form x_1^{4k}, $k \leqq q/4$. We have obviously

$$\dim_C L \geqq \frac{q}{4} - 1. \tag{11.6.20}$$

We take the intersection of the varieties L and W. By Lemma 1.3.11, in view of (11.6.19) and (11.6.20), for the dimension of $W \cap L = W_1$, we have

$$\dim_C W_1 \geqq \frac{q}{20} - 1. \tag{11.6.21}$$

In view of (11.6.7), we then have $\dim W_1 \geqq 2$.

Hence, there exists a polynomial of the single variable x_1

$$B_0(x_1) = b_0 x_1^{4k_0} + b_1 x_1^{4k_1} + \cdots + b_s x_1^{4k_s}$$

where the b_j are complex numbers, none of them zero, $s \geqq 1$, $k_0 > k_1 > \cdots > k_s$ and $4k_0 \leqq q$. This polynomial is uncorrelated with A_0, so that we have in particular

$$EA_0^2 B_0^t = EA_0^2 \cdot EB_0^t \tag{11.6.22}$$

for all integers $t \geqq 0$. We set $\beta = |b_1/b_0|$. The form A_0 has real coefficients and depends essentially on x_1. We write it in the form

$$A_0(\mathbf{x}) = x_1^{h_0} a'_0(x_2, \ldots, x_n) + x_1^{h_1} a'_1(x_2, \ldots, x_n) + \cdots + x_1^{h_m} a'_m(x_2, \ldots, x_n)$$

where $h_0 > 0$, and $h_0 > h_1 > \cdots > h_m$; the $a'_j(x_2, \ldots, x_n)$ are forms of appropriate degrees. We then have

$$A_0^2 = x_1^{2h_0} a_0(x_2, \ldots, x_n) + x_1^{r_1} a_1(x_2, \ldots, x_n) + \cdots + x_1^{2h_m} a_d(x_2, \ldots, x_n) \tag{11.6.23}$$

where, obviously,

$$a_0 = (a'_0)^2 \geqq 0; \qquad 2h_0 > r_1 > \cdots > 2h_m \geqq 0. \tag{11.6.24}$$

We consider the expression

$$EX_1^{2h_0} a_0(X_2, \ldots, X_n) B_0^t = Ea_0(X_2, \ldots, X_n) E(X_1^{2h_0} B_0^t)$$

where, in view of (11.6.24),

$$Ea_0(X_2, \ldots, X_n) > 0. \tag{11.6.25}$$

We consider the asymptotic behavior of $E(X_1^{2h_0} B_0^t)$ as $t \to \infty$. For this we recall the expression for even order moments of the standard normal distribution.

$$EX_1^{2r} = \frac{2^{r/2}\Gamma[(r+1)/2]}{\sqrt{\pi}} \tag{11.6.26}$$

Starting from the expression $B_0(x_1)$, we note that there exists a positive number $T_0 = T_0(b_0, \ldots, b_s)$ such that, for $|x_1| \geqq T_0$, we have, recalling that $\beta = |b_1/b_0|$,

$$|b_0|(x_1^{4k_0} - \beta x_1^{4k_1}) \leqq |B_0(x_1)| \leqq |b_0|(x_1^{4k_0} + \beta x_1^{4k_1}).$$

Hence, for any $h \leqq h_0$,

$$\left.\begin{aligned} |EX_1^{2h}B_0(\mathbf{X})| &\geqq |b_0|^t E\{X_1^{2h}(X_1^{4k_0} - \beta X_1^{4k_1})^t\} - T_1^t \\ |EX_1^{2h}B_0(\mathbf{X})| &\leqq |b_0|^t E\{X_1^{2h}(X_1^{4k_0} + \beta X_1^{4k_1})^t\} + T_1^t \end{aligned}\right\} \tag{11.6.27}$$

where $T_1 = T_1(T_0)$ is a constant. We then have

$$EX_1^{2h}(X_1^{4k_0} - \beta X_1^{4k_1})^t = E\Big\{X_1^{2h}\Big(X_1^{4k_0t} - \beta t X_1^{4k_0(t-1)+4k_1} + \beta^2 \frac{t(t-1)}{2} X_1^{4k_0(t-2)+8k_1} + \cdots \pm \beta^t X_1^{4k_1t}\Big)\Big\}.$$

The first term gives us the expression

$$EX_1^{4k_0t+2h} = \frac{2^{2k_0t+h}\Gamma(2k_0 t + h)}{\sqrt{\pi}} \tag{11.6.28}$$

and the $(m+1)$-th term has the estimate

$$\beta^m \frac{t(t-1)\cdots(t-m+1)}{m!} EX_1^{4k_0t-4m+2h} \leqq \frac{\beta^m T_2\, E(X_1^{4k_0t+2h})}{t^m}$$

as seen from (11.6.28), where T_2 is a constant $\geqq 1$. Thus, we see from (11.6.27) that

$$\left.\begin{aligned} |E(X_1^{2h}B_0^t)| &\geqq |b_0|^t EX_1^{4k_0t+2h}\left(1 - \frac{2\beta T_2}{t}\right) - (\beta T_1)^t \\ |E(X_1^{2h}B_0^t)| &\leqq |b_0|^t EX_1^{4k_0t+2h}\left(1 + \frac{2\beta T_1}{t}\right) + (\beta T_2)^t \end{aligned}\right\} \tag{11.6.29}$$

Keeping (11.6.28) in view, we set $h = h_0$ in the first of the inequalities (11.6.29), and, in the second, take $2h$ equal to the (even) exponent of X_1 in the expression (11.6.23)—obviously, there are no terms with odd exponents for X_1 in (11.6.23). Then the asymptotic relation given by (11.6.28) will contradict the relations (11.6.22). Thus Theorem 11.6.1 stands proved.∎

CHAPTER 12

Characterizations of Sequential Estimation Plans

12.1 FORMULATION OF THE PROBLEM

The present chapter is of an isolated character in relation to the preceding ones; it is concerned with problems of characterizing sequential estimation plans and not distribution functions.

The connection between properties of statistical estimators and the corresponding parent distributions was considered in Chapter 7. But the estimation plans considered there were those with constant sample size. In this chapter, we consider plans which are essentially sequential estimation plans.

Interest in such plans has grown in recent times (cf., for instance, [23, 148]). However, it must be noted that even the simplest problems in this area, as for instance the problem of optimal unbiased estimation of the probability in the classical Bernoulli scheme for various types of loss functions, have not been solved till now in a satisfactory fashion.

Hence it is natural to consider questions of sequential estimation, first for the simplest stochastic processes with parameters which arise or are introduced in a natural manner. We shall consider here four types of such processes: (1) Binomial, (2) Multinomial, (3) Poisson, and (4) Wiener processes, the last with unknown velocity of drift; we shall study questions of unbiased estimation of the natural parameters of these processes and of functions of these parameters. The rule of sequential estimation is determined by the stopping rule (plan) and the estimator given at the moment of stopping.

In problems of this type, we shall consider only Markovian stopping times. We give the necessary definitions for a process of type (1), viz., Binomial; the other processes are treated similarly.

We shall interpret a Binomial random walk on the lattice of integer points in the first quadrant, in the usual way; starting from the origin of coordinates, we take one step to the right with probability p or one step upwards with probability $q = 1 - p$. In the multinomial case, there will be $(k - 1)$ unknown probabilities $p_1, \ldots, p_{k-1}$ with $\sum_1^{k-1} p_i \leqq 1$, and we shall consider the n-dimensional lattice of integer points; if $p_k = 1 - p_1 - \cdots - p_{k-1}$, then a step of exactly one unit in the ith direction is taken with probability p_i, $1 \leqq i \leqq k$.

If $X(n)$ is the coordinate point (phase point) at moment n in our random walk, then the points $X(0), X(1), \ldots, X(n)$ will form a trajectory. The Markovian moment (rule) of stopping this is a random variable τ, taking only nonnegative integral values and independent of the future, i.e., the set $[\tau = n]$ is measurable with respect to the σ-algebra formed by the variables $X(0), \ldots, X(n)$. If $X_\tau = X(\tau)$ is the point (phase point) at moment τ, then the pair (X_τ, τ) form a sufficient statistic for our processes of types 1 and 2 (cf. [78]).

In considering unbiased estimators of functions $g(p)$ of the parameter p in the Binomial case, under convex loss functions, we may restrict ourselves to statistics of the form $f(X_\tau, \tau)$ in view of Theorem 7.1.2.

The estimation plan is given by the Markovian stopping rule, but, as we shall see later, in many cases, very simple and apparently superficial properties of the Markovian stopping moment τ characterize the plan completely.

There exists a well-developed theory due to Hunt [159], Dynkin [27], Shiryaev [166], and others, relating to the choice of the Markovian moment τ such that for a given function f, $Ef(X_\tau)$ is maximized (for fairly wide classes of Markov processes). Such optimization is not always possible, but in case it is not, the concept of ε-optimization can be introduced (for small values of $\varepsilon > 0$) and considered.

The theory referred to above permits the construction of solutions of similar optimization problems in terms of excessive majorants. This can be done usually for plans of "first entry," i.e., for stopping plans on first entry into a given set of points in the phase space as well as for optimal plans.

But in questions of unbiased sequential estimation, the problem is significantly more complicated.

Suppose we want to estimate a given function $g(p)$; p is the parameter of a Binomial process. A loss function W has also been given to us, and we wish to minimize $E_p W[f(X_\tau, \tau) - g(p)]$ for unbiased estimators $f(X_\tau, \tau)$, where E_p denotes the mathematical expectation under the value p of the parameter, under certain conditions on the Markovian stopping moment τ, say $E_p \tau \leqq M$, where M is a given number; or we wish to minimize

$$E_p[\alpha W(f(X_\tau, \tau) - g(p)) + \beta\tau]$$

where α and β are positive constants.

But now the property of unbiasedness makes the statistic $f(X_\tau, \tau)$ depend on the plan S of estimation itself, and not merely on the values of X_τ and τ. To illustrate this fact, we refer to an interesting theorem due to Girschick, Mosteller, and Savage [13].

We require the following definitions (cf. [13, 78]).

A *first entry plan* S is determined by its boundary ∂S; we stop at the moment of first entry into the boundary ∂S.

A plan S is said to be *closed* if the boundary is reached in finite time, i.e., if $P_p(\tau < \infty) = 1$. (P_p denotes the probabilities calculated under the value p for the parameter.)

A point will be called *transient* if, starting from the origin of coordinates, we can reach and pass it with positive probability. A point is a *boundary point* if it belongs to the boundary set ∂S, and an *inaccessible point* if it is neither a transient nor a boundary point. It is clear that all inaccessible points can be deleted from the boundary of the plan as inessential, since it is impossible to reach them. In what follows, we shall consider only plans without inaccessible points; transient and boundary points will be called *accessible*.

A plan S will be said to be finite and of size n if $P_p(\tau \leqq n) = 1$ and $P_p(\tau = n) > 0$, and we shall write: size $(S) = n$. It is necessary to note that this property in contrast to the property of being closed: $P_p(\tau < \infty) = 1$, does not depend on the set of possible values of the parameter p.

We shall denote by $Q(S)$ the set of values of p for which a plan S is closed. S will be called *complete* if the sufficient statistic (X_τ, τ) is complete in the sense of Definition 7.1.5. Thus S is complete for the set G of p-values if, for every measurable function f such that $E_p f(X_\tau, \tau)$ exists and is equal to zero for all $p \in G$, we have $P_p[f(X_\tau, \tau) = 0] = 1$ for all $p \in G$. S will be called *boundedly complete* for a set of p-values if the completeness condition formulated above holds for bounded functions $f(X_\tau, \tau)$.*

We shall consider only Markovian moments (rules) of stopping τ, for which $P_p(\tau < \infty) = 1$ for all values of p in the given parameter space. Suppose we are given a boundary $B = \partial S$ of integer points in the first quadrant and it is desired to find an unbiased estimator for the function $g(p) = p^\xi(1-p)^\eta$ where the point (ξ, η) is accessible. If $X_\tau = (x_\tau, y_\tau)$ is the stopping point at moment τ, we must form the statistic

$$f = \frac{K_{\xi,\eta}(X_\tau)}{K_{0,0}(X_\tau)} \tag{12.1.1}$$

where $K_{0,0}(X_\tau)$ is the number of possible trajectories connecting the origin to the point X_τ, and $K_{\xi,\eta}(X_\tau)$ is the number of possible trajectories from the point (ξ, η) to X_τ.

* Recently, R. A. Zaidman proved that all Markovian stopping rules which are complete are given solely by plans of first entry.

Such an estimator is easily obtained by means of the wellknown process of projecting any unbiased estimator of $g(p)$ above onto a sufficient statistic.

We note that $f \leqq 1$, since every trajectory, giving rise to the number $K_{\xi,\eta}(X_\tau)$, can be extended to a trajectory giving rise to the number $K_{0,0}(X_\tau)$.

We see that the statistic f depends on the plan S, and not merely on the values of X_τ and τ, so that the optimization theory indicated above ("Rao-Blackwellization") cannot be immediately applied. As a result of this situation, plans which are "best" in various senses, in problems of sequential estimation, are not always plans of first entry. But there exist some important cases when this is so (cf. [23]), for example, in some asymptotic problems of unbiased sequential estimation. Hence we shall study here plans of the type of first entry and their relation to the distribution of values of the Markovian stopping moment τ.

Consider a complete plan S and a parametric function $g(p)$; obviously, there exists at most one unbiased estimator of the form $f(X_\tau, \tau)$ for $g(p)$. Thus, the problem of unbiased estimation of a parametric function using estimates based on complete plans reduces to the choice of the plan and the general theory of optimization cannot be applied.

But other tools of estimation theory, in particular the Rao-Cramer information inequality, can be very effective and useful, for the case of the quadratic loss function, in problems relating to this area (cf. [23, 148]).

The form of the best unbiased estimator for a given plan S—*vide* (12.1.1)—is fairly complicated for the case of sequential estimation in comparison with estimation under a fixed sample size, and this makes the corresponding optimization problem fairly difficult. Satisfactory solutions of the corresponding optimization problems (not connected directly with characterization problems) have been obtained so far only in very special cases (cf. [148]).

These estimators can be simplified (they even become linear in some cases) if certain restrictions are imposed on the distribution of the Markovian moment τ. For example, we might consider only plans S for which $E_p \tau = \text{const.}$ for all values of $p \in Q(S)$; this would generalize plans of constant sample size.

Surprisingly, in many cases, even very mild conditions imposed on the distribution of τ restrict very substantially the plans obeying these conditions, especially in the case of plans of first entry. We shall discuss such phenomena for stochastic processes of the four types mentioned earlier.

12.2 BINOMIAL PROCESSES

Before discussing characterization problems for sequential estimation plans connected with a Binomial process, we recall some useful theorems due to Lehmann and Stein, and De Groot ([78, 23]). We shall consider only first entry plans.

We call a closed plan S *minimal* if the deletion of any of the boundary points destroys the closedness. We call a plan S *simple* if the transient points lying on every straight line of the form $x + y = k$ (k a positive integer) form an interval (are not separated by the boundary points).

The Lehmann-Stein theorem [78] runs as follows.

THEOREM 12.2.1. If $Q(S) \supset [0, p_0]$, $p_0 > 0$, then a set of necessary and sufficient conditions for the plan S to be complete is

(1) the plan is simple; and

(2) the plan in minimal.

Proof: *Vide* [78].∎

For finite plans, an elegant variant of these conditions was given by De Groot [23]. If the first entry plan S is finite, then $Q(S) = [0, 1]$. In fact, we have

$$\sum_{X_\tau \in \partial S} K_{0,0}(X_\tau) p^{x_\tau} q^{y_\tau} = 1 \tag{12.2.1}$$

where $K_{0,0}(X_\tau)$ is, as in section 12.1, the number of trajectories connecting the origin with the boundary point X_τ. If the set ∂S is bounded and size $(S) = n$, then $x_\tau + y_\tau \leqq n$, and thus the left member of (12.2.1) is a polynomial in p of degree at most n. If (12.2.1) holds for $n + 1$ distinct values of p, then it is valid for all values of p.

In this case also, the concepts of completeness and bounded completeness coincide.

De Groot's theorem has the form

THEOREM 12.2.2. In order that a plan S of size n be complete, it is necessary and sufficient that

$$\#(\partial S) = n + 1 \tag{12.2.2}$$

where $\#(\partial S)$ denotes the number of elements of the boundary.

Proof: *Vide* the work of De Groot [23].∎

We now turn to the characterization of first entry plans S by the behavior of $E_p \tau$. The behavior of this quantity is very fundamental for the theory of sequential estimation under quadratic loss in view of the following important theorem due to De Groot.

THEOREM 12.2.3. If the parametric function $g(p)$ is unbiasedly and efficiently—in the sense of the Rao-Cramer information inequality—estimable at the point $p = p_0$, then $g(p)$ must be of the form

$$g(p) = a(p - p_0)E_p \tau + b \tag{12.2.3}$$

where a and b are arbitrary constants.

Now, following the Note [28], we can formulate the following characterization theorem.

THEOREM 12.2.4. A complete finite plan S of size n is completely determined by the values of $E_p\tau$ for any $(n+1)$ distinct values of p.

Remarks. The proof is fairly complicated. Before proving it, we consider some of its consequences. We see that a complete finite plan is determined by the values of $E_p\tau$, in view of the theorem, so that we can associate various plans with various functions of the type (12.2.3) which are efficiently estimated at the point p, and conversely. We shall further "enumerate" these plans and thus enumerate the types of functions (12.2.3).

Theorem 12.2.4 turns out to be fairly deep. In the particular case thereof: $E_p\tau = M$ for $p \in [0, 1]$, where M is a natural number, the theorem shows that, among complete plans, fixing $E_p\tau$ merely means confining ourselves to plans of constant sample size M.

If we also fix the variance of the Markovian moment, then this assertion holds for all closed plans without inaccessible points as easily shown.

In fact, let $E_p\tau = M_1$ and $\mathrm{Var}_p(\tau) = M_2$ be constants. For $p = 0$, we see that $E_0\tau = y_0$ where y_0 lies on ∂S and on the y-axis; similarly, $E_1\tau = x_0$. Thus $x_0 = y_0 = M_1$ is a natural number. Further,

$$E_p(\tau^2) = \mathrm{Var}_p(\tau) + (E_p\tau)^2 = M_2 + M_1^2.$$

It is clear that $E_0(\tau^2) = x_0^2 = E_1(\tau^2) = y_0^2$, whence $M_2 + M_1^2 = x_0^2$. Hence, $\mathrm{Var}(\tau) = x_0^2 - x_0^2 = 0$, so that $\tau = M_1$ with probability one, and the plan S has constant sample size M_1.

Proof of the Theorem 12.2.4. Since $E_p\tau$ is a polynomial in p of degree at most $(n+1)$, its values at $(n+1)$ distinct points p_i determine its values for all p. We now have to establish some lemmas on finite complete closed first entry plans. Let us denote the set of these plans by $\mathscr{P}$. These plans satisfy in particular the condition of completeness (Theorem 12.2.1). We turn our attention here to the condition: if P and Q are two transient points of the plan, lying on the straight line K_c: $x + y = c$, then all other integer points on K_c lying between P and Q are also transient. Suppose we call this condition the *D-condition.*

LEMMA 12.2.1. The boundary ∂S of a plan of size n, $S \in \mathscr{P}$, has the form $\partial S = B_1 \cup B_2 \cup B_3$, where $B_i \cap B_j = \phi$ if $i \neq j$, and B_1 is a "segment" of the straight line $x + y = n$ (consisting of integer-points in the first quadrant); B_2 and B_3 are the parts of the boundary connecting respectively the y-axis and the x-axis to K_n. Further, on every integer-valued vertical line to the left of B_1, B_2 has exactly one point, and on every integer-valued horizontal line below B_1, B_3 has exactly one point. B_2 and B_3 may be empty (*vide* Figure 1).

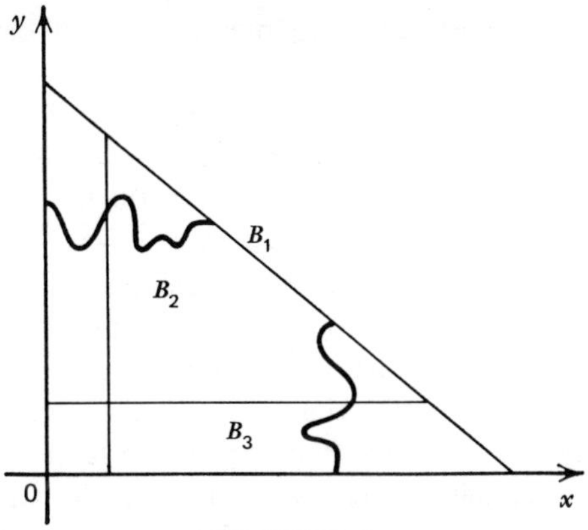

FIGURE 1.

***Proof*:** Since size $(S) = n$, the intersection $B_1 \cap K_n$ is nonempty. Then, condition (D) implies that B_1 is a "segment," and that the set of all transient points of S is simply connected and so is the boundary ∂S. The part $\partial S - B_1$ of the boundary consists of two disjoint parts B_2 and B_3, connecting B_1 with respectively the y-axis and the x-axis.

Consider B_2. On every integer-valued vertical line to the left of B_1, B_2 has at least one point P (for instance, the highest accessible point on this vertical). We shall show that such a point on the vertical is unique in B_2. Towards this end, consider a plan S' of the same size as S, obtained from S by deleting from ∂S the part B_3 and replacing it by the segment B_3' of the diagonal K_n (*vide* Figure 2). We assert that the plan S' belongs to $\mathscr{P}$ (as does plan S). It is clear that it is closed, finite (even of size n) and minimal. Its completeness

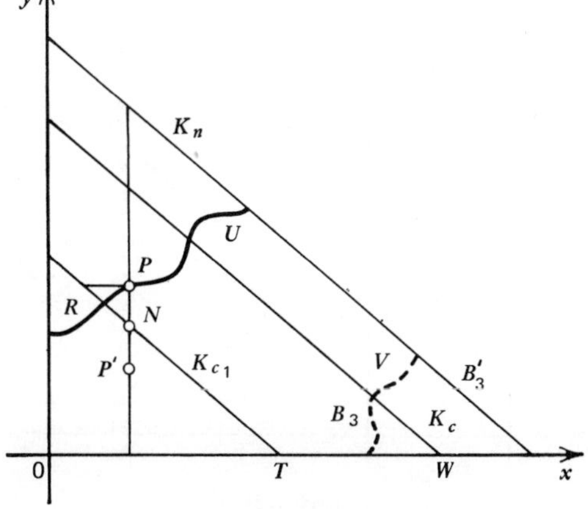

FIGURE 2.

follows from the minimality and the preservation of condition (D). If in plan S, the transient points on the line K_c form the segment UV, then in plan S', a segment VW is adjoined to the segment UV, so that a segment UW is formed and property (D) is preserved.

Now we take an integer-valued vertical line to the left of B_1 and the highest boundary point P on it. Consider the points $R = P - (1, 0)$ and $N = P - (0, 1)$. At least one of these is a transient point, otherwise P would be inaccessible. Consider the diagonal K_c which contains the points R and N. Let T be its point of intersection with the x-axis, so that T is a transient point in plan S'. Hence, in view of condition (D), the whole of the segment NT consists of transient points. Hence, if a point P' exists on the same vertical but below P (cf. Figure 2) and belongs to B_2, we can delete it without violating the closedness of S'. But then S' will not be minimal, leading to a contradiction. Thus, P is the unique boundary point on the vertical. A similar argument holds for B_3 (we may also consider the plan S^* symmetric to S', obtained by taking reflections in the bisector $y = x$). Thus, Lemma 12.2.1 is proved.▌

We see that B_2 can be given by the equation: $y = f(x)$, where f is a single- and integer valued function of the integer variable x; a similar representation holds for B_3.

LEMMA 12.2.2.

$$f(x+1) \geqq f(x) - 1 \tag{12.2.4}$$

and a similar inequality holds for B_3.

***Proof*:** If $f(x+1) - f(x) < -1$, plan S cannot be closed in view of Lemma 12.2.1, since (cf. Figure 3), from the transient point $(x, f(x) - 1)$, we can pass to the other side of the boundary.▌

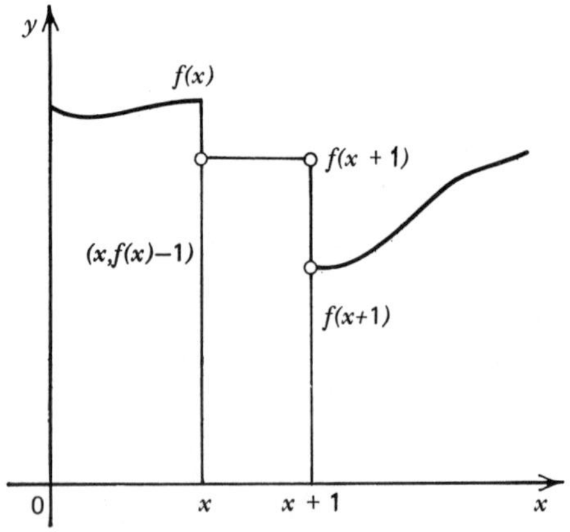

FIGURE 3.

LEMMA 12.2.3. Every finite plan S with boundary ∂S satisfying the conditions of Lemma 12.2.1 and condition (12.2.4) belongs to $\mathscr{P}$.

Proof: Consider a closed plan S with boundary $\partial S = B_1 \cup B_2 \cup B_3$, the decomposition being as described in the statement of Lemma 12.2.1. We now verify the closedness of plan S, showing that B_2 stops trajectories going upwards and B_3 stops trajectories going to the right. Consider B_2. The vertical trajectory going from the origin along the y-axis is obviously stopped by B_2. Suppose some trajectory does not intersect B_2 up to the vertical $x = x_0 \geqq 0$; we shall show that it cannot pass through B_2 up to the vertical $x = x_0 + 1$ (if it has not stopped earlier). In fact, let the final point of the trajectory on the vertical $x = x_0$ be (x_0, y_0), $y_0 = f(x_0)$; if the trajectory reaches the vertical $x = x_0 + 1$, its final point will be $(x_0 + 1, y_0)$; $y_0 \leqq f(x_0) - 1 \leqq f(x_0 + 1)$ in view of (12.2.4) so that the boundary point $(x_0 + 1, f(x_0 + 1))$ does not permit its passing above through B_2. Thus the required property of B_2 is established by induction; the corresponding property of B_3 is proved similarly.

The minimality of our plan is obvious. We have now to verify property (D) to convince ourselves that the plan is complete.

Suppose (D) is not satisfied. Then there exists a diagonal K_m (*vide* Figure 4) on which transient points P and R are separated by a point Q which is either a boundary point or an inaccessible point. Obviously, $m < n$, so that the point Q must be either not below B_2 or not to the left of B_3. Suppose for definiteness that it is not below B_2 (*vide* Figure 4). Let x_1 and x_2 be the abcissae of P and Q. Then $f(x_2) \leqq m - x_2$ since Q is not below B_2 and $f(x_1) > m - x_1$ since P is below B_2. Hence $-f(x_1) < x_1 - m$ and $f(x_2) - f(x_1) < x_1 - x_2$ which contradicts (12.2.4). A similar argument applies to B_3, so that condition (D) is proved and plan S is complete. Thus, $S \in \mathscr{P}$ and our lemma stands proved. ∎

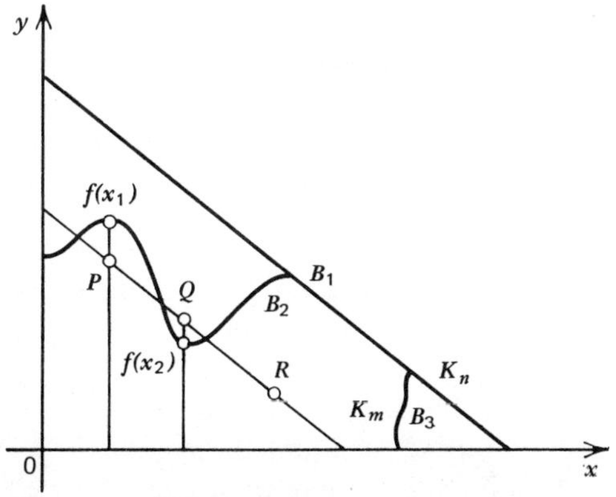

FIGURE 4.

LEMMA 12.2.4. Let $S \in \mathscr{P}$ and ∂S have points on the x-axis and the y-axis at distances of a and b, respectively, from the origin. If $a \leqq b$, then, there is exactly one point of ∂S on the horizontal line $y = 1$ (similarly, if $a \geqq b$, then the vertical line $x = 1$ contains exactly one point of ∂S).

***Proof*:** Let $P = (a, 0)$ and $Q = (0, b)$ be points on ∂S and on the axes, and $a \leqq b$. Consider the horizontal line $y = 1$. Let R be the point of ∂S lying on this line and having maximum abscissa, and W another boundary point on the same horizontal (*vide* Figure 5). Then $x_W \leqq a - 2$, otherwise the point R would be inaccessible, as easily seen. Now consider the point $U = (x_W + 1, 0)$

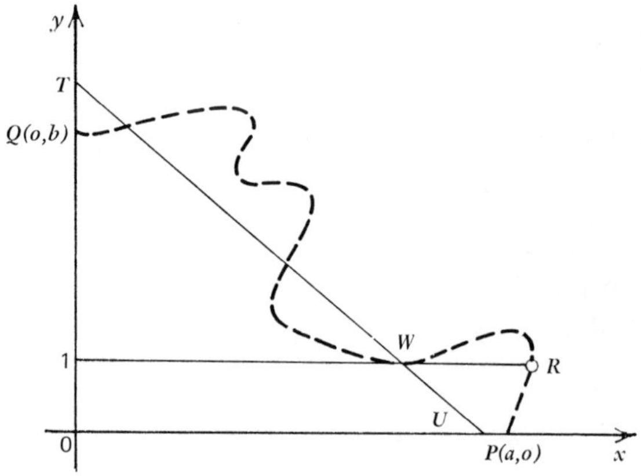

FIGURE 5.

on the x-axis, and draw the straight line WU. Let T be its point of intersection with the y-axis. We see that U is a transient point since $x_U = x_W + 1 \leqq a - 2 + 1 = a - 1 < a$, and W is a boundary point. But then, in view of condition (D), T cannot be a transient point and hence is either inaccessible or a boundary point. We thus have (*vide* Figure 5) $b = y_Q \leqq y_T = x_U < a$, and consequently $b < a$, a contradiction. Hence W cannot exist. The lemma follows.▌

For the proof of Theorem 12.2.4, we introduce yet another concept, of "pseudo-plans." Consider a first entry plan $S \in \mathscr{P}$ and suppose that part of its boundary ∂S, together with the quadrant $x \geqq x_0$, $y \geqq y_0$, forms a new plan S_{x_0, y_0} (*vide* Figure 6).

The probability that a trajectory reaches the axes $x = x_0$ $(y \geqq y_0)$ and $y = y_0 (x \geqq x_0)$ depends on (x_0, y_0) and the parameter p. We denote it by $Q(p, x_0, y_0)$. Let ξ be a point on one of these axes: $\xi = (x_0, y)$ where $y \geqq y_0$ or $\xi = (x, y_0)$ where $x \geqq x_0$.

If we denote by $P(p, \xi)$ the probability of reaching these axes for the first time at the point ξ, then we can calculate the conditional probability of getting first to the point ξ if we find ourselves on the axes of the plan S_{x_0, y_0}.

$$\nu_p(\xi) = P(p, \xi)/Q(p, x_0, y_0)$$

This is a measure concentrated on the axes of the plan S_{x_0, y_0} (inclusive of the boundary points). If such a measure $\nu_p(\xi)$ is given, we can compute the mean stopping time $E_{\nu, p}(\tau)$ for the new plan S_{x_0, y_0}, at which the trajectories leave from the point ξ on the axes with probability $\nu_p(\xi)$. If a trajectory leaves from a point of the boundary ∂S, we set $\tau = 0$ and consider it as immediate stopping.

We now define a pseudo-plan as a finite first entry plan, equipped with a probability measure ν_p concentrated on the transient and boundary points on both its axes; this measure may depend on p.

Further, the conditions below must be satisfied:

(a) $\nu_p(0, 0) \geqq C > 0$ for all p, where C is a constant which is independent of p, but may depend on S:

(b) as $p \to 0$, $\nu_p(x, 0) = o(p^x)$; and

(c) as $p \to 1$, $\nu_p(0, y) = o(q^y)$.

We shall denote the class of such measures $\nu_p(.,.)$ by $\mathcal{N}$. Obviously, a plan of the usual kind is a pseudo-plan with $\nu_p(0, 0) = 1$ (such a measure ν_p obviously belongs to $\mathcal{N}$). Consider now a pseudo-plan S such that it has a boundary point $(0, a)$ on the vertical $x = 0$, and one and only one boundary point $(1, y_e)$ on the vertical $x = 1$. We shall construct a new pseudo-plan S^*, taking $x = 1$ and $y = 0$ as its axes. We define a new measure ν_p^*, at the transient and boundary points of its axes, induced by the measure ν_p. We shall show that $\nu_p^* \in \mathcal{N}$ so that we have in fact obtained a pseudo-plan S^* from the pseudo-plan S.

We verify that conditions (a)–(c) are satisfied.

Let M denote the event that the trajectories under plan S reach the vertical $x = 1$ or start from the horizontal line $y = 0$, $x \geqq 1$. Then

$$P(M) = 1 - \sum_{y=0}^{a} \nu_p(0, y)q^{a-y}$$

and, as $p \to 1$, $P(M) \to 1$, as follows from (c); as $p \to 0$, $P(M) \geqq \nu_p(0, 0)p \geqq Cp$, and $P(M) = \sum_{x=1}^{b} \nu_p(x, 0) + \sum_{y=0}^{a} \nu_p(0, y)(1 - q^{a-y}) \leqq C_1 p$, where $(b, 0)$ is a boundary point of the pseudo-plan S, in view of condition (b).

We now verify condition (a) for S^*. $\nu_p^*(1, 0) = [\nu_p(0, 0) \cdot p + \nu_p(1, 0)]/P(M)$ $\geqq \nu_p(0, 0)/C_1 = C_2$ where C_2 is a new constant.

Condition (b) is verified as follows: as $p \to 0$, $x \geqq 1$, $\nu_p^*(x, 0) = \nu_p(x, 0)/P(M)$ $\leqq \nu_p(x, 0)/Cp = o(p^{x-1})$ in view of condition (b) for the pseudo-plan S.

Finally, as $p \to 1$, $y \geqq 1$, we have $\nu_p^*(1, y) = (1/P(M)) \sum_{i=0}^{y} \nu_p(0, i)pq^{y-i} =$ $o(q^y)$ in view of condition (c) for the plan S, verifying (c) for S^*.

Thus, our construction indeed yields a new pseudo-plan S^*.

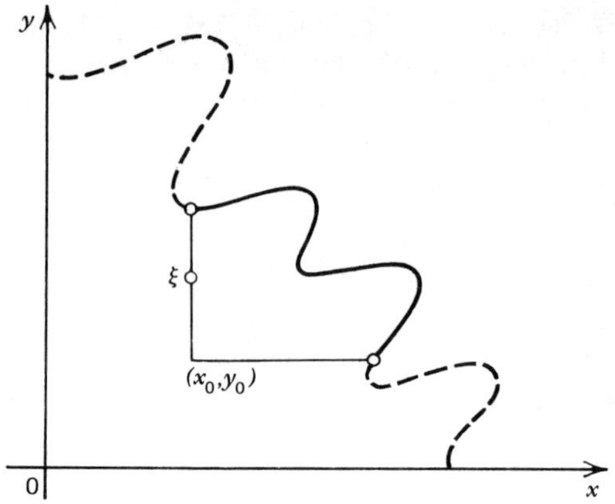

FIGURE 6.

We note that the plan S_{x_0, y_0} (*vide* Figure 6) is also a pseudo-plan, if the conditional probability of a trajectory leaving from a given point of the axes, given that it has reached the axes, is taken as v_p.

Let a closed pseudo-plan S be given, with boundary points on the axes: $(0, a)$ and $(b, 0)$, such that there is only one boundary point $(1, c)$ on the vertical $x = 1$ (cf. Figure 7).

We must assume that $c - a \geqq -1$, otherwise the pseudo-plan S will not be closed.

We now prove a lemma.

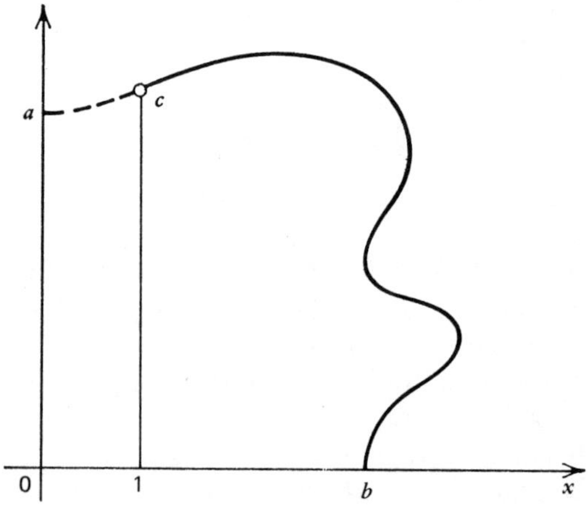

FIGURE 7.

LEMMA 12.2.5. Let S' be a pseudo-plan, constructed from the original pseudo-plan S by deleting the y-axis and forming on the axes the measure v'_p induced by the measure $v_p \in \mathcal{N}$ of S. Then $v'_p \in \mathcal{N}$ and $E_p(\tau \mid S')$ is completely determined by $E_p(\tau \mid S)$, p, and the value of a (cf. Figure 7).
Note: For any pseudo-plan S, $E_p(\tau \mid S)$ denotes the mean stopping time for a given p.
Proof: We first give the explicit expression for the measure v'_p. Let the measure v_p be given by the probabilities $v_p(x, 0)$, $x = 0, \ldots, b$ and $v_p(0, y)$, $y = 1, \ldots, a$. The probability that a trajectory does not reach the vertical $x = 1$ is

$$\sum_{y=0}^{a} v_p(0, y)q^{a-y} \qquad (c \geqq a - 1, \text{ as noted above}). \tag{12.2.5}$$

The probability that a trajectory first reaches the point $(1, y)$, $y = 1, 2, \ldots, a - 1$, is

$$\sum_{z=0}^{y} v_p(0, z)q^{y-z}p, \tag{12.2.6}$$

that it passes the point $(1, 0)$ is

$$v_p(0, 0)p, \tag{12.2.7}$$

that it leaves from $(1, 0)$ is

$$v_p(1, 0), \tag{12.2.8}$$

and that it leaves from $(x, 0)$ is, for $x = 1, 2, 3, \ldots$

$$v_p(x, 0). \tag{12.2.9}$$

From the above formulas, we find that the conditional measure v'_p for the pseudo-plan S' is completely determined by v_p, p, and a.

We now have

$$E_p(\tau \mid S) = P_p(G)E_p(\tau \mid S, G) + (1 - P_p(G))E_p(0 \mid S')$$

where G is the event that a trajectory never reaches the axes of the plan S' and does not originate from its axes. In this case, the trajectory must originate from the y-axis and go along it; the probability of this event is given by (12.2.5), remembering that $c - a \geqq -1$, and depends only on v_p, p, and a. The conditional expectation $E_p(\tau \mid S, G)$ is given by the sum

$$\sum_{y=0}^{a} v_p(0, y)q^{a-y}(a - y)$$

divided by (12.2.5), and this expression depends only on v_p, p, and a. Thus, $E_p(\tau \mid S')$ is determined by $E_p(\tau \mid S)$, p, and a, as required to prove.∎

We can now prove Theorem 12.2.4, which we shall establish in a form which is a generalization to the case of pseudo-plans.

THEOREM 12.2.5. Let S be a pseudo-plan, constructed from a complete, finite plan of size n, endowed with a probability measure $\nu_p \in \mathcal{N}$ concentrated on the set of transient and boundary points of its axes. Then its boundary ∂S is completely determined by the values of $E_p \tau$ at any $n+1$ distinct points.

Remark. It follows that the boundary ∂S of an ordinary plan of size n is completely determined by the values of $E_p \tau$ at $(n+1)$ distinct points.

***Proof*:** We shall proceed by induction on the size n of the plan. Suppose that two such plans A and A' are given, with $\nu_p = \nu'_p$, where ν_p and ν'_p are the corresponding measures, and that $E_p(\tau | A) = E_p(\tau | A')$. Let first size $(A) =$ size $(A') = 1$. Then the boundaries ∂A and $\partial A'$ consist of the two points $(1, 0)$ and $(0, 1)$. $E_p(\tau | A) = E_p(\tau | A') = \nu_p(0, 0) = \nu'_p(0, 0)$ and this is the only possible value of $E_p(\tau | S)$ for all such plans; the fact that plans A and A' coincide now follows from the fact that they are both of size one. Let now the theorem be true for all pseudo-plans of size $\leqq n_0$; we shall show that it is valid for pseudo-plans of size $\leqq (n_0 + 1)$. Let A and A' be two such pseudo-plans, with boundary points on the y-axis given by $(0, a)$ and $(0, a')$ respectively. As $p \to 0$, keeping in view the continuity of $\nu_p(\cdot)$ in p and the estimate for ν_p as $p \to 0$, we obtain

$$E_p(\tau | A) = E_p(\tau | A') = E_{\nu_p}(a - y) + O(p) = E_{\nu_p}(a' - y) + O(p)$$

where E_{ν_p} is the expectation under measure ν_p, whence it follows that $a = a'$. Similarly, if $(b, 0)$ and $(b', 0)$ respectively are boundary points on the x-axis of the two plans, then, letting $p \to 1$, we obtain $b = b'$.

Let now $a \leqq b$(as we may take without loss of generality). By Lemma 12.2.4, both plans will have exactly one boundary point each on the vertical $x = 1$, which we take to be $(1, c)$ and $(1, c')$, respectively. Further, we have from Lemma 12.2.2 that

$$c - a \geqq -1; \qquad c' - a \geqq -1.$$

We now construct a new pair of pseudo-plans A'' and A''' by deleting the vertical line $x = 1$ from the pseudo-plans A and A' (cf. Figure 7) and constructing new conditional probabilities for the emergence of trajectories from the axes. These conditional measures will coincide, as follows from the proof of Lemma 12.2.5 (as indicated above, the boundary points correspond to trajectories stopping at the moment of emergence, i.e., generally not emerging from the axes).

By Lemma 12.2.5, we have $E_p(\tau | A'') = E_p(\tau | A''')$, since obviously size (A''), size (A''') are both $\leqq n_0$. By induction, $\partial A'' = \partial A'''$, and the points $(0, a)$ and $(0, a')$, as we have seen, coincide. If $a > b$, we interchange the roles of the x- and y-axes and proceed in the same way. The theorem stands proved.▌

We now turn to functions which are estimable efficiently at $p = p_0$: *vide* formula (12.2.3). Since the constants a and b are arbitrary, we can call the set of such functions as the type of efficiently estimable functions. By Theorem 12.2.3, the number of types of functions $g(p)$ which can be estimated efficiently at a given point p_0 with the help of complete plans of size n is equal to the number of these plans. Hence a formula enumerating these plans is useful.

THEOREM 12.2.6. Complete plans of size n have exactly one boundary point on the x- and y-axes. Let us denote the distances from the origin of these points by a and b respectively, and the number of complete plans of size n, with these points as boundary points on the axes, by $A_n(a, b)$. If $a \leqq b$, then we have

$$A_n(a, b) = \sum_{i=a-1}^{n-1} A_{n-1}(i, b-1). \tag{12.2.10}$$

Remark. We note that the condition $a \leqq b$ does not result in any loss of generality, since there exists an obvious symmetry among (complete) plans of size n: $A_n(a, b) = A_n(b, a)$.

***Proof*:** By definition, we set $A_n(0, 0) = 0$. It is easily seen that $A_n(n, n) = 1$ (then only a plan of constant sample size n can be complete). Now consider a plan S of size n with boundary points $(0, a)$ and $(b, 0)$, with $a \leqq b$, on the axes. By Lemma 12.2.5, it will have exactly one boundary point P on the vertical $x = 1$, and, by Lemma 12.2.2, $y_P \geqq a - 1$. Let us now transfer the y-axis to the position $x = 1$ and delete the previous axis $x = 0$. By Lemma 12.2.3, we obtain a new plan $S' \in \mathscr{P}$ of size $n - 1$ with boundary points on its axes at the points $(0, y_P)$ and $(a - 1, 0)$. This means that the left-hand side of (12.2.10) is not greater than the right. We now take a plan S' of size $n - 1$ with boundary points on the axes at $(0, a')$ and $(b - 1, 0)$, where $a' \geqq a - 1$, and extend it by introducing a new y-axis located one unit to the left of the former y-axis, endow it with a boundary point at height a, to obtain by Lemma 12.2.4 a plan of size n with boundary points on the axes at $(0, a)$ and $(b, 0)$. Thus, the right side of (12.2.10) is not greater than its left side, and our relation stands proved.▌

We now turn to infinite complete plans. We can make the following assertion.

THEOREM 12.2.7. For infinite complete plans, closed for certain values of $p \neq 0, 1$, one of the following statements holds.

(a) The plan S is closed for $p \in [0, p_0)$ for some $p_0 \in (0, 1)$ and not closed for $(p_0, 1]$: thus, $Q(S) = [0, p_0)$ or $[0, p_0]$. The boundary contains exactly one point on every vertical, generating an integer-valued function f defined on the set of nonnegative integers. If $\overline{\lim}_{n\to\infty} f(n)/n = a$, then $p_0 = 1/(1 + a)$.

(b) (Dual of (a)): $Q(S) = [p_0, 1]$ or $(p_0, 1]$. ∂S has exactly one point on every horizontal. The number p_0 is determined similarly to (a).

(c) S is closed for values of p near the origin and near unity. ∂S consists of two disjoint parts of types (a) and (b) above respectively.

Proof: Consider a minimal complete infinite plan S. Condition (D) must hold. We take the n-truncated plan, obtained from S by deleting from ∂S all points where $x + y > n$ and introducing as new boundary points all the transient points on the diagonal $x + y = n$ (cf. Figure 8).

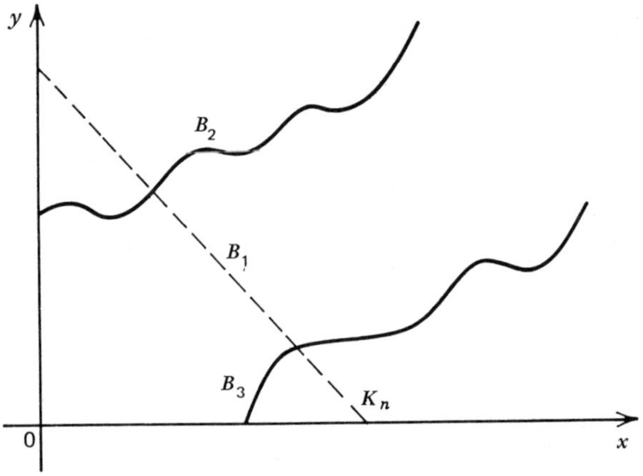

FIGURE 8.

We thus obtain a finite plan S_n of size n. In view of the minimality of S, S_n will also be minimal. Condition (D) will obviously be satisfied for S_n, so that S_n is a complete plan. Hence, according to Lemmas 12.2.1–3, the boundary ∂S_n will consist of three disjoint parts B_1, B_2, and B_3, where B_1 will lie along the diagonal $x + y = n$. B_2 and B_3 may be empty. As $n \to \infty$, B_2 and B_3 form the former boundary ∂S. By the above lemmas, B_2 (B_3) has exactly one point on every integer-valued vertical (horizontal) to the left of (below) B_1. Thus Theorem 12.2.5 is proved except for the assertion on the value of p_0. We now turn to this.

We have to show that a plan S of type (a) has the property that $Q(S) \supset [0, p_0)$ and $Q(S) \cap (p_0, 1] = \phi$. Let $p < p_0$; we shall show that S is closed. Let us denote $(1/p_0) - 1$ by $a > 0$. For $p < p_0$, $(q/p) = (1/p) - 1 > (1/p_0) - 1 = a$. Consider the regression line $L_1: y = (q/p)x$ and the line $L_2: y = (a + \varepsilon)x$, where $\varepsilon > 0$, $a + \varepsilon < (q/p)$. By the definition of a, there exists an infinite sequence $x_1, x_2, \ldots$ such that the boundary points $(x_i, f(x_i))$ will lie below L_2. If S is not closed for the value p, then we can find with positive probability, an infinite trajectory not intersecting B_2, and hence lying below B_2. In particular, this happens for points with the abscissas x_i. But this contradicts the law of the iterated logarithm, according to which, with probability one, a trajectory can leave the neighborhood, of a well-known type, of the line L_1 situated above L_2 only a finite number of times. Hence $p \in Q(S)$.

In the same way, we prove that, for $p > p_0$, the plan S is not closed. In fact, consider the two straight lines: L_1 given by $y = (q/p)x$ and L_2 by $y = (a - \varepsilon)x$ where $0 < \varepsilon < 1 + a - (1/p)$, where the line L_2 lies below L_1, and, beginning with some abscissa $x = x_0$, lies strictly below ∂S. Again, by the law of the iterated logarithm, there exists with positive probability a trajectory lying below L_2, and thus not meeting the boundary.

A dual assertion holds for plans of type (b), and similarly for those of type (c). This proves Theorem 12.2.7.▌

The above theorem leads us to a theorem on characterization of infinite complete plans through the values of $E_p \tau$.

THEOREM 12.2.8. An infinite complete plan S is determined by the values of $E_p \tau$. If $E_p \tau$ is bounded in the neighborhood of $p = 0$ and in the neighborhood of $p = 1$, we have a plan of type (c); if it is bounded only in the neighborhood of $p = 0$, then we have a plan of type (a), with a dual assertion for plans of type (b).

***Proof*:** The second assertion, regarding the types, follows from the proof of the preceding theorem.

If we have two plans of type (a), a successive comparison and deletion of the verticals $x = 0$, $x = 1$, ... shows that the boundaries of the two plans coincide; the same is true for plans of type (*b*). For plans of type (c), we may proceed as in the case of plans of finite size, introducing pseudo-plans and successively deleting vertical and horizontal lines to show that the boundaries coincide.▌

It is a curious fact that some types of first entry plans are determined by very mild conditions imposed on the moments of τ for an arbitrary given individual value of p, say $p = p_0$.

Such for example are inverse sampling plans—plans involving sampling up to the appearance of exactly $k(\geqq 1)$ events, each occurring with probability $q = 1 - p$. If $p = p_0 < 1$, we have, as is well-known

$$E_{p_0} \tau = k/q_0, \qquad (E_{p_0} \tau)' = k/q_0^2, \qquad E_{p_0} \tau^2 = (k^2 + kp_0)/q_0^2 \qquad (12.2.11)$$

where the prime denotes the derivative at $p = p_0$. We have the following.

THEOREM 12.2.9. Let $p_0 \in (0, 1)$ be a given parametric value and suppose for a plan S without inaccessible points and closed in the neighborhood of p_0, we have

$$E_{p_0} \tau = k/q_0, \qquad (E_{p_0} \tau)' = k/q_0^2, \qquad E_{p_0} \tau^2 \leqq (k^2 + kp_0)/q_0^2. \qquad (12.2.12)$$

Then S in an inverse sampling plan.

Proof: The plan is given to be closed in the neighborhood of $p = p_0$. Let $K_{x,y}$ be the number of trajectories connecting the origin with the point $(x, y) \in \partial S$. We have the obvious equality

$$\sum_{(x,y)\in\partial S} K_{x,y} q_0^y (1-q_0)^x = 1 \tag{12.2.13}$$

which also holds in the neighborhood of q_0. According to (12.2.12),

$$\sum_{(x,y)\in\partial S} K_{x,y}(x+y) q_0^y (1-q_0)^x = k/q_0 \tag{12.2.14}$$

$$\sum_{(x,y)\in\partial S} K_{x,y}(x+y)[(y/q_0) - (x/(1-q_0))] q_0^y (1-q_0)^x = k/q_0^2 \tag{12.2.15}$$

$$\sum_{(x,y)\in\partial S} K_{x,y}(x+y)^2 q_0^y (1-q_0)^x \leqq (k^2 + k(1-q_0))/q_0^2. \tag{12.2.16}$$

In view of (12.2.16), we can differentiate the identity (12.2.13) with respect to q at q_0, to obtain

$$\sum_{(x,y)\in\partial S} K_{x,y}[y/q_0 - x/(1-q_0)] q_0^y (1-q_0)^x = 0.$$

Hence $E_{p_0}(Y/q_0 - X/(1-q_0)) = 0$. In view of the fact that

$$X + Y = \tau \tag{12.2.17}$$

we have hence $E_{p_0} Y = qE_{p_0} \tau$ (which is clearly a special case of the well-known Wald identity in sequential analysis, derived by us here directly). Then we find from (12.2.14) that

$$E_{p_0} Y = k \tag{12.2.18}$$

for $(X, Y) \in \partial S$. We now show that

$$\operatorname{Var}_{p_0} Y = 0 \tag{12.2.19}$$

which will prove our theorem. We have from (12.2.9) that

$$E_{p_0}(Y - q\tau)\tau = -k(1-q_0)/q_0$$

whence

$$E_{p_0}(\tau Y) = q_0 E_{p_0} \tau^2 - k(1-q_0)/q_0.$$

Differentiating the identity (12.2.13) once again with respect to q at q_0, we hence obtain

$$E_{p_0}(Y - q_0\tau)^2 = k(1-q_0).$$

Hence, according to (12.2.16),

$$\begin{aligned} E_{p_0} Y^2 &= k(1-q_0) + 2q_0 E_{p_0} Y\tau - q_0^2 E_{p_0} \tau^2 \\ &= -k(1-q_0) + q_0^2 E_{p_0} \tau^2 \leqq -k(1-q_0) + k^2 + k(1-q_0) = k^2. \end{aligned}$$

Further, $\mathrm{Var}_{p_0}\,\tau = E_{p_0}\tau^2 - (E_{p_0}\tau)^2 \leqq k^2 - k^2 = 0$ (cf. (12.2.12)). This proves (12.2.19).

We see that the boundary ∂S of plan S lies entirely on the horizontal line $y = k$. Since the plan is closed for $p = p_0$, ∂S must evidently contain all integer points $y = k$, proving the theorem.∎

Theorem 12.2.9 is a (somewhat refined) version of the following.

THEOREM 12.2.10. Let the boundary of a plan S, closed in a neighborhood of $p_0 \in (0, 1)$, be made up of m line segments and n points. Let $k = 2m + 4n$, and the $k(k + 1)/2$ expectations and their derivatives below exist.

$$(E_{p_0}\tau^s)^{(r)}, \qquad r + s \leqq k, \qquad s \geqq 1. \tag{12.2.20}$$

Then all the straight lines and points forming the boundary ∂S are determined by the values (12.2.20). Note: Since S is closed in the neighborhood of p_0, this will usually determine the plan S itself (as in the case of Theorem 12.2.9). In what follows, we discuss some examples.

***Proof*:** Let the straight lines supporting segments of ∂S be given by the equations

$$a_j x + b_j y + c_j = 0, \qquad j = 1, \ldots, m$$

and the points which are isolated integer points in ∂S be $(\alpha_j, \beta_j), j = 1, \ldots, n$. We set up the expression

$$\Pi = \Pi(x, y) = \prod_{j=1}^{m} (a_j x + b_j y + c_j) \prod_{j=1}^{n} [(x - \alpha_j)^2 + (y - \beta_j)^2]. \tag{12.2.21}$$

If we show that

$$E_{p_0}(\Pi) = 0, \qquad \mathrm{Var}_{p_0}(\Pi) = 0 \tag{12.2.22}$$

then it would follow that $\Pi = 0$ with probability one (under p_0) and ∂S in fact lies on a set of lines and points defined by the equation $\Pi(x, y) = 0$.

$\Pi(x, y)$ given by (12.2.21) is a polynomial in x and y of degree $k = m + 2n$. To calculate $\mathrm{Var}_{p_0}(\Pi)$, we have to compute $E_{p_0}(\Pi^2)$, Π^2 being a polynomial of degree $2k$. We have further, $\tau = X + Y$, and

$$E_{p_0}\tau^s = \sum_{(x, y) \in \partial S} K_{x,y}(x + y)^s p_0^x (1 - p_0)^y \tag{12.2.23}$$

where $K_{x,y}$ as usual is the number of trajectories connecting the origin to the point $(x, y) \in \partial S$.

Differentiating the expression (12.2.23) with respect to p at p_0, we find

$$(E_{p_0}\tau^s)' = E_{p_0}\{[X/p_0 - Y/(1 - p_0)](X + Y)^s\}, \qquad (X, Y) \in \partial S.$$

Multiplying both sides of this relation by $p_0(1 - p_0)$, we find

$$E_{p_0}(X + Y)^s(X - p_0\tau) = p_0(1 - p_0)(E_{p_0}\tau^s)'.$$

Hence, the left side is determined by the values of $(E_{p_0}\tau^s)'$ and, generally, as is easily seen, the expression

$$E_{p_0}[(X+Y)^{s_0}(X-p_0\tau)^{r_0}]$$

is determined by the values of $(E_{p_0}\tau^s)^{(r)}$, $r+s \leqq r_0+s_0$. In addition to relation (12.2.20), we also have the identity $E_{p_0}\tau^0 = E_{p_0}1 = 1$, which also can be differentiated with respect to p at p_0.

We consider (12.2.22) for plan S: the left side of these relations is defined by expectations of the form $E_{p_0}X^{m_1}Y^{m_2}$, $m_1+m_2 \leqq 2k = 2m+4n$. We shall show that these expectations are uniquely determined by the values (12.2.20). For this purpose, we introduce the linear transformation

$$\tau = x+y; \qquad \xi = q_0 x - p_0 y$$

with determinant -1. We note that $\xi = x - p_0\tau$. We see that the monomials $x^{m_1}y^{m_2}$ are expressible as linear combinations of the monomials $\tau^s\xi^r$, $r+s \leqq m_1+m_2$ with coefficients depending on r and s, and the same is true of the corresponding expectations. Since $E_{p_0}\tau^{s_0}\xi^{r_0}$ can be expressed in terms of $(E_{p_0}\tau^s)^{(k)}$, $k+s \leqq s_0+r_0$, we see that $E_{p_0}\Pi$ and $E_{p_0}\Pi^2$ must coincide with the same expressions for the plan S, and consequently, $\Pi = 0$ for $p = p_0$. Since Π does not depend on p_0, it vanishes for all values of p, proving the theorem. ▌

We derive some consequences of this theorem. In Theorem 12.2.9, we considered an inverse sampling plan S; ∂S consisted of integer points on the horizontal line $y = n$; in the formulation of Theorem 12.2.10, we have to set $m = 1$, $n = 0$, $k = 2m = 2$. Thus, the three quantities

$$E_{p_0}\tau, \qquad (E_{p_0}\tau)', \qquad \text{and } E_{p_0}\tau^2 \tag{12.2.24}$$

define the line $y = n$ and the inverse sampling plan.

We also consider a "stoping plan" with boundary ∂S of the type $y = kx + b$, where $k \geqq 0$, $b > 0$ are integers, or the line symmetric with it: $y = (1/k)x - b$. The first of these is of type (a) and the second of type (b), in the terminology of Theorem 12.2.7. In view of Theorem 12.2.10, the three expressions (12.2.24) determine a sloping plan. In fact, they determine the line $y = kx + b$ (or $y = (1/k)x - b$). Since the plan S is closed in the neighborhood of p_0, all the integer points on the above line and in the positive quadrant belong to ∂S; thus, the values (12.2.24) determine the plan (the condition relating to $E_{p_0}\tau^2$ can be weakened further, as in Theorem 12.2.9).

Other plans of interest in this context are plans having a boundary made up of two or more parallel lines; the existence of several lines here is necessary to guarantee the closedness of the plans in some interval. Such, for example, will be a plan S with boundary ∂S made up of the integer points on the two lines: $y = (1/2)x + 3$, $y = (1/2)x + 4$.

12.3 MULTINOMIAL PROCESS

We consider a multinomial process with probability of a step in the direction of the i-th axis being p_i. The definitions concerning first entry plans are similar to those in section 12.2 for the Binomial process. In this case, however, simple geometrical conditions for completeness are not known (cf. [28]).

We discuss here only an analogue of Theorem 12.2.10 and other simple facts.

First we consider a finite plan without inaccessible points and such that

$$E_{\mathbf{p}}\tau = M_1, \qquad E_{\mathbf{p}}\tau^2 = M_2 \tag{12.3.1}$$

where M_1 and M_2 are constants and $\mathbf{p} = (p_1, \ldots, p_n)$ is an arbitrary element of the simplex $p_1 + \cdots + p_n = 1$, $p_i \geqq 0$ for all i. We can make the assertion that M_1 and M_2 are integers, and $M_2 = M_1^2$ and the plan S is one of constant size with the boundary ∂S: $\tau = x_1 + \cdots + x_n = M_1$.

In fact, since S is a finite plan, it must have on each of the x_i-axes exactly one point of the boundary set; let the distance of such a point on the x_i-axis be a_i from the origin. Taking $\mathbf{p} = (0, \ldots, 0, 1, 0, \ldots, 0)$ with the 1 in the i-th place, in (12.3.1), we see that $a_i = M_1$ for all i, so that M_1 is an integer and all the a_i have a common value, M_1. Further, $\tau = x_1 + \cdots + x_n$, $E_p\tau^2 = a_i^2 = M_1^2$, so that $M_2 = M_1^2$. Hence $\operatorname{Var}_p \tau = 0$, $E_p\tau = M_1$, proving our assertion.

We now turn to a more complicated proposition, the analogue of Theorem 12.2.10.

Let a plan S be closed in the neighborhood of the point $\mathbf{p}_0 = (p_{01}, \ldots, p_{0n})$ in the simplex of probabilities, and let ∂S consist of a given number of intersections of linear subspaces $R^k \subset R^n$. Then we can assert that the linear subspaces whose intersections support the boundary ∂S are defined by expressions of the form

$$D^{r_1 \cdots r_{k-1}} E_{p_0} \tau^s \tag{12.3.2}$$

where $D^{r_1 \cdots r_{k-1}}$ is the symbol of differentiation r_i times with respect to p_i at the point $\mathbf{p}_0$, and $r_1 + \cdots + r_{k-1} + s \leqq c$, c being defined by the numbers and types of the subspaces under consideration; it is clear that we assume the existence of the expression (12.3.2).

As in the proof of Theorem 12.2.10, it is sufficient to show that the quantities (12.3.2) define the moments $E_{\mathbf{p}_0} X_1^{m_1} \cdots X_n^{m_n}$ for $(X_1, \ldots, X_n) \in \partial S$.

For brevity, we consider only the case $n = 4$. We have $\tau = x_1 + x_2 + x_3 + x_4$. As in the proof of Theorem 12.2.10, we note that the quantities (12.3.2) determine the quantities $E_{p_0}\xi_1^{m_1}\cdots\xi_4^{m_4}$, for $m_1+\cdots+m_4 \leqq r_1+r_2+r_3+s$, where

$$\begin{aligned}
\xi_1 &= \tau = x_1+x_2+x_3+x_4\\
\xi_2 &= \quad x_1(1-p_1-p_2-p_3)-p_1x_4\\
\xi_3 &= \quad x_2(1-p_1-p_2-p_3)-p_2x_4\\
\xi_4 &= \quad x_3(1-p_1-p_2-p_3)-p_3x_4.
\end{aligned}$$

Setting $\lambda_i = p_i/(1-p_1-p_2-p_3)$ and introducing the new variables $\xi_1' = \xi_1$, $\xi_2' = \xi_2\lambda_2/p_2$; $\xi_3' = \xi_3\lambda_3/p_3$, and $\xi_4' = \xi_4\lambda_4/p_4$, we note that the quantities $E_{p_0}(\xi_1')^{m_1}\cdots(\xi_4')^{m_4}$ determine the quantities $E_{p_0}\xi_1^{m_1}\cdots\xi_4^{m_4}$, and hence the row vector $\boldsymbol{\xi}^T = (\xi_1, \ldots, \xi_4)$ can be replaced by the row vector $(\boldsymbol{\xi}')^T = (\xi_1', \ldots, \xi_4')$ in our subsequent arguments. If $\mathbf{x}^T = (x_1, \ldots, x_4)$, then we have the linear transformation $\boldsymbol{\xi}' = \boldsymbol{\Lambda}\mathbf{x}$, where

$$\boldsymbol{\Lambda} = \begin{pmatrix} 1 & 1 & 1 & 1\\ 1 & 0 & 0 & -\lambda_1\\ 0 & 1 & 0 & -\lambda_2\\ 0 & 0 & 1 & -\lambda_3 \end{pmatrix}.$$

It is easily seen that $\det(\Lambda) = -(1+\lambda_1+\lambda_2+\lambda_3) = -(1-p_1-p_2-p_3)^{-1} \neq 0$, for points inside the simplex. Thus (12.3.2) indeed determines $E_{p_0}X_1^{m_1}\cdots X_4^{m_4}$ and our assertion is proved.

We now consider some examples.

Some plans S for a multinomial process easily reduce to plans for a Binomial process; such are plans of first entry, with boundary ∂S of the form $x_i = k$. But sloping plans do not reduce to binomial plans.

Consider the first "octant" and the plane

$$a_1x_1+\cdots+a_nx_n = b. \tag{12.3.3}$$

In order for such a plane to be the boundary set of a closed plan S, it must not admit the passage of trajectories of the multinomial process through it. It is easily seen that the numbers a_i can be chosen to be integers in such a case (multiplying both sides of (12.3.3) by one and the same number if necessary for the purpose). Under our conditions, the following result holds.

THEOREM 12.3.1. (Malyshev)* If $b > 0$ and the a_i $(i = 1, \ldots, n)$ are integers, with the greatest common factor $(a_1, \ldots, a_n, b) = 1$, then, in order for the plane (12.3.3) to be the boundary ∂S of a first entry plan S, it is necessary and sufficient that $a_i \leqq 1$ for all i and that $a_i = 1$ for at least one value of i.

* This theorem was intimated to the authors by Prof. A. V. Malyshev in January 1969.

We shall not present a proof of this theorem, which involves the geometry of numbers.

It follows that there exist infinitely many sloping multinomial plans. For example, the plane

$$x_1 + x_2 + \cdots + x_{n-2} - 3x_{n-1} - 7x_n = 31$$

will define such a plan S. $Q(S)$ will contain the neighborhood of some point inside the simplex. If $\mathbf{p}_0$ is an element of $Q(S)$, then the quantities $D^{r_1 \cdots r_{n-1}}(E_{p_0} \tau^s)$ for $r_i = 0, 1$; $s = 1, 2$, will determine the sloping plan S.

12.4 POISSON PROCESS

We shall now consider a homogeneous Poisson process $X_\tau(w)$:

$$P[X_\tau(w) = k] = (\lambda\tau)^k e^{-\lambda\tau}/k!$$

$k = 0, 1, 2, \ldots$; $\tau \geqq 0$, $\lambda > 0$. For estimating a parametric function $g(\lambda)$, we shall consider a sequential first entry plan S with boundary ∂S defined on the set $U = N \times T$, where N is the set of all nonnegative integers and T is the set of positive numbers. Suppose that ∂S consists of a finite or countably infinite number of points or segments. We suppose that the end points of these segments and the isolated boundary points do not have any point of accumulation.

We shall make use of the definitions in section 12.1 concerning first entry plans S; in particular, such a plan S will be said to be finite (or bounded) if ∂S lies wholly within a bounded subset of $U = N \times T$. Consider a point u of U. Let $k(u) \in N$ and $t(u) \in T$ be the components of the point u of the process to be characterized. Let $\mathscr{B}$ be the σ-algebra of subsets of U, generated by the points of N and the segments closed to the right, of T.

The following result, first obtained by Trybula [148], follows as a special case from Lemma 1.6.1.

LEMMA 12.4.1. In the case of a boundary ∂S without limit points, of the type described above, there exists a countably additive measure μ_S defined on $\mathscr{B}$ and independent of λ such that, for every λ

$$P_\lambda[(X_\tau(w), \tau(w)) \in A] = \int_A \lambda^{x(u)} e^{-\lambda t(u)} \mu_S(du) \tag{12.4.1}$$

where A is any member of $\mathscr{B}$, contained in ∂S, and $(X_\tau(w), \tau(w))$ is a point in ∂S.

In [148], Trybula also extended the arguments of De Groot on efficiently estimable functions at a given parameter point in the case of a Poisson process.

If (X_τ, τ) is a point on ∂S, then a function $g(\lambda)$ is efficiently estimable at λ_0 if and only if it has the form

$$g(\lambda) = a(\lambda - \lambda_0)E_\lambda \tau + b \tag{12.4.2}$$

where a and b are constants.

Hence the study of the functions $E_\lambda \tau$ and their relation to sequential estimation plans S is of some interest.

Since we shall use the notion of complete plans, we establish a result relating to the completeness of finite plans. We shall study only plans without inaccessible points and with a boundary of the form considered above. We shall denote by $\mathscr{S}$ the class of such first entry type plans.

THEOREM 12.4.1. In order that a finite, minimal, closed plan $S \in \mathscr{S}$ be boundedly complete, it is necessary and sufficient that

$$\text{meas.}\ (\partial S) = T_0 \tag{12.4.3}$$

where meas. (∂S) is the Lebesgue measure of the boundary and T_0 is the least upper bound of possible stopping moments.

Proof: Since plan S is finite and belongs to $\mathscr{S}$, its boundary ∂S must consist of only a finite number of points and segments. If meas. $(\partial S) < T_0$, then the plan will obviously not be closed, since trajectories can pass through ∂S with positive probability. Further, T_0 must be equal to the t-coordinate of the right-most point of ∂S. Let now S be complete and closed. From what has been said above, it follows that we must have meas. $(\partial S) \geqq T_0$; we shall show that the inequality sign cannot hold, so that condition (12.4.3) is necessary. Let, then, meas. $(\partial S) > T_0$, if possible. Then, obviously, the projections of two segments of ∂S on the t-axis must overlap. Let us have two segments

$$x = k_1, \quad t \in [t_1, t_2] \qquad \text{and} \qquad x = k_2, \quad t \in [t_1, t_2]$$

lying on ∂S, where $k_1 \neq k_2$ are integers. In view of (12.4.1), in order to show that the plan S is not complete, it suffices to construct two continuous nonzero functions ϕ_1 and ϕ_2 such that

$$\lambda^{k_1} \int_{t_1}^{t_2} \phi_1(t)e^{-\lambda t}\, dt = \lambda^{k_2} \int_{t_1}^{t_2} \phi_2(t)e^{-\lambda t}\, dt.$$

Let $k_2 > k_1$ and $k_2 - k_1 = k > 0$. Then, it is sufficient that

$$\int_{t_1}^{t_2} \phi_1(t)e^{-\lambda t}\, dt = \lambda^k \int_{t_1}^{t_2} \phi_2(t)e^{-\lambda t}\, dt.$$

Let ϕ_1 be a smooth function, vanishing along with its first derivative at t_1 and t_2; then, integrating by parts, we have

$$\int_{t_1}^{t_2} \phi_1(t)e^{-\lambda t}\, dt = (1/\lambda) \int_{t_1}^{t_2} e^{-\lambda t}\phi_1'(t)\, dt.$$

If $\phi_1', \ldots, \phi_1^{(k-1)}$ have the same properties as ϕ_1, we see that we can take $\phi_2 = \phi_1^{(k)}$ and obtain the desired equality, so that the plan S cannot be complete. This proves the necessity part.

Suppose now that meas. $(\partial S) = T_0$; we shall show the bounded completeness of the plan S. Let this not be so. Then we must have

$$\sum_{s=1}^{L} \lambda^{k_2} \int_{t_{1s}}^{t_{2s}} e^{-\lambda t}\phi_s(t)\,dt + \sum_{m=1}^{M} \lambda^m e^{-\lambda t_m} k_m = 0 \tag{12.4.4}$$

where L and M are integers, k_m are real numbers, the segments $[t_{1s}, t_{2s}]$ of the t-axis do not overlap, and the ϕ_s are integrable with respect to Lebesgue measure over $[t_{1s}, t_{2s}]$, $s = 1, \ldots, L$.

Evidently, we must have $L \geqq 1$. We multiply the relation (12.4.4) by $e^{\lambda M}$ and differentiate M times. We obtain a relation of the form (12.4.4) with M replaced by $M - 1$ and the monomials λ^k replaced by polynomials $Q_k(\lambda)$. Let us multiply that relation by $e^{\lambda(M-1)}$ and differentiate $M - 1$ times. Continuing in this manner, we obtain finally the relation

$$\sum_{s=1}^{L} Q_s(\lambda) \int_{t_{1s}}^{t_{2s}} e^{-\lambda t}\phi_s(t)\,dt = 0$$

where the Q_s are polynomials. We assume further that $t_{2,s-1} = t_{1,s}$; $s = 2, 3, \ldots, L$. Hence we deduce that the entire function of λ given by

$$\phi(\lambda) = \int_{t_{1L}}^{t_{2L}} e^{-\lambda t}\phi_L(t)\,dt \tag{12.4.5}$$

has the estimate

$$\phi(\lambda) = O[\exp\{|\lambda|(t_{2,L-1} + \varepsilon)\}]$$

for any $\varepsilon > 0$. Hence, by Lemma 1.4.1 (the Paley-Wiener theorem), we must have

$$\phi(\lambda) = \int_{-t_{2,L-1}}^{t_{2,L-1}} e^{-\lambda t}\psi(t)\,dt$$

where

$$\psi \in L_2(-t_{2,L-1}, \quad t_{2,L-1}) = L_2(-t_{1,L}, t_{1,L}).$$

Comparing this with (12.4.5), we see that we must have $\phi_L(t) = 0$ a.e., $\psi(t) = 0$ a.e., in view of the uniqueness of the representation as a Fourier integral, leading to a contradiction and proving the theorem. ∎

We remark that the theorem is not true for closed plans which are not finite. Consider for instance a plan with constant time $T = T_0$; here ∂S consists of all points of the form $(\tau = T_0, X_\tau = k)$, $k = 0, 1, 2, \ldots$. It is easily seen that this plan is complete, and meas. $(\partial S) = 0$.

We shall now establish an analogue of Theorem 12.2.3, which enables us to characterize parametric functions which are efficiently estimable at a given parametric point.

THEOREM 12.4.2. A finite plan $S \in \mathscr{S}$ is fully determined by the values of $E_\lambda \tau$ in an arbitrarily small segment of values of λ.

Remark. We note that a finite plan $S \in \mathscr{S}$ is simply a complete, closed, finite plan, whose boundary ∂S consists of a finite number of points and segments.

***Proof*:** In the case of a finite plan, in view of formula (12.4.1), $E_\lambda \tau$ is an entire function of λ and hence is fully determined by its values on an arbitrarily small segment of values of λ.

We shall represent the plan S on the axes (t, x); ∂S consists of a finite number of horizontal segments with integer ordinates $x = i$, $i = 1, \ldots, k$; their projections on the t-axis do not overlap. Let us call k the *height of the plan.* We shall call the other important parameter (the number of segments n in ∂S) the *order* of the plan. We have $n \geqq 1$, $k \geqq 1$. If $n > 1$, then $k > 1$. A plan $S \in \mathscr{S}$ of order n and height k will be called an *(n, k)-plan.*

For the proof of Theorem 12.4.2, we have to generalize the notion of a plan, as was done in section 12.2 for the Binomial processes.

We shall call an (n, k)-plan a *pseudo-plan* if it is endowed with a family of probability measures ν_λ, $\lambda \in (0, \infty)$, on the set of all transient and boundary points of the t-axis and x-axis. These measures must satisfy the conditions below: as $\lambda \to 0$,

(a_1) for any interval $U_r = [0, r)$ of the t-axis, we must have $\nu_\lambda(U_r) \geqq C_r$, where $C_r > 0$ and is independent of λ, but possibly dependent on the plan S;

(a_2) $\nu_\lambda\{(0, x)\} = o(\lambda^x)$, $x = 0, 1, \ldots$;

as $\lambda \to \infty$,

(b_1) $\nu_\lambda(U_r) \geqq C_r \lambda^{-k}$, where k is an integer constant which may depend on the plan S;

(b_2) for any interval $V_t = (t, \infty)$ of the t-axis, there exists an integer constant l, depending only on the plan S, such that

$$\nu_\lambda(V_t) = O(\lambda^l e^{-t\lambda}).$$

We shall denote the family of probability measures of such a type by $\mathscr{N}$. As was done in the Binomial case, we verify similarly that these properties are preserved when we pass from the pseudo-plan S to a new pseudo-plan S^*. For this we shall consider two types of constructing a new pseudo-plan.

(1) S^* is obtained from S as follows: the left boundary segment of S^* has the form $x = x_0$, $0 \leqq t \leqq t_0$, and the next boundary segment is situated on the horizontal $x = x_1$, $x_1 < x_0$; then, if $x_0 - x_1 > 1$, it is possible that boundary points (t_0, x), $x_1 < x < x_0$, x integer, are present.

(2) The plan S has a boundary point $(t_0, 0)$ on the t-axis. It is deleted for S^*, and the horizontal $x = 1$ is chosen as the (new) axis of the plan S^*.

We shall verify that in case (1) or case (2), the measures v_λ^*, induced by the measures v_λ of plan S, belong to $\mathcal{N}$.

Let us start with case (1). Denote by M the event that the trajectories of plan S reach the axes of plan S^*, i.e., that the trajectories do not come to a stop on the segment $[0, t_0]$ or at the integer points (t_0, x) of the vertical $t = t_0$, for $x > x_1$.

Let A denote a Borel subset of $X \times T$ and B a Borel subset of the axes (i.e., of $X \cup 0t$).

Let $R_\lambda(A \mid B)$ be the probability of first entry into A starting from B, under the measure v_λ. If $B = X \cup 0t$, then we shall simply write the above as $R_\lambda(A)$. We shall check that $v_\lambda^* \in \mathcal{N}$.

Let $\lambda \to 0$. We have to verify conditions (a_1) and (a_2).

(a_1) For the point $(t_0, 0)$, we have

$$v_\lambda^*\{(t_0, 0)\} \geqq \frac{1}{P(M)} \int_0^{t_0} e^{-\lambda(t_0 - t)}\, dv_\lambda(t).$$

As $\lambda \to 0$, $P(M) \to 1$ and the above expression will be asymptotically $v_\lambda(U_{t_0}) \geqq C_{t_0}$.

(a_2) For $0 < x < x_1$,

$$v_\lambda^*\{(t_0, x)\} = \frac{1}{P(M)} \left[\sum_{i=0}^{x} R_\lambda\{(t_0, x) \mid (0, i)\} v_\lambda\{(0, i)\} \right.$$
$$\left. + \int_0^{t_0} R_\lambda\{(t_0, x) \mid (t, 0)\}\, dv_\lambda\{(t, 0)\} \right] = o(\lambda^x)$$

as easily checked, starting from elementary properties of a Poisson process.

Let $\lambda \to \infty$. We shall verify conditions (b_1) and (b_2) for v_λ^*. We have

$$P(M) = \sum_{i=0}^{x_1} v_\lambda\{(0, i)\} \sum_{j=i}^{x_1} R\{(t_0, j) \mid (0, i)\}$$
$$+ \int_0^{t_0} \sum_{j=0}^{x_1} R_\lambda\{(t_0, j) \mid (t, 0)\}\, dv_\lambda\{(t, 0)\} + v_\lambda(V_{t_0})$$
$$= \sum_{i=0}^{x_1} v_\lambda\{(0, i)\} O(e^{-t_0\lambda} \lambda^{x_1 - i})$$
$$+ \int_0^{t_0} e^{-(t - t_0)\lambda} \sum_{j=0}^{x_1} \frac{[(t_0 - t)\lambda]^j}{j!}\, dv_\lambda\{(t, 0)\} + v_\lambda(V_{t_0}).$$

Integration by parts easily leads to the estimate

$$P(M) \leqq C_1 \lambda^{x_1 + l + 1} e^{-t_0\lambda}.$$

On the other hand,

$$P(M) \geqq \sum_{j=0}^{x_1} R\{(t_0, j) \mid U_{(1/2)t_0}\}$$

$$= \int_0^{t_0/2} e^{-(t_0-t)\lambda} \sum_{j=0}^{x_1} \frac{[(t_0-t)\lambda]^j}{j!} \, dv_\lambda\{(t, 0)\}$$

$$\geqq C_2 e^{-t_0\lambda} \lambda^{x_1} v_\lambda(U_{(1/2)t_0}) \geqq C_3 e^{-t_0\lambda} \lambda^{x_1-k}$$

in view of condition (b_1) for S; here and below, the C_i are positive constants ($i = 1, 2, \ldots$).

To check condition (b_1) for S^* now,

$$v_\lambda^*\{(t_0, 0)\} \geqq \frac{1}{P(M)} \int_0^{t_0} e^{-(t_0-t)\lambda} \, dv_\lambda\{(t, 0)\}$$

$$\geqq \frac{1}{P(M)} e^{-t_0\lambda} v_\lambda(U_{t_0})$$

$$\geqq \frac{1}{C_1} \lambda^{-x_1-l-1} e^{t_0\lambda} C_{t_0} \lambda^{-k} \geqq C_{t_0}^* \lambda^{-k^*}$$

where C_t^* and k^* denote new constants obtained for the plan S^*. For any interval $U_t^* = [t_0, t_1)$, we have $v_\lambda^*(U_t^*) \geqq v_\lambda^*\{(t_0, 0)\} \geqq C_{t_0}^* \lambda^{-k^*}$, which proves condition (b_1) for plan S^*.

As for condition (b_2) for S^*, this follows from

$$v_\lambda^*(V_t) = \frac{1}{P(M)} v_\lambda(V_t) \leqq \frac{C_4}{C_3} e^{t_0\lambda} \lambda^{-x_1+k+l} e^{-t\lambda}$$

$$= O(\lambda^{l^*} e^{-(t-t_0)\lambda})$$

where l^* denotes a new positive constant.

We pass on to case (2). Let M be the event that the trajectories of the plan S reach the axes of plan S^*. This is equivalent to the event that the trajectories do not stop at the point $(t_0, 0)$.

Let $\lambda \to 0$. Then we have for plan S

$$P(M) \geqq \int_0^{t_0} (1 - e^{-\lambda(t_0-t)}) \, dv_\lambda\{(t, 0)\} \sim \lambda \int_0^{t_0} (t_0 - t) \, dv_\lambda\{(t, 0)\}$$

$$\geqq \lambda \int_0^{t_0/2} (t_0 - t) \, dv_\lambda\{(t, 0)\} > C_1 \lambda v_\lambda(U_{(1/2)t_0}) > C_1 C_{(1/2)t_0} \lambda = C_2 \lambda$$

in view of condition (a_1) for plan S. On the other hand, if $(0, a)$ is a boundary point of S, then

$$P(M) = \sum_{i=1}^{a} v_\lambda\{(0, i)\} + \int_0^{t_0} (1 - e^{-(t_0-t)\lambda}) \, dv_\lambda\{(t, 0)\} \leqq C_3 \lambda.$$

Now we can establish properties (a_1) and (a_2) for S^*.

(a_1) Let U_t^* be the interval $(0, t)$ lying on the horizontal $x = 1$. Then,

$$\nu_\lambda^*(U_t^*) \geqq \frac{1}{P(M)} \int_0^t (1 - e^{-\lambda(t-\xi)})\, d\nu_\lambda\{(\xi, 0)\}$$

$$\sim \frac{\lambda}{P(M)} \int_0^t (t - \xi)\, d\nu_\lambda\{(\xi, 0)\} \geqq \frac{1}{C_3} \int_0^{t/2} (t - \xi)\, d\nu_\lambda\{(\xi, 0)\}$$

$$\geqq C_4\, \nu_\lambda(U_{(1/2)t}) \geqq C_4\, C_{(1/2)t} = C_t^*$$

where C_t^* is a new constant, for the plan S^*, depending on t.

(a_2) $\nu_\lambda^*\{(0, x)\} = (1/P(M))\, \nu_\lambda\{(0, x)\} = o(\lambda^{x-1})$ for $x \geqq 1$.

Let now $\lambda \to \infty$. Then $P(M) \to 1$. We verify that conditions (b_1) and (b_2) are satisfied for S^*.

(b_1) We have

$$\nu_\lambda^*(U_t^*) \geqq \frac{1}{P(M)} \int_0^t (1 - e^{-\lambda(t-\xi)})\, d\nu_\lambda\{(\xi, 0)\}$$

$$\geqq \frac{1}{P(M)} \int_0^{t/2} (1 - e^{-\lambda(t-\xi)})\, d\nu_\lambda\{(\xi, 0)\}$$

$$\geqq C_1 \nu_\lambda(U_{(1/2)t}) \geqq C_1 C_{(1/2)t} \lambda^{-k} = C_t^* \lambda^{-k}$$

where C_t^* is yet another constant depending on t and the plan S^*.

(b_2) Let V_t^* be the interval (t, ∞) on the line $x = 1$. Then, as is easily seen,

$$\nu_\lambda^*(V_t^*) = \frac{1}{P(M)} \left[\int_0^t e^{-\lambda(t-\xi)}(1 - e^{-\lambda(t_0-t)})\, d\nu_\lambda\{(\xi, 0)\} \right.$$

$$\left. + \int_t^{t_0} (1 - e^{-\lambda(t_0-\xi)})\, d\nu_\lambda\{(\xi, 0)\} \right]$$

$$\leqq \int_t^{t_0} d\nu_\lambda\{(\xi, 0)\} + \int_0^t e^{-\lambda(t-\xi)}\, d\nu_\lambda\{(\xi, 0)\}.$$

Integrating by parts and applying property (b_2) of plan S, we easily arrive at the estimate

$$\nu_\lambda^*(V_t^*) = 0(\lambda^{l+1} e^{-\lambda t}).$$

This completes the proof of all our assertions.▮

Theorem 12.4.2 is a special case of the following.

THEOREM 12.4.3. If two pseudo-plans S and S' of types (n, k) and (n', k') have identical measures $\nu_\lambda = \nu'_\lambda$ and mean stopping times: $E_\lambda \tau = E_\lambda \tau'$, then $\partial S = \partial S'$.

Note: The coincidence of the measures ν_λ and ν'_λ means that the measures have a common support, contained in the intersection of the sets of transient and boundary points on the axes of the plans and coincide on it.

Proof: We prove this by a two-fold induction—on n as well as k. Firstly, let $n = n' = 1$, so that we have pseudo-plans of types $(1, k)$ and $(1, k')$ (cf. Figure 9). The boundary of the $(1, k)$-plan consists of the segment: $x = k$; $t \in [0, T]$ and the boundary points $(i, t_i), 0 \leqq t_1 \leqq \cdots \leqq t_{n-1} = T$. The boundary points have to be so located in order for the plan to be closed.

Let $k = 1$, and ν_λ be given on the segment $x = 0$, $t \in [0, T]$, and at the point $(0, 1)$. Thus we have a $(1, 1)$-pseudo-plan (cf. Figure 10). This plan is defined by the value of T. For every λ, $E_\lambda \tau$ is a strictly increasing function of T.

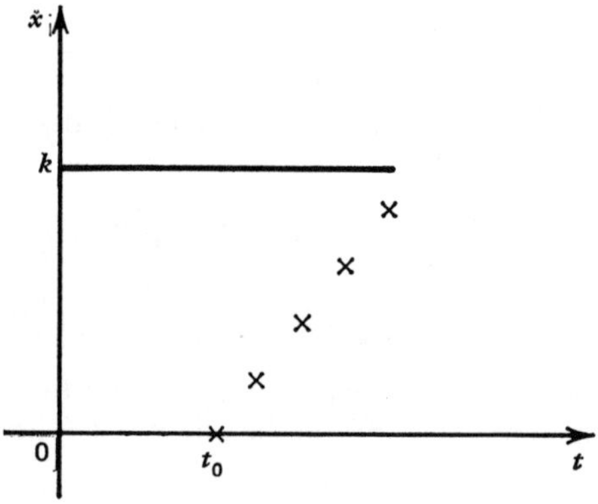

FIGURE 9.

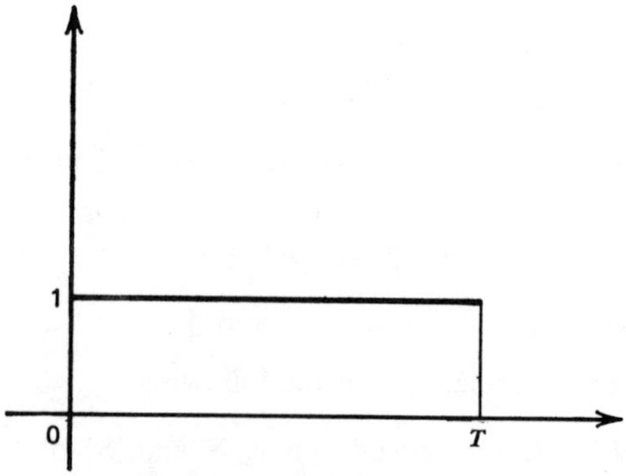

FIGURE 10.

Hence, for given λ, $E_\lambda \tau$, as a function of T must be uniquely determined. Thus, (1, 1)-pseudo-plans with given measures ν_λ on the axes are uniquely determined.

Let now our assertion be assumed proved for all $(1, k)$-pseudo-plans with $k \leqq k_0$. We shall prove it for $k = k_0 + 1$. First, we show that the points $(t_0, 0)$ and $(t_0', 0)$ (cf. Figure 9) coincide: $t_0 = t_0'$. We have $E_\lambda \tau = E_\lambda \tau'$, where τ and τ' are the respective stopping times for the two pseudo-plans. In view of the fact that $\nu_\lambda \in \mathcal{N}$, we obtain for $\lambda \to 0$,

$$E_{\nu_\lambda}(t_0 - \xi) = E_{\nu_\lambda}(t_0' - \xi) + o(\lambda)$$

where E_{ν_λ} denotes the expectation with respect to the measure ν_λ, and ξ is a random variable corresponding to this measure. Hence $t_0 = t_0'$, so that the "bases" of our pseudo-plans: the segments $x = 0$, $t \leqq t_0$ and $x = 0$, $t \leqq t_0'$, coincide.

We now consider two new pseudo-plans, obtained by deleting the common base of the plans S and S'. Let M be the event that a trajectory does not stop at the point $(t_0, 0)$. It is clear that $P(M) > 0$ for $\lambda > 0$, and $P(M)$ is the same for both S and S'. We shall assume that $k \geqq 2$, $k' \geqq 2$ (the justification for this assumption will be given later).

The horizontal $x = 1$ is free of boundary segments and has only one boundary point of each of the plans. We obtain two new pseudo-plans $\tilde{S}$ and $\tilde{S}'$; ν_λ will induce for both the plans one and the same conditional measure, given M; call it $\tilde{\nu}_\lambda$. If $\tilde{E}_\lambda(\tau)$ and $\tilde{E}_\lambda(\tau')$ be the expectations of the stopping times of the pseudo-plans S and S', then we have

$$\begin{aligned} E_\lambda(\tau) &= P(M^c) E_{\nu_\lambda}[(t_0 - \xi) e^{-\lambda(t_0 - \xi)} \mid M^c] + P(M) \tilde{E}_\lambda(\tau) \\ &= P(M^c) E_{\nu_\lambda}[(t_0 - \xi) e^{-\lambda(t_0 - \xi)} \mid M^c] + P(M) \tilde{E}_\lambda(\tau') \qquad (12.4.6) \end{aligned}$$

where ξ is a random point on the t-axis distributed according to the measure $(\nu_\lambda \mid M^c)$. Hence $\tilde{E}_\lambda(\tau) = \tilde{E}_\lambda(\tau')$, while the heights of $\tilde{S}$ and $\tilde{S}'$ do not exceed k_0. By our assumption, $\partial \tilde{S} = \partial \tilde{S}'$, whence $\partial S = \partial S'$, as required to prove.

We now turn to the proof of Theorem 12.4.2, through induction on n and k. As we shall see below, the theorem can be proved for all pseudo-plans of types $(n, 2)$, $(n', 2)$. Taking this fact for granted, let us assume that Theorem 12.4.2 has been proved for all (n, k)-pseudo-plans with $n \leqq n_0$ and for $n = n_0 + 1$, $k \leqq k_0$ (note that there are no plans with $n > 1$, $k = 1$).

Let S and S' be two pseudo-plans of types (n, k) and (n', k') respectively, with $n, n' \leqq n_0 + 1$ and $k, k' \leqq k_0 + 1$. Since our pseudo-plans are closed, the boundaries contain one and only one segment each, abutting on the x-axis; let these be respectively $x = m$, $t \in (0, l)$ and $x = m'$, $t \in (0, l')$. Further they have boundary points $(t_0, 0)$ and $(t_0', 0)$ on the bases. Arguing as earlier, we show that $m = m'$. Both the pseudo-plans have one and the same measure ν_λ with support on the axes.

Let L be the event that a trajectory starts on the x-axis; X a random variable, distributed over the set of integer points on the x-axis according to the conditional measure $(\nu_\lambda | L)$. If we leave from the point $(0, x)$, $x \leqq m$, then we have for S

$$\lim_{\lambda \to \infty} \lambda E_\lambda(\tau | x) = m - x, \text{ i.e., } \lambda E_\lambda(\tau | x) = m - x + o(1) \text{ as } \lambda \to \infty$$

and similarly, for S' and $x \leqq m'$

$$\lambda E_\lambda(\tau' | x) = m' - x + o(1) \qquad \text{as} \qquad \lambda \to \infty.$$

Further, $P(L) \to 1$ as $\lambda \to \infty$, and $\lambda E_\lambda \tau = \lambda E_\lambda \tau'$. Hence, m, m', X being bounded, we have as $\lambda \to \infty$,

$$P(L) E\nu_\lambda(m - x | L) - P(L) E(m' - x | L) = o(1).$$

Hence $m = m'$. Thus, as asserted above, $k \geqq 2$ and $k' \geqq 2$. Now consider the segments $x = m$, $t \in (0, l)$ and $x = m$, $t \in (0, l')$. Suppose $l \leqq l'$. We shall compare the abscissa t_0 of the boundary point on the base with the number l. We distinguish the two possible cases: (a) $t_0 \leqq l$, and (b) $t_0 > l$.

Consider case (a). Let $t_0 \leqq l$. We take $k \geqq 2$. Then $m > 1$, for, if $m = 1$, then (cf. Figure 11) our pseudo-plan will contain some (1, 1)-pseudo-plan, to which we can add a segment of inaccessible points, which is impossible. We now assert that boundary points cannot exist on the interval $x = 1$, $0 \leqq t < t_0$ (cf. Figure 12).

In fact, in view of the completeness of our pseudo-plans, a segment of boundary points cannot lie on the interval $x = 1$, $0 < t < l$. If P is an isolated boundary point of S or S', with $P = (t_1, 1)$, $t_1 < t_0$, the points on the horizontal $x = 1$ to the right of P would be accessible, and P can be deleted without destroying the closedness of S or S', so that S and S' would no longer be minimal. Thus, S has a boundary point $Q = (\tilde{t}_1, 1)$, $\tilde{t}_1 \geqq t_0$. The points to its right on $x = 1$ are inaccessible, so that Q is an isolated boundary point.

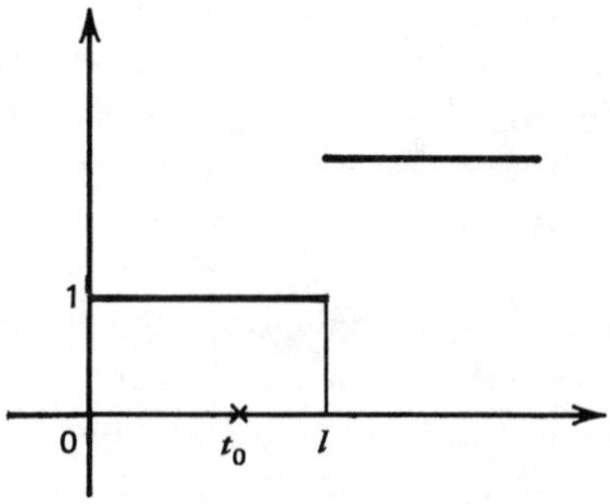

FIGURE 11.

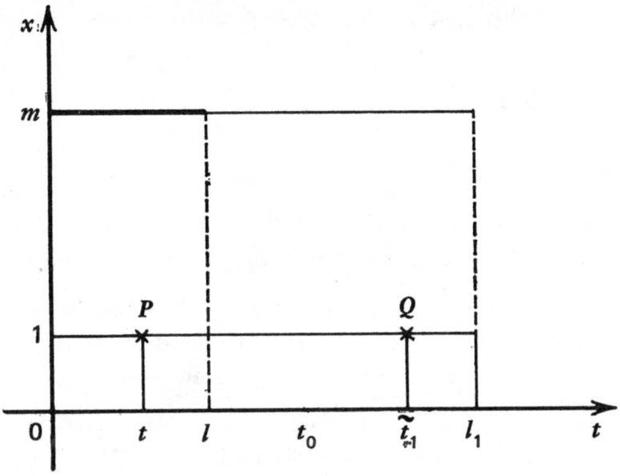

FIGURE 12.

The same is true of the plan S'; this pseudo-plan has an isolated boundary point $Q' = (\tilde{t}'_1, 1)$, $\tilde{t}'_1 \geqq t_0$.

We now delete the base $x = 0$ from both plans. We thus obtain two new plans $\tilde{S}$ and $\tilde{S}'$ of smaller height. The measure ν_λ with support on the axes of our pseudo-plans now induces a new conditional measure $\tilde{\nu}_\lambda$ on the new axes given the event that the trajectories do not stop at $(t_0, 0)$.

As before, we show that the coincidence of the (original) expectations $E_\lambda \tau$ implies that of the expectations $E_\lambda(\tilde{\tau})$ of the pseudo-plans $\tilde{S}$ and $\tilde{S}'$ with the measure $\tilde{\nu}_\lambda$. By our assumption, $\partial\tilde{S} = \partial\tilde{S}'$ and this settles case (a).

As for case (b), let $t_0 > l$. We assert that if then there exists an isolated boundary point $P = (t_i, i)$, then $t_i > l$. In fact, suppose such a point, with $t_i < l$ exists (cf. Figure 13). Then, $i \leqq m$, because, if $i > m$, P would be inaccessible.

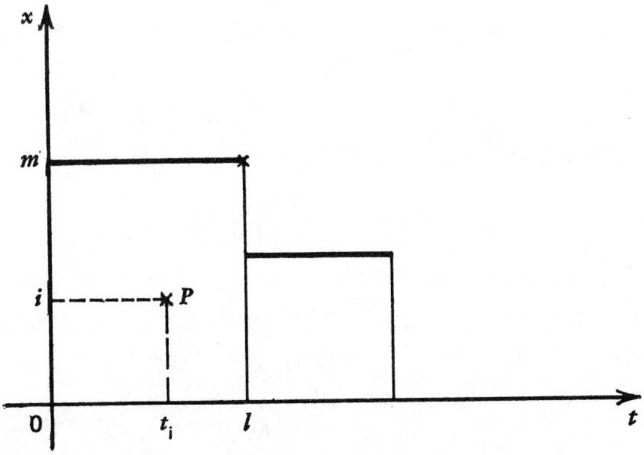

FIGURE 13.

Further, points on the horizontal $x = i$, sufficiently close to P, must be inaccessible, since, otherwise, P could be deleted from ∂S without destroying the closedness of the plan. In view of the completeness of the pseudo-plans, projections of the segments of the boundary on the t-axis cannot overlap, and there must be an isolated boundary point $Q = (t_{i-1}, i-1)$, where we must have $t_{i-1} \leqq t_i$ from the closedness of the plan. Similarly, we prove the existence of an isolated boundary point $(t_{i-2}, i-2)$ with $t_{i-2} \leqq t_{i-1}$, and so on. Hence, $l > t_i \geqq t_{i-1} \geqq \cdots \geqq t_0$, so that $l > t_0$, contrary to our assumption (b). Thus, $t_i \geqq l$. The boundary ∂S, in addition to the segment $x = m$, $t \in [0, l]$ must contain at least one more segment (since $n > 1$), say, the segment $x = r$, $t \in (l, q)$. We can now repeat the preceding argument, to show that, for $i = 0, 1, \ldots, \min(m, r)$, there is no isolated boundary point of the form (i, l). Hence $t_i > l$ for $i = 0, 1, \ldots, \min(m, r)$. The same is true for the plan S'. Hence the boundary ∂S, for $t \in [0, l + \varepsilon]$ ($\varepsilon > 0$ sufficiently small) must be of one of two types (cf. Figures 14a and 14b).

If $r < m - 1$, then the points (s, l), $s = r + 1, \ldots, m - 1$, must be isolated boundary points. For the plan S' (for which $l' \geqq l$) we have the same alternatives. If $l' > l$, we must have $r' = m$, since $m' = m$, and the boundary segment $x = m$, $t \in (0, l')$ contains the segment $x = m$, $t \in (0, l)$.

Now we shift the x-axis of both plans to the right, to the position $t = l$, and delete the segment $x = m$, $t \in (0, l)$, and if necessary the boundary points (s, l) for $s = r + 1, \ldots, m - 1$ or (s, l) for $s = r' + 1, \ldots, m - 1$ of both the pseudo-plans. We obtain the boundaries of new pseudo-plans $\tilde{S}$ and $\tilde{S}'$. To obtain the new pseudo-plans, we shall show that $r = r'$ and that the measure ν_λ induces one and the same conditional measure $\tilde{\nu}_\lambda$ for both plans.

Let M be the event that a trajectory stops on the segment $x = m$, $t \in (0, l)$, i.e., the event $\tau < l$, since we have $t_0 > l$.

Let $N(N')$ be the event that at time $t = l$, we have $x(l) \leqq r - 1$ ($x(l) \leqq r' - 1$); $\tilde{\nu}_\lambda$ and $\tilde{\nu}'_\lambda$ are taken to be the conditional measures induced by the measure ν_λ on the axes of the plans S and S' given the events N and N'.

We then have

$$\begin{aligned} E_\lambda \tau = E_\lambda \tau' &= P_\lambda(M) E\nu_\lambda(\tau \mid M) + P_\lambda(M^c) \cdot l + P_\lambda(N) \cdot E\tilde{\nu}_\lambda(\tau) \\ &= P_\lambda(M) E\nu_\lambda(\tau \mid M) + P_\lambda(M^c) \cdot l + P_\lambda(N') E\tilde{\nu}_{\lambda'}(\tau'). \end{aligned} \tag{12.4.7}$$

This gives

$$\lambda P_\lambda(N) E\tilde{\nu}_\lambda(\tau) = \lambda P_\lambda(N') E\tilde{\nu}_{\lambda'}(\tau). \tag{12.4.8}$$

We deduce hence that $r = r'$.

Let $P_i(\lambda)$, $0 \leqq i \leqq m - 1$, denote the probability that a trajectory first reaches the point (l, i); this is obviously the same for both the plans since $m = m'$. Let $r \neq r'$; for definiteness, let $r > r'$. Then,

$$P_\lambda(N) = \sum_{i=0}^{r-1} P_i(\lambda) \geqq \sum_{i=0}^{r'-1} P_i(\lambda) = P_\lambda(N'). \tag{12.4.9}$$

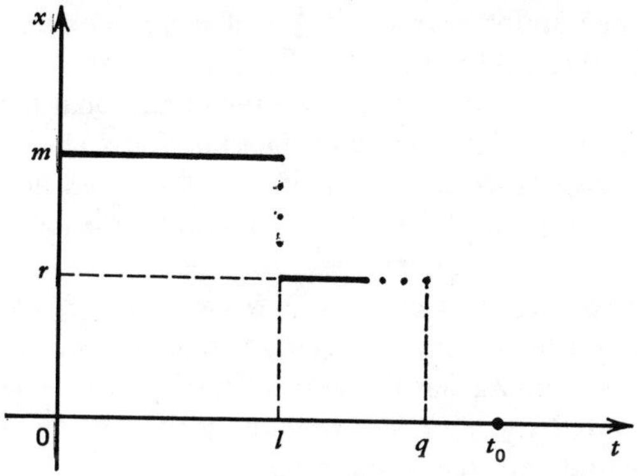

FIGURE 14a.

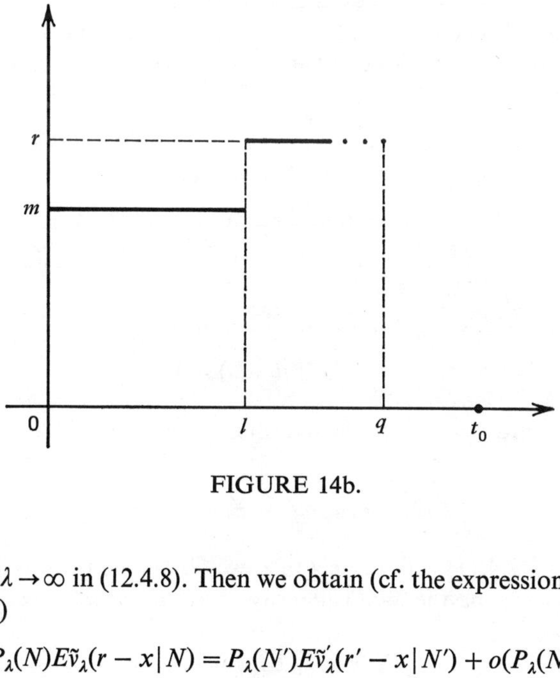

FIGURE 14b.

Now, we let $\lambda \to \infty$ in (12.4.8). Then we obtain (cf. the expression given earlier for $\lambda E_\lambda(\tau | x)$)

$$P_\lambda(N) E\tilde{v}_\lambda(r - x | N) = P_\lambda(N') E\tilde{v}'_\lambda(r' - x | N') + o(P_\lambda(N)).$$

Hence,

$$\sum_{i=0}^{r-1} P_i(\lambda)(r - i) = \sum_{i=0}^{r'-1} P_i(\lambda)(r' - i) + o\left(\sum_{i=0}^{r-1} P_i(\lambda)\right),$$

$$(r - r') \sum_{i=0}^{r'-1} P_i(\lambda) + \sum_{r'}^{r-1} (r - i) P_i(\lambda) = o\left(\sum_{i=0}^{r-1} P_i(\lambda)\right)$$

as $\lambda \to \infty$, so that we cannot have $r - r' \geqq 1$. Thus, $r = r'$, and the structure of the boundary in the neighborhood $t \in (0, l + \varepsilon)$ is one and the same for S and S', for sufficiently small $\varepsilon > 0$. Then the conditional measures $\tilde{v}_\lambda$ and $\tilde{v}'_\lambda$ under conditioning by N are one and the same for $\tilde{S}$ and $\tilde{S}'$, and $\tilde{E}_\lambda(\tau)$ is also one and the same. Since the order of the plans does not exceed n_0, it follows from our assumption that $\partial\tilde{S} = \partial\tilde{S}'$, so that $\partial S = \partial S'$, as required to prove.

Now it remains to consider the case $k = 2$. In case (a): $t_0 \leqq l$, there can be no boundary points on the horizontal $x = 1$, but then $m = 2$ and $n = 1$; however, the case $n = 1$ and arbitrary k has been considered earlier. Hence, in the case $k = 2$, only the alternative (b): $t_0 > l$ remains, and this can be investigated as indicated above and with the same results. ∎

The theorem just proved can be extended to other types of processes, in particular to the well-known Erlangian process, for which the time between successive unit jumps has a gamma distribution.

An analogue of Theorem 12.2.8 holds for the Poisson process. We shall establish here only the analogue of its simplest corollary, namely, Theorem 12.2.7.

THEOREM 12.4.4. If the plan S is closed and does not have inaccessible points, and, for some value λ_0 of the parameter λ, we have:

$$E_{\lambda_0}\tau = k_0/\lambda_0, \qquad E'_{\lambda_0}\tau = -k_0/\lambda_0^2, \qquad \text{and} \qquad \operatorname{Var}_{\lambda_0}\tau \leqq k_0/\lambda_0^2 \tag{12.4.10}$$

(where the prime stands for the derivative), then k_0 is a positive integer and S is an inverse sampling plan.

Proof: In view of the representation (12.4.1), we have

$$\int_{\partial S} \lambda^x e^{-\lambda t}\mu_s(du) = 1 \tag{12.4.11}$$

where $x = x(u)$ and $t = t(u)$ are the coordinates of a boundary point. Further, in view of (12.4.10),

$$\int_{\partial S} \lambda_0^x e^{-\lambda_0 t} t\mu_s(du) = E_{\lambda_0}\tau = k_0/\lambda_0. \tag{12.4.12}$$

Differentiating (12.4.11) with respect to λ, we obtain the identity $E_\lambda x = \lambda E_\lambda \tau$, which turns out to be the Wald identity for this special case; hence, from (12.4.10),

$$E_{\lambda_0}x = \lambda_0 E_{\lambda_0}\tau = k_0. \tag{12.4.13}$$

Now, we differentiate (12.4.11) once again and also (12.4.12) at the point $\lambda = \lambda_0$. This yields, after some simple calculations (cf. section 12.2) the relations

$$E_{\lambda_0}x^2 = E_{\lambda_0}\tau^2 - k_0^2 \leqq k_0^2 + k_0^2 - k_0^2 = k_0^2,$$

$$\operatorname{Var}_{\lambda_0}x = E_{\lambda_0}x^2 - (E_{\lambda_0}x)^2 \leqq k_0^2 - k_0^2 = 0, \qquad \text{or } \operatorname{Var}_{\lambda_0}(x) = 0.$$

In view of (12.4.13), the boundary ∂S lies on the horizontal $x = k_0$. Then $k_0 > 0$ must be an integer and ∂S must coincide with the horizontal $x = k_0$, since, otherwise, the plan will not be closed.∎

12.5 WIENER PROCESS

We now turn to the Wiener process given by

$$X(t) = \xi(t) + \lambda t \tag{12.5.1}$$

where $\xi(t)$ is a standard Brownian motion and λ is the velocity of drift, which it is required to estimate. For this purpose, we use a first entry plan S with boundary ∂S consisting of the smooth curves $x = f_1(t)$ and $x = f_2(t)$, $f_1(t) < 0 < f_2(t)$ for $t \leqq T_0$ and a vertical segment in $t = T_0$. The class of such plans will be denoted by $\mathscr{S}$. They are obviously finite (bounded).

The transition probability density for the process (12.5.1) is

$$p_\lambda(t, x) = \frac{1}{\sqrt{2\pi t}} \exp\left[-\frac{(x - \lambda t)^2}{2t}\right].$$

A sufficient statistic for λ will be $(X_\tau(w), \tau(w))$, where τ is the stopping time and w is the elementary event. We have now to obtain the analogue of formula (12.4.1) giving the distribution of the sufficient statistic on the boundary ∂S. For the very special case of constant f_1 and f_2, such a formula was given by Trybula in [148]. In the general case, Lemma 1.6.1. enables us to write

$$P[X_\tau(w), \tau(w) \in A] = \int_A \frac{p_\lambda(t, x)}{p_0(t, x)} \mu_s(du) \tag{12.5.2}$$

where A is a Borel set of points u, $t = t(u)$ and $x = x(u)$ are the values of the coordinates of the sufficient statistic on the boundary ∂S.

We have further

$$\frac{p_\lambda(t, x)}{p_0(t, x)} = \exp\left[-\frac{1}{2}\lambda(2x - \lambda t)\right] \tag{12.5.3}$$

Now we can establish the following result.

THEOREM 12.5.1. Let the boundary ∂S of a plan S of the type indicated above satisfy an algebraic equation of degree k (for instance, consist of a finite number of straight line segments). Then the same equation is satisfied for any plan of the same type for which the $k(2k + 1)$ quantities

$$(E_{\lambda_0} \tau^s)^{(r)}, \qquad s + r \leqq 2k,\ s \geqq 1 \tag{12.5.4}$$

for some given value λ_0, coincide with the same quantities for the original plan.

***Proof*:** In fact, let our equation have the form

$$\Pi(x, t) = 0. \tag{12.5.5}$$

If we show that, for the given values of the quantities (12.5.4), we have the relations

$$E_{\lambda_0}[\Pi(X_\tau, \tau)] = 0, \qquad \mathrm{Var}_{\lambda_0}[\Pi(X_\tau, \tau)] = 0 \tag{12.5.6}$$

then the equation (12.5.5) will certainly be satisfied for (X_τ, τ). Further, to compute (12.5.6), we have to form $\Pi(X_\tau, \tau)$ and $(\Pi(X_\tau, \tau))^2$ and apply formulas (12.5.2) and (12.5.3).

The polynomial $(\Pi(X_\tau, \tau))^2$ has degree $2k$. In view of (12.5.2) and (12.5.3), we have

$$E_{\lambda_0}\tau^s = \int_{\partial S} \exp\left[-\frac{1}{2}\lambda_0(2x - \lambda_0 t)\right] \cdot t^s \mu_s(du). \tag{12.5.7}$$

Differentiating this expression with respect to λ at λ_0,

$$E_{\lambda_0}\left\{\left[-\frac{1}{2}(2X - \lambda_0 \tau) - \frac{1}{2}\lambda_0 \tau\right]\tau^s\right\} = (E_{\lambda_0}\tau^s)'.$$

Thus we can express $E_{\lambda_0}(X\tau^s)$ by means of $(E_{\lambda_0}\tau^s)'$ and $E_{\lambda_0}\tau^{s+1}$. To express $E_{\lambda_0}(X^2\tau^s)$, we must differentiate once again, and it is easily seen that $E_{\lambda_0} X^a \tau^b$ for $a + b \leqq 2k$ can be expressed by means of (12.5.4). This leads to the relations (12.5.6), which proves the theorem. ∎

CHAPTER 13

Other Characteristic Properties of Distributions

In this chapter, we shall deal with various characteristic properties of distributions which are of interest from the point of view of mathematical statistics.

13.1 DISTRIBUTIONS WITH MINIMAL FISHERIAN AMOUNT OF INFORMATION

The Fisherian amount of information for the family of distributions $\{P_\theta, \theta \in \Delta\}$ on the space $(\mathscr{X}, \mathscr{A})$ where Δ is an interval of the real axis, with densities $dP_\theta/d\mu = p(x, \theta)$ with respect to some measure μ which dominates the family, was defined in section 7.1 as

$$I(\theta) = E_\theta\left(\frac{\partial \log p(X, \theta)}{\partial \theta}\right)^2 = \int_{\mathscr{X}} \left(\frac{\partial \log p(x, \theta)}{\partial \theta}\right)^2 p(x, \theta)\, d\mu$$

under the assumption that the right hand side exists. In estimation theory, the role of the amount of information is given by the Rao-Cramer inequality (7.1.2).

In this section, we shall show that in the class of distributions depending on a shift parameter (and satisfying certain smoothness conditions) and having given variance, the minimum Fisherian amount of information is attained by the normal distributions. This is thus yet another characteristic property of the normal law; it illustrates yet again the fact that the shift parameter is estimated "worst of all" in the case of samples from a normal population (it is useful to compare the results of this section with those of Chapter 7). The gamma distribution is similarly distinguished among all distributions on R^1_+ depending on a scale parameter.

Consider a family of distributions on R^1, depending on a shift parameter θ and given by the densities $p(x-\theta)$ with respect to Lebesgue measure. We assume that

(i) p is continuously differentiable;
(ii) $\int x^2 p(x)\,dx < \infty$, and
(iii) $|x|p(x) \to 0$ as $|x| \to \infty$.

The Fisherian amount of information for the family of densities $p(x-\theta)$ is equal to

$$I_p(\theta) = \int \left(\frac{\partial \log p(x-\theta)}{\partial \theta}\right)^2 p(x-\theta)\,dx$$

where the integrand above is taken as zero if $p(x-\theta) = 0$ for some value(s) of x. It is easily seen that

$$I_p(\theta) = \int_{[p(x)>0]} \left(\frac{p'}{p}\right)^2 p\,dx = I_p(0) = I_p \quad \text{(say).} \tag{13.1.1}$$

THEOREM 13.1.1. In the class of all densities with given variance σ^2 and satisfying conditions (i)–(iii) above, $\min_p I_p$ is attained by the normal distributions.

***Proof*:** Without loss of generality, we may take $\int xp(x)\,dx = 0$. We set $A^+ = \{x\colon p(x) > 0\}$; in view of the continuity of p, A^+ is an open set and hence $A^+ = \bigcup_n \Delta_n$, where $\{\Delta_n\}$ is a sequence of disjoint open intervals. Integration by parts gives

$$\int_{\Delta_n} xp'(x)\,dx = -\int_{\Delta_n} p(x)\,dx$$

whence

$$\int_{A^+} xp'(x)\,dx = -1. \tag{13.1.2}$$

By the Cauchy-Schwarz-Bunyakovski inequality, we have from the above

$$1 = \left(\int_{A^+} xp'(x)\,dx\right)^2 = \left(\int_{A^+} x\,\frac{p'(x)}{p(x)}\,p(x)\,dx\right)^2$$

$$\leqq \int_{A^+} \left(\frac{p'(x)}{p(x)}\right)^2 p(x)\,dx \cdot \int_{A^+} x^2 p(x)\,dx$$

whence $I_p \geqq 1/\sigma^2$, where, moreover, the equality is attained when and only when, for some constant c,

$$\frac{p'(x)}{p(x)} = cx \text{ a.e. on } A^+ \tag{13.1.3}$$

(with respect to Lebesgue measure). In view of the continuity of p', we conclude that $p'(x)/p(x) = cx$, for $x \in A^+$. Integrating (13.1.3), we obtain

$$p(x) = A \exp \frac{cx^2}{2} \qquad \text{for } x \in A^+. \tag{13.1.4}$$

But $A \exp (cx^2/2) \neq 0$ for real x; hence, the continuity of p, together with (13.1.4), gives $p(x) = A \exp (cx^2/2)$ for all x, where $c = -1/\sigma^2$, $A = 1/(\sqrt{2\pi}\sigma)$. The theorem stands proved. ▮

Let now a family of distributions be given on the half line R_+^1, depending on a scale parameter $\sigma \in R_+^1$ and given by the densities $(1/\sigma)p(x/\sigma)$ with respect to Lebesgue measure. Suppose that p satisfies conditions (i) and (ii) above, and, in addition,

(iv) $x^2 p(x) \to 0$ as $x \to \infty$, $xp(x) \to 0$ as $x \to 0$.

The Fisherian amount of information for the family of densities $(1/\sigma)p(x/\sigma)$ is equal to

$$\begin{aligned} I_p(\sigma) &= \int_{[p(x/\sigma)>0]} \left(-\frac{1}{\sigma} - \frac{x}{\sigma^2} \frac{p'(x/\sigma)}{p(x/\sigma)} \right)^2 \frac{1}{\sigma} p\left(\frac{x}{\sigma}\right) dx \\ &= \frac{1}{\sigma^2} \left\{ \int_{A^+} \left(\frac{xp'(x)}{p(x)} \right)^2 p(x)\, dx - 1 \right\} = \frac{1}{\sigma^2} I_p(1) = \frac{I_p}{\sigma^2} \end{aligned}$$

where A^+ is the same set as above. The following result about such a family is due to Kagan [61].

THEOREM 13.1.2. In the class of all densities with given moments $\alpha_i = \int_0^\infty x^i p(x)\, dx$, $i = 1, 2$, satisfying the conditions (i), (ii), and (iv), $\min_p I_p$ is attained by the gamma distributions.

***Proof*:** As in the derivation of (13.1.2), integration by parts gives

$$\int_{A^+} xp'(x)\, dx = -1, \qquad \int_{A^+} x^2 p'(x)\, dx = -2\alpha_1. \tag{13.1.5}$$

We multiply the first of these relations by μ, the second by $-\lambda$, and add, to obtain

$$\int_{A^+} x(\mu - \lambda x)p'(x)\, dx = (2\lambda\alpha_1 - \mu).$$

Applying the Cauchy-Schwarz-Bunyakovski inequality, we get

$$\begin{aligned} (2\lambda\alpha_1 - \mu)^2 &= \left\{ \int_{A^+} \frac{xp'(x)}{p(x)} (-\lambda x + \mu) p(x)\, dx \right\}^2 \\ &\leqq \int_{A^+} \left(\frac{xp'(x)}{p(x)} \right)^2 p(x)\, dx \int_{A^+} (\mu - \lambda x)^2 p(x)\, dx. \end{aligned}$$

Hence, for all λ and μ,

$$I_p \geqq \frac{(-2\lambda\alpha_1 + \mu)^2}{\int_{A^+} (\mu - \lambda x)^2 p(x)\, dx} = \frac{(-2\lambda\alpha_1 + \mu)^2}{(\lambda^2\alpha_2 - 2\lambda\mu\alpha_1 + \mu^2)}$$

where, for given λ and μ, equality holds when and only when, for some constant c, the following relation

$$\frac{xp'(x)}{p(x)} = c(\mu - \lambda x) \tag{13.1.6}$$

is fulfilled almost everywhere in A^+(in the sense of Lebesgue measure). In view of the continuity of p', the above relation is then satisfied for all x in A^+. We absorb the constant c in λ and μ, and integrate (13.1.6) to obtain

$$p(x) = ke^{-\lambda x}x^{\mu}, \qquad x \in A^+. \tag{13.1.7}$$

Since $e^{-\lambda x}x^{\mu} \neq 0$ for $x \in R^1_+$, the continuity of p guarantees the validity of (13.1.7) for all $x \in R^1_+$. Setting $\mu = \nu - 1$, and determining k from the normalizing conditions, we arrive at the usual form of the gamma density

$$p(x) = \frac{\lambda^{\nu}}{\Gamma(\nu)} e^{-\lambda x}x^{\nu - 1}, \qquad x > 0. \tag{13.1.8}$$

It is easily seen that the density (13.1.8) will have the given moments α_1, α_2 if we take $\lambda = \alpha_1/(\alpha_2 - \alpha_1^2))$, $\nu = \alpha_1^2/(\alpha_2 - \alpha_1^2)$. Thus, in the class of distributions saitsfying the conditions of Theorem 13.1.2, $\min_p I_p$ is attained by the density (13.1.8) with the indicated values for λ and ν.▌

13.2 DISTRIBUTIONS WITH MAXIMUM ENTROPY

13.2.1. Inequalities from Information Theory

We formulate here two standard lemmas which will be used for characterizing distributions with maximum entropy. Proofs can be found in [131], p. 58.

LEMMA 13.2.1. Let $\sum a_i$ and $\sum b_i$ be convergent series of positive terms such that $\sum a_i \geqq \sum b_i$. Then

$$\sum a_i \log (b_i/a_i) \leqq 0. \tag{13.2.1}$$

LEMMA 13.2.2. Let f and g be nonnegative functions integrable with respect to the measure μ, and S the set where $f > 0$. If $\int_S (f - g)\, d\mu \geqq 0$, then

$$\int_S f \log (f/g)\, d\mu \geqq 0 \tag{13.2.2}$$

with equality if and only if $f = g$ a.e. $[\mu]$.

13.2.2. Characterization through Maximum Entropy

Let X be an r.v. taking values in a given interval (a, b), and probability density function p. The entropy of X is by definition

$$-\int_a^b p(x) \log p(x)\, dx. \tag{13.2.3}$$

In the case of a discrete distribution, the integration is replaced by summation, and the entropy given by

$$-\sum_{x \in S} p(x) \log p(x) \tag{13.2.4}$$

where the sum extends over the whole of the discrete set of values of X. We shall show that many well-known distributions are distinguished by the fact that, among suitable classes, they attain the maximum entropy.

We first establish a general theorem.

THEOREM 13.2.1. Let X be an r.v. with density

$$p(x) > 0 \quad \text{for } x \in (a, b) \quad \text{and} \quad p(x) = 0 \text{ otherwise.} \tag{13.2.5}$$

Let, further, $h_1, \ldots$ be integrable functions on (a, b) satisfying for given constants $g_1, \ldots,$ the conditions

$$\int_a^b h_i(x) p(x)\, dx = g_i, \qquad i = 1, 2, \ldots. \tag{13.2.6}$$

Then the maximum entropy is attained by distributions with the density of the form

$$p(x) = \exp\{a_0 + a_1 h_1(x) + a_2 h_2(x) + \cdots\} \tag{13.2.7}$$

(and only by them), if there exist $a_0, a_1, \ldots$ such that the above density satisfies the conditions (13.2.5) and (13.2.6).

***Proof*:** Let p have the form (13.2.7), and q be another density satisfying the conditions (13.2.5) and (13.2.6). Using the inequality (13.2.2), we obtain

$$\begin{aligned} -\int q \log q \leqq -\int q \log p &= -\int q(a_0 + a_1 h_1 + \cdots) \\ &= -(a_0 + a_1 g_1 + \cdots). \end{aligned}$$

We see from (13.2.8) that $-(a_0 + a_1 g_1 + \cdots)$ is a fixed upper bound for $-\int q \log q$, if the density q satisfies conditions (13.2.5) and (13.2.6). The upper bound is attained when q is chosen according to (13.2.7), if only it is possible to find suitable constants $a_0, a_1, \ldots$. ▮

From the general theorem above, we obtain a series of results, tabulated below.

Set of values of the r.v.	Restrictions	Density function corresponding to the maximum entropy
$(0, 1)$	—	$p(x) = 1$ (Uniform)
$(0, 1)$	$E \log X = g_1$ $E \log (1 - X) = g_2$	$p(x) = \dfrac{x^{m-1}(1-x)^{n-1}}{B(m, n)}$ (Beta)
$(0, \infty)$	$EX = g_1$	$p(x) = ae^{-ax}$ (Exponential)
$(0, \infty)$	$EX = g_1$ $E \log X = g_2$	$p(x) = \dfrac{a^p}{\Gamma(p)} x^{p-1} e^{-ax}$ (Gamma)
$(-\infty, \infty)$	$EX = g_1$ $EX^2 = g_2$	$p(x) = \dfrac{1}{\sigma\sqrt{2\pi}} e^{-(x-\mu)^2/2\sigma^2}$ (Normal)
$(-\infty, \infty)$	$E\|X\| = g_1$	$p(x) = \dfrac{1}{2} ae^{-a\|x\|}$ (Laplace)

Theorem 13.2.1 can be generalized to random vectors, to obtain the following characterization of the multivariate normal distribution.

THEOREM 13.2.2. Let $\mathbf{X}$ be a p-variate random vector with expectation vector $\boldsymbol{\mu}$ and nondegenerate dispersion matrix $\mathbf{A}$, and let $p(\mathbf{x})$ be the density of $\mathbf{X}$. The maximum entropy of $\mathbf{X}$ is attained by the normal distribution with density function corresponding to $(\boldsymbol{\mu}, \mathbf{A})$ and only by that distribution.

13.3 CHARACTERIZATION OF DISTRIBUTIONS THROUGH THE FORM OF THE MAXIMUM LIKELIHOOD ESTIMATOR

Let $f(x, \theta)$ be the density of the r.v. X, depending on a parameter $\theta \in \Omega \subset R^1$. For given θ, let $S_\theta = \{x: f(x, \theta) > 0\}$. Set $S = \bigcup_{\theta \in \Omega} S_\theta$ and $S^n = S \times S \times \cdots \times S$. Let $(X_1, \ldots, X_n)$ be a random sample of observations on X. Then the *maximum likelihood estimator* (MLE) $\hat{\theta}_n$ of θ is defined as a function from S^n into Ω, satisfying for all $\theta \in \Omega$ the condition

$$\prod_{i=1}^{n} f(x_i, \hat{\theta}_n) \geqq \prod_{i=1}^{n} f(x_i, \theta). \tag{13.3.1}$$

In this section, we deal with characterizations of distributions by the requirement that the MLE be a given function of the observations.

The following result is due to Teicher [145].

THEOREM 13.3.1. Let $\{F(x-\theta),\ \theta \in R^1\}$ be a family of absolutely continuous distributions on R^1, depending on a shift parameter θ. Suppose that the density $f(x) = F'(x)$ is lower semicontinuous at $x = 0$. If the sample mean $\bar{X}$ is the MLE for θ for $n = 2$, $n = 3$, then F is a normal d.f. with zero mean.

***Proof*:** Let $\bar{X} = (\sum_{i=1}^{n+1} X_i)/(n+1)$. Then, it follows from (13.3.1) under the conditions of the theorem, that for, $n = 1, 2$, and all real $x_1, \ldots, x_{n+1}$ and θ,

$$\prod_{i=1}^{n+1} f(x_i - \bar{x}) \geqq \prod_{i=1}^{n+1} f(x_i - \theta)$$

whence

$$\prod_{i=1}^{n+1} f(y_i) \geqq \prod_{i=1}^{n+1} f(y_i - \theta) \tag{13.3.2}$$

for $n = 1, 2$, and all real θ and $y_1, \ldots, y_{n+1}$ such that $\sum_{i=1}^{n+1} y_i = 0$.

Taking $n = 1$, $y_1 = -y_2 = y$ in (13.3.2), we have

$$f(y)f(-y) \geqq f(y-\theta)f(-y-\theta) \qquad \text{for all real } y, \theta. \tag{13.3.3}$$

We note that $f(0) = 0$ would imply that $f \equiv 0$. Suppose then that for some a, $f(a) = \infty$. Then, according to (13.3.3), for any $y \in R^1$, either $f(y)f(-y) = \infty$ or $f(2y + a) = 0$. For a density f, the former condition cannot be fulfilled on a set of positive Lebesgue measure. Consequently, the density f satisfying (13.3.2) must be positive at the origin and finite everywhere (so that products of the type $0 \cdot \infty$ do not appear in (13.3.2)).

Let $h = \log f$; h can possibly take the value $-\infty$. Then it follows from (13.3.2) that, for $n = 1, 2$ and all real $y_1, \ldots, y_n, \theta$,

$$\sum_{1}^{n} h(y_i) + h\left(-\sum_{1}^{n} y_i\right) \geqq \sum_{1}^{n} h(y_i - \theta) + h\left(-\sum_{1}^{n} y_i - \theta\right). \tag{13.3.4}$$

In particular, we have from (13.3.4) that

$$nh(y) + h(-ny) \geqq nh(y-\theta) + h(-ny-\theta)$$

which, for $n = 1$, gives

$$h(y) + h(-y) \geqq h(y-\theta) + h(-y-\theta). \tag{13.3.5}$$

We note that if θ and y_i are replaced by $-\theta$ and $-y_i$ in (13.3.4), then, adding the resulting inequality to (13.3.4), we obtain the following result: if $g(y)$ satisfies (13.3.4), then so does $h(y) = g(y) + g(-y)$. Hence we first consider symmetric solutions of (13.3.4), for which the inequality takes the form

$$\sum_1^n h(y_i) + h\left(\sum_1^n y_i\right) \geqq \sum_1^n h(y_i - \theta) + h\left(\sum_1^n y_i + \theta\right). \tag{13.3.6}$$

Similarly, (13.3.5) becomes

$$2h(y) \geqq h(y - \theta) + h(y + \theta) \qquad \text{for all } \theta,\ y. \tag{13.3.7}$$

We set $A = \{y\colon y > 0,\ h(y) = -\infty\}$ and $c = \inf A$. Taking $n = 2$, $y_1 = y_2 = (1/2)c_m$, $\theta = -(1/4)c_m$ in (13.3.6), we obtain

$$2h(c_m/2) + h(c_m) \geqq 3h(3c_m/4).$$

Choosing a sequence $\{c_m\}$ of elements of A with $\{c_m\} \downarrow c$, we will have $h(3c_m/4) = -\infty$, $m = 1, 2, \ldots$. For $c > 0$, we thus obtain a contradiction, while for $c = 0$ the condition of lower semicontinuity of f at the origin is violated.

Thus, h is a measurable, everywhere finite, and, by (13.3.7), a concave function. It follows hence that h is a continuous concave function.

Let $\mathscr{D}$ denote the complement of the (at most denumerable) set of points at which h is not differentiable and q denote the derivative of h. Then q is monotone and defined for all $x \in \mathscr{D}$. Since, according to (13.3.6), $\max_\theta[\sum_1^n h(y_i - \theta) + h(\sum_1^n y_i + \theta)]$ is attained for $\theta = 0$, we have

$$-\sum_1^n q(y_i) + q\left(\sum_1^n y_i\right) = 0 \tag{13.3.8}$$

for all y_i in $\mathscr{D}$ such that $y_i \in \mathscr{D}$, $n = 1, 2$. For $n = 2$, (13.3.8) leads to

$$q(y_1) + q(y_2) = q(y_1 + y_2) \qquad \text{if } y_1, y_2, y_1 + y_2 \text{ all} \in \mathscr{D}. \tag{13.3.9}$$

Let C denote the class of nonnegative measurable everywhere-finite functions on R^1, lower semicontinuous at the origin, and nonzero on a set of positive Lebesgue measure. Let C' be the subclass of C consisting of functions vanishing nowhere. Since the only solution of the functional equation (13.3.9) of Cauchy's is $q(y) = cy$, the only symmetric functions $f \in C$, satisfying (13.3.2)—and which are necessarily in C'—are such that

$$h(y) = \log f(y) = -cy^2 + d \qquad \text{for } y \in \mathscr{D}$$

and so, by continuity, for all $y \in R^1$.

Suppose now that $f \in C$ satisfies (13.3.2). According to (13.3.3), $f(y)f(-y)$ is nonzero on a set of positive measure (it is only necessary to set $\theta = -y$ to obtain this); it is clear that $f(y)f(-y)$ is symmetric, belongs to C and as noted above, satisfies (13.3.2). Thus, $f(y)f(-y) \in C'$ whence $f(y) \in C'$.

Further, for some real c and d,

$$g(y) = \log f(y) = -(1/2)(cy^2 - d) + b(y)$$

where $b(y)$ is an odd function. In fact, according to the preceding, $g(y) + g(-y) = -cy^2 + d$ for some c and d, and it follows hence that $b(y) = g(y) + (1/2)(cy^2 - d)$ is an odd function.

Substituting the expression for $g(y)$ in (13.3.5), we obtain

$$c\theta^2 \geqq b(y - \theta) - b(y + \theta) \qquad \text{for all } y \text{ and } \theta. \qquad (13.3.10)$$

Replacing y by $-y$ and θ by $-\theta$ in (13.3.10) and combining the result with (13.3.10), we have

$$|b(y - \theta) - b(y + \theta)| \leqq c\theta^2 \qquad \text{for all } y, \theta \qquad (13.3.11)$$

whence $c \geqq 0$ and then we successively obtain that $b(y)$ is differentiable, constant, and equals zero.

Consequently, the only solutions of (13.3.2) belonging to C are the functions $f(x) = \exp(-(1/2)cx^2 + d)$, and the theorem is proved.▮

The following result is due to Ghosh and Rao, [21].

THEOREM 13.3.2. Let $\{F(x - \theta),\ \theta \in R^1\}$ be the same family as in the statement of the preceding theorem. If the sample median is the MLE for θ for $n = 4$, then $F'(x) = f(x) = (1/2)a \exp(-a|x|)$ (a distribution with such a density is called a Laplace distribution).

***Proof*:** We consider the sample values $-y_1 \leqq -y \leqq 0 \leqq y \leqq y_2$, whose median is zero. For such a sample, according to the conditions of the theorem,

$$f(-y_1)f(-y)f(y)f(y_2) \geqq f(-y_1 - \theta)f(-y - \theta)f(y - \theta)f(y_2 - \theta) \qquad (13.3.12)$$

for all $\theta \in R^1$, whence

$$f(-y)f(y) \geqq f(-y - \theta)f(y - \theta). \qquad (13.3.13)$$

The last relation is the same as (13.3.3) of Theorem 13.3.1; hence $f(0) > 0$, and f is finite everywhere. Further, $g(y) = f(y)f(-y)$ satisfies (13.3.12) and (13.3.13), so that $\log g(y)$ is a concave function.

Since f is lower-semicontinuous at the origin and $f(0) > 0$, $f(y) > \varepsilon$ for y sufficiently small. Suppose that $g(y) = 0$ for some $y > 0$ and let c be the exact lower bound of those $y > 0$ for which $g(y) = 0$. Then $c > 0$. It is possible to choose sufficiently small y_1, y_2, y and θ such that $g(y_1) = 0$ and $0 < y_1 + \theta < c$, $0 < y + \theta < c$, $0 < y - \theta < c$, and $0 < y_2 - \theta < c$. But, for such a choice, inequality (13.3.12) will not be valid for the function g (instead of f) if it is noted that g is symmetric. Hence $g(y) > 0$ for all $y \in R^1$ and $h = \log g$ is a continuous concave function.

Let us denote by $\mathscr{D}$ the set on which h' exists; the complementary set is at most denumerable. If $-y_1$, $-y$, y, y_2 belong to $\mathscr{D}$ and satisfy (13.3.12), then we derive from (13.3.12) and (13.3.13) that

$$h'(-y)+h'(y)=0; \qquad h'(-y_1)+h'(-y)+h'(y)+h'(y_2)=0.$$

Hence

$$h'(-y_1)+h'(y_2)=0 \qquad \text{for all } y_1, y_2 \in \mathscr{D}, y_1, y_2>0.$$

Since g is symmetric, $h'(-y_1)=-h'(y_1)$ and we have $h'(y_1)=h'(y_2)$ for all y_1, $y_2 \in \mathscr{D}$ with y_1, $y_2>0$ Hence h' is constant on the set $\{y \in \mathscr{D},\ y>0\}$, and hence $h'(y)=$ const. for all $y \geqq 0$ (taking the right-hand derivative at the origin). Hence

$$h(y)=\begin{cases} b-ay & \text{for } y \geqq 0 \\ b-a|y| & \text{for } y<0 \end{cases} \tag{13.3.14}$$

But, if $y>0$ and θ is sufficiently small, we obtain from (13.3.13)

$$-ay \geqq -a(y+\theta)-\log f(y+\theta)+\log f(y-\theta)$$

or

$$-a\theta \leqq \log f(y+\theta)-\log f(y-\theta).$$

Substituting $-\theta$ in place of θ, we obtain

$$a\theta \leqq \log f(y-\theta)-\log f(y+\theta)$$

or

$$-a\theta \geqq \log f(y+\theta)-\log f(y-\theta).$$

Hence

$$(\log f(y+\theta)-\log f(y-\theta))/\theta=-a$$

so that $(d/dy)\log f(y)=-(1/2)a$, which gives $f(y)=k_1e^{-(1/2)ay}$ for $y>0$. But we have from (13.3.14) that $f(y)f(-y)=ke^{-ya}$, whence $f(-y)=k_2e^{-(1/2)ay}$ for $y>0$. Setting $y=0$ in (13.3.13), we find that $k \geqq k_i^2$ for $i=1, 2$. But $k^2=(k_1k_2)^2$ in view of the continuity of g at the origin. Thus, $k_1=k_2=k^{1/2}$, and $f(y)=$ const. $e^{-(1/2)a|y|}$ for all real y. The theorem is proved.∎

The following result is due to Teicher [145].

THEOREM 13.3.3. Let $\{F(x/\sigma),\ \sigma>0\}$ be a family of absolutely continuous distributions on the half line R_+^1, depending on a scale parameter σ. Suppose that the density $F'(x)=f(x)$ satisfies the conditions

(i) f is continuous on $(0, \infty)$;

(ii) $\lim_{y \downarrow 0} f(\lambda y)/f(y)=1$ for all $\lambda>0$.

If the sample mean $\overline{X}$ is the MLE for σ for all $n = 1, 2, \ldots$, then f is the exponential density

$$f(x) = \begin{cases} e^{-x} & \text{for } x > 0 \\ 0 & \text{for } x \leqq 0. \end{cases}$$

Proof: The joint density function of $X_1, \ldots, X_n$ is

$$\sigma^{-n} \prod_1^n f\left(\frac{x_i}{\sigma}\right).$$

If $\hat{\sigma}$ is the MLE, then, for all $\sigma > 0$,

$$\hat{\sigma}^{-n} \prod_1^n f\left(\frac{x_i}{\hat{\sigma}}\right) \geqq \sigma^{-n} \prod_1^n f\left(\frac{x_i}{\sigma}\right). \tag{13.3.15}$$

We set $y_i = x_i/\hat{\sigma}$, $\lambda = \hat{\sigma}/\sigma$. If $\hat{\sigma}(x_1, \ldots, x_n)$ is a homogeneous function of degree one, then we have from (13.3.15) that

$$\prod_1^n f(y_i) \geqq \lambda^n \prod_1^n f(\lambda y_i) \tag{13.3.16}$$

for all $\lambda > 0$ and $y_1, \ldots, y_n$ satisfying the condition

$$\hat{\sigma}(y_1, \ldots, y_n) = 1. \tag{13.3.17}$$

If $\hat{\sigma} = \overline{X}$, then condition (13.3.17) reduces to

$$\sum_1^n y_i = n. \tag{13.3.18}$$

In (13.3.16), we set $y_i = k/m$ for $1 \leqq i \leqq m$ and $y_i = (n-k)/(n-m)$ for $m + 1 \leqq i \leqq n$, where k, m, n are positive integers such that $k < m < n$; we have

$$f^m\left(\frac{k}{m}\right) f^{n-m}\left(\frac{n-k}{n-m}\right) \geqq \lambda^n f^m\left(\lambda \frac{k}{m}\right) f^{n-m}\left(\frac{\lambda(n-k)}{n-m}\right). \tag{13.3.19}$$

Let $k/m \to a$, $m/n \to c$. Then, for all $\lambda > 0$ and all $0 < c < 1$, $0 < a < 1$, we have

$$f^c(a) f^{1-c}\left(\frac{1-ac}{1-c}\right) \geqq \lambda f^c(\lambda a) f^{1-c}\left(\frac{\lambda(1-ac)}{(1-c)}\right). \tag{13.3.20}$$

If now there exists a sequence $\{a_n\} \to 0$ with $f(a_n) > 0$ for all n, then from (13.3.20) and the condition (ii) of the theorem it follows that

$$f(y) \geqq \lambda^y f(y) \qquad \text{for } y \geqq 1, \lambda > 0. \tag{13.3.21}$$

Thus, if f vanishes for some $y \geqq 1$, then it is identically zero. Further, by (13.3.16), if f vanishes for some y in $(0, 1)$, then f will vanish identically in $(1, \infty)$.

Let us now suppose that there is no sequence $\{a_n\} \to 0$ with $f(a_n) > 0$ for all n, i.e., in some interval $(0, \delta)$, $f(y) \equiv 0$. Since $f(y) \not\equiv 0$ in R^1_+, we obtain from (13.3.16)—taking $y_i = 1$—that $f(1) > 0$. Hence $\delta < 1$, and we may assume that $(0, \delta)$ is the maximal interval in which f vanishes identically. We have from (13.3.20) that

$$0 = \lambda f^c(\lambda\delta) f^{1-c}\left(\frac{\lambda(1-\delta c)}{(1-c)}\right) \tag{13.3.22}$$

for all $\lambda > 0$, $0 < c < 1$. But the continuity of f implies that $f(y) > 0$ in the interval $(1 - \varepsilon, 1 + \varepsilon)$ for sufficiently small $\varepsilon > 0$. Hence, setting $\lambda = (1 - \varepsilon)/\delta$ in (13.3.22), we have

$$f\left(\frac{(1-\varepsilon)(1-\delta c)}{\delta(1-c)}\right) = 0$$

for $0 < c < 1$. Hence $f[(1 - \varepsilon)/\delta] = 0$ for sufficiently small $\varepsilon > 0$.

Let now $k + 1$ be an arbitrary integer $> (1 - \varepsilon)/\delta$. We have from (13.3.16) that

$$0 \geqq f^{n-k}\left(\frac{\lambda(1-\varepsilon)}{\delta}\right) f^k\left\{\frac{\lambda[k+1-(1-\varepsilon)\delta^{-1}]}{k}\right\}.$$

Setting here $\lambda = \delta/(1 - \varepsilon)$, we obtain

$$f\left\{\left[\frac{\delta}{k(1-\varepsilon)}\right][k+1-(1-\varepsilon)\delta^{-1}]\right\} = 0$$

for all sufficiently large k and small ε, whence $f[\delta/(1 - \varepsilon)] = 0$ for sufficiently small $\varepsilon > 0$, contradicting the assumption about the maximality of δ.

Thus a density f satisfying (13.3.16) does not vanish anywhere on R^1_+. Setting $h = \log f$, we write (13.3.21) in the form

$$\frac{h(y) - h(\lambda y)}{y} \geqq \log \lambda, \qquad y \geqq 1, \lambda > 0. \tag{13.3.23}$$

Replacing λ by $1/\lambda$ in (13.3.23) and combining the result obtained with (13.3.23), we have

$$0 \geqq h(\lambda y) - 2h(y) + h(y/\lambda) \qquad \text{for } y \geqq 1, \lambda > 0. \tag{13.3.24}$$

This means that the function $H(y) = h(e^y)$ is concave for $y \geqq 0$ and hence h is differentiable on $(1, \infty)$ with the possible exception of a countable set $\mathscr{D}$. From (13.3.23), we have for $\lambda < 1$ and $y \geqq 1$,

$$\frac{h(\lambda y) - h(y)}{y(\lambda - 1)} \geqq \frac{\log \lambda}{1 - \lambda} \geqq \frac{h(y/\lambda) - h(y)}{\lambda y[(1/\lambda) - 1]}$$

whence, letting $\lambda \uparrow 1$, we deduce that $h'(y) = -1$ on $(1, \infty) - \mathscr{D}$. Then, by continuity,

$$h(y) = c - y \qquad \text{for } y \geqq 1. \tag{13.3.25}$$

Further, taking $y_i < 1$ for $i = 1, \ldots, r < n$ and $y_i > 1$ for $i > r$ in (13.3.16) and using (13.3.18) and (13.3.25), we find that for all $\lambda > 0$, $r < n$ and $y_i \in (0, 1)$ for $1 \leqq i \leqq r$,

$$\sum_{i=1}^{r} [h(y_i) - h(\lambda y_i) + (1 - \lambda)y_i] \geqq n[\log \lambda + 1 - \lambda]. \tag{13.3.26}$$

Setting $r = 1$ in (13.3.26), we obtain for $0 < y < 1$

$$\frac{h(y) - h(\lambda y)}{n} \geqq \log \lambda + (1 - \lambda)\left(1 - \frac{y}{n}\right). \tag{13.3.27}$$

But, for $0 < x < 1$, and for all λ sufficiently close to 1 and larger than 1, $\log \lambda + x(1 - \lambda) > 0$. Thus, it follows from (13.3.27) that h is monotone decreasing in (0, 1).

Let now $h = h_1 + h_2$, where h_2 is absolutely continuous and h_1 is singular, i.e., $h_1'(y) = 0$ almost everywhere in (0, 1) and $h_1(y) \equiv 0$ for $y > 1$.

Taking $r = 1$ again in (13.3.26), we obtain for $0 < y < 1$ and $\lambda < 1$,

$$\frac{h(\lambda y) - h(y) + (\lambda - 1)y}{\lambda - 1} \geqq n\left(\frac{\log \lambda}{1 - \lambda} + 1\right). \tag{13.3.28}$$

Hence $y(h_2'(y) + 1) \geqq 0$ almost everywhere in (0, 1). Similarly, replacing λ by $1/\lambda$ in (13.3.26), we obtain the reverse inequality. Hence $h_2'(y) = -1$ almost everywhere in (0, 1), which reduces to $h_2(y) = c_1 - y$.

Using this result in (13.3.28), we obtain

$$\limsup_{\lambda \to 1} \frac{h_1(\lambda y) - h_1(y)}{y(\lambda - 1)} \geqq \limsup_{\lambda \uparrow 1} \frac{h_1(\lambda y) - h_1(y)}{y(\lambda - 1)} \geqq 0 \tag{13.3.29}$$

for all y in (0, 1); hence $h_1(y) \equiv 0$. Also, by continuity, $c_1 = c$.

Thus, for $y \in R_+^1$, $f(y) = ae^{-y}$ and since f is zero outside R_+^1, $a = 1$. The theorem stands proved. ▮

THEOREM 13.3.4. (cf. [145]). Let $\{F(x/\sigma), \sigma > 0\}$ be a family of absolutely continuous distributions on the real line depending on the scale parameter σ. Suppose that the density $F'(x) = f(x)$ satisfies the conditions

(i) f is continuous on R^1;

(ii) $\lim_{y \to 0} f(\lambda y)/f(y) = 1$ for all $\lambda > 0$.

If now $(\sum_1^n X_i^2/n)$ is the MLE for σ^2 for all n, where $X_1, \ldots, X_n$ constitute a random sample of size n, then F is the normal law $N(0, 1)$.

***Proof*:** Here (13.3.17) reduces to

$$\sum_1^n y_i^2 = n. \tag{13.3.30}$$

We set $y_i = \pm(k/m)^{1/2}$ for $1 \leqq i \leqq m$; $y_i = \pm[(n-k)/(n-m)]^{1/2}$ for $m < i \leqq n$; then, similarly to (13.3.20), we obtain the inequality

$$f^c(a)f^{1-c}\left(\pm\left(\frac{1-a^2c}{1-c}\right)^{1/2}\right) \geqq \lambda f^c(\lambda a)f^{1-c}\left(\pm\left(\frac{1-a^2c}{1-c}\right)^{1/2}\right) \tag{13.3.31}$$

valid for $\lambda > 0$, $|a| \leqq 1$, $0 < c < 1$. Arguments similar to those used in the proof of Theorem 13.3.3 show that f does not vanish, and it then follows from (13.3.31) that

$$\frac{h(y) - h(\lambda y)}{y^2} \geqq \log \lambda, \qquad |y| \geqq 1, \qquad \lambda > 0.$$

As before, h is differentiable, but this time in the set $A: \{y: |y| \geqq 1\}$, with the possible exception of an at most countable set of points $\mathscr{D}^*$. Proceeding in the same way as in the proof of Theorem 13.3.3, we find that $h'(y) = -y$ for $y \in A - \mathscr{D}^*$ and hence $h(y) = c - (1/2)y^2$ for $|y| \geqq 1$. The analogue of (13.2.26) is the inequality

$$\sum_1^r \left[h(y_i) - h(\lambda y_i) + \frac{(1-\lambda^2)y_i^2}{2}\right] \geqq n\left[\log \lambda + \frac{1}{2}(1-\lambda^2)\right] \tag{13.3.32}$$

where $|y| < 1$ and $r < n$. Hence it follows that h decreases in (0, 1) and increases in $(-1, 0)$. Again, using arguments similar to those used in the proof of Theorem 13.3.3, we arrive at the conclusion that $h(y) = c - (1/2)y^2$ for $|y| < 1$ also, whence $f(y) = \text{const.} \exp(-(1/2)y^2)$, where, of course, the constant $= (2\pi)^{-1/2}$. ∎

13.4 CHARACTERIZATION OF DISTRIBUTIONS THROUGH VARIOUS PROPERTIES OF THE CONDITIONAL DISTRIBUTION OF ONE STATISTIC GIVEN ANOTHER

13.4.1 The Symmetry of the Conditional Distribution of One Linear Form Given Another, and the Normal Law

In Theorem 3.1.1, it was shown that the independence of two linear forms in independent r.v.'s is a characteristic property of the normal law. Here, we prove a result fairly close to it, due to Heyde [160], in which the independence of the linear forms is replaced by the symmetry of the conditional distribution of one of them given the other.

THEOREM 13.4.1. Let $X_1, \ldots, X_k$ be independent r.v.'s and a_j, b_j, $j = 1, \ldots, k$ be nonzero constants satisfying the conditions $a_i b_i^{-1} + a_j b_j^{-1} \neq 0$ for any i, j, $i \neq j$. If the conditional distribution of $\sum_1^k a_j X_j$ given $\sum_1^k b_j X_j$ is symmetric, then the r.v.'s X_j are normal (possibly degenerate). The parameters of the distributions are connected as follows; if the c.f. of X_j be $\exp(itA_j - B_j t^2)$, with A_j real and $B_j \geqq 0$, then $\sum_1^k A_j a_j = 0$ and $\sum_1^k B_j a_j b_j = 0$.

***Proof*:** We begin remarking that the conditional c.f.

$$E\left[\exp\left(it\sum_1^k a_j X_j\right) \middle| \sum_1^k b_j X_j\right]$$

must be real-valued, so that, by Lemma 1.1.1,

$$E\left\{\left[\exp\left(it\sum_1^k a_j X_j\right) - \exp\left(-it\sum_1^k a_j X_j\right)\right]\exp\left(is\sum_1^k b_j X_j\right)\right\} = 0$$

for all real s and t. Thus, if f_j is the c.f. of X_j, then in view of the independence of the X_j, we have

$$\prod_{j=1}^k f_j(a_j t + b_j s) = \prod_{j=1}^k f_j(-a_j t + b_j s). \tag{13.4.1}$$

Suppose now that all the $a_j b_j^{-1}$ are distinct, and define the following functions of u in some neighborhood of the origin.

$$\phi_{1,j}(u) = \log f_j(b_j u), \qquad \phi_{2,j}(u) = -\log f_j(b_j u); \qquad 1 \leqq j \leqq k.$$

Then, we obtain from (13.4.1) that

$$\sum_1^k [\phi_{1,j}(s + b_j^{-1} a_j t) + \phi_{2,j}(s - b_j^{-1} a_j t)] = 0$$

for sufficiently small s and t. Since we have assumed that $a_i b_i^{-1} \neq -a_j b_j^{-1}$ for any i and j, $i \neq j$, the conditions of Lemma 1.5.1 are satisfied, and, therefore, in some neighborhood of the origin

$$\log f_j(t) = P_j(t), \qquad 1 \leqq j \leqq k \tag{13.4.2}$$

where the P_j are polynomials (of degree $\leqq 2k$). Hence, by Marcinkiewicz' theorem, the P_j are polynomials of degree $\leqq 2$, and the X_j are normal (possibly degenerate).

The case where some of the $a_j b_j^{-1}$ coincide is studied in the same way as in Theorem 3.1.1.

If now

$$f_j(t) = \exp(iA_j t - B_j t^2), \qquad 1 \leqq j \leqq k \tag{13.4.3}$$

substituting the expression (13.4.3) in (13.4.1) and regrouping the terms, we obtain

$$2it \sum_{1}^{k} A_j a_j - 4ts \sum_{1}^{k} B_j a_j b_j = 0$$

for all t and s, so that $\sum_1^k A_j a_j = 0$, $\sum_1^k B_j a_j b_j = 0$.∎

Remark. Theorem 13.4.1 is a special case of the general Theorem 10.3.11 proved in Chapter 10. We choose $Z_1 = -Z_2$, and Z_1, Z_2, Z_3 to be single linear functions in Theorem 10.3.11.

13.4.2 The Poisson Distribution in a Binomial Ruin Model

Let X be a discrete r.v. with $P[X = n] = p_n$, n running over the nonnegative integers. Let $d(r \mid n)$ be the probability that the value n of the r.v. X is reduced by some ruin process to the value r. If Y denotes the resulting r.v., then

$$P[Y = s] = \sum_{n=s}^{\infty} p_n \, d(s \mid n). \tag{13.4.4}$$

We introduce the conditional probabilities

$$P[Y = s \mid \text{unruined}] = \frac{p_s \, d(s \mid s)}{\sum_{s=0}^{\infty} p_s \, d(s \mid s)},$$

$$P[Y = s \mid \text{ruined}] = \frac{\sum_{n=s+1}^{\infty} p_n \, d(s \mid n)}{\sum_{s=0}^{\infty} \sum_{n=s+1}^{\infty} p_n \, d(s \mid n)}. \tag{13.4.5}$$

THEOREM 13.4.2. Let X be a Poisson r.v. and the ruin process be binomial i.e., $d(r \mid n) = \binom{n}{r} \pi^r \phi^{n-r}$, $\phi = 1 - \pi$. Then,

$$P[Y = s] = P[Y = s \mid \text{unruined}] = P[Y = s \mid \text{ruined}]. \tag{13.4.6}$$

***Proof*:** The probability of the compound event $[Y = s, \text{unruined}]$ is given by

$$P[Y = s, \text{ unruined}] = \frac{e^{-\lambda} \lambda^s \pi^s}{s!}$$

so that

$$P[Y = s \mid \text{unruined}] = \frac{e^{-\lambda}(\lambda^s \pi^s / s!)}{\sum_0^{\infty} e^{-\lambda} (\pi\lambda)^s / s!}$$

$$= e^{-\pi\lambda} (\pi\lambda)^s / s!$$

Similarly,

$$P[Y = s \mid \text{ruined}] = e^{-\pi\lambda} (\pi\lambda)^s / s!$$

and consequently the unconditional probability $P[Y = s]$ is equal to the same value. Formula (13.4.6) is proved.∎

We now investigate whether the property expressed by the formula (13.4.6) is characteristic of the Poisson distribution of the original r.v. and the Binomial ruin process. The main result in this direction is due to Rao and Rubin [132].

THEOREM 13.4.3. Let X be a discrete r.v. with values 0, 1, ... and $d(r|n)$ the probability that the value n of X is reduced by a ruin process to the value r. Let Y denote the resulting r.v., also taking the values 0, 1, ... Suppose that

$$P[Y = s|\text{ruined}] = P[Y = s|\text{unruined}]. \tag{13.4.7}$$

Then,

(i) if for all n, $d(r|n) = \binom{n}{r}\pi^r\phi^{n-r}$, where π and $\phi = 1 - \pi$ are fixed, then X is a Poisson r.v.;

(ii) if X is a Poisson r.v. with parameter λ and (13.4.7) is satisfied for all λ, then the ruin process is Binomial.

***Proof*:** (i) Let $G(t)$ be the generating function of X. If the original r.v. X is subjected to a Binomial ruin process, i.e., $d(r|n) = \binom{n}{r}\pi^r\phi^{n-r}$, then the generating function of Y, the resulting random variable, is

$$\sum_{r=0}^{\infty}\sum_{n=r}^{\infty} p_n\binom{n}{r}\pi^r\phi^{n-r}t^r = G(\phi + \pi t).$$

Further,

$$P[Y = r|\text{unruined}] = \frac{p_r\pi^r}{\sum p_s\pi^s} \tag{13.4.8}$$

and the generating function of the conditional distribution given by (13.4.8) is

$$\frac{\sum p_r\pi^r t^r}{(\sum p_s\pi^s)} = \frac{G(\pi t)}{G(\pi)}.$$

If $P(Y = r) = P(Y = r|$ unruined), then we have the functional equation

$$G(\phi + \pi t) = \frac{G(\pi t)}{G(\pi)}. \tag{13.4.9}$$

Setting $x = \pi t$, we reduce (13.4.9) to the form

$$G(x) = G(\pi)G(x + \phi) = [G(\pi)]^s G(x + s\phi); \qquad s = 0, 1, 2, \ldots. \tag{13.4.10}$$

Hence G is defined for all real x and further

$$G(x) = [G(\pi)]^{-s}G(x - s\phi) \tag{13.4.11}$$

for all positive integers s. Let now $G(x) = H(x)\cdot e^{-[x\log G(\pi)]/\phi}$. Then, it follows from (13.4.10) that

$$H(x) = H(x \pm s\phi), \qquad s = 0, 1, 2, \ldots$$

so that H is periodic with period ϕ.

The next step consists in showing the absolute monotonicity of the function G on $(-\infty, 0]$: in other words, we shall show that

$$G^{(k)}(x) \geqq 0 \qquad \text{for } -\infty < x \leqq 0; \qquad k = 1, 2, \ldots$$

where $G^{(k)}$ denotes the k-th derivative of G. We have from (13.4.11) that

$$G(x - s\phi) = [G(\pi)]^s G(x); \qquad G^{(k)}(x - s\phi) = [G(\pi)]^s G^{(k)}(x). \quad (13.4.12)$$

But, for $x \geqq 0$, $G^{(k)}(x)$ is nonnegative and hence we have from (13.4.12) that

$$G^{(k)}(y) \geqq 0 \qquad \text{for } -s\phi < y \leqq 0, \qquad s = 0, 1, \ldots.$$

Thus, G is absolutely monotone on $(-\infty, 0]$ and hence, by the well-known theorem due to Bernstein (cf. for instance [150], p. 162), there exists a distribution function $\alpha(t)$ such that

$$G(x) = G(0) \int_0^\infty e^{xt}\, d\alpha(t).$$

Setting $G(x) = e^{\lambda x} H(x)$, we have

$$e^{\lambda x} H(x) = H(0) \int_0^\infty e^{xt}\, d\alpha(t). \qquad (13.4.13)$$

Substituting $x = -\phi$ and $x = -2\phi$ successively in the above, and noting that $H(0) = H(-\phi) = H(-2\phi)$, we obtain

$$e^{-\lambda\phi} = \int_0^\infty e^{-\phi t}\, d\alpha(t), \qquad e^{-2\lambda\phi} = \int_0^\infty e^{-2\phi t}\, d\alpha(t)$$

whence $(Ee^{-\phi t})^2 = Ee^{-2\phi t}$, where the expectation is taken with respect to the d.f. $\alpha(t)$. This means that the d.f. of the variable $e^{-\phi t}$ is a degenerate one, with its point of increase at $t = \lambda$. Then we have from (13.4.13) that

$$e^{\lambda x} H(x) = H(0) e^{\lambda x}$$

i.e., $H(x) = H(0) = \text{constant}$. But $H(0) = e^{-\lambda}$ since $G(1) = 1$; hence

$$G(x) = e^{\lambda(x-1)} \qquad (13.4.14)$$

which is the generating function of a Poisson distribution.

(ii) We follow Srivastava and Srivastava [140] for this part. We have from (13.4.7)

$$e^{-\lambda} \sum_{n=r}^{\infty} \frac{\lambda^n}{n!} d(r \mid n) = \frac{e^{-\lambda} \lambda^r\, d(r \mid r)/r!}{\sum_{j=0}^{\infty} e^{-\lambda} \lambda^j\, d(j \mid j)/j!} \qquad (13.4.15)$$

this equation being satisfied for $\lambda > 0$ and $r = 0, 1, \ldots$.

Setting $r = 0$ and equating the coefficients of λ^n, we get

$$\frac{1}{n!} = \sum_{j=0}^{n} \frac{d(0 \mid n-j)}{(n-j)!} \cdot \frac{d(j \mid j)}{j!}. \qquad (13.4.16)$$

The generating function for the sequence $\{1/n!\}$ is e^t, so that we obtain from (13.4.16) that

$$e^t = a(t) \cdot b(t) \tag{13.4.17}$$

where $a(t)$ and $b(t)$ are the generating functions of the sequences

$$\{d(0 \,|\, n - j)/(n - j)!\} \text{ and } [d(j|j)/j!\}$$

respectively.

We have from (13.4.17) that

$$e^{t-1} = a(t)b(t)e^{-1}.$$

But e^{t-1} is the generating function of the Poisson distribution with parameter $\lambda = 1$; then it follows from Raikov's theorem [118] that $a(t)b(t)e^{-1}$ is the product of the generating functions of two Poisson distributions, so that for some π, $0 \leqq \pi \leqq 1$, $a(t)e^{-(1-\pi)}$ and $b(t)e^{-\pi}$ are the generating functions of Poisson distributions. It follows that

$$a(t) = e^{(1-\pi)t} \qquad \text{and} \qquad b(t) = e^{\pi t}. \tag{13.4.18}$$

We deduce from (13.4.18) that $d(j|j) = \pi^j$ and $d(0|j) = (1 - \pi)^j$. Substituting the values $d(0|0) = 1$ and $d(n|n) = \pi^n$ in (13.4.15), we obtain

$$\sum_{n=r}^{\infty} \frac{\lambda^n}{n!} d(r|n) = \frac{\lambda^r \pi^r e^{\lambda(1-\pi)}}{r!} \tag{13.4.19}$$

the above relation being valid for all $\lambda > 0$. Equating the coefficients of λ^{r+s}, we have

$$\frac{d(r|r+s)}{(r+s)!} = \frac{\pi^r(1-\pi)^s}{r!\, s!},$$

i.e.,

$$d(r|n) = \binom{n}{r} \pi^r (1-\pi)^{n-r} \qquad \text{if } n \geqq r. \blacksquare$$

Remark. If, in Theorem 13.4.3, X is a discrete r.v. taking the values c, $c + 1, \ldots$ instead of $0, 1, \ldots$, then it is (only) necessary to replace the phrase "Poisson r.v." by "Poisson r.v. truncated to the left at the point $c - 1$" in the statement of the theorem.

13.4.3 Discrete Distributions

We first cite a general result from Patil and Seshadri [109] for discrete distributions and derive from it characterizations of the Binomial, Poisson, Negative Binomial, and Geometric distributions.

THEOREM 13.4.4. Let X and Y be independent discrete r.v.'s and $c(x; x+y) = P[X = x | X + Y = x + y]$. If

$$\frac{c(x+y, x+y)c(0, y)}{c(x, x+y)c(y, y)} = \frac{h(x+y)}{h(x)h(y)}$$

where h is a nonnegative function, then

$$f(x) = f(0)h(x)e^{ax}, \; g(y) = g(0)k(y)e^{ay}$$

where

$$0 < f(x) = P[X = x], \qquad 0 < g(y) = P[Y = y], \qquad k(y) = \frac{h(y)c(0, y)}{c(y, y)}.$$

***Proof*:** If $r(x + y) = P[X + Y = x + y]$, then

$$f(x)g(y) = r(x+y)c(x, x+y) \tag{13.4.20}$$

whence we have

$$\frac{f(x+y)f(0)}{f(x)f(y)} = \frac{c(x+y, x+y)c(0, y)}{c(x, x+y)c(y, y)} = \frac{h(x+y)}{h(x)h(y)} \tag{13.4.21}$$

under the conditions of the theorem.

If $\phi(x) = f(x)/[f(0)h(x)]$, then (13.4.21) reduces to the Cauchy functional equation $\phi(x+y) = \phi(x)\phi(y)$, whose solution has the form $\phi(x) = e^{ax}$, a being a constant. Thus, $f(x) = f(0)h(x)e^{ax}$. Then, using (13.4.20), we obtain

$$g(y) = \frac{r(y)c(0, y)}{f(0)} = \frac{c(0, y)}{f(0)} \frac{f(y)g(0)}{c(y, y)} = g(0)k(y)e^{ay}$$

where $k(y) = h(y)c(0, y)/c(y, y)$. ▮

Corollary 1. If the conditional distribution of X for given $X + Y$ is hypergeometric with parameters m and n, then X and Y have Binomial distributions with parameters (m, θ) and (n, θ) respectively.

***Proof*:** Under the conditions given, we have

$$c(x, x+y) = \frac{\binom{m}{x}\binom{n}{y}}{\binom{m+n}{x+y}}.$$

Then $h(x) = \binom{m}{x}$ and $k(y) = \binom{n}{y}$; in this case,

$$f(x) = \binom{m}{x} f(0)e^{ax} \qquad \text{and} \qquad g(y) = \binom{n}{y} g(0)e^{ay}.$$

Let $e^a = \alpha$; then, $f(0) = (1+\alpha)^{-m}$ and $g(0) = (1+\alpha)^{-n}$. Hence, letting $\theta = \alpha/(1+\alpha)$,

$$f(x) = \binom{m}{x}\theta^x(1-\theta)^{m-x}, \qquad g(y) = \binom{n}{y}\theta^y(1-\theta)^{n-y}.$$

Corollary 2. If the conditional distribution of X given $X+Y$ is Binomial with the same parameter θ for all values of $X+Y$, then X and Y are Poisson r.v.'s.

***Proof*:** In this case,

$$c(x, x+y) = \binom{x+y}{x}\theta^x(1-\theta)^y = \left(\frac{\theta}{1-\theta}\right)^x (1-\theta)^{x+y}.$$

Hence, $h(x) = 1/x!$ and $k(y) = [(1-\theta)/\theta]^y/y!$. Thus,

$$f(x) = \frac{f(0)e^{ax}}{x!} \qquad \text{and} \qquad g(y) = g(0)\left(\frac{1-\theta}{\theta}\right)^y \frac{e^{ay}}{x!}$$

Setting $\lambda = e^a$ and $\mu = (1-\theta)e^a/\theta$, we have

$$f(x) = \frac{e^{-\lambda}\lambda^x}{x!} \qquad \text{and} \qquad g(y) = \frac{e^{-\mu}\mu^y}{y!}. \blacksquare$$

Corollary 3. If the conditional distribution of X given $X+Y$ is negative hypergeometric with parameters m and n for all values of $X+Y$, then X and Y have negative Binomial distributions with parameters (m, θ) and (n, θ) respectively.

***Proof*:** In this case

$$c(x, x+y) = \binom{x+y}{x}\frac{B(m+x, n+y)}{B(m, n)}$$

and it is easily seen that

$$h(x) = \frac{\Gamma(m+x)}{x!} \qquad \text{and} \qquad k(y) = \frac{\Gamma(n+y)}{y!}.$$

Then

$$f(0) = \frac{(1-\lambda)^m}{\Gamma(m)} \qquad \text{and} \qquad g(0) = \frac{(1-\lambda)^n}{\Gamma(n)}$$

where $\lambda = e^a$. Thus,

$$f(x) = \binom{x+m-1}{m-1}\lambda^x(1-\lambda)^m, \qquad g(y) = \binom{y+n-1}{n-1}\lambda^y(1-\lambda)^n. \blacksquare$$

Corollary 4. If the conditional distribution of X given $X + Y$ is a discrete uniform distribution for all values of $X + Y$, then X and Y have one and the same geometric distribution.

***Proof*:** This is a special case of Corollary 3 for $m = n = 1$. Then $h(x) = k(y) = 1$. Thus, $f(x) = \lambda^x(1 - \lambda)$ and $g(y) = \lambda^y(1 - \lambda)$.▮

13.5 DETERMINATION OF THE PARENT DISTRIBUTION THROUGH THE DISTRIBUTION OF CERTAIN STATISTICS

Sometimes it becomes necessary to test hypotheses concerning whether a distribution belongs to a given class of distributions (normal, gamma, Poisson, etc.), on the basis of a certain number of small samples. Then the parameters of the distributions might vary from sample to sample, and therefore, for testing hypotheses of the type indicated, one has to use statistics whose distribution eliminates in a certain sense these parameters (cf., Petrov [111]).

In this case, there arises the analytical problem of reconstructing the parent or original distribution from that of such a statistic. There are a series of isolated, interesting results in this area, which we present below, and fairly general results due to Prokhorov obtained under stronger restrictions; we begin with these results, following [117].

Let $(\mathscr{X}, \mathscr{A})$ be a measurable space; $\mathscr{X}$ is the set of elementary events, and $\mathscr{A}$ a σ-algebra. Let $\mathscr{P}$ be a set of probability distributions defined on it. Let $(\mathscr{Y}, \mathscr{B})$ be another measurable space, and a statistic $Y: \mathscr{X} \to \mathscr{Y}$ be a measurable mapping of $(\mathscr{X}, \mathscr{A})$ into $(\mathscr{Y}, \mathscr{B})$. Under this mapping, every distribution P induces on $\mathscr{B}$ a corresponding distribution of the statistic Y, which we denote by Q_P^Y. The statistic Y must possess the following properties.†

(1) Its distribution should not depend on the parameter(s) of $P \in \mathscr{P}$ (so that we may denote it simply as $Q_{\mathscr{P}}^Y$).

(2) From its distribution, it is possible to reconstruct the class of distributions $\mathscr{P}$, i.e., if P^* is a distribution on $\mathscr{A}$, and $Q_{P^*}^Y = Q_{\mathscr{P}}^Y$, then $P^* \in \mathscr{P}$. Often, requirement (2) is weakened by means of restrictions on the class of distributions P^*, so that we have the requirement

(2a) If $P^* \in \mathscr{P}^* \supset \mathscr{P}$, and $Q_{P^*}^Y = Q_{\mathscr{P}}^Y$, then $P^* \in \mathscr{P}$. Here $\mathscr{P}^*$ is a class containing $\mathscr{P}$.

We consider some examples to illustrate this general postulation. Let $(\mathscr{X}, \mathscr{A})$ be the Euclidean space R^n with its Borel σ-algebra; the distribution P will correspond to a random sample, being given by the probability density:

$$p(x_1, \theta) \ldots p(x_n, \theta) \tag{13.5.1}$$

where θ belongs to a parameter space Θ.

† If Y satisfies (1), it is called an ancillary statistic. If in addition (2) or (2a) is satisfied, it may be called a sufficient ancillary.

Example 1. (shift parameter: additive type) Let $p(x, \theta) = p(x - \theta)$ for θ real. We may take the Euclidean space R^{n-1} with its Borel σ-algebra as the space $(\mathscr{Y}, \mathscr{B})$, and for the statistic Y, the $(n - 1)$-dimensional vector

$$\mathbf{Y} = (x_1 - x_n, \ldots, x_{n-1} - x_n)$$

or any single-valued invertible function thereof, for instance,

$$\mathbf{Y}' = (x_1 - \bar{x}, \ldots, x_n - \bar{x}).$$

In the work [67] of Kovalenko's, presented below in the generalized form due to Prokhorov (Theorem 13.5.1), it was shown that for $n \geqq 3$, the distribution of $\mathbf{Y}$ determines the c.f. $f(t) = \int e^{itx} p(x)\, dx$ to within a factor e^{ict} (which clearly corresponds to a shift) in every interval where f does not vanish. In particular, if $f(t) \neq 0$ for any t, then, for $n \geqq 3$, the statistic Y satisfies the requirements (1) and (2). The same is true if f is analytic in the neighborhood of the origin, since we may take $f(t) \neq 0$ in that neighborhood and then it is uniquely defined by its values on this neighborhood, since it turns out to be in fact analytic in the neighborhood of the entire real axis (cf., for instance, [85]).
Example 2. (a linear transformation) Let $\theta = (a, \sigma)$ where a is real and $\sigma > 0$. Let

$$p(x, \theta) = \frac{1}{\sigma} p\left(\frac{x - a}{\sigma}\right). \tag{13.5.2}$$

Here, it is natural to take as $\mathbf{Y}$ the $(n - 2)$-dimensional statistic $\mathbf{Y} = (y_1, \ldots, y_n)$, with $y_k = (x_k - \bar{x})/S$, where $S^2 = \sum_1^n (x_k - \bar{x})^2$, $S > 0$. Then, we have

$$\sum_1^n y_k = 0, \qquad \sum_1^n y_k^2 = 1.$$

There is a common group-theoretic property involved in all these examples. We start from a group $\sum$ of one-one (or almost one-one) transformations g of the sample space into itself, under which the elements x and gx simultaneously belong to, or do not belong to, $\mathscr{P}$. If the maximal invariant group of $\sum$ is taken as Y, then, for $P \in \mathscr{P}$, Y possesses property (1). The question of when Y has the property (2) is, generally speaking, a very difficult problem of analytical statistics.

We first give a generalization of a result from [67] on samples from a multivariate population.

Let the family $\mathscr{P}$ be given by the formula (13.5.1), where $\mathbf{x}_j$ is the s-dimensional vector $(x_j^{(1)}, \ldots, x_j^{(s)})$ and

$$p(\mathbf{x}, \boldsymbol{\theta}) = p(\mathbf{x} - \boldsymbol{\theta}) \tag{13.5.3}$$

$\boldsymbol{\theta}$ being a vector in a k-dimensional subspace π_k of the Euclidean space R^s.

Without loss of generality, we may take π_k to be given by the relations

$$\theta_{k+1} = \cdots = \theta_s = 0. \tag{13.5.4}$$

We shall assume that the density $p(\mathbf{x}, \mathbf{0})$ satisfies the well-known condition due to Cramer:

$$\int_{R^s} e^{(\mathbf{h}, \mathbf{x})} p(\mathbf{x}, \mathbf{0})\, d\mathbf{x} < \infty \tag{13.5.5}$$

for values of h lying in a neighborhood of the origin in R^s (*vide* [46] as regards Cramer's condition and its use in the theory of limit theorems).

We also note that the integral in (13.5.5) may not be convergent if we rcplace $p(\mathbf{x}, \mathbf{0})$ by $p(\mathbf{x}, \boldsymbol{\theta})$, for the values of $\boldsymbol{\theta}$ indicated above. We now establish the following

THEOREM 13.5.1. Let $(\mathbf{x}_1, \ldots, \mathbf{x}_n)$ be a random sample of size $n \geqq 3$ from a multivariate distribution of the type (13.5.3) obeying the condition (13.5.5), and let the condition (13.5.4) be satisfied. We set

$$\mathbf{x}_j' = (x_j^{(1)}, \ldots, x_j^{(k)}, 0, \ldots, 0).$$

Then the statistic $\mathbf{y} = (\mathbf{y}_1, \mathbf{y}_2,)$ where $\mathbf{y}_1 = \mathbf{x}_1 - \mathbf{x}_3'$, $\mathbf{y}_2 = \mathbf{x}_2 - \mathbf{x}_3'$, satisfies requirements (1) and (2).

Proof: Let $\mathbf{t} = (t^{(1)}, \ldots, t^{(s)})$, $\boldsymbol{\tau} = (\tau^{(1)}, \ldots, \tau^{(s)})$, and let $\mathbf{t}'$ and $\boldsymbol{\tau}'$ be defined with respect to $\mathbf{t}$ and $\boldsymbol{\tau}$ in the same way as $\mathbf{x}_j'$ with respect to $\mathbf{x}_j$. Let

$$f(\mathbf{t}) = E \exp i(\mathbf{t}, \mathbf{x}_j).$$

We see that $\mathbf{y}_1 - \mathbf{y}_2 = \mathbf{x}_1 - \mathbf{x}_2$, so that the c.f. of $\mathbf{x}_1 - \mathbf{x}_2$, i.e., $|f(\mathbf{t})|^2$ is uniquely defined in terms of the statistic $(\mathbf{y}_1, \mathbf{y}_2)$. For the determination of $f(\mathbf{t})$, we have to study the behavior of its argument arg $f(\mathbf{t})$ in the neighborhood of the origin.

Let $f_1(\mathbf{t})$ and $f_2(\mathbf{t})$ be two distinct c.f.'s for the $\mathbf{x}_j$ such that $\mathbf{y}$ has the same distribution. Denote the known c.f. of the statistic $\mathbf{y}$ by $\psi_0(\mathbf{t}, \boldsymbol{\tau})$. We have

$$\begin{aligned}\psi_0(\mathbf{t}, \boldsymbol{\tau}) &= E \exp [i(\mathbf{t}, \mathbf{x}_1 - \mathbf{x}_3') + i(\boldsymbol{\tau}, \mathbf{x}_2 - \mathbf{x}_3')] \\ &= f_j(\mathbf{t}) f_j(\boldsymbol{\tau}) \overline{f_j(\mathbf{t}' + \boldsymbol{\tau}')}, \qquad j = 1, 2. \end{aligned} \tag{13.5.6}$$

Let $\delta > 0$ be so small that in a δ-neighborhood of the origin the functions $f_j(\mathbf{t})$, $j = 1, 2$ do not vanish, and let $\mathbf{t}$, $\boldsymbol{\tau}$, and $\mathbf{t} + \boldsymbol{\tau}$ lie in this neighborhood. Then, $\psi_0(\mathbf{t}, \boldsymbol{\tau})$ does not vanish; let $S_0(\mathbf{t}, \boldsymbol{\tau})$ be the principal value of arg $\psi_0(\mathbf{t}, \boldsymbol{\tau})$ in this set, and $A_j(\mathbf{t})$ the principal value of arg $f_j(\mathbf{t})$, $j = 1, 2$. All these functions are continuous. For the indicated values of $\mathbf{t}$ and $\boldsymbol{\tau}$, we have

$$A_j(\mathbf{t}) + A_j(\boldsymbol{\tau}) - A_j(\mathbf{t}' + \boldsymbol{\tau}') = S_0(\mathbf{t}, \boldsymbol{\tau}). \tag{13.5.7}$$

We consider the corresponding homogeneous equation in the continuous functions, for $a(\mathbf{t}) = A_1(\mathbf{t}) - A_2(\mathbf{t})$, we have

$$a(\mathbf{t}) + a(\boldsymbol{\tau}) - a(\mathbf{t}' + \boldsymbol{\tau}') = 0 \tag{13.5.8}$$

with $a(\mathbf{0}) = 0$.

We find, on setting $\boldsymbol{\tau} = \mathbf{0}$ or $\mathbf{t} = \mathbf{0}$, $a(\mathbf{t}) = a(\mathbf{t}')$ and $a(\boldsymbol{\tau}) = a(\boldsymbol{\tau}')$, so that $a(\mathbf{t}') + a(\boldsymbol{\tau}') = a(\mathbf{t}' + \boldsymbol{\tau}')$.

In view of the continuity of the functions, we obtain the well-known solution

$$a(\mathbf{t}') = \sum_{j=1}^{k} c_j t^{(j)}$$

where the c_j are numerical coefficients.

Since $a(\mathbf{t}) = a(\mathbf{t}')$, we have

$$A_1(\mathbf{t}) - A_2(\mathbf{t}) = \sum_{j=1}^{k} c_j t^{(j)},$$

$$f_1(\mathbf{t}) = f_2(\mathbf{t}) \exp\left(i \sum_{j=1}^{k} c_j t^{(j)}\right). \tag{13.5.9}$$

In view of the analyticity of f_1 and f_2, this equality holds for all values of t. Thus the theorem is proved.∎

We note that if $f(\mathbf{t}) \neq 0$ for any $\mathbf{t}$, then (13.5.9) holds even without the condition (13.5.5).

We now turn to the case of a scale parameter in a multivariate population.

Let the family $\mathscr{P}$ be given by the formula (13.5.1) for an s-dimensional vector $\mathbf{x}_j$ with scale parameter θ.

$$p(\mathbf{x}, \theta) = \theta^{-s} p(\mathbf{x}/\theta). \tag{13.5.10}$$

Let $\mathbf{x} = (\mathbf{x}_1, \ldots, \mathbf{x}_n)$ be a random sample from (13.5.10).

$$\mathbf{x}_j = (x_j^{(1)}, \ldots, x_j^{(s)}).$$

Consider the $2s$-dimensional statistic

$$\mathbf{V}_j = \{\ln |x_j^{(1)}|, \ldots, \ln |x_j^{(s)}|, \operatorname{sign} x_j^{(1)}, \ldots, \operatorname{sign} x_j^{(s)}\}.$$

It belongs to the additive type, where $\theta_1 = \ln \theta$ plays the role of the shift parameter (the last s coordinates of $\mathbf{V}_j$ do not involve the shift parameter). Further, Cramer's condition (13.5.5) will be satisfied if $p(\mathbf{x}, 1)$ is bounded and satisfies Cramer's condition, since then, it would follow from the existence of $E \exp(h|\mathbf{x}|)$ for sufficiently small real values of h that $E \exp(h \ln |\mathbf{x}|)$ exists. This leads us to the following theorem.

THEOREM 13.5.2. Let $p(\mathbf{x}, 1)$ be bounded and satisfy Cramer's condition. Then the statistic

$$\mathbf{y} = (\mathbf{V}_1 - \mathbf{V}_3', \mathbf{V}_2 - \mathbf{V}_3')$$

where $\mathbf{V}_3'$ is defined in relation to $\mathbf{V}_3$ in accordance with the principle in Theorem 13.5.1 (replacing s by $2s$ and k by s), has the properties (1) and (2).

The proof of this theorem thus reduces to an immediate application of Theorem 13.5.1.

We turn to the case of a general linear transformation (cf. Example 2). Here $\boldsymbol{\theta} = (a, b)$, a real and b positive, and $p(x, \boldsymbol{\theta}) = b^{-1}p[(x - a)/b]$, preserving the notation of Example 2. We first consider the case of a symmetric density $p(x)$. Then we have

THEOREM 13.5.3. If p is symmetric, bounded and satisfies Cramer's condition, then, for $n \geqq 6$, the statistic

$$\mathbf{y}' = (y_1', y_2') = \left[\left(\frac{y_4 - y_3}{y_2 - y_1}\right)^2, \left(\frac{y_6 - y_5}{y_2 - y_1}\right)^2\right]$$

has properties (1) and (2a) if $\mathscr{P}'$ be the class of symmetric densities.

***Proof*:** We have (cf. Example 2)

$$y_1' = \left(\frac{x_4 - x_3}{x_2 - x_1}\right)^2 \quad \text{and} \quad y_2' = \left(\frac{x_6 - x_5}{x_2 - x_1}\right)^2.$$

Let p' be a symmetric density different from p and such that $Q_{p'}^{\mathbf{y}'} = Q_p^{\mathbf{y}}$. From the facts that p satisfies Cramer's condition and is bounded, we easily see that $\ln y_1'$ and consequently $\ln (x_2 - x_1)^2$ satisfy Cramer's condition (for p as well as for p').

Now consider the sample of size 3 of i.i.d.r.v.'s

$$\ln (x_2 - x_1)^2, \qquad \ln (x_4 - x_3)^2, \qquad \text{and} \qquad \ln (x_6 - x_5)^2$$

and apply Theorem 13.5.1 to it. We see that the distribution of $\ln (x_2 - x_1)^2$ can be reconstructed from that of $\mathbf{y}'$ to within a shift parameter, and that of $(x_2 - x_1)^2$ can be, to within a scale parameter. Since $x_2 - x_1$ has a symmetric distribution, its distribution can be reconstructed then to within a scale parameter. For a symmetric density p' the distribution of x_1 can be reconstructed from that of $x_2 - x_1$ to within a shift parameter, proving the theorem. We note that if p is normal, then the symmetry of p' is not required, since the distribution of x_1 can be reconstructed from that of $x_2 - x_1$. ∎

If we omit the requirement of symmetry, then the following result can be established.

THEOREM 13.5.4. Let p be bounded and satisfy Cramer's condition. Then, for $n \geqq 9$, the statistic

$$\mathbf{y}'' = (\mathbf{y}_1'', \mathbf{y}_2'')$$

where

$$\mathbf{y}_1'' = \left(\ln\left|\frac{y_3 - y_1}{y_9 - y_7}\right|, \ln\left|\frac{y_2 - y_1}{y_8 - y_7}\right|, \operatorname{sign}(y_3 - y_1), \operatorname{sign}(y_2 - y_1)\right)$$

$$\mathbf{y}_2'' = \left(\ln\left|\frac{y_6 - y_4}{y_9 - y_7}\right|, \ln\left|\frac{y_5 - y_4}{y_8 - y_7}\right|, \operatorname{sign}(y_6 - y_4), \operatorname{sign}(y_5 - y_4)\right)$$

has the properties (1) and (2).

***Proof*:** We note that the distribution of the vector $(x_3 - x_1, x_2 - x_1)$ belongs to the multiplicative type.

In terms of a sample of size 3 from the given two-dimensional distribution, i.e., $(x_3 - x_1,\ x_2 - x_1)$, $(x_6 - x_4,\ x_5 - x_4)$, $(x_9 - x_7,\ x_8 - x_7)$, this multiplicative type can be uniquely reconstructed from the distribution of our statistic. Again, from the distribution of $(x_3 - x_1,\ x_2 - x_1)$, we reconstruct the additive type of distribution which x_1 has, and the theorem is proved. ∎

From among the theorems of a more or less general character concerning the reconstruction of the distribution of the members of a random sample from the distribution of a statistic, we present here some theorems from the work of Linnik [84].

In the work [117] presented above, essentially we considered the reconstruction of the original distribution of the sample from that of a vector statistic. In [84], such reconstruction from the distribution of a scalar statistic $Q(x_1, \ldots, x_n)$ of the type of a definite tube statistic of dimension n is considered (cf. section 4.1). Problems of this type may be considered as generalizations of the classical moment problem.

In fact, let us assume that the distribution $F(x) = P[x_i \leqq x]$ has moments of all orders: $\alpha_k = \int x^k\, dF(x)$ and let $Q = n\bar{x} = (x_1 + \cdots + x_n)$. Q will also have moments of all orders; it is easily seen that, computing the moments of Q (under the known distribution of Q), we can successively find all the α_k. Therefore, in the class of laws F having moments of all orders and satisfying the condition of unique solution for the moment problem, the distribution of Q will determine F.

We may take any linear form $Q = \sum_1^n a_i X_i$ with positive coefficients, in place of $n\bar{X}$. If negative coefficients are also admitted, then, in general, the distribution of Q will not uniquely define the parent distribution; thus, for $n = 2$, $Q = X_1 - X_2$, on the basis of the distribution of the statistic Q, the law $F(x + a)$, for any a, is not distinguishable from $F(x)$.

Let a continuous statistic Q be given; we call the set of all points $(x_1, \ldots, x_n)$ satisfying the condition

$$Q(x_1, \ldots, x_n) = c \tag{13.5.11}$$

a *level surface* for Q. In what follows, we shall consider only such statistics Q for which the level surfaces (13.5.11) are star shaped and piecewise-smooth (made up of a finite number of smooth pieces), or are empty.

The statistic Q will be called *definite* if the level surface (13.5.11) is, for any c, located in a bounded subset of the sample space, and indefinite in the contrary case.

The behavior of nonlinear indefinite statistics and the connection between their distributions and the parent distributions present significant difficulties in studying. In our work, we shall consider only definite statistics. However, generally speaking, such statistics will not uniquely determine the parent distribution.

Take, for instance, $n = 1$, and a distribution concentrated on the interval $(-\pi/2, \pi/2)$ and having the density function $(1/\sqrt{2}) \cos(x/2)$ thereon, as the parent distribution. Let $Q = X^2$. Here the sample space is the segment $[-\pi/2, \pi/2]$ and the statistic is definite thereon. At the same time, for any sufficiently small ε_j and any integers k and N, we have

$$(1/\sqrt{2}) \int_{-\pi/2}^{\pi/2} Q^k \left[\cos(x/2) + \sum_{j=1}^{N} \varepsilon_j \sin jx\right] dx = (1/\sqrt{2}) \int_{-\pi/2}^{\pi/2} Q^k(\cos(x/2))\, dx$$

where, the expression within the brackets (on the left) is the probability density for a random variable taking values in $[-\pi/2, \pi/2]$.

The situation, however, changes if we consider symmetric r.v.'s with mean value zero. Under fairly general conditions, their distributions can be reconstructed from that of the statistic Q. Here we shall examine such cases.

We introduce some definitions, for formulating the results obtained in this direction. We consider the class of functions $y(x)$ which are continuous throughout the real line. We say that a set of such functions form a class $\{D\}$ on the segment $[\alpha, \beta]$ (the values $\mp\infty$ are admissible for α and β), if , in every finite segment $[a, b]$ contained in $[\alpha, \beta]$, the difference between two functions of the class $\{D\}$

$$y_1 - y_2 \qquad \text{with} \qquad y_1 \in \{D\},\ y_2 \in \{D\}$$

has only a finite number of zeros or is identically zero.

The difference $y_1 - y_2$ remains unchanged if all the functions of the class are "shifted" by the addition of one and the same continuous function h; hence every continuous function belongs to some class $\{D\}$.

The class of polynomials form a class $\{D\}$; so do all functions analytic in the neighborhood of the segment $[\alpha, \beta]$; quasianalytic functions constitute yet another example.

According to Mandelbrojt [100], with a given sequence $\{m_n\}$ of positive numbers, we associate the class of all functions $y(x)$, having derivatives of all orders on the finite segment $[\alpha, \beta]$, for each of which there exists a constant $K > 0$ such that

$$|y^{(n)}(x)| \leqq K^n m_n \qquad (n = 1, 2, \ldots). \tag{13.5.12}$$

Call it the class $C\{m_n\}$. This class will be called *quasianalytic* (Δ), if it follows from the relations

$$y_1, y_2 \in C_{\{m_n\}}; \qquad y_1^{(n)}(x_0) = y_2^{(n)}(x_0) \qquad \text{for } n = 0, 1, \ldots \tag{13.5.13}$$

that

$$y_1 \equiv y_2 . \tag{13.5.14}$$

A necessary and sufficient condition for the quasi-analyticity of $C\{x_n\}$ is given by Carleman's theorem (cf. Mandelbrojt [100], p. 31).

It is easily seen that a class $C\{x_n\}$, quasi-analytic (Δ) on every finite segment $[a, b] \subset [\alpha, \beta]$, will be a class $\{D\}$ on $[\alpha, \beta]$. To see this, we set up the difference of two functions of the class $y_1 - y_2$. If it has an infinite number of zeros on $[a, b]$, then any accumulation point x_0 of these zeros will also be such a zero. By Rolle's theorem, between two zeros of $y_1 - y_2$ will be contained a zero of $y_1' - y_2'$, and the set of zeros of this difference will also have the same accumulation point x_0. The same holds for $y_1^{(n)} - y_2^{(n)}$ so that $y_1^{(n)}(x_0) = y_2^{(n)}(x_0)$ for $n = 1, 2, \ldots$ so that we must have $y_1 \equiv y_2$.

To obtain examples which are derivable from the class of infinitely differentiable functions (apart from trivial examples obtained by a "shift" of all functions of the class by a continuous function without a derivative), we may adopt the method of superposition of functions.

Let g be a continuous function defined on the whole of R^1 and such that, for any g_0, the equation $g(x) = g_0$ has only a finite number of solutions. Consider also some class of functions of the type $\{D\}$ and the new class of continuous functions $\{y(g(x))\}$ formed from them. If on a finite segment $[a, b]$, the difference $y_1(g(x)) - y_2(g(x))$ of two functions of the new class has an infinite number of zeros, then, by the property of the class $\{D\}$ we must have $g(x_1) = g(x_2) = \cdots$ for an infinite set of (distinct) numbers $x_1, x_2, \ldots,$ which is impossible, so that the new class is also a class of the type $\{D\}$.

We now introduce a more general type of classes $\{D_{pw}\}$, We shall say that a set of continuous functions $y(x)$ on $[\alpha, \beta]$ form a class of the type $\{D_{pw}\}$, $\{D\text{-piecewise}\}$, if every finite segment $[a, b] \subset [\alpha, \beta]$ can be split up into a finite number of segments $[a_m, b_m]$ on each of which $y(x)$ belongs to a class of type $\{D\}$.

We shall consider below classes of symmetric probability densities $y(x)$ defined on the whole real line and belonging to a class of type $\{D_{pw}\}$. We note that $\{D_{pw}\}$ is, generally speaking, not a class of type $\{D\}$.

We now formulate the results that have been obtained.

THEOREM 13.5.5. Let an even continuous probability density function y be given in the segment $[-\alpha, \alpha]$, which possibly vanishes only at the end points of this segment. Let $Q(x_1, \ldots, x_n)$ be a definite, piecewise-smooth, star shaped, homogeneous, positive measurable function. Then the distribution of this statistic, induced by the measure corresponding to $y(x)$, determines $y(x)$ in the class of even densities of the type $\{D_{pw}\}$ containing $y(x)$.

THEOREM 13.5.6. Let a probability density y, given on the segment $[-\alpha, \alpha]$, belong to a class of type $\{D_{pw}\}$ and have an at most denumerable set of zeros on that segment. Then the distribution of a statistic $Q(x_1, \ldots, x_n)$, of the type indicated in Theorem 13.5.5, determines the density $y(x)$ in its class.

We now proceed to formulate two other results.

We shall say, as in Theorem 13.5.5, that the distribution of the statistic $Q = Q(x_1, \ldots, x_n)$ is *induced by a parent population* with law of distribution $F(x)$, if it corresponds to a random sample $(x_1, \ldots, x_n)$ from this population. Further, the probability density $y(x)$ will be said to be *completely symmetric* if a random variable Z corresponding to it can be represented in the form

$$Z = Z_1 - Z_2 \tag{13.5.15}$$

where Z_1 and Z_2 are identically distributed r.v.'s having a density which belongs to $\{D_{pw}\}$.

THEOREM 13.5.7. Let a completely symmetric density $y(x)$ be holomorphic at all points of the real axis, and let the distribution corresponding to it of a definite statistic $Q(x_1, \ldots, x_n)$, of the type indicated in Theorem 13.5.5, be given. Then this distribution determines $y(x)$ in the class of all completely symmetric densities of bounded variation in the neighborhood of the origin.

In particular, if $y(x)$ is a normal density with zero mean, we have

THEOREM 13.5.8. A normal density with zero mean is determined by the distribution of a statistic of the type indicated above, among all densities which are completely symmetric and of bounded variation in the neighborhood of the origin.

We now proceed to prove the four theorems stated above in order.

***Proof*:** Let $Q = Q(x_1, \ldots, x_n)$ be a definite statistic with piecewise-smooth level surfaces and homogeneous with degree $\mu > 0$, so that

$$Q(tx_1, \ldots, tx_n) = t^{\mu} Q(x_1, \ldots, x_n) \tag{13.5.16}$$

for all $t > 0$. Let us replace the statistic Q by $Q_1 = |Q|^{1/\mu}$. This will be of the same type as Q, and, besides, nonnegative and of degree 1.

Let the distribution of Q_1 be induced by a parent distribution having a continuous probability density $y(x)$. Then, as easily seen, Q_1 will also have a density function defined and continuous on $x \geqq 0$. This follows from the definiteness of the statistic and the piecewise smoothness of its level surfaces.

Denoting the probability density of the distribution of the statistic Q_1 by $\gamma(\rho)$, so that $\gamma(\rho) = 0$ for $\rho < 0$, we find

$$P[Q_1 < R] = \int_0^R \gamma(\rho)\, d\rho.$$

We may now take the statistic Q_1, whose distribution is determined by that of Q, as the original one (instead of Q), and denote it again by the same symbol Q. The problem of interest to us is the reconstruction of $y(x)$ from $\gamma(\rho)$, from the point of view of uniqueness of solution.

We now set up the basic integro-functional equation of the problem. It consists simply in expressing the probability density $\gamma(\rho)$ of Q in terms of the density $y(x)$ in spherical coordinates. We now introduce spherical coordinates in the sample space:

$$\begin{aligned}
x_1 &= \rho \cos \phi_1 \\
x_2 &= \rho \sin \phi_1 \cos \phi_2 \\
&\cdots\cdots\cdots\cdots \\
x_{n-1} &= \rho \sin \phi_1 \sin \phi_2 \cdots \cos \phi_{n-1} \\
x_n &= \rho \sin \phi_1 \sin \phi_2 \cdots \sin \phi_{n-1}
\end{aligned}$$

$\rho \geqq 0$, $0 \leqq \phi_i \leqq \pi$ for $1 \leqq i \leqq n-2$; $0 \leqq \phi_{n-1} \leqq 2\pi$. The Jacobian of the transformation is $\rho^{n-1}T_0(\phi)$, where

$$T_0(\phi) = \sin^{n-2}\phi_1 \sin^{n-3}\phi_2 \cdots \sin \phi_{n-2}. \tag{13.5.17}$$

Let us write $\phi = (\phi_1, \ldots, \phi_{n-1})$, $d\phi = d\phi_1, \ldots, d\phi_{n-1}$ and Φ = the domain of variation of ϕ.

We write the equation of the level surface $Q = 1$ in the form

$$x_1 = g_1(\phi), \ldots, x_n = g_n(\phi). \tag{13.5.18}$$

In view of the homogeneity of Q, with degree 1, the equation of the star shaped level surface $Q = \rho$ will be

$$x_1 = \rho g_1(\phi), \ldots, x_n = \rho g_n(\phi). \tag{13.5.19}$$

Then the basic integro-functional equation can be written in the form

$$\int_\Phi y(\rho g_1(\phi))y(\rho g_2(\phi)) \cdots y(\rho g_n(\phi))\rho^{n-1}T_0(\phi)\, d\phi = \gamma(\rho) \tag{13.5.20}$$

for any $\rho \geqq 0$. Setting $\gamma(\rho)/\rho^{n-1} = \gamma_0(\rho)$, we have

$$\int_\Phi y(\rho g_1(\phi)) \cdots y(\rho g_n(\phi)) T_0(\phi)\, d\phi = \gamma_0(\rho). \tag{13.5.21}$$

The above holds for any $\rho > 0$; here $\gamma_0(\rho)$ is a given function; the question is about the reconstruction of $y(x)$ from it and the uniqueness thereof.

We shall take $y(x)$ to be the density of a symmetric law, and hence an even function. Let

$$y(x) = z(x^2). \tag{13.5.22}$$

Then (13.5.21) takes the form

$$\int_\Phi z(\rho^2 g_1^2(\phi)) \cdots z(\rho^2 g_n^2(\phi)) T_0(\phi)\, d\phi = \gamma_0(\rho). \tag{13.5.23}$$

Let $r = \rho^2$; $\rho = \sqrt{r} > 0$; $\gamma_0(\rho) = \gamma_1(r)$; $\psi_i(\phi) = g_i^2(\phi)$. Then the above can be rewritten in the form

$$\int_\Phi z(r\psi_1(\phi)) \cdots z\,(r\psi_n(\phi)) T_0(\phi)\, d\phi = \gamma_1(r). \tag{13.5.24}$$

Here $\psi_i(\phi) \geqq 0$ for $i = 1, \ldots, n$ and r can take arbitrary positive values; $\gamma_1(r)$ is a given function of r.

If the basic segment $[-\alpha, \alpha]$ on which $y(x)$ is given and satisfies (13.5.21) can be embedded in an open set of the complex variable $v = x + iy$ such that $y(v)$ is analytic at every point of this set, then in the class of functions of such a type, $y(v)$ is determined uniquely. We have $y(v) = z(v^2)$ in the neighborhood of $v = 0$, and setting $v^2 = r$, we arrive at (13.5.24). In view of $\psi_i(\phi) \geqq 0$, $T_0(\phi) \geqq 0$, we uniquely determine $z(0)$; noting that the right side will also be analytic in the above open set of the v-plane, we differentiate with respect to r successively and set $r = 0$ to successively determine $z'(0)$, $z''(0)$, ..., and hence the function $y(x)$ for $|x| \leqq \alpha$.

We shall now consider more general classes of solutions. Only even solutions $y(x)$ will be considered, so that it is sufficient to study only (13.5.24). Let

$$\psi_0 = \sup_{i,\,\phi} \psi_i(\phi). \tag{13.5.25}$$

We have $\psi_0 > 0$ and for all values of ϕ

$$0 < r\psi_i(\phi) \leqq r\psi_0. \tag{13.5.26}$$

Let (13.5.21) have a solution $y(x)$ concentrated in the segment $[-\alpha, \alpha]$, continuous and nonvanishing in the interval $(-\alpha, \alpha)$. In this case $z(x) \neq 0$ for x in $(-\sqrt{\alpha}, \sqrt{\alpha})$. Let $y(x)$ be contained in some class of the type $\{D_{pw}\}$;

as remarked in section 13.3, $z(x)$ will be contained in the same class. Suppose $z_1(x)$ from the class is a solution of (13.5.24). We set up the difference $\eta(x) = z_1(x) - z(x)$.

Since (13.5.24) determines $z(0) = z_1(0)$, $\eta(0) = 0$. We investigate the behavior of $\eta(x)$ for $x \neq 0$. $\eta(x)$ is continuous; taking $z_1 \not\equiv z$, we have $\eta \not\equiv 0$.

Let $x_0 \geqq 0$ be the largest number satisfying the following conditions: $\eta(x) = 0$ for $0 \leqq x \leqq x_0$, and for any $\Delta > 0$, however small, there exists a point ξ in the segment $[s_0, x_0 + \Delta]$ at which $\eta(\xi) \neq 0$.

We set, choosing some $\Delta \in (0, 1)$,

$$r_0 = x_0/\psi_0 \qquad \text{and} \qquad r_1 = (x_0 + \Delta)/\psi_0. \tag{13.5.27}$$

We consider the functions $\eta(r\psi_i(\phi))$ for $0 \leqq r \leqq r_1$, $i = 1, \ldots, n$. Obviously $\eta(r\psi_i(\phi)) = 0$ for $0 \leqq r \leqq r_0$. Thus, only the behavior of $\eta(r\psi_i(\phi))$ in the set $r_0 \leqq r \leqq r_1$ is of importance. Setting $\prod(z, r, \phi) = \prod_{i=1}^n z(r\psi_i(\phi))$, we consider the difference

$$\prod(z_1, r, \phi) - \prod(z, r, \phi) = \sigma_1(r, \phi) + \sigma_2(r, \phi)$$

where

$$\sigma_1(r, \phi) = \sum \eta(r\psi_i(\phi)) \prod_{k \neq i} z(r\psi_k(\phi)) \tag{13.5.28}$$

$$\begin{aligned} \sigma_2(r, \phi) = &\sum_{\substack{i_1, i_2 \\ i_1 \neq i_2}} \eta(r\psi_{i_1}(\phi))\eta(r\psi_{i_2}(\phi)) \prod_{k \neq i_1, i_2} z(r\psi_k(\phi)) \\ &+ \sum_{\substack{i_1, i_2, i_3 \\ i_1 \neq i_2 \neq i_3 \neq i_1}} \eta(r\psi_{i_1}(\phi))\eta(r\psi_{i_2}(\phi))\eta(r\psi_{i_3}(\phi)) \prod_{k \neq i_1, i_2, i_3} z(r\psi_k(\phi)) + \cdots. \end{aligned} \tag{13.5.29}$$

Substituting two solutions z and z_1 in (13.5.24) and subtracting one of the resulting expressions from the other, we find

$$\int_\Phi \sigma_1(r, \phi)T_0(\phi)\, d\phi = -\int_\Phi \sigma_2(r, \varphi)T_0(\phi)\, d\phi. \tag{13.5.30}$$

This relation holds for any $r \geqq 0$, but we shall consider it only in the segment

$$I_\Delta = [r_0, r_1]. \tag{13.5.31}$$

$y(x)$ and $y_1(x)$ belong to a class of the type $\{D_{pw}\}$, and hence $z(x)$ and $z_1(x)$ will belong to a class of the same type. Thus, in every sufficiently small segment, $\eta(x) = z_1(x) - z(x)$ either vanishes identically or has only a finite number of zeros. We now establish a lemma.

LEMMA 13.5.1. For sufficiently small Δ in (13.5.31) and $r \in I_\Delta$, $\eta(r\psi_i(\phi))$ will be of constant sign (or will vanish identically) for all i and all ϕ.

***Proof*:** For $r \in I_\Delta$,

$$\eta(r\psi_i(\phi)) = \eta\left(\frac{x_0\psi_i(\phi)}{\psi_0} + \frac{\theta\Delta\psi_i(\phi)}{\psi_0}\right), \qquad \theta \in [0, 1].$$

Further, $0 \leqq \psi_i(\phi) \leqq \psi_0$. Thus, the set of values of $r\psi_i(\phi)$ is contained wholly in the segment $[0, x_0 + \Delta]$. Further, $\eta(x) = 0$ for $0 \leqq x \leqq x_0$, so that the set of those values of $r\psi_i(\phi)$ for which $\eta(r\psi_i(\phi))$ is nonzero, is contained in the segment $[x_0, x_0 + \Delta]$. We have $\eta(x_0) = 0$; we know that, however small be Δ, there is a point x in the above segment where $\eta(x) \neq 0$. From the properties of the classes of type $\{D_{pw}\}$, it follows that, for sufficiently small Δ, $\eta(x)$ will preserve its sign throughout the segment $[x_0, x_0 + \Delta]$, vanishing only at $x = x_0$. This proves the lemma.▌

We note that if $\eta(x)$ is positive, then (13.5.30) is possible only for $\sigma_1(r, \phi)$ and $\sigma_2(r, \phi)$ equal to zero almost everywhere, and if the solution $z(x)$ is nonzero for $0 \leqq x \leqq x_0 + \Delta$, then the relation $\eta(x) \equiv 0$ is trivial and the solution $z(x)$ is unique. We may take the sign of $\eta(x)$ to be positive, replacing $\eta(x)$ by $-\eta(x)$ and $z(x)$ by $z_1(x)$ if necessary. But then it would be necessary to assume that any other solution $z_1(x) \neq 0$ for $0 \leqq x \leqq x_0 + \Delta$, which is not desirable.

We write

$$\eta_\Delta = \sup\,[|\eta(x)| : x_0 \leqq x \leqq x_0 + \Delta].$$

Suppose $\eta_\Delta \neq 0$ and we shall show that this leads to a contradiction. According to what has been said above, we assume that $z(x) \neq 0$ for $0 \leqq x < \alpha^2$; taking $\Delta < (\alpha^2 - x_0)/2$, we obtain

$$c_1 \leqq z(x) \leqq c_2 \tag{13.5.32}$$

(here and below, as usual, $c_1, c_2, \ldots$ will denote positive constants).

We take some term of the sum (13.5.29) containing the factor $\eta(r\psi_i(\phi))$, multiply it by $T_0(\phi)$ and integrate with respect to ϕ. The absolute value of the expression obtained does not exceed

$$c_2^{n-2}\eta_\Delta \cdot \int_\Phi |\eta(r\psi_i(\phi))|\, T_0(\phi)\, d\phi \tag{13.5.33}$$

The sum $\sigma_2(r, \phi)$ has at most 2^n terms, so that we conclude from (13.5.33) that

$$\left|\int_\Phi \sigma_2(r, \phi) T_0(\phi)\, d\phi\right| = 2^n c_2^{n-2} \eta_\Delta \sum_{i=1}^{n} \int_\Phi |\eta(r\psi_i(\phi))|\, T_0(\phi)\, d\phi. \tag{13.5.34}$$

On the other hand, we have from (13.5.28) and (13.5.32) that

$$\left|\int_\Phi \sigma_1(r, \phi) T_0(\phi)\, d\phi\right| \geqq c_1^{n-1} \sum_{i=1}^{n} \int_\Phi |\eta[r\psi_i(\phi)]|\, T_0(\phi)\, d\phi. \tag{13.5.35}$$

For sufficiently small Δ, η_Δ is as small as desired, so that (13.5.34) and (13.5.35) contradict (13.5.30), if only the relations below are not satisfied.

$$\int_\Phi |\eta(r\psi_i(\phi)|\, T_0(\phi)\, d\phi = 0 \qquad \text{for } i = 1, \ldots, n. \tag{13.5.36}$$

Setting $i = i_0$, $\phi = \phi_0$, so that $\psi_i(\phi_0) = \psi_0$, we see that $\eta(r\psi_{i_0}(\phi)) \neq 0$ for r and ϕ sufficiently close to r_1 and ϕ_0. This contradicts (13.5.36), and proves Theorem 13.5.5.▌

We turn to Theorem 13.5.6. We shall consider even solutions $y(x)$ of the equation (13.5.21) belonging to a class of the type $\{D_{pw}\}$ and further having an at most denumerable set of zeros in the domain of definition. We introduce the functions $z(x) = y(\sqrt{|x|})$ and $z_1(x) = y_1(\sqrt{|x|})$, as before, corresponding to the two solutions $y(x)$ and $y_1(x)$ assumed. With the earlier notations, we choose Δ so small that the difference $z_1(x) - z(x)$ preserves the same sign on the segment $[x_0, x_0 + \Delta]$, and set $\eta(x) = \pm[z_1(x) - z(x)]$, where the sign is so chosen that $\eta(x) > 0$. We may take this sign to be positive, renaming the solutions if necessary. Turning to (13.5.30), we find that

$$\int_\Phi \sigma_1(r, \phi) T_0(\phi)\, d\phi = 0. \tag{13.5.37}$$

Arguing in the same way as in respect of (13.5.36), choosing an index i_0 and a closed set $\Omega(r, \phi)$ such that $\eta(r\psi_{i_0}(\phi)) \geqq \lambda_1 > 0$ in that set, we find from (13.5.28) that

$$\int_{\Omega(r,\phi)} \prod_{k \neq i_0} z(r\psi_k(\phi)) T_0(\phi)\, d\phi = 0. \tag{13.5.38}$$

The integrand is everywhere continuous, and nonzero everywhere except at a set of (r, ϕ)-points of measure zero, in $\Omega(r, \phi)$. Thus, we have a contradiction to (13.5.38), and $\eta(x) \equiv 0$ everywhere. This proves Theorem 13.5.6.▌

Turning to the proof of Theorem 13.5.7, let $f(t)$ be a completely symmetric density which is of bounded variation in the neighborhood of the origin; we have

$$f(t) = \int_{-\infty}^{\infty} p(t - u) p(u)\, du$$

where p and f, as probability densities, belong to $L_1(-\infty, \infty)$. Further, $p \in L_2(-\infty, \infty)$. If π is the c.f. of the density p, then

$$|\pi(u)|^2 = \int_{-\infty}^{\infty} f(t) e^{itu}\, dt.$$

We now apply Theorem 23 of [146], p. 58. If $f \in L_1(-\infty, \infty)$ and is of bounded variation in the neighborhood of the point $t = x$, then

$$(\tfrac{1}{2})[f(x + 0) + f(x - 0)] = (1/2\pi) \lim_{\lambda \to \infty} \int_{-\lambda}^{\lambda} e^{-ixu}\, du \int_{-\infty}^{\infty} f(t) e^{iut}\, dt.$$

We set $x=0$, so that the conditions of the above theorem are satisfied. Then,

$$f(0)=(1/2\pi)\lim_{\lambda\to\infty}\int_{-\lambda}^{\lambda}|\pi(u)|^2\,du$$

whence

$$f(0)=(1/2\pi)\int_{-\infty}^{\infty}|\pi(u)|^2\,du.$$

Thus, $|\pi(u)|^2\in L_1(-\infty,\infty)$ and we then have

$$f(x)=(1/2\pi)\int_{-\infty}^{\infty}e^{-ixu}|\pi(u)|^2\,du.$$

In view of the evenncss of f, we have

$$f(x)=(1/2\pi)\int_{-\infty}^{\infty}e^{ixu}|\pi(u)|^2\,du$$

the integral converging absolutely, so that f is a positive definite function.

Let $y(x)$ be a completely symmetric solution of (13.5.21) which is also analytic on the entire real axis (for instance, a normal density with zero mean). Obviously, it will be of bounded variation in the neighborhood of $x=0$.

We embed $y(x)$ in the class of densities $f(x)$ which are completely symmetric and of bounded variation in the neighborhood of $x=0$. We shall show the uniqueness of the solution $y(x)$ of (13.5.21) in this class.

Let $f(x)$ be another solution of (13.5.21) in this class. Set up the difference $\eta_0(x)=f(x)-y(x)$. If this difference vanishes at infinitely many points in every neighborhood of the origin, then, by Corollary 1 to Lemma 1.2.1, in view of the positive definiteness of $f(x)$ and $y(x)$, they must coincide, and the uniqueness of the solution follows. Further, $\eta_0(0)=0$. Suppose now that there exists a neighborhood $[-\delta,\delta]$ of the origin, in which $\eta_0(x)$ vanishes only for $x=0$. Introducing as before the functions $z(x)=y(\sqrt{|x|})$ and $z_1(x)=f(\sqrt{|x|})$ on the segment $[0,\delta^2]$, and using the fact that $\eta(x)=z_1(x)-z(x)$ preserves the same sign on this segment, we show, on the basis of the above arguments and the fact that $y(x)\neq 0$ in some neighborhood of the origin, as seen from the representation of $f(t)$ as a convolution, that $\eta_0(x)\equiv 0$ in some segment $[0,\delta_1]$ and consequently in $[-\delta_1,\delta_1]$. But then $f(x)=y(x)$ for x in $[-\sqrt{\delta_1},\sqrt{\delta_1}]$ and, in view of the Corollary 1 to Lemma 1.2.1, these functions coincide on the whole axis. This proves Theorem 13.5.7. Theorem 13.5.8 is a special case thereof.∎

If the original random variable is a random m-dimensional vector, then, generally speaking, for its determination, the knowledge of the distributions of m distinct definite statistics is required. Under fairly general conditions, an analogue of equation (13.5.20) can be established. The class in which the solutions are sought can be generalized as follows. If $y(x_1, \ldots, x_m)$ is the probability density function of the vector $\mathbf{X}$, then, for fixed $x_1, \ldots, x_m$, the function $y(xx_1, \ldots, xx_m)$ of the single variable x is considered, and the conditions introduced above are imposed on that functions, for all possible values of $x_1, \ldots, x_m$.

It is possible to obtain results similar to Theorems 13.5.6 and 13.5.7 in respect of indefinite statistics of certain types if some conditions are imposed on the rate of divergence to infinity of the level surfaces. Besides, in some cases, it is possible to replace investigations in the neighborhood of the origin $(0, \ldots, 0)$ by investigations near the point at infinity; then the behavior of the level surfaces near the planes of coordinates turns out to be essential.

We now present some results of a special kind, on the reconstruction of the parent distribution from that of a statistic. Despite their specialized character, they are distinguished by the comparative mildness of the conditions imposed.

We go back to Example 2 (linear transformations), keeping to that notation. Consider a random sample $(X_1, \ldots, X_n)$ of size $n \geqq 6$. The statistics Y_j, $j = 1, \ldots, n$ are distributed on the $(n-2)$-dimensional sphere Φ:

$$y_1 + \cdots + y_n = 0, \; y^2 + \cdots + y_n^2 = 1. \tag{13.5.39}$$

If the X_j are normal, then, as is well-known, the vector $(Y_1, \ldots, Y_n)$ is uniformly distributed on the above sphere (cf., for instance, [71], p. 421]. Since normal distributions have bounded densities and satisfy Cramer's condition (13.5.5), it follows from Theorem 13.5.3 (for observations on a normal law, the requirement of symmetry of the density function can be omitted) that the uniform distribution of our statistic on the above sphere implies the normality of the original distribution. This result was established by Zinger [30] using a different method, in 1956. Then, Zinger and Linnik [40] considerably strengthened this result. It turned out that, for characterizing normality, the uniformity of distribution of the statistic $(Y_1, \ldots, Y_n)$ on the entire sphere Φ was not needed, only the equality of the values of the density function of this distribution at a finite number of points on the sphere was. Precisely, we have:

THEOREM 13.5.9. Let $(X_1, \ldots, X_n)$ be a random sample from a one-dimensional distribution, having a continuous density function f. If at least one triplet of points y', y'', y''' can be found on the sphere Φ, which are of the form

$$\begin{aligned} y' &= (y_1, y_2, y_3, y_4, y_5, y_6, \ldots, y_n) \\ y'' &= (y_1, y_2, y_3, y_1, y_2, y_3, \ldots, y_n) \\ y''' &= (y_4, y_5, y_6, y_4, y_5, y_6, \ldots, y_n) \end{aligned} \tag{13.5.40}$$

where, further,

$$\begin{aligned} y_1 + y_2 + y_3 &= y_4 + y_5 + y_6, \\ y_1^2 + y_2^2 + y_3^2 &= y_4^2 + y_5^2 + y_6^2 \end{aligned} \tag{13.5.41}$$

but

$$(y_1, y_2, y_3) \neq (y_4, y_5, y_6)$$

at which the density function of the distribution of $(Y_1, \ldots, Y_n)$ on the sphere Φ turns out to have the same value, then the original distribution is normal.

***Proof*:** It is easily seen that the probability density function of $\mathbf{Y} = (Y_1, \ldots, Y_n)$ on the sphere Φ can be represented in the form

$$\phi(\mathbf{y}) = c \int_{-\infty}^{\infty} dt \int_0^{\infty} s^{n-2} f(t + sy_1) \cdots f(t + sy_n)\, ds \tag{13.5.42}$$

where $c > 0$ is a constant.

We take three points of the form (13.5.40), at which ϕ has the same value. It follows immediately from (13.5.42) that

$$\int_{-\infty}^{\infty} dt \int_0^{\infty} s^{n-2}[f(t + sy_1)f(t + sy_2)f(t + sy_3) - f(t + sy_4)f(t + sy_5)f(t + sy_6)]^2\, ds = 0. \tag{13.5.43}$$

We take some point t_0 such that $f(t_0) \neq 0$. In view of the continuity of f, there exists a neighborhood $|t - t_0| < \delta$ such that $f(t) \neq 0$ there. Let s and t be such that

$$0 \leqq s < s_0 = \delta/(2 \max |y_j|), \qquad |t - t_0| < (1/2)\delta. \tag{13.5.44}$$

It will then follow from (13.5.43) that, for the chosen values of s and t,

$$f(t + sy_1)f(t + sy_2)f(t + sy_3) = f(t + sy_4)f(t + sy_5)f(t + sy_6). \tag{13.5.45}$$

In the neighborhood of the point t_0 concerned, $\ln f$ is a continuous function. Let, for $|t| < \delta$,

$$\phi(t) = \ln f(t_0 + t).$$

Under condition (13.5.44), equation (13.5.45) can be written in the form

$$\sum_{j=1}^{6} \varepsilon_j \phi(t + sz_j) = 0; \qquad \varepsilon_j = 1 \quad \text{for } j \leqq 3, \qquad \varepsilon_j = -1 \quad \text{for } j > 3. \tag{13.5.46}$$

We shall show that the only continuous solution of (13.5.46) is a quadratic polynomial.

We multiply (13.5.46) by $(u - t)^2$ and integrate with respect to t from 0 to u ($|u| < (1/2)\delta$), obtaining

$$\sum_1^6 \varepsilon_j \int_0^u (u - t)^2 \phi(t + sy_j)\, dt = 0 \tag{13.5.47}$$

or

$$\sum_1^6 \varepsilon_j \int_0^{u+sy_j} (u + sy_j - t)^2 \phi(t)\, dt = \sum_1^6 \varepsilon_j \int_0^{sy_j} (u + sy_j - t)^2 \phi(t)\, dt. \tag{13.5.48}$$

The left side of (13.5.48) is thrice differentiable with respect to s. Since the right side is a quadratic polynomial in u, its coefficients can be thrice differentiated with respect to s. We so differentiate, and set $s = 0$, to obtain

$$\left(\sum_1^6 \varepsilon_j y_j^3\right) \phi(u) = Au^2 + Bu + c \tag{13.5.49}$$

where A, B, and C are constants. Recalling that $(y_1, y_2, y_3) \neq (y_4, y_5, y_6)$, we then have from (13.4.41) that $\sum_1^6 \varepsilon_j y_j^3 \neq 0$ after which it follows from (13.5.49) that ϕ is a quadratic polynomial.

Thus, in the neighborhood of each point t where $f(t) \neq 0$, $f = \exp \phi$ where ϕ is a quadratic. Thus f is continuous and it then follows that f is a normal density function on the whole of the real line, and the theorem stands proved.∎

The conditions of this theorem can be further weakened (cf. [40]).

Following essentially the method of the work [30], Ghurye [18] has proved a series of special results, which can apparently be derived also by Prokhorov's method (cf. the proof of Theorem 13.5.4). We shall confine ourselves to the formulation of one such result and a brief indication of the application to it of the above method.

THEOREM 13.5.10. Let $(X_1, \ldots, X_n)$ be a random sample from a population with a continuous density function, and let $Y_i = (X_i - T)/W$, $i = 1, \ldots, n$, where T and W are satistics to be indicated below. If the vector $\mathbf{Y} = (Y_1, \ldots, Y_n)$ is distributed uniformly over the surface $W = 1$, $S = 0$, then

(1) if $n \geqq 5$, $T = \min(X_1, \ldots, X_n)$, $W = \sum_1^n (X_i - T)$, there exist $\sigma > 0$ and θ such that X_i have the exponential density $(1/\sigma) \exp[-(x - \theta)/\sigma]$ with support $x \geqq \theta$;

(2) if $n \geqq 4$, $T = \min(X_1, \ldots, X_n)$, $W = \max(X_1, \ldots, X_n) - T$, then there exist $\sigma > 0$ and θ such that X_i is distributed uniformly over $(\theta, \theta + \sigma)$.

***Proof*:** The application of Prokhorov's method presented above to proving this result would consist in setting up and studying the statistics

$$(Y_i - Y_j)/(Y_k - Y_m) = (X_i - X_j)/(X_k - X_m). \blacksquare$$

New results of interest concerning characterizations of the symmetric stable laws through the behavior of the types of the distributions were recently obtained by Kovalenko [68].

13.6 CHARACTERIZATION OF THE EXPONENTIAL DISTRIBUTION THROUGH THE PROPERTIES OF ORDER STATISTICS

We consider the exponential distribution with density function

$$p(x; \theta, A) = \begin{cases} \theta^{-1} \exp[-(x-A)/\theta] & \text{for } x \geqq A \\ 0 & \text{otherwise} \end{cases} \tag{13.6.1}$$

where $\theta > 0$. If $X_1 \leqq X_2 \leqq \cdots \leqq X_n$ be a "variational series" constructed on the basis of a random sample of size n from the population (13.6.1), then the statistics X_1 and $\sum_1^n (X_i - X_1)$, which are estimators of A and θ respectively, are independent. In this section, we consider characterizations of the exponential law through the independence of the above statistics, and other similar problems; the general picture strongly resembles the situation of a random sample from a normal population $N(\mu, \sigma^2)$, where the independence of the estimators $\bar{X}$ and S^2 of the parameters μ and σ^2 respectively characterizes the normal law. The material of this section is based on the papers Govindarajulu [14], Tanis [144], and Ferguson [153].

THEOREM 13.6.1. Let $X_1 \leqq \cdots \leqq X_n$ be the variational series from a random sample of size $n \geqq 2$ from a population with d.f. F such that $F(0) = 0$. Let $U_1 = X_1$, $U_2 = X_2 - X_1$, ..., $U_n = X_n - X_1$. The r.v. U_1 and the vector $(U_2, \ldots, U_n)$ are independent if and only if $F(X) = 1 - e^{-ax}$ for $x > 0$ (and zero otherwise), for some $a > 0$.

Proof: From elementary considerations, it follows that the probability element of $(U_2, \ldots, U_n)$ given $U_1 = u_1$ is

$$\begin{cases} (n-1)! \prod_{i=2}^{n} \dfrac{dF(u_i + u_1)}{[1 - F(u_1)]} & \text{for } 0 \leqq u_2 \leqq \cdots \leqq u_n, \\ 0 & \text{otherwise} \end{cases} \tag{13.6.2}$$

If $F(x) = 1 - e^{-ax}$ for $x > 0$, then

$$\frac{dF(u_i + u_1)}{1 - F(u_1)} = ae^{-au_i}\, du_i \tag{13.6.3}$$

and, since the set of values of the variables $u_2, \ldots, u_n$ in (13.6.2) does not involve u_1, the independence of U_1 and $(U_2, \ldots, U_n)$ follows, showing the sufficiency of the conditions of the theorem.

For the necessity part, we note that the independence of U_1 and $(U_2, \ldots, U_n)$ implies that of each of the factors in the product (13.6.2) in regard to U_1. Hence

$$\int_0^w \frac{d_v F(v + u_1)}{1 - F(u_1)} = \frac{F(w + u_1) - F(u_1)}{1 - F(u_1)} \tag{13.6.4}$$

does not depend on u_1. Proceeding to the limit as $u_1 \to 0+$, and remembering that $F(0) = 0$, we find that

$$\frac{F(u_1 + w) - F(u_1)}{1 - F(u_1)} = F(w) \tag{13.6.5}$$

i.e.,

$$G(u_1)G(w) = G(u_1 + w), \qquad \text{where} \qquad G(x) = 1 - F(x). \tag{13.6.6}$$

It is well-known that the only measurable solution of (13.6.6) is $G(x) = e^{-ax}$ for a real, and it follows from probabilistic considerations that $a > 0$.∎

THEOREM 13.6.2. Under the conditions of Theorem 13.6.1, U_1 and $\sum_1^n U_t$ are independent if and only if $F(x) = 1 - e^{-ax}$ for $x > 0$ and 0 otherwise, for some $a > 0$.

Proof: The sufficiency is a consequence of Theorem 13.6.1. To prove the necessity, we note that the c.f. of the conditional distribution of the sum $\sum U_i = \sum (X_i - X_1)$ for given $U_1 \equiv X_1 = x_1$, is

$$E[e^{it\,\Sigma(X_i - X_1)} | X_1 = x_1] = \left\{\int_{x_1}^{\infty} \frac{e^{it(x_2 - x_1)}\,dF(x_2)}{1 - F(x_1)}\right\}^{n-1}. \tag{13.6.7}$$

In view of the independence of X_1 and $\sum_1^n (X_i - X_1)$, the right side of (13.6.7) does not involve x_1. But then

$$\int_{x_1}^{\infty} e^{it(x_2 - x_1)} \frac{dF(x_2)}{1 - F(x_1)} \tag{13.6.8}$$

does not depend on x_1. But (13.6.8) is the c.f. of the conditional distribution of $X_2 - X_1$ given $X_1 = x_1$ where $X_1 \leqq X_2$ is the variational series for a sample of size 2. Using Theorem 13.6.1 with $n = 2$, we obtain the desired result.∎

THEOREM 13.6.3. Let $X_1 \leqq \cdots \leqq X_n$ be the variational series for a random sample from a population with d.f. $F(x)$, satisfying the condition of Theorem 13.6.1. The r.v. X_i and the r.v. $X_{j+1} - X_j$, $i \leqq j \leqq n - 1$, are independent if and only if $F(x) = 1 - e^{-ax}$ for $x > 0$, zero otherwise, for some $a > 0$.

Proof: The proof is similar to that of Theorem 13.6.1. We write down the probability element of the conditional distribution of the pair (X_j, X_{j+1}) given $X_i = x_i$ and then use the fact that the conditional distribution of $X_{j+1} - X_j$ given $X_i = x_i$, does not depend on x_i.∎

THEOREM 13.6.4. Let $X_1 \leqq \cdots \leqq X_n$ be the variational series for a random sample of size n from a population with continuous d.f. F. If for some $m < n$

$$E(X_m | X_{m+1} = x) = ax - b \tag{13.6.9}$$

then the d.f. F has the following form, to within a shift and a change of scale:

(i) $F(x) = e^x$ for $x < 0$ if $a = 1$;
(ii) $F(x) = x^\theta$ for $0 < x < 1$ if $0 < a < 1$; and
(iii) $F(x) = (-x)^\theta$ for $x < -1$ if $a > 1$,

where $\theta = a/[m(1-a)]$.

Proof: The function $R(x)$, defined by (13.6.10) below, is the mathematical expectation of the maximum of the observations in a random sample of size m truncated the point x, and is one version of the regression function of X_m given $X_{m+1} = x$.

$$R(x) = \begin{cases} \int_{-\infty}^{x} \dfrac{y\, d[F(y)]^m}{[F(x)]^m} & \text{if } F(x) > 0 \\ x & \text{if } F(x) = 0. \end{cases} \tag{13.6.10}$$

We first show that $R(x)$ is a continuous and nondecreasing function. Further, $R(x_1) < R(x_2)$ if $F(x_1) < F(x_2)$.

Since $F(x)$ and $\int_{-\infty}^{x} y\, d[F(y)]^m$ are continuous functions, it is clear that $R(x)$ is continuous everywhere except possibly at the point $c = \inf\{x: F(x) > 0\}$. But, since for $x > c$, $R(x)$ is the mathematical expectation of a r.v. taking values between c and x, $R(x) \to c$ as $x \to c + 0$. Obviously, $R(x)$ increases in $x \leqq c$ and $R(x) > R(c)$ for $x > c$. Suppose $c < x_1 < x_2$. Then, setting $F(x_1) = F_1$ and $F(x_2) = F_2$, we have

$$\begin{aligned}[R(x_2) - R(x_1)](F_2^m - F_1^m) &= F_1^m \int_{-\infty}^{x_2} y\, d[F(y)]^m - F_2^m \int_{-\infty}^{x_1} y\, d[F(y)]^m \\ &\geqq F_1^m x_1 (F_2^m - F_1^m) - (F_2^m - F_1^m) x_1 F_1^m = 0\end{aligned}$$

where the last inequality is strict if $F_1 < F_2$. Thus, $R(x)$ is continuous and nonincreasing. It follows that if $E(X_m \mid X_{m+1} = x) = ax - b$, then $a > 0$.

We now turn to the proof of the assertion of the theorem.

We have from (13.6.10) that for almost all x (according to the measure corresponding to the d.f. F)

$$\int_{-\infty}^{x} y\, d[F(y)]^m = (ax - b)[F(x)]^m. \tag{13.6.11}$$

Let the numbers c and d be such that $\{x: 0 < F(x) < 1\} = (c, d)$, c and d may be infinite. There does not exist an interval (c_1, d_1) with $c < c_1 < d_1 < d$, on which F is constant. In fact, on such an interval, the left side of (13.6.11) is constant, and the right side strictly increases, and both of them are continuous functions. Therefore, (13. 6.11) is true for all x in (c, d).

The left side of (13.6.11) can be written as

$$\int_{-\infty}^{x} y\, d[F(y)]^m = x[F(x)]^m - \int_{-\infty}^{x} [F(y)]^m\, dy. \tag{13.6.12}$$

The existence of the integral on the left follows from that of the integral on the right. Let $H(x) = \int_{-\infty}^{x} [F(y)]^m \, dy$; then, for all x, $H'(x) = [F(x)]^m$. Now, (13.6.11) can be written in the form

$$\frac{d}{dx} \log H(x) = \frac{1}{(1-a)x + b} \quad \text{for all } x \text{ in } (c, d). \tag{13.6.13}$$

We obtain the solution of this differential equation in the three cases indicated in the statement of the theorem.

(i) $a = 1$. Since $\log H(x)$ increases in (c, d), we must have $b > 0$. Integrating (13.6.13), we find $H(x) = ke^{x/b}$, whence $[F(x)]^m = (k/b)e^{x/b}$. Obviously, $c = -\infty$ and $d < \infty$. Since $F(d) = 1$, it follows that $F(x) = \exp[(x-d)/mb]$ for $x < d$. We have consequently, to within a shift and a change of scale, the d.f. indicated in (i).

(ii) $0 < a < 1$. We integrate (13.6.13) to obtain

$$H(x) = k[(1-a)x + b]^{1/(1-a)} \qquad \text{for } x \in (c, d).$$

In this case, c and d are finite and the conditions $F(c) = 0$, $F(d) = 1$ give $F(x) = [(x-c)/(d-c)]^{\theta}$ for $c < x < d$, where $\theta = [a/m(1-a)]$ and $c = -b/(1-a)$.

(iii) $a > 1$. Integration of (13.6.13) gives

$$H(x) = k[b - (a-1)x]^{1/(a-1)} \qquad \text{for } x \in (c, d).$$

In this case, $c = -\infty$ and $d < \infty$. The condition $F(d) = 1$ leads to $F(x) = [(k-x)/(k-d)]^{\theta}$ for $x < d$, where $\theta = -a/[m(a-1)]$ and $k = b/(a-1) > d$.

The theorem stands completely proved.▌

13.7 IDENTICAL DISTRIBUTION OF A MONOMIAL AND A LINEAR STATISTIC

In Chapter 2, we studied characterizations of distributions through the property of equidistribution of two linear statistics of a random sample. Here we consider some special cases of this problem, when one of the linear forms degenerates into a monomial, i.e., has only one nonzero coefficient. For instance, if $X_1, \ldots, X_n$ are i.i.d.r.v.'s, then what can be said about the distribution of the X_i if X_i has the same distribution as $\bar{X}$? It is well-known that if the X_i have a Cauchy distribution, then X_i and $\bar{X}$ are identically distributed. The direct converse of this assertion is not true; however, the Cauchy distribution is characterized by the identity of distribution of X_i and $\bar{X}$ for two values of n, say, n_1 and n_2, such that $\log n_1/\log n_2$ is irrational (cf. Theorem 13.7.1).

We first formulate a general result, which follows from Theorems 5.4.2–5.4.4.

THEOREM 13.7.1. Let $X_1, \ldots, X_n$ be nondegenerate i.i.d.r.v.'s. Let $a \neq 0$ be a real number such that aX_1 and $a_1 X_1 + \cdots + a_n X_n$ are equidistributed. Then $|a| > |a_i|$ for all i, and X_i follows a generalized stable law of one of the types described in Theorem 5.4.2, depending on the root λ of the equation $\sum_1^n |a_i|^\lambda = |a|^\lambda$ and the nature of the vector $(-\log|b_1|, \ldots, -\log|b_n|)$, where $b_i = a_i/a$.

Proof: Let us denote by f the c.f. of X_i. Then, the identity of distribution of aX_i and $a_1 X_1 + \cdots + a_n X_n$ means that

$$f(at) = f(a_1 t) \cdots f(a_n t) \tag{13.7.1}$$

for all real t. Hence, setting $b_j = a_j/a$ for all j, we have

$$f(t) = \prod_1^n f(b_i t) \tag{13.7.2}$$

or, letting $\phi(t) = f(t)f(-t)$, we have

$$\phi(t) = \prod_1^n \phi(|b_i| t). \tag{13.7.3}$$

We shall show that $|b_i| < 1$ for all i. If $|b_i| = 1$, then we see from (13.7.3) that ϕ will be the c.f. of a degenerate r.v. If $|b_i| > 1$, then $\phi(t) \leqq \phi(|b_i| t)$, whence

$$\phi(t) \geqq \phi(t/|b_i|) \geqq \cdots \geqq \phi(t/|b_i|^k) \to 1 \qquad \text{as} \qquad k \to \infty$$

which also contradicts the assumption that the X_i are nondegenerate. Hence we must have $|b_i| < 1$. Then (13.7.2) becomes a particular case of equation (5.4.31) considered in the course of the proof of Theorem 5.4.2, corresponding to $\gamma_i = 1$ for all i. Hence, all the assertions of that theorem are applicable to our present case. ▮

We derive a characterization of symmetric stable laws, and in particular, the Cauchy distribution, from Theorem 13.7.1.

THEOREM 13.7.2. Let $X_1, \ldots, X_n$, with $n \geqq 2$, be nondegenerate i.i.d.r.v.'s.

(i) If the statistics X_i and $a_1 X_1 + \cdots + a_n X_n$ are equidistributed, then

(a) X_i follows a normal law if (and only if) $\sum a_i^2 = 1$;

(b) X_i follows a Cauchy law if $\sum |a_i| = 1$ and at least one pair of the numbers $-\log|a_i|$ $(1 \leqq i \leqq n)$ are mutually incommensurable; and

(c) X_i has a symmetric stable distribution with exponent $0 < \lambda < 1$ or $1 < \lambda < 2$, if $\sum |a_i|^\lambda = 1$, there is at least one negative number among the a_i and at least one pair of the numbers $-\log|a_i|$ are mutually incommensurable.

(ii) If X_i and $\bar{X}$ are equidistributed for two values of n, say n_1 and n_2, such that $\log n_1/\log n_2$ is irrational, then X_i has a Cauchy distribution.

(iii) Let $a_1, \ldots, a_n$ be such that $|a_i| < 1$ for all i, $\sum a_i^2 \neq 1$, and the numbers $-\log |a_i|$ are mutually commensurable, and $\rho_1 > 0$ is such that $m_j = -\log |a_j|/\rho_1$ are positive integers with their greatest common factor $=1$; similarly, let $(b_1, \ldots, b_n)$ be another sequence of n numbers different from $(a_1, \ldots, a_n)$ with the same properties but corresponding to a different number $\rho_2 > 0$. Suppose that the statistics X_i, $\sum a_j X_j$ and $\sum b_j X_j$ are identically distributed. Then, if ρ_1/ρ_2 is irrational, and either among the numbers $a_1, \ldots, a_n$ and $b_1, \ldots, b_n$, there is at least one negative number, or $\sum |a_i| = 1, \sum |b_i| = 1$, then X_i follows a symmetric stable law.

***Proof*:** (i) The c.f. $f(t)$ of X_i satisfies the equation

$$f(t) = \prod_{i=1}^{n} f(a_i t) \tag{13.7.4}$$

which is a special case of (5.4.31) of Theorem 5.4.2. Let λ be the unique real root of the equation $\sum |a_i|^\lambda = 1$.

If $\lambda = 2$, then f is the c.f. of a normal law, as shown in part (iii) of Theorem 5.4.2. This proves (ia).

If at least one pair of the numbers $-\log |a_j|$ is incommensurable and $\lambda \neq 2$, then f is the c.f. of an i.d. law. Then, as has been shown earlier—cf. (5.4.32)—the Levy functions M and N in the Levy representation $L(\mu, \sigma^2, M, N)$ for $\log f$ have the form

$$M(u) = \xi |u|^{-\lambda}, \qquad N(u) = -\eta u^{-\lambda}$$

where ξ and η are nonnegative real numbers with $\xi + \eta > 0$. But then f is the c.f. of a stable law, and for $\lambda = 1$, we have the explicit representation (see (1.1.11))

$$\log f(t) = i\mu t - c|t|\{1 + i(2b/\pi)(t/|t|) \log |t|\}. \tag{13.7.5}$$

Substituting from (13.7.5) into (13.7.4), we find that $b = 0$, so that $f(t) \exp(-i\mu t)$ is the c.f. of a Cauchy distribution, proving (ib).

To prove (ic), we note that under the given conditions, in the solution (5.4.32) we must have $\xi = \eta$, leading to a symmetric stable law.

To prove (ii), we have to consider the functional equation

$$f(t) = [f(t/n)]^n \qquad \text{for real } t \tag{13.7.6}$$

which is a particular case of (5.3.7). If (13.7.6) is satisfied for $n = n_1$, $n = n_2$, then in the Levy representation for $\log f$, the Levy functions M and N have the form (5.3.3) with $\lambda = 1$, where the functions ξ and η from (5.3.3) have two periods $\rho_1 = \log n_1$ and $\rho_2 = \log n_2$. Since ρ_1/ρ_2 is irrational, the set $\{m_1\rho_1 + m_2\rho_2\colon\ m_1 \text{ and } m_2 \text{ integers}\}$ is everywhere dense in R^1 and $\xi(m_1\rho_1 + m_2\rho_2) = \xi(0)$. The right-continuity of the function ξ then implies that $\xi \equiv \xi(0)$ and similarly $\eta \equiv \eta(0)$. Thus,

$$M(u) = c_1/|u| \qquad \text{and} \qquad N(u) = -c_2/u$$

where $c_1 \geqq 0$, $c_2 \geqq 0$, and $c_1 + c_2 > 0$. Thus f is the c.f. of a stable law with $\lambda = 1$. Using the explicit representation (13.7.5) above, and substituting it in (13.7.6) for $n = n_1$ and $n = n_2$, we find that $b = 0$, i.e., the stable law differs only by a shift from a symmetric stable law with $\lambda = 1$, i.e., a Cauchy law. In case (iii), we have two equations

$$f(t) = \prod_1^n f(a_i t) = \prod_1^n f(b_i t) \tag{13.7.7}$$

and the solution of either of them has the form (5.4.33) or (5.4.34), where the corresponding functions ξ and η are periodic with two periods ρ_1 and ρ_2. Using arguments similar to those for part (ii) above, we establish that ξ and η are constant functions, and thus f is a stable c.f. Substituting then from (1.1.11) into (13.7.7), we find that the stable law is also symmetric under the conditions given on the a_i, b_i ▮

We now consider a generalization of Theorem 13.7.1 to the case of a denumerable number of r.v.'s.

THEOREM 13.7.3. Let $X_1, X_2, \ldots$ be nondegenerate, i.i.d.r.v.'s and $\{a_j\}$ a sequence of real constants such that the series $\sum a_j X_j$ converges with probability one. If aX_j and $\sum a_j X_j$ are equidistributed, then $|a| > |a_j|$ for all j, and X_j follows a generalized stable law of one of the types described in Theorems 5.6.2 and 5.6.3, depending on the root λ of the equation $\sum |a_j|^\lambda = |a|^\lambda$, and the nature of the infinite vector $(-\log |b_1|, -\log |b_2|, \ldots)$, where $b_j = a_j/a$.

Proof: Let f denote the c.f. of the X_j. The basic assumption of the theorem means that

$$f(at) = \prod_1^\infty f(a_j t). \tag{13.7.8}$$

We show as in Theorem 13.7.1 that $|a| > |a_j|$ for all j, and (13.7.8) can be written in the form

$$f(t) = \prod_1^\infty f(b_j t) \tag{13.7.9}$$

where $|b_j| < 1$ for all j. But (13.7.9) is a special case of (5.6.7), considered in Theorem 5.6.3, corresponding to $\gamma_j = 1$ for all j. Hence, the results of Theorems 5.6.2 and 5.6.3 apply to this situation. ▮

13.8 OPTIMALITY CRITERIA BASED ON THE SAMPLE MEAN, AND THE NORMAL LAW

Let $(X_1, \ldots, X_n)$ be a random sample from a population with d.f. $F(x - \theta)$. We wish to test the hypothesis $H_0 : \theta = 0$ against the composite alternative $H_1 : \theta > 0$. It is well-known (cf. for instance [71], p. 580) that, if F is the d.f.

of a normal law, then a test with the critical region $\{\bar{x} > c_\alpha\}$ where c_α is chosen according to the given level α, is uniformly most powerful (optimal) among all criteria with level $\leqq \alpha$, for all α. Following the note by Morozenski [104], we show below that for a continuous d.f. F with

$$\int x^2 \, dF(x) < \infty \tag{13.8.1}$$

the normality of the X_i follows from the optimality, for all α, of the criterion $\{\bar{x} > c_\alpha\}$. Apparently, this result can be proved by various methods; that of the note [104] appears to us to be of interest in view of the connection which is established in the process between point estimation and testing of hypotheses. For simplicity, we shall assume that $\int x \, dF(x) = 0$.

THEOREM 13.8.1. Let $(X_1, \ldots, X_n)$ be a random sample of size $n \geqq 3$ from a population with continuous d.f. $F(x-\theta)$, satisfying (13.8.1). If, for every $\alpha \in (0, 1)$, the test with critical region $\{\bar{x} > c_\alpha\}$, where $c_\alpha = \max\{c: P(\bar{X} > c \mid H_0) = \alpha_\alpha\}$, is optimal in the class of criteria of level $\leqq \alpha$ for H_0 against H_1, then F is a normal d.f.

***Proof*:** Let $t = t(x_1, \ldots, x_n)$ be the Pitman estimator for θ under quadratic loss (cf. section 7.3). We set $H(u) = P_0(\bar{X} < u)$ and $G(u) = P_0(t < u)$. To avoid certain technical complications, we assume that G is continuous. Let $c_\alpha = c_1(\alpha)$, and $c_2(\alpha) = \max\{c: P(t > c \mid H_0) = \alpha\}$. It follows from the optimality of the test criterion $\{\bar{x} > c_1(\alpha)\}$ that

$$P_\theta(\bar{X} > c_1(\alpha)) \geqq P_\theta(t > c_2(\alpha))$$

i.e.,

$$H(c_1(\alpha) - \theta) \leqq G(c_2(\alpha) - \theta) \qquad \text{for } \alpha \in (0, 1), \theta > 0. \tag{13.8.2}$$

LEMMA 13.8.1. The function $c_2(\alpha) - c_1(\alpha)$ is nonincreasing on (0, 1).

***Proof*:** Let $0 < \beta < \gamma < 1$. Let $\theta = c_1(\beta) - c_1(\gamma)$, and use (13.8.2). We have

$$1 - \gamma = G(c_2(\gamma)) = H(c_1(\gamma)) = H(c_1(\beta) - \theta) \leqq G(c_2(\beta) - \theta)$$

whence

$$c_2(\gamma) \leqq c_2(\beta) - \theta = c_2(\beta) - c_1(\beta) + c_1(\gamma)$$

and

$$c_2(\beta) - c_1(\beta) \geqq c_2(\gamma) - c_1(\gamma). \blacksquare$$

LEMMA 13.8.2. Let $\int_0^1 c^2(x)\, dx < \infty$; set $R(x) = \int_0^x c(t)\, dt$. If the function $c^2(x)$ is monotone on some interval $(0, \delta)$ (on $(\delta, 1)$) then $\lim_{x \to 0+} c(x)R(x) = 0$ ($\lim_{x \to 1-0} c(x)R(x) = 0$, under the additional condition $R(1) = 0$).

***Proof*:** $|R(x)| \leq \int_0^x |c(t)|\, dt \leqq A\sqrt{x}$ for some $A > 0$. Hence $|c(x)R(x)| \leqq A\sqrt{x}|c(x)|$ and we need only consider the case of $c^2(x)$ monotonically decreasing on $(0, \delta)$. If $\phi(x) = xc^2(x)$ does not tend to zero as $x \to 0+$, then an $\varepsilon > 0$ and a sequence $\{a_n\} \downarrow 0$ with $\phi(a_n) \geqq \varepsilon$ exist. We may take it that $a_{n+1} \leqq (1/2)a_n$. Consider the function on $(0, \delta)$, given by

$$\psi(x) = \begin{cases} 0 & \text{for } x \in (a_1, \delta) \\ \varepsilon/a_n & \text{for } x \in (a_{n+1}, a_n). \end{cases}$$

Since $c^2(x)$ is monotone decreasing, for all x in $(0, \delta)$ we have

$$c^2(x) = \phi(x)/x \geqq \psi(x).$$

Hence $\int_0^\delta c^2(x)\, dx \geqq \int_0^\delta \psi(x)\, dx = \sum_1^\infty (\varepsilon/a_n)(a_n - a_{n+1}) = \infty$, contrary to the condition that $\int_0^1 c^2(x)\, dx < \infty$. Thus, $\lim_{x\to 0+} c(x)R(x) = 0$. It is shown similarly that $\lim_{x\to 1-0} c(x)R(x) = 0$ if $R(1) = 0$.∎

We continue with the proof of the theorem. We have from (13.8.1) that $\text{Var}_0(\bar{X}) = \int u^2\, dH(u) < \infty$, and *a fortiori* $\text{Var}_0(t) = \int u^2\, dG(u) < \infty$. According to [66], p. 241,

$$\text{Var}_0(\bar{X}) = \int_0^1 c_1^2(\alpha)\, d\alpha, \qquad \text{Var}_0(t) = \int_0^1 c_2^2(\alpha)\, d\alpha,$$

$$\int_0^1 c_1(\alpha)\, d\alpha = \int x\, dH(x) = 0; \qquad \int_0^1 c_2(\alpha)\, d\alpha = 0.$$

Hence

$$\text{Var}_0(t) - \text{Var}_0(\bar{X}) = \int_0 [c_2^2(\alpha) - c_1^2(\alpha)]\, d\alpha. \tag{13.8.3}$$

We introduce the function $R(t) = \int_0^t [c_1(\alpha) + c_2(\alpha)]\, d\alpha$. In view of the monotone decreasing nature of $c_1(\alpha)$ and $c_2(\alpha)$ and the condition $R(1) = 0$, we have $R(t) \geqq 0$ for all t in $(0, 1)$. Integrating (13.8.3) by parts—it is necessary to integrate over an interval $(a, b) \subset (0, 1)$ and then proceed to the limit as $a \to 0$, $b \to 1$, and to apply Lemma 13.8.2—we have

$$\begin{aligned} \text{Var}_0(t) - \text{Var}_0(\bar{X}) &= [\{c_2(\alpha) - c_1(\alpha)\}R(\alpha)]_0^1 - \int_0^1 R(\alpha)\, d[c_2(\alpha) - c_1(\alpha)] \\ &= -\int_0^1 R(\alpha)\, d[c_2(\alpha) - c_1(\alpha)] \geqq 0 \end{aligned}$$

according to Lemmas 13.8.1 and 13.8.2. But, as we saw in Theorem 7.3.1, $\text{Var}_0(\bar{X}) \geqq \text{Var}_0(t)$, where the equality holds only for normal variables X_i. The theorem stands proved.∎

13.9 CHARACTERIZATION OF THE NORMAL LAW THROUGH A PROPERTY OF THE NONCENTRAL CHI-SQUARE DISTRIBUTION

In the theory of test criteria involving the chi-square distribution, an important role is played by this fact: if $X_1, \ldots, X_n$ are i.i.d.r.v.'s following a normal distribution $N(0, \sigma^2)$, then the distribution of the noncentral chi-square statistic $\chi^2 = \sum_1^n (X_i + a_i)^2$ depends on $a_1, \ldots, a_n$ only through the non-centrality parameter $\sum_1^n a_i^2$. Following the note [58], we shall show below that this property is characteristic of the normal law.

THEOREM 13.9.1. If, for i.i.d.r.v.'s $X_1, \ldots, X_n$, with $n \geqq 2$, the distribution of the statistic $\chi^2 = \sum_1^n (X_i + a_i)^2$ depends only on $\sum_1^n a_i^2$, where the a_i are real, then the X_i are normal.

***Proof*:** Consider the function

$$\phi(a) = E \exp [-(X_1 + a)^2]. \tag{13.9.1}$$

Obviously ϕ is positive-valued and twice differentiable everywhere. By the conditions of the theorem

$$Ee^{-\Sigma_1^n (X_i + a_i)^2} = \prod_1^n Ee^{-(X_i + a_i)^2} = \prod_1^n \phi(a_i) = \psi\left(\sum_1^n a_i^2\right). \tag{13.9.2}$$

Let $h(a) = \log \phi(a)$ and $H(a) = \log \psi(a)$. We obtain from (13.9.2) that

$$\sum_1^n h(a_i) = H\left(\sum_1^n a_i^2\right). \tag{13.9.3}$$

Differentiating with respect to a_1 and then with respect to a_2, we will have for all $a_1, \ldots, a_n$ that

$$H''\left(\sum_1^n a_i^2\right) = 0.$$

Thus $H(a) = c_1 a + c_2$ (for $a > 0$) and from (13.9.1) and (13.9.3) we then have

$$\phi(a) = \int e^{-(x+a)^2}\, dF(x) = e^{c_1 a^2 + c_3} \tag{13.9.4}$$

where $F(x) = P(X_i \leqq x)$. Now, let $dG = e^{-x^2}\, dF$. Then (13.9.4) can be re-written as

$$\int e^{-2ax}\, dG(x) = e^{c_4 a^2 + c_3}.$$

By the uniqueness theorem for the Laplace transform, it then follows that

$$dG(x) = c_5 e^{c_6 x^2}\, dx$$

for some c_5 and c_6, and hence that F is the d.f. of a normal law, proving the theorem.▮

13.10 CHARACTERIZATIONS OF THE WIENER PROCESS

13.10.1 Stochastic Integrals

In this subsection, we present some properties of random processes, needed below for characterization purposes. We shall denote a random process by $\{X(t): t \in T\}$ or simply by $\{X(t)\}$ when the domain of variation of t is clear.

Continuous process: A process $\{X(t)\}$ will be said to be *continuous in probability* at t if, as $\tau \to 0$, $X(t+\tau) - X(t)$ tends to zero in probability. We shall use the following notation in this context.

$$p\text{-}\lim_{\tau \to 0} [X(t+\tau) - X(t)] = 0. \tag{13.10.1}$$

The process will be said to be *continuous in the mean* at t, if

$$\lim_{\tau \to 0} E[X(t+\tau) - X(t)]^2 = 0. \tag{13.10.2}$$

The process will be said to be continuous (in some sense) in the interval $[A, B]$, if it is continuous at each point of this interval (in that sense).

Homogeneous process: $\{X(t)\}$ will be called *homogeneous* if the d.f. of the increment $X(t+\tau) - X(t)$ depends only on τ, and not on t.

Process with independent increments: $\{X(t)\}$ is called a process with *independent increments* if its increments over nonoverlapping intervals (except possibly for common endpoints) are independent.

We cite without proof a series of properties of a continuous (in probability) homogeneous process with independent increments.

LEMMA 13.10.1. Let $\{X(t)\}$ be a continuous (in probability) homogeneous process with independent increments; denote the c.f. of the difference $X(t+\tau) - X(t)$ by $f(u; \tau)$. Then, $f(u; \tau)$ is a continuous function of τ.

LEMMA 13.10.2. Under the same conditions, $f(u; \tau)$ is an infinitely divisible c.f., and further $f(u, \tau) = [f(u, 1)]^{\tau}$.

Mean value: If $\mu(t) = EX(t)$ exists for every t, then $\mu(t)$ is called the *mean value function* (or simply the mean value) of the process.

Covariance function: The function

$$\Gamma(s, t) = E[X(t)X(s)] - \mu(t)\mu(s) \tag{13.10.3}$$

(if it exists) is called the *covariance function* of the process.

Wiener process: A homogeneous process $W(t)$ with independent increments, defined for $t \geqq 0$, is called a *Wiener* (or a *Brownian motion*) *process* if $W(t) - W(s)$ has a normal distribution $N(0, \sigma^2|t-s|)$ and $W(0) = 0$. In this case, $\Gamma(s, t) = \sigma^2 \min(s, t)$. We call $V(t)$ a *Wiener process with mean value function* $\mu(t)$, if $V(t) = \mu(t) + W(t)$, where $W(t)$ is a Wiener process (with zero mean). If $\mu(t)$ is a linear function of t, then we say that $V(t)$ is a Wiener process with linear mean value function.

Stochastic integrals: Let $\alpha(.)$ be a nonconstant, nondecreasing, right-continuous function, defined on a compact interval $[a, b]$ with $\alpha(a) = A$ and $\alpha(b) = B$, and $X(t)$ a continuous (in probability) homogeneous process with independent increments, defined on $[A, B]$. Let g be a continuous function on $[a, b]$. The random variable

$$Y_g = \int_a^b g(t)\, dX[\alpha(t)] \tag{13.10.4}$$

is defined as

$$\underset{\|\Delta\| \to 0}{p\text{-}\lim} \sum_{r=0}^{n-1} g(t_r^*)\{X[\alpha(t_{r+1})] - X[\alpha(t_r)]\} \tag{13.10.5}$$

where Δ: $t_0(=a) < t_1 < \cdots < t_n(=b)$ is an arbitrary subdivision of $[a, b]$, $\|\Delta\| = \max_r |t_{r+1} - t_r|$ and $t_r^* \in [t_r, t_{r+1}]$. The definition is meaningful in that the limit (13.10.5) does not depend on the subdivision Δ and the choice of the points t_r^* given the subdivision. Y_g is called a *stochastic integral.* We formulate two lemmas which enable us to compute the c.f.'s of stochastic integrals.

Let $f(u; \tau) = E \exp\{iu[X(t + \tau) - X(t)\}$, $A \leqq t < t + \tau \leqq B$,

$$f(u) = [f(u; \tau)]^{1/\tau} \qquad \text{and} \qquad \phi(u) = \ln f(u). \tag{13.10.6}$$

LEMMA 13.10.3. Let $X(t)$ be a continuous (in probability) homogeneous process with independent increments, g a function continuous on $[a, b]$ and Y_g the stochastic integral (13.10.4). Then the c.f. ψ_g of Y_g is given by

$$\log \psi_g(u) = \int_a^b \phi[ug(t)]\, d\alpha(t) \tag{13.10.7}$$

ϕ being defined by (13.10.6).

LEMMA 13.10.4. Let $X(t)$ be a continuous homogeneous process with independent increments; $g(t)$ and $h(t)$ continuous functions on $[a, b]$ and Y_g, Y_h random variables defined in terms of the functions g and h according to (13.10.4). Then the joint c.f. $\psi_{g,h}$ of Y_g and Y_h is given by

$$\log \psi_{g,h}(u, v) = \int_a^b \phi[ug(t) + vh(t)]\, d\alpha(t)$$

where ϕ is defined by (13.10.6).

We may also consider stochastic integrals defined as limits in the mean, instead of as limits in probability as was done in formula (13.10.5). The existence of such integrals would require imposing certain conditions on the covariance function of the process $X(t)$. We shall consider here only the stochastic integral (13.10.4) defined by the formula (13.10.5).

13.10.2 Characterization Theorems

We cite here three different characterizations of the Wiener process.

THEOREM 13.10.1. (identical distribution of a stochastic integral and an r.v.) Let $X(t)$ be a continuous (in probability) homogeneous process with independent increments, defined on $[A, B]$ and g a continuous function on $[a, b]$, satisfying one of the conditions:

(i) $|g(t)| < 1$ for all t in $[a, b]$ and g has a finite number of zeros on $[a, b]$; or

(ii) $|g(t)| > 1$ for all t in $[a, b]$; it follows then that g maintains a constant sign on $[a, b]$.

Suppose that for some (and hence for all) integer $n \geqq 1/(B-A)$, Y_g has the same distribution as the sum of n independent r.v.'s each distributed as $X(t+1/n) - X(t)$, $A \leqq t < t + 1/n \leqq B$. Then $X(t)$ is a Wiener process with linear mean value function if and only if

$$\int_a^b g^2(t)\, d\alpha(t) = 1. \tag{13.10.8}$$

Further, in that case,

$$\int_a^b g(t)\, d\alpha(t) = 1 \tag{13.10.9}$$

or the mean value function is identically zero.

***Proof*:** The "only if" assertion and (13.10.9) follow from the fact that the condition of equidistribution given in the statement of the theorem is equivalent to the equation

$$\int_a^b \phi[ug(t)]\, d\alpha(t) = \phi(u) \qquad \text{for all real } u. \tag{13.10.10}$$

We take up the sufficiency. Let (13.10.8) and (13.10.10) be satisfied. Set $\theta(u) = \phi(u) + \phi(-u)$; then, we have from (13.10.10)

$$\int_a^b \theta[ug(t)]\, d\alpha(t) = \theta(u) \qquad \text{for all real } u. \tag{13.10.11}$$

We now obtain from (13.10.8), setting $c(t) = |g(t)|$, and

$$F(u, t) = \theta[uc(t)] - [c(t)]^2\theta(u)$$

that

$$\int_a^b F(u, t)\, d\alpha(t) = 0. \tag{13.10.12}$$

Suppose first that g satisfies condition (i). Fix $u \neq 0$. If $a_1 < a_2 < \cdots < a_n$ and, possibly $a_0 = a$ and $a_{n+1} = b$ are zeros of $g(t)$, then $F(u, a_r) = 0$ and (13.10.12) can be rewritten as

$$\sum_{r=0}^{n} \int_{[a_r,\, a_{r+1}]} F(u, t)\, d\alpha(t) = 0. \tag{13.10.13}$$

In the sum of the left above, we may ignore those r for which $\int_{[a_r,\, a_{r+1}]} d\alpha(t) = 0$, so that (13.10.13) implies the existence of an $r = r(u)$, with $0 \leqq r \leqq n$, such that

$$\int_{[a_r,\, a_{r+1}]} d\alpha(t) > 0, \qquad \int_{[a_r,\, a_{r+1}]} F(u, t)\, d\alpha(t) \geqq 0.$$

Since F is continuous in t, it follows from the above inequalities that $F(u, t)$ cannot be strictly negative on $[a_r, a_{r+1}]$, so that a t_u with $c(t_u) > 0$ and $F(u, t_u) \geqq 0$ exists.

Similarly, a t_u^* with $c(t_u^*) > 0$ and $F(u, t_u^*) \leqq 0$ exists. Thus, to every $u \neq 0$, there correspond a $\xi_u = uc(t_u)$ and $\xi_u^* = uc(t_u^*)$, where $0 < |\xi_u| < |u|$, $0 < |\xi_u^*| < |u|$, such that, if $\eta(u) = \theta(u)/u^2$ (for $u \neq 0$), then

$$\eta(u) \geqq \eta(\xi_u), \qquad \eta(u) \leqq \eta(\xi_u^*).$$

Hence, as in Theorem 5.6.1, Part (ii), it is seen that $\eta(u)$ is constant, so that $\theta(u) = -cu^2$, and by Cramer's theorem, f is a normal c.f. Relation (13.10.9) follows from (13.10.10).

Let us now take up the case of g satisfying condition (ii); then g maintains a constant sign, and $|g|$ is bounded below by a constant $\gamma > 1$ and also bounded above. In this case, it follows directly from (13.10.12) that, for every $u \neq 0$, there exist t_u in $[a, b]$ and t_u^* in $[a, b]$ such that $F(u, t_u) \geqq 0$, $F(u, t_u^*) \leqq 0$, where, of course, $c(t_u)$ as well as $c(t_u^*) \geqq \gamma$. Thus, for every fixed $u > 0$, there exists a sequence $u = u_0 < u_1 < \cdots \to \infty$ such that $\eta(u_{k+1}) \geqq \eta(u_k)$ for all k, and a sequence $u = u_0^* < u_1^* < \cdots \to \infty$ such that $\eta(u_{k+1}^*) \leqq \eta(u_k^*)$ for all k. But, since f is an i.d.c.f., $\lim_{u \to \infty} \eta(u) = -\sigma^2$, where $(1/2)\sigma^2$ corresponds to the normal component in the Levy representation for $\log f$. Then we have from what we have seen above that $\eta(u) \leqq -\sigma^2 \leqq \eta(u)$ for all $u > 0$, so that $\theta(u) = -\sigma^2 u^2$. It follows from Cramer's theorem that f is a normal c.f. As before, (13.10.9) follows from (13.10.10).

THEOREM 13.10.2. (independence of stochastic integrals) Let $X(t)$ be a continuous (in probability) homogeneous process with independent increments given on $[A, B]$, and g and h be continuous on $[a, b]$, at least one of them nonvanishing and the other nonvanishing on a set of positive α-measure. Then the stochastic integrals Y_g and Y_h are independent if and only if

(i) $\{X(t)\}$ is a Wiener process with linear mean value function, and

(ii) $\int_a^b gh\, d\alpha = 0$ if $X(t)$ is a nondegenerate process.

***Proof*:** The sufficiency part is easily proved. As for the necessity, if Y_g and Y_h are independent, we have the basic relation

$$\int_a^b \phi(ug + vh)\, d\alpha = \int_a^b \phi(ug)\, d\alpha + \int_a^b \phi(vh)\, d\alpha. \tag{13.10.14}$$

For definiteness, let us take g to be nonvanishing on $[a, b]$, so that h is nonvanishing at least on a set of positive α-measure on $[a, b]$. Then g preserves a constant sign, and we may assume that $g(t) \geqq g_0 > 0$. Multiply both sides of (13.10.14) by $(x - u)$, $x > 0$, and integrate with respect to u from 0 to x. Then, making an easily justified interchange of the order of integration, we obtain

$$\int_a^b g^{-2} \left[\int_{vh}^{xg+vh} (xg + vh - w)\phi(w)\, dw \right] d\alpha - \int_a^b g^{-2} \left[\int_0^{xg} (xg - w)\phi(w)\, dw \right] d\alpha$$
$$= (1/2)x^2 \phi(vh) \int_a^b d\alpha.$$

The left side is differentiable with respect to v (under the integral sign) for any fixed x; hence, so must the right side be. After such differentiation, we have

$$\int_a^b g^{-2} h \left[\int_{vh}^{xg+vh} \phi(w)\, dw \right] d\alpha = x^2 A(v) + xB(v).$$

Again, from the differentiability with respect to v of the left side, for all x, it follows that both the functions A and B are differentiable. So differentiating, setting $v = 0$ and $\theta(u) = \phi(u) + \phi(-u)$, we obtain for some real c (noting that $\int_a^b h^2\, d\alpha > 0$):

$$\int_a^b h^2 g^{-2} [\theta(xg) + cx^2 g^2]\, d\alpha = 0.$$

Now two possibilities arise.

Case 1. There exists a sequence $\{x_n\} \to 0$ such that, for every n, the function $G_n(t) = \theta[x_n\, g(t)] + cx_n^2 g^2(t)$ changes sign as t varies from a to b; let then $\{t_n\}$ be such that $G_n(t_n) = 0$. Let $u_n = x_n\, g(t_n)$, so that $\{u_n\} \to 0$ (remembering that g is bounded), and $\theta(u_n) + cu_n^2 = 0$ for all n. By Corollary 2 to Lemma 1.2.1, we must then have $\theta(u) = -cu^2$, so that f is a normal c.f.

Case 2. There exists a $\delta > 0$ such that for every x with $|x| < \delta$, the function $\theta[xg(t)] + cx^2 g^2(t)$ preserves its sign as t varies from a to b. Let S_h denote the subset of $[a, b]$ where $h(t) \neq 0$. Then, for any x with $|x| < \delta$, we have $\theta[xg(t)] + cx^2 g^2(t) = 0$ for all $t \in S_h$ with the possible exception of a set of points (depending on x) of α-measure zero. Thus, to every x with $|x| < \delta$, there corresponds at least one t_x in S_h such that

$$\theta[xg(t_x)] + cx^2 g^2(t_x) = 0.$$

Choosing a sequence $\{x_n\} \to 0$ and using the boundedness of g, we arrive at the conclusion that f is the c.f. of a normal law as we did in Case 1. It follows from (13.10.14) that $\int_a^b gh\, d\alpha = 0$ if $\{X(t)\}$ is a nondegenerate process. ∎

THEOREM 13.10.3. (equidistribution of two stochastic integrals) Let $\{X(t)\}$ be a continuous (in probability) homogeneous process with independent increments, defined on $[A, B]$, and g and h continuous on $[a, b]$ and such that $\max|g(t)| \neq \max|h(t)|$ there. Suppose $X(t)$ has moments of all orders. Then the stochastic integrals Y_g and Y_h are equidistributed if and only if

(i) $X(t)$ is a Wiener process with linear mean value function,
(ii) $\int_a^b g\,d\alpha = \int_a^b h\,d\alpha$ unless the mean value function is identically zero, and
(iii) $\int_a^b g^2\,d\alpha = \int_a^b h^2\,d\alpha$.

Proof: The equidistribution of Y_g and Y_h is equivalent to the relation

$$\int_a^b \phi[ug(t)]\,d\alpha(t) = \int_a^b \phi[uh(t)]\,d\alpha(t) \qquad \text{for all real } u. \tag{13.10.15}$$

Setting $\theta(u) = \phi(u) + \phi(-u)$, rewrite (13.10.15) in the form

$$\int_a^b \theta[ug(t)]\,d\alpha(t) = \int_a^b \theta[uh(t)]\,d\alpha(t). \tag{13.10.16}$$

From the existence of moments of all orders for $X(t)$ follows the differentiability under the integral sign in relation (13.10.16). Differentiating $2k$ times, we obtain

$$\int_a^b [g(t)]^{2k}\theta^{(2k)}[ug(t)]\,d\alpha(t) = \int_a^b [h(t)]^{2k}\theta^{(2k)}[uh(t)]\,d\alpha(t). \tag{13.10.17}$$

We set $u = 0$ here to obtain

$$\theta^{(2k)}(0)\int_a^b \{[g(t)]^{2k} - [h(t)]^{2k}\}\,d\alpha(t) = 0. \tag{13.10.18}$$

But the relation

$$\int_a^b \{[g(t)]^{2k} - [h(t)]^{2k}\}\,d\alpha(t) = 0$$

can hold only for a finite number of values of k, since, otherwise, we would have, by Lemma 1.6.3, $\max|g| = \max|h|$ on $[a, b]$. Thus we obtain from (13.10.18) that $\theta^{(2k)}(0) = 0$ for all $k \geqq k_0$. Since θ is an even function, $\theta^{(2k+1)}(0)$ also $= 0$ for all k. Thus $\theta^{(s)}(0) = 0$ for all $s \geqq 2k_0$, so that $\theta(u)$ is a polynomial of degree at most $2k_0$ in the neighborhood of the origin, and standard arguments show that $\theta(u)$ is a polynomial of degree two, and the same is true of $\phi(u)$ as well, by Cramer's theorem. Hence we have the theorem. ∎

CHAPTER 14

Unsolved Problems

In this chapter, some problems will be formulated, which appear to the authors to be of interest and importance in the circle of problems with which this work is connected.

CHAPTER 2

1. Describe all the possible distributions of i.i.d. random vectors $\mathbf{X}_1, \ldots, \mathbf{X}_n$, for which the linear statistics $\mathbf{A}_1\mathbf{X}_1 + \cdots + \mathbf{A}_n\mathbf{X}_n$ and $\mathbf{B}_1\mathbf{X}_1 + \cdots + \mathbf{B}_n\mathbf{X}_n$, with $\mathbf{A}_i$, $\mathbf{B}_i$ nonsingular square matrices, are identically distributed.

2. Construct the corresponding theory of equidistributed linear forms on algebraic structures.

3. Develop a theory of identically distributed order statistics of the form $\max((X_1 - a)/A_1, \ldots, (X_n - a)/A_n)$, outlined in section 2.6.

CHAPTER 3

1. Extend the theorem of Ghurye and Olkin [19] to a denumerable number of random vectors.

2. Obtain conditions for the independence of suitable algebraic functions of independent r.v.'s $V_1, \ldots, V_n$, on the basis of the addition theorem for the Weierstrass function $\mathscr{P}(z)$, similar to what was done in section 3.4 in respect of the function tanh z.

3. Extend the theorem of Ghurye and Olkin [19] to Hilbert space valued and more general random variables.

CHAPTER 4

1. Let $\{X_i\}$ be a sequence of independent (and, to start with, identically distributed) r.v.'s, $\{a_i\}$ and $\{b_i\}$ be two sequences of real numbers, and τ a

Markovian stopping time (cf. Chapter 12 also). Construct the "linear forms" with a random number of summands

$$L_1 = \sum_1^{\tau} a_i X_i \qquad \text{and} \qquad L_2 = \sum_1^{\tau} b_i X_i.$$

Investigate conditions for the independence of L_1 and L_2. Under what conditions on τ and the sequences $\{a_i\}$ and $\{b_i\}$ would the independence of L_1 and L_2 imply the normality of the X_i?

2. Generalize Theorems 4.7.3 and 4.7.4 to the case of $\overline{X}$ and other symmetric and nonsymmetric polynomial statistics being independent, and also relax the requirement on the sample size (in this context, *vide* [41]).

3. Find conditions for the independence of a quadratic form $Q(X_1, \ldots, X_n)$ and a linear form $L(X_1, \ldots, X_n)$ in independent r.v.'s.

4. Is it possible to replace the condition of continuity imposed on $y(x)$ in Anosov's theorem (section 4.9) by a weaker condition?

5. Relax the condition on the density $y(x)$ in Theorem 4.10.1.*

6. Extend the investigation of independence of linear forms with random coefficients.

7. Extend the investigation of independence of indefinite tube statistics.

CHAPTER 5

1. Let $\mathbf{X}_1$ and $\mathbf{X}_2$ be independent random vectors, and $\mathbf{A}_1$, $\mathbf{A}_2$, $\mathbf{B}_1$, $\mathbf{B}_2$ constant matrices. Describe all possible distributions for the vectors $\mathbf{X}_1$ and $\mathbf{X}_2$ if it is known that for the statistics $\mathbf{Y}_1 = \mathbf{A}_1\mathbf{X}_1 + \mathbf{A}_2\mathbf{X}_2$ and $\mathbf{Y}_2 = \mathbf{B}_1\mathbf{X}_1 + \mathbf{B}_2\mathbf{X}_2$, the relation $E(\mathbf{Y}_1 | \mathbf{Y}_2) = \mathbf{0}$ is satisfied.

2. Let $\mathbf{X}_1, \ldots, \mathbf{X}_n$ be independent random vectors with finite dispersion matrices, $\mathbf{Y}_1 = \mathbf{A}_1\mathbf{X}_1 + \cdots + \mathbf{A}_n\mathbf{X}_n$ and $\mathbf{Y}_2 = \mathbf{B}_1\mathbf{X}_1 + \cdots + \mathbf{B}_n\mathbf{X}_n$, where the $\mathbf{A}_i$, $\mathbf{B}_i$ are constant matrices. Consider the relation

$$E(\mathbf{Y}_1 | \mathbf{Y}_2) = \mathbf{c} + d\mathbf{Y}_2, \qquad \text{disp}\,(\mathbf{Y}_1 | \mathbf{Y}_2) = \mathbf{D}$$

where $\mathbf{c}$ is a vector, d a scalar, and $\mathbf{D}$ a matrix of constants ("disp." standing for "dispersion matrix"). Under what conditions on $\mathbf{A}_i$, $\mathbf{B}_i$ will it follow from the above relation that the $\mathbf{X}_i$ are normal?

3. Consider problems 1 and 2 above for Hilbert space valued random variables.

CHAPTER 6

1. Let $X_1, \ldots, X_n$ be i.i.d.r.v.'s, $Z_i = (X_i - \overline{X})/S$ where $S^2 = \sum_1^n (X_i - \overline{X})^2$. What can be said about the distribution of the X_i if $E(\overline{X} | Z_1, \ldots, Z_n) = 0$?

* Recently, L. B. Klebanov removed all restrictions on $y(x)$.

2. Let Z be a normal r.v. and X related to Z according to

$$E(Z^k \mid X) = X^k + Q_{k-1}(X)$$

for $k = 0, 1, \ldots$, where Q_{k-1} is a polynomial of degree $\leqq k - 1$. Does it follow hence that X is also normal? Note that if this assertion were true, then Cramer's theorem on the independent components of a normal law (also being normal) would easily follow from it.

Apparently, other formulations are possible, where the distribution of one r.v. is required to determine that of another "closely connected" with the former.

3. Let $X_1, \ldots, X_n$ be i.i.d. variables. Find all distributions for which $E(\overline{X} \mid S^2) =$ constant, where $\overline{X}$ and S^2 are as defined in (1).

CHAPTER 7

1. Describe the d.f.'s F for which the Pitman estimator of the shift parameter, under quadratic loss, has the form $\overline{X} + Q_k(X_1 - \overline{X}, \ldots, X_n - \overline{X})$, where Q_k is some polynomial of degree $\leqq k$.

2. Study conditions for the admissibility of $\overline{X}$ as an estimator of the shift parameter, given a random sample $(X_1, \ldots, X_n)$, under a loss function $W(\tilde{\theta} - \theta)$ of a general type. In particular, if $W(u) \geqq 0$ is a symmetric convex function with $W(0) = 0$, then is the admissibility of $\overline{X}$ characteristic of the normal law?

This problem reduces to the following. Consider the class of regular estimators $\tilde{\theta} = \overline{X} + \phi(X_1 - \overline{X}, \ldots, X_n - \overline{X}) = \overline{X} + \phi(\mathbf{Y})$. If $\overline{X}$ is admissible, then

$$\min_{\phi} E_0 W(\overline{X} + \phi(\mathbf{Y})) = E_0 W(\overline{X}). \tag{1}$$

On the other hand,

$$E_0 W(\overline{X} + \phi(\mathbf{Y})) = E_0 E_0[W(\overline{X} + \phi(\mathbf{Y})) \mid \mathbf{Y}]$$

and for (1) to be satisfied, it is necessary that for almost all $\mathbf{y}$

$$\min_{\phi} E_0[W(\overline{X} + \phi(\mathbf{y}) \mid \mathbf{Y} = \mathbf{y}] = E_0[W(\overline{X}) \mid \mathbf{Y} = \mathbf{y}]. \tag{2}$$

In the same way as (7.9.3) was derived from (7.9.1), we obtain the following functional equation from (2) under certain smoothness conditions:

$$\int_{-\infty}^{\infty} w(u) \prod_{1}^{n} f(u + y_i)\, du = 0 \qquad \text{if} \sum_{1}^{n} y_i = 0 \tag{3}$$

where $w(u) = W'(u)$, f is the probability density function of the X_i. Under what conditions on f will a normal density function be the unique solution of (3)?

3. *Estimation of a shift parameter in the presence of a nuisance scale parameter.* Let $(X_1, \ldots, X_n)$ be a random sample from a population with d.f. $F[(x-\theta)/\sigma]$, where $\theta \in R^1$ is a structural parameter, and $\sigma \in R^1_+$ a nuisance parameter. In investigating the admissibility of $\bar{X}$ as an estimator for θ under quadratic loss, it is natural to consider a class of estimators of the form $\tilde{\theta} = \bar{X} + \phi(\mathbf{Y})$ where $\mathbf{Y}$ is as defined above, and $\phi(\mathbf{Y})$ is a homogeneous function of the first degree of homogeneity. For such estimators,

$$E_{\theta,\sigma}(\tilde{\theta} - \theta)^2 = \sigma^2 E_{0,1}\tilde{\theta}^2$$

so that, for the admissibility of $\bar{X}$, it is necessary that

$$E_{0,1}(\bar{X}^2) = \min E_{0,1}[\bar{X} + \phi(\mathbf{Y})]^2 \tag{4}$$

where the minimum is taken over all functions ϕ of the first degree of homogeneity. Such ϕ can be represented in the form $\phi(\mathbf{y}) = \phi(y_1, \ldots, y_n) = s\phi(y_1/s, \ldots, y_n/s) = s\phi(z_1, \ldots, z_n) = s\phi(\mathbf{z})$. Conversely, every function $s\phi((\mathbf{z})$ has the first degree of homogeneity as a function of $\mathbf{y}$. Hence (4) is equivalent to

$$E_{0,1}(\bar{X}^2) = \min_{\phi(\mathbf{z})} E_{0,1}[\bar{X} + s\phi(\mathbf{z})]^2. \tag{5}$$

Under certain smoothness conditions, (5) leads to the following equation

$$\int_0^\infty v^{n-1}\, dv \int_{-\infty}^\infty u \prod_1^n f(u + z_i v)\, du = 0, \qquad \text{if } \sum_1^n z_i = 0,\ \sum_1^n z_i^2 = 1. \tag{6}$$

Is the normal density the only solution of (6)? An equation, similar to (6), is obtained without difficulty for the problem of estimation of a shift parameter in the presence of a nuisance scale parameter, for loss functions other than quadratic. In this area, only the Laplacian loss function (Zinger, Kagan) has been studied so far.

4. Investigate the problems similar to 2 and 3 above for a scale parameter (naturally, the shift parameter will be the nuisance parameter then).

5. Extend the class of processes for which the normality of the process would follow from the admissibility of the optimal linear estimator of the shift parameter (see section 7.8).

6. For what d.f.'s F will the Pitman estimator for the parameter θ (under quadratic loss), constructed from a random sample $(X_1, \ldots, X_n)$, for the population $F((x-\theta)/\sigma)$, be independent of σ? (For such F, it is not necessary to know σ, in order to estimate θ in an "optimal manner.") The analytical problem consists in clarifying conditions under which $E_{0,1}(\bar{X} \mid \mathbf{Y})$ has the first degree of homogeneity as a function of $\mathbf{Y}$. Note that such a property is possessed by the uniform distribution as well as by the normal; it is possible that, for given values of the first two moments, the above property is characteristic of some one-parameter family of d.f.'s F (see section 7.10).

CHAPTER 8

1. Construct a moment theory of sufficient subspaces. In particular, describe in terms of the first $2k$ moments all families of d.f.'s $F(x, \theta)$ for which a nontrivial $L_k^{(2)}$-sufficient subspace exists.

2. In the note [55], families of distributions $\{P_\theta;\ \theta \in \Theta\}$ given by the densities $p(x;\ \theta)$ with respect to some measure μ, where

$$p(\mathbf{x};\ \theta) = R(T(\mathbf{x});\ \theta) \sum_1^m d_i(\theta) r_i(\mathbf{x}) \tag{7}$$

were introduced. For such families, $T(\mathbf{x})$ is, generally speaking, not a sufficient statistic if $m > 1$; however, in the construction of similar zones for instance, $T(\mathbf{x})$ plays a role similar to that of a sufficient statistic. Describe all (regular) one-dimensional distributions, some "power" of which has the form (7) with a nontrivial $T(\mathbf{x})$. The conjecture here is that the densities of such distributions must have the form

$$f(x;\ \theta) = \sum_1^k a_i(\theta) s_i(x) \cdot \exp \sum_1^s c_i(\theta) \phi_i(x).$$

3. Remove the condition $\int_0^\infty x^\delta\, dF(x) < \infty$ for some $\delta > 0$ from Theorem 8.5.4.*

CHAPTER 9

1. Examine the order of stability in Theorems 9.2.1 and 9.3.1; it is very likely that it can be considerably improved in the latter case.

2. Investigate the stability of the characterization of the normal law through the properties of independence of linear statistics (Chapter 3) and the identical distribution of linear statistics (Chapter 2).

3. Examine the stability of the characterizations given by Theorems 13.5.1–13.5.4 and 13.5.9.

4. Investigate the stability of the characterizations of sequential estimation plans through the properties of Markovian stopping times.

CHAPTER 10

1. Let $\mathbf{X} = \mathbf{A}\mathbf{Y}$ be a random vector, having linear structure. In Theorem 10.3.10, it was shown that $\mathbf{X}$ can be represented in the form $\mathbf{X} = \mathbf{X}_1 + \mathbf{X}_2$, where $\mathbf{X}_1$ and $\mathbf{X}_2$ are independent random vectors, $\mathbf{X}_2$ is normally distributed and $\mathbf{X}_1$ is degenerate or a nonnormal random vector with a unique linear structure. When is such a representation unique?

* This has been achieved in a recent paper by D. Kelkar and T. Matthes, *Ann. Math. Statist.*, **41**, (1970).

CHAPTER 11

1. Does there exist a nonlinear (measurable) transformation $T:R^1 \to R^1$ which transforms two normal distributions with different means into the same normal distribution? (D. Basu)

2. Classify rational transformations of a standard normal random sample into another such.

3. Is it possible to generalize Theorem 11.1.3 on nonlinear meromorphic transformations from C^n to C^1 of a complex normal vector for more general analytic transformations?

4. Is the distribution of a nonhomogeneous polynomial statistic based on a normal sample determined by its moments?

5. Is the U-conjecture true for pairs of nonhomogeneous polynomial statistics based on a sample of size 2?

6. The U-conjecture under the conditions of Theorem 11.6.1 corresponds to the assertion: $\mathcal{N} - \mathcal{Z} = \phi$; Theorem 11.6.1 asserts merely that dim $(\mathcal{N} - \mathcal{Z}) \leqq \dim(\mathcal{N}) - q/6$. Is it possible to obtain in the same way a closer estimate for the dimension of $\mathcal{N} - \mathcal{Z}$?

7. Is it possible to establish an analogue of Theorem 11.6.1 for nonhomogeneous polynomial statistics P and Q?

8. Indicate an upper bound for dim $(\mathcal{N} - \mathcal{Z})$ in the case when the degrees of P and Q are not widely different.

CHAPTER 12

1. Is the Markovian stopping time τ characterized in the case of a multinomial process with parameter $\mathbf{p} = (p_1, \ldots, p_k)$ by the function $\phi(\mathbf{p}) = E_{\mathbf{p}}(\tau)$ corresponding to a complete plan?

2. Consider a Binomial process with parameter p. For what closed (and, in general, not complete) first entry plans is $E_{\mathbf{p}}(\tau)$ a constant?

3. Extend the results of section 12.2 to nonrandom sampling schemes.

4. Carry over the results of section 12.4 to stationary ordinary flows with limited after-effect.

CHAPTER 13

1. Let the d.f. F satisfy the conditions of section 13.1 and let its first $2k$ moments be given: $\mu_r = \int x^r \, dF(x)$, $1 \leqq r \leqq 2k$. For which of such F is $\min_F I$ attained where I is the Fisherian amount of information for the family $\{F(x - \theta), \theta \in R^1\}$? Possibly, this problem is linked to Problem 1 of Chapter 7.

2. Consider the preceding problem for the family of d.f.'s $\{F(x/\sigma), \sigma \in R^1_+\}$ depending on a scale parameter.

3. Relax the conditions of Theorems 13.5.1–13.5.4; in particulár, replace Cramer's condition by a weaker one.

4. Establish the analogue of Theorem 13.5.9 for the gamma distribution and for distributions with density of the form $\exp[-P(x)]$ where $P(x)$ is a nonnegative polynomial.

5. Let $(X_1, \ldots, X_n)$ be a random sample from a population with d.f. $F(x - \theta)$. The hypothesis $H_0 : \theta = 0$ is to be tested against the alternative $H_1 : \theta \neq 0$. If the test with critical region $|\bar{x}| > c$ is admissible in the class of criteria of given level, then does the normality of F follow hence?

6. Let $X_1, \ldots, X_n$ be i.i.d. r.v.'s and $\bar{X} = (X_1 + \cdots + X_n)/n$. Find necessary and sufficient conditions for the distribution of $\bar{X}$ to uniquely specify the parent distribution.

7. Let $X_1, \ldots, X_n$ be i.i.d. r.v.'s. What can be said about the distribution of X_1 if it is known that $\sum X_i^2$ has a chi-square distribution, that $\sum (X_i - \bar{X})^2$ has a chi-square distribution.

8. Let $X_1, \ldots, X_n$ be i.i.d. as normal variables, and T be a statistic with chi square (central or non-central) on n.d.f. What function of X_i is T?

9. Extend the results of Addendum A to Hilbert space valued random variables.

ADDENDUM A

Further Characterizations of Probability Laws Through Properties of Linear Functions of Random Variables

A.1 INTRODUCTION

This chapter describes some further work on characterizations of probability distributions since the Russian edition went to press.

Let $\mathbf{X}$ be a vector random variable with m components $X_1, \ldots, X_m$, and let

$$Z_i = a_{i1}X_1 + \cdots + a_{im}X_m, \qquad i = 1, \ldots, p$$

be $p < m$ linear functions; these p relations may be represented in matrix notation as $\mathbf{Z} = \mathbf{AX}$. The distribution of $\mathbf{Z}$ is of course completely determined if that of $\mathbf{X}$ is given. In this chapter, we address ourselves to the converse problem of characterizing the distribution of $\mathbf{X}$ given some properties of $\mathbf{Z}$. Some special cases of this problem have been studied by Kotlarski ([171, 172, 173]).

A surprising result is that when the m components of $\mathbf{X}$ are independently distributed and $m \leqq p(p+1)/2$, the joint distribution of p suitably chosen linear functions of the components of $\mathbf{X}$ determines the distributions of each of the m components, apart from a change of location. Thus, if there are 210 independent random variables, it is enough to know the joint distribution of 20 suitably chosen linear functions thereof to find the distribution of each of the 210 variables to within a location parameter (vide Rao [176]).

We also consider problems where the components of $\mathbf{X}$ may be dependent, and obtain generalizations of all the previous results on characterizations of the multivariate normal distribution.

A.2 SOME IMPORTANT LEMMAS

We quote for ready reference a few lemmas on solutions to certain functional equations, some of which have been proved in Chapter 1.

LEMMA A.2.1. Let $\psi_1, \ldots, \psi_n$ be continuous complex-valued functions of a real variable. If there exist distinct nonzero real numbers $c_1, \ldots, c_n$ such that

$$\psi_1(t + c_1 u) + \cdots + \psi_n(t + c_n u) = A(t \mid u) + B(u \mid t) \tag{A.2.1}$$

where $A(x \mid y)$ and $B(x \mid y)$ are, for every fixed real y, polynomials in x of degrees at most a and b respectively, then the ψ_i are all polynomials of degree at most $(a + b + n)$.

Corollary A.2.1. If in (A.2.1)

$$A(t \mid u) = A(u) \qquad \text{and} \qquad B(u \mid t) = B(t) \tag{A.2.2}$$

where A and B are given to be continuous functions, then the ψ_i, A, and B are all polynomials of degree at most n.

LEMMA A.2.2. (Rao [128]) If the right-hand side of (A.2.1) is of the form

$$A(t) + B(u) + P_k(t, u) \tag{A.2.3}$$

where A and B are continuous functions and P_k is a polynomial of degree k, then the ψ_i, A, and B are all polynomials of degree at most max (n, k).

LEMMA A.2.3. (Rao [176]) If the right-hand side of (A.2.1) consists only of $P_k(t, u)$ as defined in (A.2.3), then the ψ_i are polynomials of degree at most max $(n - 2, k)$.

We shall extend the results of Lemma A.2.1 by considering ψ_i as a function of $\boldsymbol{\alpha}_i' \mathbf{t}$, where $\boldsymbol{\alpha}_i$ is a given p-vector and $\mathbf{t}$ is a p-vector variable. We write the equation in the form

$$\psi_1(\boldsymbol{\alpha}_1'\mathbf{t}) + \cdots + \psi_n(\boldsymbol{\alpha}_n'\mathbf{t}) = \sum_1^p A_i(t_i) + P_k(\mathbf{t}) \tag{A.2.4}$$

where t_i is the i-th component of $\mathbf{t}$ and P_k is a k-th degree polynomial in $\mathbf{t}$.

LEMMA A.2.4. Let $\psi_1, \ldots, \psi_n$, $A_1, \ldots, A_p$ be complex-valued continuous functions and let the functional equation (A.2.4) hold for all values of $\mathbf{t}$ in a neighborhood of the origin. Further, let $\boldsymbol{\alpha}_i$ be not proportional to $\boldsymbol{\alpha}_j$ for $i \neq j$ or to any unit vector (a vector with only one nonzero component). Then the ψ_i and A_j are all polynomials of degree at most max (n, k) in an interval around the origin.

We shall obtain more precise estimates for the degrees of the polynomials ψ_i by imposing further restrictions on the $\boldsymbol{\alpha}_i$.

Let us denote by $\mathbf{A}$ the matrix with $\boldsymbol{\alpha}_1, \ldots, \boldsymbol{\alpha}_n$ as its columns and similarly by $\mathbf{B}$ the matrix with columns $\boldsymbol{\beta}_1, \ldots, \boldsymbol{\beta}_n$, and define the operation (cf. definition (1.5.16))

$$\mathbf{A} \odot \mathbf{B} = (\boldsymbol{\alpha}_1 \otimes \boldsymbol{\beta}_1 \vdots \cdots \vdots \boldsymbol{\alpha}_n \otimes \boldsymbol{\beta}_n) \tag{A.2.5}$$

where $\otimes$ denotes the Kronecker product. The matrix $\mathbf{A} \odot \mathbf{A}$ is of order $p^2 \times n$. We use the notation

$$(\mathbf{A} \odot)^s \mathbf{A} = \mathbf{A} \odot \mathbf{A} \odot \cdots \odot \mathbf{A} \text{ (involving the symbol } \odot \ s \text{ times).}$$

It is easy to see that

$$R[(\mathbf{A} \odot)^s \mathbf{A}] \geqq R[(\mathbf{A} \odot)^r \mathbf{A}] \qquad \text{if } s \geqq r \tag{A.2.6}$$

where $R(X)$ denotes the rank of matrix X.

LEMMA A.3.5. [176] Consider the equation

$$\psi_1(\boldsymbol{\alpha}_1'\mathbf{t}) + \cdots + \psi_n(\boldsymbol{\alpha}_n'\mathbf{t}) = 0 \tag{A.2.7}$$

where the ψ_i are as in (A.2.4) and $\boldsymbol{\alpha}_i$ is not proportional to $\boldsymbol{\alpha}_j$ for $i \neq j$. Let s be the smallest integer such that

$$R[(\mathbf{A} \odot)^s \mathbf{A}] = n. \tag{A.2.8}$$

Then the ψ_i are polynomials of degree at most s.

Note that we do not impose the condition that $\boldsymbol{\alpha}_i$ is not proportional to any unit vector in the statement of Lemma A.2.5.

A.3 CHARACTERIZATION OF PROBABILITY LAWS: THE UNIVARIATE CASE

Let X_1, X_2, X_3 be three independent real random variables and let

$$Z_1 = X_1 - X_2, \qquad Z_2 = X_2 - X_3. \tag{A.3.1}$$

If the characteristic function (c.f.) of the pair (Z_1, Z_2) does not vanish, it was shown by Kotlarski ([171, 172]) and Pakshirajan and Mohun [174] that the distribution of (Z_1, Z_2) determines that of X_1, X_2, X_3 up to a change of location. We prove the same result by considering two general linear forms

$$Z_1 = a_1 X_1 + a_2 X_2 + a_3 X_3; \qquad Z_2 = b_1 X_1 + b_2 X_2 + b_3 X_3 \tag{A.3.2}$$

such that $a_i{:}b_i \neq a_j{:}b_j$ for $i \neq j$, and also investigate the general case of p linear forms in n independent variables.

THEOREM A.3.1. (Rao) Let X_1, X_2, X_3 be three independent random variables and Z_1, Z_2 as defined in (A.3.2). If the c.f. of (Z_1, Z_2) does not vanish, then the distribution of (Z_1, Z_2) determines the distributions of the X_i up to a change of location.

Proof: Let ϕ_i be the c.f. of X_i and $\xi_i = \log \phi_i$ be the *second characteristic* (s.c.) of X_i (i.e., the principal branch of the logarithm of the c.f.). Let f_i and s_i be an alternative c.f. and s.c. of X_i, respectively. Define $\psi_i = \xi_i - s_i$.

Computing the c.f.'s of (Z_1, Z_2) for the two choices of the distributions of X_1, X_2, X_3, and equating the s.c.'s we obtain

$$\psi_1(a_1 t + b_1 u) + \psi_2(a_2 t + b_2 u) + \psi_3(a_3 t + b_3 u) = 0 \tag{A.3.3}$$

valid for all real t and u. In view of the conditions on the coefficients a_i, b_i given in (A.3.2), the equation (A.3.3) can take only one of the three forms.

$$\text{(i)}\quad \psi_1(t + c_1 u) + \psi_2(t + c_2 u) + \psi_3(t + c_3 u) = 0,$$

with c_1, c_2, c_3 and zero all distinct.

$$\text{(ii)}\quad \psi_1(t + c_1 u) + \psi_2(t + c_2 u) = A(t), \tag{A.3.4}$$

with c_1, c_2 and zero all distinct.

$$\text{(iii)}\quad \psi_1(t + cu) = A(t) + B(u),\ c \neq 0.$$

In each case, applying the results of Lemma A.2.2 and Lemma A.2.3, we see that ψ_i must be linear, $i = 1, 2, 3$, thus proving the theorem. ▮

THEOREM A.3.2. [176] Let $X_1, \ldots, X_n$ be n independent random variables and

$$Z_1 = a_1 X_1 + \cdots + a_n X_n \quad \text{and} \quad Z_2 = b_1 X_1 + \cdots + b_n X_n \tag{A.3.5}$$

be two linear functions that $a_i{:}b_i \neq a_j{:}b_j$ for $i \neq j$. If the c.f. of (Z_1, Z_2) is specified and does not vanish anywhere, and ϕ_i, f_i be two alternative possible c.f.'s of X_i, then

$$\phi_i(t) = f_i(t) \exp[P_{n-2}(t)] \tag{A.3.6}$$

where P_k denotes a polynomial of degree at most k.

***Proof*:** This result is obtained by proceeding as in the proof of Theorem A.3.1 and applying the result of Lemma A.2.3. An interesting consequence of Theorem A.3.2 is given by Corollary A.3.1 below. ▮

Corollary A.3.1. If $n = 4$ in (A.3.5), then the distribution of X_i is determined up to a convolution with a normal variable.

We have seen that the distribution of two (suitable) linear functions of n independent random variables specifies the distribution of each individual variable up to a change of location when $n = 3$, up to convolution with a normal variable when $n = 4$, and so on. We now pose the following question: what is the smallest number of linear functions of n independent random variables whose joint distribution specifies the distribution of each variable up to a change of location, up to convolution with a normal variable, and so on? Theorem A.3.3 provides the answer.

Let $X_1, \ldots, X_n$ be n independent random variables and let

$$Z_i = a_{i1} X_1 + \cdots + a_{in} X_n; \qquad i = 1, \ldots, p \tag{A.3.7}$$

be p linear functions. Let us denote the (row) vector $(a_{1i}, \ldots, a_{pi})$ by $\boldsymbol{\alpha}_i'$ and the matrix with $\boldsymbol{\alpha}_i$ as its i-th column by $\mathbf{A}$. We assume that the c.f. of $(Z_1, \ldots, Z_p)$ does not vanish anywhere.

THEOREM A.3.3. [176] Let $Z_1, \ldots, Z_p$ be p linear functions of n independent variables $X_1, \ldots, X_n$ as defined in (A.3.7). Further, let $\boldsymbol{\alpha}_i$ be not proportional to $\boldsymbol{\alpha}_j$ for $i \neq j$ and the integer s be such that $R[(\mathbf{A}\odot)^s\mathbf{A}] = n > R[(\mathbf{A}\odot)^{s-1}\mathbf{A}]$. Then the c.f. of each X_i is determined up to a factor $\exp[P_{i,s}(t)]$ where $P_{i,s}$ is a polynomial of degree at most s.

Proof: This result is proved by an application of Lemma A.2.5. ▮

Theorem A.3.3 provides the basic result for answering the questions raised. We shall find the minimum value of p for given n to determine the c.f. of X_i up to a factor of the form $\exp(c_i t)$. In such a case we must have $s=1$, i.e., $R(\mathbf{A} \odot \mathbf{A}) = n$. Since the number of independent rows in $\mathbf{A} \odot \mathbf{A}$ can be at most $p(p+1)/2$, the minimum value of p for given n is such that

$$p(p-1)/2 < n \leqq p(p+1)/2.$$

The minimum values of p for a few values of n are as follows.

n	2	3	4	5	6	7	8	9	10
min p	2	2	3	3	3	4	4	4	4 .

To show that min p is attainable we shall give a particular choice of the coefficients of the p linear forms. Let $\mathbf{I}_s$ denote the identity matrix of order s and $\mathbf{R}_{ts}$ denote the $(t+s) \times (s-1)$ matrix with the first t rows null, $(1, 1, \ldots, 1)$ as the $(t+1)$-th row, and the rows of $\mathbf{I}_{s-1}$ as the last $(s-1)$ rows. Let us construct the partitioned matrix

$$\mathbf{A} = (\mathbf{I}_p \vdots \mathbf{R}_{0,p}\vdots \ \ldots \ \vdots\mathbf{R}_{p-2,2}).$$

It is easy to see that $R(\mathbf{A} \odot \mathbf{A}) = p(p+1)/2$ so that the maximum number of independent variables can be $p(p+1)/2$ when p linear forms are considered.

Corollary A.3.2. [176] Let $\mathbf{A}$ be a $p \times \{p(p+1)/2\}$ matrix such that $R(\mathbf{A} \odot \mathbf{A}) = p(p+1)/2$ and $\mathbf{X}$ be a vector of $p(p+1)/2$ independent random variables X_i, $i = 1, \ldots, p(p+1)/2$. Then the joint distribution of just the p functions which are the components of $\mathbf{Z} = \mathbf{AX}$ determines the distributions of the individual X_i up to changes of location.

A.4 SOME GENERAL FUNCTIONAL EQUATIONS

Let ϕ_i be a continuous complex-valued function of a real p_i-vector variable, $i = 1, \ldots, s$. We denote by $C_a(\mathbf{u}|\mathbf{t})$ a polynomial of degree a in $\mathbf{u}$ given $\mathbf{t}$, where $\mathbf{u}$ and $\mathbf{t}$ are vector variables of possibly different dimensions; by $D_b(\mathbf{t}|\mathbf{u})$ a polynomial of degree b in $\mathbf{t}$ given $\mathbf{u}$; and by $P_k(\mathbf{t}, \mathbf{u})$ a polynomial of degree k in $\mathbf{t}$ and $\mathbf{u}$. All the functions and the functional equations considered here need be defined only in a neighborhood of the origin in appropriate spaces, unless otherwise stated. Naturally the solutions obtained are valid only in a neighborhood of the origin. We denote the rank of a matrix $\mathbf{X}$ as before by $R(\mathbf{X})$.

THEOREM A.4.1. (Khatri and Rao [170]). Let ϕ_i, $i = 1, \ldots, s$, $C_a(\mathbf{u}|\mathbf{t})$, $D_b(\mathbf{t}|\mathbf{u})$ and $P_k(\mathbf{t}, \mathbf{u})$ be as defined above, and $\mathbf{A}_i$ and $\mathbf{B}_i$, $i = 1, \ldots, s$, be matrices of orders $p \times p_i$ and $m \times p_i$ respectively, such that

$$\sum_1^s \phi_i(\mathbf{A}_i'\mathbf{t} + \mathbf{B}_i'\mathbf{u}) = C_a(\mathbf{u}|\mathbf{t}) + D_b(\mathbf{t}|\mathbf{u}) + P_k(\mathbf{t}, \mathbf{u}). \tag{A.4.1}$$

(i) If $R(\mathbf{A}_i) = R(\mathbf{B}_i) = p_i$, $i = 1, \ldots, s$, then $C_a(\mathbf{u}|\mathbf{t})$ and $D_b(\mathbf{t}|\mathbf{u})$ are polynomials in $\mathbf{t}$ and $\mathbf{u}$ of degree $\leqq \max(k, s+a+b)$.

(ii) If $C_a(\mathbf{u}|\mathbf{t}) = C(\mathbf{t})$, $D_b(\mathbf{t}|\mathbf{u}) = D(\mathbf{u})$ and $R(\mathbf{A}_i) = R(\mathbf{B}_i) = p_i$ for $i = 1, \ldots, s$, then $C(\mathbf{t})$ and $D(\mathbf{u})$ are polynomials in $\mathbf{t}$ and $\mathbf{u}$ respectively, of degree $\leqq$ max (k, s).

(iii) If $C_a(\mathbf{u}|\mathbf{t}) = 0$ and $R(\mathbf{A}_i) = p_i$, $i = 1, \ldots, s$, then $D_b(\mathbf{t}|\mathbf{u})$ is a polynomial in $\mathbf{t}$ and $\mathbf{u}$ of degree $\leqq$ max $(k, s + b - 1)$.

(iv) If $R(\mathbf{A}_i) = p_i$, $i = 1, \ldots, s$, then $D_b(\mathbf{t}|\mathbf{u})$ is a polynomial in t and u of degree $\leqq$ max $(k + s + a + b)$. (Note that a similar statement cannot be made about $C_a(\mathbf{u}|\mathbf{t})$ when no assumption is made on $R(\mathbf{B}_i)$.)

***Proof*:** The results are established by following the method of differences as in the proof of Theorem 1.5.9, with appropriate modifications to suit the unequal dimensions of the arguments of the different functions in (A.4.1). ▮

THEOREM A.4.2. (Khatri and Rao [170]). Let ϕ_i be functions as in Theorem A.4.1 and $\mathbf{C}_i$ be an $m \times p_i$ matrix of rank p_i such that

$$\phi_1(\mathbf{C}_1'\mathbf{t}) + \cdots + \phi_s(\mathbf{C}_s'\mathbf{t}) = P_k(\mathbf{t}). \tag{A.4.2}$$

If

$$R(\mathbf{C}_i \vdots \mathbf{C}_j) = p_i + p_j, \qquad i = 1, \ldots, r; \qquad j = r + 1, \ldots, s \tag{A.4.3}$$

then $\phi_1(\mathbf{C}_1'\mathbf{t}) + \cdots + \phi_r(\mathbf{C}_r'\mathbf{t})$ and $\phi_{r+1}(\mathbf{C}_{r+1}'\ \mathbf{t}) + \cdots + \phi_s(\mathbf{C}_s'\mathbf{t})$ are polynomials in $\mathbf{t}$ of degree at most max $(k, s - 2)$.

***Proof*:** Let us take

$$\mathbf{t} = \mathbf{H}_{11}\mathbf{u} + \mathbf{H}_{21}\mathbf{v} \tag{A.4.4}$$

with $\mathbf{H}_1 = (\mathbf{H}_{11} \vdots \mathbf{H}_{21})$ nonsingular and

$$\mathbf{C}_s'\mathbf{H}_{11} = \mathbf{I} \qquad \text{and} \qquad \mathbf{C}_s'\mathbf{H}_{21} = \mathbf{0} \tag{A.4.5}$$

where $\mathbf{I}$ is of order p_s. We define

$$\mathbf{C}_{i,1} = \mathbf{H}_1'\mathbf{C}_i, \qquad \mathbf{B}_{i,1} = \mathbf{H}_{2,1}'\mathbf{C}_i \tag{A.4.6}$$

$$\phi_i^{(1)}(\mathbf{C}_{i,1}'\mathbf{w}) = \phi_i(\mathbf{C}_{i,1}'\mathbf{w} + \mathbf{B}_{i,1}'\mathbf{h}_1) - \phi_i(\mathbf{C}_{i,1}'\mathbf{w}) \tag{A.4.7}$$

$$P_{k-1}^{(1)}(\mathbf{w}) = P_k(\mathbf{H}_1\mathbf{w} + \mathbf{H}_{21}\mathbf{h}_1) - P_k(\mathbf{H}_1\mathbf{w}) \tag{A.4.8}$$

where $\mathbf{w}' = (\mathbf{u}' \vdots \mathbf{v}')$. From (A.4.2), by taking the first difference with increment $\mathbf{h}_1$ for $\mathbf{v}$ and writing $\mathbf{t}$ for $\mathbf{w}$, we obtain

$$\sum_{i=1}^{s-1} \phi_i^{(1)}(\mathbf{C}_{i,1}'\mathbf{t}) = P_{k-1}^{(1)}(\mathbf{t}). \tag{A.4.9}$$

Equation (A.4.9) is of the same form as (A.4.2) with the same conditions $R(\mathbf{C}_{i,1}) = p_i$ and $R(\mathbf{C}_{i,1} \vdots \mathbf{C}_{j,1}) = p_i + p_j$; $i = 1, \ldots, r$; $j = r + 1, \ldots, s - 1$, in view of the condition (A.4.3), but with one function (ϕ_s) eliminated. We also note that $\phi_i^{(1)}(\mathbf{t})$ as a function of $\mathbf{h}_1$ is a p_i-vector function since $R(\mathbf{B}_{i,1}) = p_i$ by (A.4.3). We then eliminate $\phi_{s-1}^{(1)}$ in the same manner to obtain a functional equation in $\phi_i^{(2)}$, $i = 1, \ldots, s - 2$ and so on. Finally, after $s - r$ eliminations, we obtain the equation

$$\sum_{i=1}^{r} \phi_i^{(s-r)}\ (\mathbf{C}_{i,s-r}'\mathbf{t}) = P_{k-s+r}^{(s-r)}(\mathbf{t}) \tag{A.4.10}$$

where $\phi_i^{(s-r)}$ and $P_{k-s+r}^{(s-r)}$ denote the $(s-r)$-th difference functions of ϕ_i and P_k involving the increment vectors $\mathbf{h}_1, \ldots, \mathbf{h}_{s-r}$ used in the different stages of the reduction, and $P_{k-s+r}^{(s-r)} = 0$ if $k < s - r$.

Equation (A.4.10) can be written in the explicit form

$$\sum_1^r \phi_i^{(s-r-1)} (\mathbf{C}'_{i,s-r}\mathbf{t} + \mathbf{B}'_{i,s-r}\mathbf{h}_{s-r}) = Q(\mathbf{t}) + P_{k-s+r+1}(\mathbf{t}, \mathbf{h}_{s-r}) \qquad \text{(A.4.11)}$$

where $\mathbf{C}_{i,s-r}$ and $\mathbf{B}_{i,s-r}$ are suitably defined as in Equations (A.4.4)–(A.4.8). Also, $R(\mathbf{C}_{i,s-r}) = R(\mathbf{B}_{i,s-r}) = p_i$, and

$$Q(\mathbf{t}) = \sum_1^r \phi_i^{(s-r-1)}(\mathbf{C}'_{i,s-r}\mathbf{t}). \qquad \text{(A.4.12)}$$

Now, we apply result (iii) of Theorem A.4.1 to show that $Q(\mathbf{t})$ is a polynomial in $\mathbf{t}$ of degree at most $\alpha = \max(k - s + r + 1, r - 1)$ given $\mathbf{h}_{s-r-1}, \ldots, \mathbf{h}_1$. Let us write this polynomial as $P_\alpha(\mathbf{H}_{s-r}\ \mathbf{t}|\mathbf{h}_{s-r-1})$. Hence

$$\sum_1^r \phi_i^{(s-r-1)}(\mathbf{C}'_{i,s-r-1}\mathbf{t}) = P_\alpha(\mathbf{t}|\mathbf{h}_{s-r-1}). \qquad \text{(A.4.13)}$$

Writing

$$Q_1(\mathbf{t}) = \sum_1^r \phi_i^{(s-r-2)}(\mathbf{C}'_{i,s-r-1}\mathbf{t}) \qquad \text{(A.4.14)}$$

we have

$$Q_1(\mathbf{t} + \mathbf{F}\mathbf{h}_{s-r-1}) = \sum_1^r \phi_i^{(s-r-2)} (\mathbf{C}'_{i,s-r-1}\mathbf{t} + \mathbf{B}'_{i,s-r-1}\mathbf{h}_{s-r-1}) \qquad \text{(A.4.15)}$$

where $\mathbf{F}' = (\mathbf{0} \vdots \mathbf{I})$ is an $(m - p_{r+2}) \times m$ matrix of rank $m - p_{r+2}$. Using (A.4.14) and (A.4.15) in (A.4.13), we get

$$Q_1(\mathbf{t} + \mathbf{F}\mathbf{h}_{s-r-1}) = Q_1(\mathbf{t}) + P_\alpha(\mathbf{t}|\mathbf{h}_{s-r-1}). \qquad \text{(A.4.16)}$$

From (A.4.16), it is easy to see that if $\mathbf{t}' = (\mathbf{u}', \mathbf{v}')$ and $\mathbf{h}_{s-r-1} = \mathbf{h}$, then

$$P_\alpha(\mathbf{u}, \mathbf{v} + \mathbf{w}|\mathbf{h} - \mathbf{w}) = P_\alpha(\mathbf{u}, \mathbf{v}|\mathbf{h}) - P_\alpha(\mathbf{u}, \mathbf{v}|\mathbf{w}) \qquad \text{(A.4.17)}$$

valid for all $\mathbf{w}(\neq \mathbf{h})$ in an appropriate region near the origin. Equating the coefficients of various powers of the u_i's and v_j's, the components of $\mathbf{u}$ and $\mathbf{v}$, it is easily shown that $P_\alpha(\mathbf{u}, \mathbf{v}|\mathbf{h})$ is a polynomial in $\mathbf{u}$, $\mathbf{v}$, and $\mathbf{h}$ of degree at most $(\alpha + 1)$ and so we write $P_\alpha(\mathbf{u}, \mathbf{v}|\mathbf{h}) = P_{\alpha+1}(\mathbf{t}, \mathbf{h})$. Using this and (A.4.15) in (A.4.16), we have

$$\sum_1^r \phi_i^{(s-r-2)}(\mathbf{C}'_{i,s-r-1}\mathbf{t} + \mathbf{B}'_{i,s-r-1}\mathbf{h}) = Q_1(\mathbf{t}) + P_{\alpha+1}(\mathbf{t}, \mathbf{h}) \qquad \text{(A.4.18)}$$

with $\mathbf{h} = \mathbf{h}_{s-r-1}$. This is similar to (A.4.11) and hence $Q_1(\mathbf{t})$ is a polynomial in $\mathbf{t}$ of degree at most $(\alpha + 1)$ given $\mathbf{h}_{s-r-2}, \ldots, \mathbf{h}_1$. Proceeding in this manner, we get finally $\sum_1^r \phi_i(\mathbf{C}'_i\,\mathbf{t})$ as a polynomial in $\mathbf{t}$ of degree at most $\alpha + s - r - 1 = \max(k, s - 2)$ and this establishes the theorem. ∎

Note: In Theorem A.4.2, the condition $R(\mathbf{C}_i) = p_i$ is unnecessary if (instead) the main condition (A.4.3) is written as

$$R(\mathbf{C}_i \vdots \mathbf{C}_j) = R(\mathbf{C}_i) + R(\mathbf{C}_j), \qquad i = 1, \ldots, r \quad \text{and} \quad j = r + 1, \ldots, s. \qquad \text{(A.4.19)}$$

However, if $R(\mathbf{C}_i) \neq p_i$, the function $\phi_i(\mathbf{C}_i'\mathbf{t})$ can be redefined as $\psi_i(\mathbf{B}_i'.\mathbf{t})$ where $\mathbf{B}_i$ is of order $m \times q_i$ and of rank q_i, so that the conditions of Theorem A.4.2 as stated are satisfied.

Corollary A.4.1. Let ϕ_i, $\mathbf{A}_i$, and $\mathbf{B}_i$ be as in Theorem A.4.1, satisfying relation (A.4.1). If $R(\mathbf{A}_i) = R(\mathbf{B}_i) = p_i$ and

$$R\begin{pmatrix} \mathbf{A}_i & \mathbf{A}_j \\ \mathbf{B}_i & \mathbf{B}_j \end{pmatrix} = p_i + p_j \qquad \text{for } i = 1, \ldots, r; \qquad j = r+1, \ldots, s \tag{A.4.20}$$

then

$$\sum_1^r \phi_i(\mathbf{A}_i'\mathbf{t} + \mathbf{B}_i'\mathbf{u}) \qquad \text{and} \qquad \sum_{r+1}^s \phi_i(\mathbf{A}_i'\mathbf{t} + \mathbf{B}_i'\mathbf{u}) \tag{A.4.21}$$

are polynomials in $\mathbf{t}$ and $\mathbf{u}$ of degree at most max $(k, s+a+b)$.

Corollary A.4.2. Let ϕ_i, $\mathbf{A}_i$, and $\mathbf{B}_i$ be as in Theorem A.4.1, satisfying relation (A.4.1) with $C_a(\mathbf{u}|\mathbf{t}) = 0$. If $R(\mathbf{A}_i) = p_i$ and (A.4.20) holds, then the two expressions given by (A.4.21) are polynomials in $\mathbf{t}$ and $\mathbf{u}$ of degree at most max $(k, s+b-1)$.

Corollary A.4.3. Let ϕ_i and $\mathbf{C}_i$ of rank p_i be as in Theorem A.4.2, satisfying relation (A.4.2). If

$$R(\mathbf{C}_i \vdots \mathbf{C}_j) = p_i + p_j \qquad \text{for all } i \neq j \tag{A.4.22}$$

then each ϕ_i is a polynomial of degree at most max $(k, s-2)$.

Corollary A.4.4. Let ϕ_i be a continuous complex q_i-vector-valued function of a p_i-vector real variable and $\mathbf{C}_i$ be an $m \times p_i$ matrix of rank p_i, $i = 1, \ldots, s$. Further, let

$$\sum_1^s \mathbf{G}_i \phi_i(\mathbf{C}_i'\mathbf{t}) = \mathbf{P}_k(\mathbf{t}) \tag{A.4.23}$$

where $\mathbf{P}_k$ is a q-vector of polynomials in m-vector $\mathbf{t}$ of degree at most k, and $\mathbf{G}_i$ is a $q \times q_i$ matrix. If the relation (A.4.3) holds, then

$$\sum_1^r \mathbf{G}_i\, \phi_i(\mathbf{C}_i'\, \mathbf{t}) \qquad \text{and} \qquad \sum_{r+1}^s \mathbf{G}_i\, \phi_i(\mathbf{C}_i'\, \mathbf{t}) \tag{A.4.24}$$

are vectors of polynomials in $\mathbf{t}$ of degree at most max $(k, s-2)$.

Corollary A.4.5. If in Corollary A.4.4, instead of condition (A.4.3), the conditions $R(\mathbf{G}_i) = q_i$ and $R(\mathbf{C}_i : \mathbf{C}_j) = p_i + p_j$ for all $i \neq j$ hold, then each ϕ_i is a vector of polynomials of degree at most max $(k, s-2)$.

We shall now give an estimate of the degree of the polynomials of Theorem A.4.2. For this purpose we introduce a matrix product which is a generalization of the product (A.2.5). Let $\mathbf{E} = (\mathbf{E}_1 \vdots \cdots \vdots \mathbf{E}_n)$ and $\mathbf{F} = (\mathbf{F}_1 \vdots \cdots \vdots \mathbf{F}_n)$ be partitions of matrices $\mathbf{E}$ and $\mathbf{F}$ and $\otimes$ denote the Kronecker product as usual. Then we define the product $\mathbf{E} \odot \mathbf{F}$ according to

$$\mathbf{E} \odot \mathbf{F} = (\mathbf{E}_1 \otimes \mathbf{F}_1 \vdots \cdots \vdots \mathbf{E}_n \otimes \mathbf{F}_n). \tag{A.4.25}$$

We denote by $(\mathbf{E}\odot)^2\mathbf{E}$ the product $(\mathbf{E} \odot \mathbf{E}) \odot \mathbf{E}$, and similarly for $(\mathbf{E}\odot)^r\mathbf{E}$.

THEOREM A.4.3. (Khatri and Rao [170]) Consider ϕ_i and $\mathbf{C}_i$ as in Theorem A.4.2 and the equation (A.4.2) with $k < r < s - 2$. Let $\mathbf{C} = (\mathbf{C}_1 \vdots \cdots \vdots \mathbf{C}_s)$. If

$$R[(\mathbf{C}\odot)^{r-1}\mathbf{C}] < \sum p_i^r \quad \text{and} \quad R[(\mathbf{C}\odot)^r\mathbf{C}] = \sum p_i^{r+1} \tag{A.4.26}$$

then each ϕ_i is a polynomial of degree at most r.

Corollary A.4.6. If $r = 1$ and $k = 1$ in Theorem A.4.3, then each ϕ_i is at most linear.

A.5 CHARACTERIZATION OF THE MULTIVARIATE NORMAL DISTRIBUTION

A.5.1 Nonuniqueness of Structural Representation

A p-vector variable $\mathbf{X}$ is said to have a linear structure if it can be represented as

$$\mathbf{X} = \mathbf{A}_1\mathbf{Y}_1 + \cdots + \mathbf{A}_m\mathbf{Y}_m \tag{A.5.1}$$

where $\mathbf{Y}_i$ is a p_i-vector variable, $\mathbf{A}_i$ is a $p \times p_i$ matrix, and the $\mathbf{Y}_i$ are independent. We may take $R(\mathbf{A}_i) = p_i$, without loss of generality. Suppose $\mathbf{X}$ admits an alternative representation

$$\mathbf{X} = \mathbf{B}_1\mathbf{Z}_1 + \cdots + \mathbf{B}_n\mathbf{Z}_n \tag{A.5.2}$$

where $\mathbf{Z}_i$ is a q_i-vector variable, $\mathbf{B}_i$ is a $p \times q_i$ matrix of rank q_i, and the $\mathbf{Z}_i$ are independent. Rao [129] has shown that when $\mathbf{Y}_i$ and $\mathbf{Z}_j$ are one-dimensional variables and no $\mathbf{B}_i$ is a scalar multiple of any $\mathbf{A}_j$, $\mathbf{X}$ has a multivariate normal distribution. We extend this result to the general case where $\mathbf{Y}_i$ and $\mathbf{Z}_j$ are possibly vector variables as defined in (A.5.1) and (A.5.2) respectively.

THEOREM A.5.1. Let $\mathbf{X}$ admit two alternative representations as given by (A.5.1) and (A.5.2). If

$$R(\mathbf{A}_i : \mathbf{B}_j) = p_i + q_j \quad \text{for all } i \text{ and } j \tag{A.5.3}$$

then $\mathbf{X}$ has a multivariate normal distribution.

***Proof*:** Let ϕ_i be the second characteristic of $\mathbf{Y}_i$ and ψ_j that of $\mathbf{Z}_j$. Then, as usual (cf. Rao [129]), we obtain the equation

$$\phi_1(\mathbf{A}_1'\mathbf{t}) + \cdots + \phi_m(\mathbf{A}_m'\mathbf{t}) - \psi_1(\mathbf{B}_1'\mathbf{t}) - \cdots - \psi_n(\mathbf{B}_n'\mathbf{t}) = 0 \tag{A.5.4}$$

in a neighborhood of the origin in R^p. Under the condition (A.5.3), applying Theorem A.4.2 to solve (A.5.4), we find that $\sum \phi_i(\mathbf{A}_i'\mathbf{t})$ is a polynomial in $\mathbf{t}$. Then $\sum \mathbf{A}_i\mathbf{Y}_i = \mathbf{X}$ has a multivariate normal distribution. Thus the theorem is proved. ∎

Theorems characterizing $\mathbf{X}$ as the sum of a multivariate nonnormal and a normal random variable, similar to those in Rao [129], [175] for the univariate case, can also be established.

A generalization of Theorem 10.3.11 on conditional distributions can also be obtained, providing an extension of Theorem 13.4.1 to random vectors.

A.5.2 Independence of Sets of Linear Functions

Let $\mathbf{X}_1, \ldots, \mathbf{X}_n$ be independent vector random variables, where $\mathbf{X}_i$ is a p_i-vector variable. Consider q sets of linear functions

$$\mathbf{Y}_i = \mathbf{A}_{i1}\mathbf{X}_1 + \cdots + \mathbf{A}_{in}\mathbf{X}_n, \qquad i = 1, \ldots, q \tag{A.5.5}$$

where $\mathbf{Y}_i$ is a k_i-vector. We shall characterize the distributions of the $\mathbf{X}_i$ under the condition that $\mathbf{Y}_1, \ldots, \mathbf{Y}_q$ are independent.

THEOREM A.5.2. Let $\mathbf{Y}_i$ and $\mathbf{X}_j$ be as defined in (A.5.5). Denote $\mathbf{A}'_j = (\mathbf{A}'_{1j} \vdots \cdots \vdots \mathbf{A}'_{qj})$ and by $\mathbf{A}'_{j(i)}$ the matrix obtained from $\mathbf{A}'_j$ by deleting the i-th partition. If $R(\mathbf{A}_{j(i)}) = p_j$ for all i and j, then $\mathbf{X}_i$ has a p_i-variate normal distribution.

***Proof*:** Let ϕ_i be the second characteristic of $\mathbf{X}_i$. Then the condition of independence of the $\mathbf{Y}_j$ gives

$$\sum_1^n \phi_j(\mathbf{A}'_{1j}\,\mathbf{t}_1 + \cdots + \mathbf{A}'_{qj}\,\mathbf{t}_q) = C_1(\mathbf{t}_1) + \cdots + C_q(\mathbf{t}_q)$$

in the vectors $\mathbf{t}_j$ in neighborhoods of the origin in respective spaces. Considering $\mathbf{t}_1$ and $\mathbf{u}' = (\mathbf{t}'_2 \vdots \ldots \vdots \mathbf{t}'_q)$ as two vector variables, we have the equation

$$\sum_1^n \phi_j(\mathbf{A}'_{1j}\,\mathbf{t}_1 + \mathbf{A}'_{j(i)}\mathbf{u}) = C_1(\mathbf{t}_1) + D_1(\mathbf{u}).$$

Applying result (ii) of Theorem A.4.1, we find that $C_1(\mathbf{t}_1)$ is a polynomial, implying that $\mathbf{Y}_1$ is multivariate normal. Similarly, $\mathbf{Y}_2, \ldots, \mathbf{Y}_q$ are all multivariate normal, and they are also independently distributed. Then $\sum \mathbf{m}'_i\mathbf{Y}_i$ is univariate normal, for arbitrary choice of the vectors $\mathbf{m}_i$. But

$$\mathbf{m}'_1\mathbf{Y}_1 + \cdots + \mathbf{m}'_q\mathbf{Y}_q = (\sum \mathbf{m}'_i\mathbf{A}_{i1})\mathbf{X}_1 + \cdots + (\sum \mathbf{m}'_i\mathbf{A}_{in})\mathbf{X}_n$$

which implies that $(\sum \mathbf{m}'_i\mathbf{A}_{ij})\mathbf{X}_j$ is univariate normal for arbitrary choice of vectors $(\mathbf{m}'_1 \vdots \ldots \vdots \mathbf{m}'_n)$. Since $\mathbf{A}_j$ has rank p_j, it follows that $\mathbf{h}'\mathbf{X}_j$ is univariate normal for any choice of the vector $\mathbf{h}$. Hence $\mathbf{X}_j$ is p_j-variate normal. ∎

Theorem A.5.2 is a generalization of a result due to Ghurye and Olkin [19] in two directions, involving as it does more than two sets of linear functions and different dimensions for the basic variables.

It may be seen that when $q = 2$, the minimum number of linear functions needed for the result of Theorem A.5.2 to be true is $2p_0$ where p_0 is the maximum among $p_1, \ldots, p_n$, the respective dimensions of the variables $\mathbf{X}_1, \ldots, \mathbf{X}_n$. Further, the number of linear functions in each set is p_0. When $q = p_0 + 1$, the minimum number of linear functions is $p_0 + 1$, with one function in each set. For a general q, the minimum number is $qp_0/(q-1)$ if it is an integer, otherwise the next higher integer. The number of linear functions in each set is $p_0/(q-1)$ if it is an integer, otherwise the integral part thereof in some sets and the next higher integer in others.

A.5.3 Regression Problems

Let $\mathbf{X}_1, \ldots, \mathbf{X}_n$ be independent random variables where $\mathbf{X}_i$ may be a p_i-vector variable. Consider the linear functions

$$\mathbf{Y} = \mathbf{A}_1\mathbf{X}_1 + \cdots + \mathbf{A}_n\mathbf{X}_n \quad \text{and} \quad \mathbf{Z} = \mathbf{B}_1\mathbf{X}_1 + \cdots + \mathbf{B}_n\mathbf{X}_n \tag{A.5.6}$$

where $\mathbf{Y}$ and $\mathbf{Z}$ are k_1- and k_2-vectors respectively. We shall characterize the distribution of the $\mathbf{X}_i$ under the condition

$$E(\mathbf{Y}|\mathbf{Z}) = \mathbf{C} \tag{A.5.7}$$

where $\mathbf{C}$ is a constant k_1-vector.

THEOREM A.5.3. Let $\mathbf{X}_1, \ldots, \mathbf{X}_n$, $\mathbf{Y}$, and $\mathbf{Z}$ be as defined in (A.5.6) and satisfy (A.5.7). If $R(\mathbf{A}_i) = p_i = R(\mathbf{B}_i)$ and $R(\mathbf{B}_i \vdots \mathbf{B}_j) = p_i + p_j$ for all $i \neq j$, then $\mathbf{X}_i$ has a p_i-vector normal distribution.

***Proof*:** Let $g_{sj}(\mathbf{t})$ denote the partial derivative of $\phi_j(\mathbf{t})$, the second characteristic of $\mathbf{X}_j$, with respect to the s-th element of $\mathbf{t}$, and let $\mathbf{g}_j(\mathbf{t}) = (\mathbf{g}_{1j}(\mathbf{t}), \ldots, g_{pj}(\mathbf{t}))'$. Then, (A.5.7) gives

$$\mathbf{A}_1 \mathbf{g}_1(\mathbf{B}_1' \mathbf{t}) + \cdots + \mathbf{A}_n \mathbf{g}_n(\mathbf{B}_n' \mathbf{t}) = \mathbf{C} \tag{A.5.8}$$

in a neighborhood of the origin. Using Corollary A.4.3 and arguing on lines similar to those of Khatri [169], we find that $\mathbf{g}_j(\mathbf{w})$ is a vector of polynomials in $\mathbf{w}$ and hence $\phi_j(\mathbf{w})$ is a polynomial in $\mathbf{w}$. This shows that $\mathbf{X}_j$ has a multivariate normal distribution. ∎

A.6 CHARACTERIZATION OF PROBABILITY LAWS THROUGH PROPERTIES OF LINEAR FUNCTIONS

Let $\mathbf{X}_1, \ldots, \mathbf{X}_n$ be independent vector random variables, with $\mathbf{X}_i$ of dimension p_i, and consider an m-vector $\mathbf{Z}$ defined by

$$\mathbf{Z} = \mathbf{C}_1\mathbf{X}_1 + \cdots + \mathbf{C}_n\mathbf{X}_n. \tag{A.6.1}$$

We examine the extent to which the distribution of the $\mathbf{X}_i$ can be specified knowing the distribution of $\mathbf{Z}$, under the assumption that the characteristic function of $\mathbf{Z}$ is nonvanishing.

THEOREM A.6.1. Let $\mathbf{X}_1, \ldots, \mathbf{X}_n$ be as defined above and let $p_{(1)}$ and $p_{(2)}$ denote the two largest values (these may be equal) among the p_i. Further, let $\mathbf{C}_i$ be an $m \times p_i$ matrix such that $m \geqq p_{(1)} + p_{(2)}$, $R(\mathbf{C}_i) = p_i$ and $R(\mathbf{C}_i \vdots \mathbf{C}_j) = R(\mathbf{C}_i) + R(\mathbf{C}_j)$ for $i \neq j$. Then the c.f. of $\mathbf{X}_i$ is determined up to a factor of the form $\exp P_{n-2}$, where P_{n-2} is a polynomial of degree at most $(n-2)$.

***Proof*:** Let ϕ_i be the difference in the second characteristics of $\mathbf{X}_i$ under two possible alternative distributions. From (A.6.1) we have the relation

$$\phi_1(\mathbf{C}_1' \mathbf{t}) + \cdots + \phi_n(\mathbf{C}_n' \mathbf{t}) = 0. \tag{A.6.2}$$

Then the result of Theorem A.6.1 follows by an application of Corollary A.4.3. ∎

Theorem A.6.2 below provides the type of statement which can be made in the general case when $\mathbf{Z}$ in (A.6.1) is an m-vector.

THEOREM A.6.2. Let the $\mathbf{X}_i$ be as before and let $r \leqq n-2$ be such that

$$R((\mathbf{C}\odot)^{r-1}\mathbf{C}) < \sum p_i^r \qquad \text{and} \qquad R((\mathbf{C}\odot)^r\mathbf{C}) = \sum p_i^{r+1} \tag{A.6.3}$$

where $\mathbf{C} = (\mathbf{C}_1 \vdots \mathbf{C}_2 \vdots \cdots \vdots \mathbf{C}_n)$; then, the c.f. of $\mathbf{X}_i$ is determined up to a factor of the form $\exp P_r$.

***Proof*:** The result above follows by an application of Theorem A.4.3. ∎

In particular, if $R(\mathbf{C} \odot \mathbf{C}) = \sum p_i^2$, then the distribution of $\mathbf{X}_i$ is determined up to a change of location. If $\mathbf{C}$ is an $m \times (p_1 + \cdots + p_n)$ matrix, it is of interest to know the smallest possible value of m for which $R(\mathbf{C} \odot \mathbf{C}) = \sum p_i^2$. When $p_i = 1$ for all i, it is shown (Corollary A.3.2) that the smallest m is such that

$$m(m-1)/2 < n \leqq m(m+1)/2 \tag{A.6.4}$$

so that the joint distribution of m suitably chosen linear functions of $m(m+1)/2$ independent random variables can determine the distribution of each variable upto a change of location. In the case of general p_i, the formula for minimum m is likely to be

$$m(m-1) < \sum_{i=1}^{n} p_i(p_i + 1) \leqq m(m+1) \tag{A.6.5}$$

which has been verified to be true in special cases.

ADDENDUM B

Characterization of Distributions Through Properties of Bayes Estimators

B.1 INTRODUCTION

We present here some new results, closely related to those of Chapter 7, on characterization of distributions. We again consider (as we did there) the problem of estimation for the parameter $\theta \in R^1$ in the set-up of direct measurements:

$$X_i = \theta + \varepsilon_i, \qquad i = 1, \ldots, n \tag{B.1.1}$$

where the ε_i are independent random variables (r.v.). We define the risk incurred in using $\tilde{\theta} = \tilde{\theta}(X_1, \ldots, X_n)$ as an estimator of θ, as the number

$$\rho(\tilde{\theta}) = \int_{-\infty}^{\infty} R(\tilde{\theta}, \theta)\, d\Pi(\theta) \tag{B.1.2}$$

where Π is a probability measure on R^1 and $R(\tilde{\theta}, \theta)$ a loss function. We restrict ourselves to the case $R(\tilde{\theta}, \theta) = E_\theta(\tilde{\theta} - \theta)^2$, though a number of results given below hold in more general situations. The choice of the expression (B.1.2) as the risk associated with the choice of $\tilde{\theta}$ may be interpreted as considering θ as an r.v. independent of the vector $(\varepsilon_1, \ldots, \varepsilon_n)$ of errors, with *a priori* distribution Π. In this case, estimating θ is predicting its value on the basis of the observations X_i.

It is well known that in case $\int \theta^2\, d\,\Pi(\theta) < \infty$, the *a posteriori* mean

$$\hat{\theta} = E(\theta \mid X_1, \ldots, X_n) = \frac{\int \theta \prod_1 f_j(x_j - \theta)\, d\Pi(\theta)}{\int \prod_1^n f_j(x_j - \theta)\, d\Pi(\theta)} \tag{B.1.3}$$

is the best estimator of θ, where f_j is the probability density function of ε_j. To construct the Bayes estimator (B.1.3), it is necessary to know the *a priori* distribution and the distributions of the errors. It is natural to pose the question: when is it possible to construct (B.1.3) with the help of the first two moments of the F_j and of Π alone? Analytically this reduces to the investigation of the relation

$$E(\theta \mid X_1, \ldots, X_n) = \hat{E}(\theta \mid X_1, \ldots, X_n) \tag{B.1.4}$$

where the conditional expectation in the wide sense, on the right-hand side of (B.1.4), is simply the best *linear* estimator of θ. In this addendum, we study the relation (B.1.4) and a number of its variants. Many of the results below are analogues to the results of Chapter 7 (non-Bayesian), but there are exceptions (see, for example, Theorem B.2.3).

We shall use the following notation.

F_j is the d.f. of ε_j; Π is the d.f. of θ (the *a priori* d.f.); f_j and π are the characteristic functions of F_j and Π respectively, $\mu_1^{(j)}$ and α_1 their first moments, and, for $k \geqq 2$, $\mu_k^{(j)}$ and α_k their k-th central moments. If the ε_j are identically distributed, we shall omit the index j.

If, for an integer $k \geqq 1$, the $(2k)$-th moments of all the F_j and of Π exist finitely, we shall denote by M_k (respectively by $M_k(\bar{X})$) the linear space of all polynomials in $X_1, \ldots, X_n$ (respectively in $\bar{X}$) of degree k, endowed with the scalar product

$$(Q, R) = E(QR) = \int E_\theta(QR)\, d\Pi(\theta)$$

where

$$E_\theta(QR) = \int_{R^n} Q(x_1, \ldots, x_n) R(x_1, \ldots, x_n) \prod_1^n dF_j(x_j - \theta).$$

We denote the projection operator into M_k by $\hat{E}(. \mid M_k)$.

B.2 CONDITION OF LINEARITY OF THE BAYES ESTIMATOR

Suppose that

$$0 < \alpha_2 < \infty, \qquad 0 < \mu_2^{(j)} < \infty, \qquad j = 1, \ldots, n. \tag{B.2.1}$$

Then the best linear estimator of θ is

$$\hat{E}(\theta \mid M_1) = \hat{\theta}_1 = c_0 + \sum_1^n c_j x_j \tag{B.2.2}$$

where

$$\left.\begin{aligned} c_0 &= \frac{(\alpha_1 - \sum_{=j1}^n (\alpha_2 \mu_1^j / \mu_2^{(j)}))}{(1 + \sum_{j=1}^n (\alpha_2/\mu_2^{(j)}))} \\ c_i &= \frac{(\alpha_2/\mu_2^{(i)})}{(1 + \sum_{j=1}^n (\alpha_2/\mu_2^{(j)}))}, i = 1, \ldots, n. \end{aligned}\right\} \tag{B.2.3}$$

THEOREM B.2.1. (Kagan and Karpov [178]). Suppose $n \geqq 2$, and (B.2.1) is satisfied. Then the relation

$$E(\theta \mid X_1, \ldots, X_n) = \hat{E}(\theta \mid M_1) \tag{B.2.4}$$

holds iff all the F_j and Π are normal distributions.

***Proof*:** The necessity is proved in the same way as in the case of Theorem 7.4.1. The sufficiency follows from the fact that if the vector $(\theta, X_1, \ldots, X_n)$ has a normal distribution, then the best predictor for θ on the basis of $X_1, \ldots, X_n$ is linear. ▮

Hence, if in the set-up (B.1.1) one of the F_j and/or Π is nonnormal, then the estimator (B.2.2) can be improved upon if the statistician knows the functions F_j, Π. In section B.3, we shall see that this improvement can be achieved if the statistician knows only a certain number of high order moments of the F_j and of Π.

Here we investigate the possibility of improving on (B.2.2) in the case of identically distributed ε_j (in this case, $\hat{\theta}_1 = c_0 + c_1 \sum_1^n X_j$) if the statistician knows the values of $\bar{X}$ and of one other linear statistic $l = \sum_1^n b_j X_j$. Let $\bar{b} = (\sum b_j)/n$, $\delta_j = b_j - \bar{b}$ and suppose that among the numbers δ_j, there is exactly one having the maximum modulus. Without loss of generality, we may then suppose that

$$|\delta_1| > |\delta_j|, \qquad j = 2, \ldots, n. \tag{B.2.5}$$

THEOREM B.2.2. ([178]). Suppose the r.v.'s ε_j are identically distributed. Then the relation

$$E\left(\theta \middle| \sum_1^n X_j, l\right) = c_0 + c_1 \sum_1^n x_j, \tag{B.2.6}$$

where the statistic l satisfies the condition (B.2.5) and the common d.f. F of the ε_j satisfies the conditions of Theorem 7.5.1, holds iff F and Π are normal d.f.'s.
***Proof*:** We may suppose $\alpha_1 = 0$. Introducing the functions $\phi = \pi'/\pi$ and $h = f'/f$, we deduce from (B.2.6)

$$\phi(nt + n\bar{b}\,\tau) = A \sum_1^n h(t + b_k \tau) \tag{B.2.7}$$

where A is a nonzero constant, and t and τ are small. Letting $t = -\bar{b}\tau$ in (B.2.7), we have

$$\sum_1^n h(\delta_j \tau) = 0 \tag{B.2.8}$$

where the δ_j satisfy (B.2.5), for small τ. This relation is investigated in the same way as (7.5.2) to obtain the normality of the ε_j and hence, in view of (B.2.7), that of θ. ▮

Note: If, in addition to (B.2.5), we suppose

$$\delta_1 \delta_j \leqq 0 \qquad \text{for } j = 2, \ldots, n \tag{B.2.9}$$

then Theorem B.2.2 remains true under the minimal condition on the ε_j, namely $\mu_2 < \infty$. The proof in this case is based on Lemma 1.5.10.

Consider now, in the case of identically distributed ε_j, the situation where the statistician knows only the value of $\sum_1^n X_j$. We shall show that if at least one of F, Π is nonnormal, then, for all values of n (with the possible exception of one value), the best linear estimator (B.2.2) can be improved upon on the basis of $\sum_1^n X_j$ alone.

THEOREM B.2.3. Let the ε_j be identically distributed and α_2, μ_2 be finite. Then the relation

$$E(\theta \,|\, \bar{X}) = \hat{E}(\theta \,|\, \bar{X}) \tag{B.2.10}$$

holds for two values of $n: n_1 < n_2$, iff both F and Π are normal.

***Proof*:** *Necessity*. We may again suppose $\alpha_1 = \mu_1 = 0$. The relation (B.2.10). for a given n, has the form

$$E(\theta \mid \bar{X}) = c_n \bar{X}, \; c_n = n\alpha_2/(n\alpha_2 + \mu_2)$$

whence we obtain

$$(1 - c_n)\pi'(nt)[f(t)]^n = n\pi(nt)f'(t)[f(t)]^{n-1}.$$

If $|nt| < \varepsilon$, ε small, then we obtain hence $\phi(\tau) = a_n h(\tau/n)$ where $a_n = c_n/(1 - c_n)$, for small τ. Making use of this relation for $n = n_1$, $n = n_2$, we obtain

$$h(t) = (n_2/n_1)h(n_1 t/n_2) \qquad \text{for small } t. \tag{B.2.11}$$

Since μ_2 is finite, $h(t) = tg(t)$, where g is continuous at the origin and satisfies the relation $g(t) = g(n_1 t/n_2)$ for small t. We conclude that $g(t) = g(0) = -\mu_2$ for small t. Thus, F and hence Π are normal.

The sufficiency part is trivial. ▮

B.3 CONDITIONS FOR THE LINEARITY OF THE POLYNOMIAL BAYES ESTIMATOR

In this section, we shall assume that

$$\alpha_{2k} < \infty, \qquad \mu_{2k}^{(j)} < \infty, \qquad j = 1, \ldots, n \tag{B.3.1}$$

for some integer k. In this case, it is natural to use as an estimator of θ the statistic

$$\hat{\theta}_k = \hat{E}(\theta \mid M_k)$$

based only on the moments of the first $2k$ orders of the F_j and of Π (not on the distributions themselves).

For a fixed value of k, $\hat{\theta}_k$ is the best polynomial estimator of θ.

THEOREM B.3.1. Under the condition (B.3.1), the relation

$$\hat{E}(\theta \mid M_k) = \hat{E}(\theta \mid M_1)$$

holds iff $\alpha_1, \ldots, \alpha_{k+1}$ and $\mu_1^{(j)}, \ldots, \mu_{k+1}^{(j)}$ coincide with the corresponding moments of some normal distributions, i.e., $\alpha_r = (r-1)!!\sigma^r$ and $\mu_r^{(j)} = (r-1)!!\sigma_j^r$ for even r and zero for odd r, $1 \leqq r \leqq k+1$.

The proof is quite similar to that of Theorem 7.5.2.

Thus, if at least one of the first $(k+1)$ moments of the *a priori* distribution Π and/or of the distributions F_j differs from the moment of the same order of a suitable normal distribution, the estimator θ_1 can be improved upon by taking θ_k based only on the high order moments of Π and the F_j.

The next result has no non-Bayesian analogue and concerns the possibility of improving on the estimator θ_1 within the class of polynomials in $\bar{X}$.

THEOREM B.3.2. Let the ε_j be identically distributed and α_{2k} and μ_{2k} be finite. The relation

$$\hat{E}(\theta \mid M_k(\bar{X})) = \hat{E}(\theta \mid \bar{X}) \tag{B.3.2}$$

holds for an infinite sequence of values of $n: n_1 < n_2 < \cdots$ iff the first $(k+1)$ moments of Π and of F coincide with the corresponding moments of some normal distributions.

***Proof*:** We assume that $\alpha_1 = \mu_1 = 0$. (B.3.2) is equivalent to the relation

$$E\left[\left(\theta - \frac{n\alpha_2}{n\alpha_2 + \mu_2}\bar{X}\right)\bar{X}^r\right] = 0, \qquad r = 0, 1, \ldots, k.$$

Passing to the limit as $n_k \to \infty$ in the above relation, we can determine the successive moments $\alpha_3, \ldots, \alpha_{k+1}$ and $\mu_3, \ldots, \mu_{k+1}$ in terms of the first two moments of Π and of F. We omit the details. Thus, if at least one of the first $(k+1)$ moments of F and/or of Π differs from the corresponding moment of a suitable normal distribution, the estimator $\hat{\theta}_1$ can be improved upon for all values of n, with the possible exception of a finite number of them, if the statistician knows only the moments α_r and μ_r, $1 \leqq r \leqq 2k$, and the value of the statistic $\hat{\theta}_1$ without knowing the individual observations X_j.

B.4 ε-BAYESIAN CHARACTER OF $\hat{\theta}_1$

We call an estimator $T = T(X_1, \ldots, X_n)$ ε-Bayesian if

$$E(T - \theta)^2 \leqq E(\hat{\theta} - \theta)^2 + \varepsilon.$$

THEOREM B.4.1. Let $\varepsilon_1, \ldots, \varepsilon_n$, $n \geqq 3$, be identically and symmetrically distributed and α_2 and μ_2 be finite. If $\hat{\theta}_1$ is a $[n\alpha_2^2\mu_2\varepsilon^2/(n\alpha_2 + \mu_2)^2]$-Bayesian estimator for θ, then

$$\sup_x |F(x) - \Phi_{12}(x)| < C[\log(1/\varepsilon)]^{-1/2}$$

where Φ_{12} is the d.f. of the normal law $N(\mu_1, \mu_2)$ and C is an absolute constant.

The proof is similar to that of Theorem 9.3.1. We have considered $n\alpha_2^2\mu_2\varepsilon^2/(n\alpha_2 + \mu_2)^2$ instead of ε to obtain an absolute bound for the closeness of F to $\Phi_{\mu 1, \mu 2}$.

BIBLIOGRAPHY

[1] Anosov, D. V. "On an integral equation arising in statistics," *Vestnik Leningrad. Univ.* **7** (1964), 151–154.

[2] Akhiezer, N. I. *The classical moment problem.* (1965), Hafner, New York.

[3] Aczel, J. *Lectures on Functional Equations and their Applications.* (1966) Academic Press, New York and London.

[4] Bahadur, R. R. "A characterization of sufficiency," *Ann. Math. Statist.* **26** (1955), 286–293.

[5] Bernstein, I. N. "The possibility of analytic continuation for certain functions," *Functional Analysis and its Applications* **2**, (1968), 92–93.

[6] Bernstein, S. N. "On a property characteristic of the normal law," *Trudy Leningrad. Polytech. Inst.* **3** (1941), 21–22 (*Collected Works* **4** (1964), 314–315).

[7] Blackwell, D. "Conditional expectation and unbiased sequential estimation," *Ann. Math. Statist.* **18**, (1947), 105–110.

[8] Van der Waerden, B. L. *Algebra*, Vol. II, 5th edition. (1970), Frederick Ungar, New York.

[9] Weber, H. *Lehrbuch der Algebra*, Vol. I, 3rd edition. (1912), Chelsea, New York.

[10] Wolfowitz, J. "The efficiency of sequential estimates and Wald's equation for sequential procedures," *Ann. Math. Statist.* **18** (1947), 215–230.

[11] Hajek, J. "On a property of normal distributions of an arbitrary stochastic process," *Czechoslovak. Math. Journal* **8** (1958), 610–618.

[12] Giri, N. "On the complex analogues of the T^2 and R^2 tests," *Ann. Math. Statist.* **35**, (1964), 664–670.

[13] Girshick, M. A., Mosteller, F. and Savage, L. J. "Unbiased estimation for certain binomial sampling," *Ann. Math. Statist.* **17** (1946), 13–23.

[14] Govindarajulu, Z. "Characterization of the exponential distribution and power-series distributions," *Skand. Aktuarie-tidskrift* **49** (1966), 132–136.

[15] Gradstein, I. C. and Ryzhik, I. M. *Tables of Integrals, Sums, Series and Products.* (1957), Martin Strauss, Berlin.

[16] Grenander, U. *Probability on Algebraic Structures.* (1963), John Wiley, New York.

[17] Goodman, J. "Statistical analysis based on a certain multivariate complex Gaussian distribution," *Ann. Math. Statist.* **34**, (1963), 152–177.

[18] Ghurye, S. G. "Characterization of some location- and scale- parameter families of distributions," *Contributions to Probability and Statistics: Essays in Honour of H. Hotelling.* (1960), Stanford University Press, Stanford, Calif., 203–205.

[19] Ghurye, S. G. and Olkin, I. "A characterization of the multivariate normal distribution," *Ann. Math. Statist.* **33** (1962), 533–541.

[20] Ghosh, J. K. "Only linear transformations preserve normality," *Sankhya, Ser-A* **31** (1969), 309–312.

[21] Ghosh, J. K. and Rao, C. R. "A note on some translation parameter families of densities for which the median is an m.l.e," *Sankhya,* **A, 33,** (1971), 91–93.

[22] Darmois, G. "Analyse générale des liaisons stochastiques," *Rev. Inst. Internationale Statist.* **21** (1953), 2–8.

[23] De Groot, M. "Unbiased binomial sequential estimation," *Ann. Math. Statist.* **30** 1(1959), 80–101.

[24] Doetsch, G. *Theorie und Anwendung der Laplace-Transformation.* (1937), Springer, Berlin.

[25] Doob, J. L. *Stochastic Processes.* (1953), John Wiley, New York.

[26] Dynkin, E. B. "Necessary and sufficient statistics for families of probability distributions," *Uspekhi Matem. Nauk* **VI** (1951), 66.

[27] Dynkin, E. B. *Markov Processes.* (1961), Pergamon Press, New York.

[28] Zaidman, R. A., Linnik, Yu, V. and Romanovskii, I. V. "Plans of sequential estimation and Markovian stopping times," *DAN SSSR* **185** (1969), 1222–1225.

[29] Zinger, A. A. "On independent samples from a normal population," *Uspekhi Matem. Nauk* **VI** (1951), 172.

[30] Zinger, A. A. "On a problem of Kolmogorov's," *Vestnik Leningrad. Univ.* **1** (1956), 53–56.

[31] Zinger, A. A. "Independence of quasi-polynomial statistics and analytical properties of distributions," *Teoriia Veroiatn. Prim.* **III** (1958), 265–284.

[32] Zinger, A. A. "On the distribution of polynomial statistics in samples from a normal population," *DAN SSSR* **149** (1963), 20–21.

[33] Zinger, A. A. "On the distribution of polynomial statistics in samples from normal populations and populations related to them," *Trudy Matem. Inst. Steklov. AN SSSR* **79** (1965), 150–159.

[34] Zinger, A. A. "On a class of limit distributions for normed sums of independent random variables," *Teoriia Veroiatn. Prim.* **X,** (1965), 672–692.

[35] Zinger, A. A. "On a problem of Gnedenko's," *DAN SSSR* **162** (1965), 1238–1240.

[36] Zinger, A. A. "On an extension of the class of stable distributions," *DAN SSSR* **173** (1967), 1255–1256.

[37] Zinger, A. A. "Investigations into Analytical Statistics and Their Applications to Limit Theorems of Probability Theory," Doctoral Dissertation, Leningrad University, 1969.

[38] Zinger, A. A. and Linnik, Yu. V. "On an analytical generalization of a theorem of H. Cramer's," *Vestnik Leningrad. Univ.* **11** (1955), 51–56.

[39] Zinger, A. A. and Linnik, Yu. V. "On a class of differential equations and its applications to some problems of regression theory," *Vestnik Leningrad. Univ.* **7** (1957), 121–130.

[40] Zinger, A. A. and Linnik, Yu. V. "On characterizations of the normal distributions," *Teoriia Veroiatn. Prim.* **IX** (1964), 692–695.

[41] Zinger, A. A. and Linnik, Yu. V. "Polynomial statistics of a normal sample," *DAN SSSR* **176** (1967), 766–767.

[42] Zinger, A. A., Kagan, A. M., and Klebanov, L. B. "The sample mean as an estimator for a shift parameter under loss functions other than quadratic," *DAN SSSR* **189** (1969), 29–30.

[43] Zinger, A. A. and Linnik, Yu. V. "Non-linear statistics and random linear-forms," *Trudy Matem. Inst. Steklov. AN SSSR* **111** (1970), 23–39.

[44] Zolotarev, V. M. "The Mellin-Stieltjes transform in probability theory," *Teoriia Veroiatn. Prim.* **II** (1957), 444–469.

[45] Zolotarev, V. M. "On the problem of stability for the decomposition of a normal law into (independent) components," *Teoriia Veroiatn. Prim.* **XIII**, (1968), 738–742.

[46] Ibragimov, I. A. and Linnik, Yu. V. *Independent and Stationarily-Connected Random Variables.* Nauka, Moscow. (1965).

[47] Eaton, M. L. "Characterization of distributions by the identical distribution of linear forms," *J. Appl. Prob.* **3** (1966), 481–494.

[48] Joshi, V. M. "Admissibility of confidence intervals," *Ann. Math. Statist.* **37** (1966), 629–638.

[49] Kawata, T. and Sakamoto, H. "On the characterization of the independence of the sample mean and the sample variance," *J. Math. Soc. Japan.* **I** (1949), 111–115.

[50] Kagan, A. M. "On the estimation theory of location parameters," *Sankhya, Ser. A* **28** (1966), 335–352.

[51] Kagan, A. M. "Partial sufficiency and unbiased estimation of polynomials in a shift paramter," *DAN SSSR* **174** (1967), 1257–1259.

[52] Kagan, A. M. and Shalaevskii, O. V. "Characterization of the normal law by the property of partial sufficiency," *Teoriia Veroiatn. Prim.* **XII** (1967) 567–569.

[53] Kagan, A. M. "Theory of estimation for families with shift-, scale- and exponential parameters," *Trudy Matem. Inst. Steklov. AN SSSR* **104** (1968), 19–87.

[54] Kagan, A. M. "On ε-admissibility of the sample mean as an estimator of location parameters," *Sankhya, Ser. A* **32** (1970), 37–40.

[55] Kagan, A. M. and Linnik, Yu. V. "A class of families of distributions admitting similar zones," *Vestnik Leningrad. Univ.* **13** (1964), 25–36.

[56] Kagan, A. M., Linnik, Yu. V. and Rao, C. R. "On a characterization of the normal law based on a property of the sample average," *Sankhya, Ser. A* **27** (1965), 405–406.

[57] Kagan, A. M. and Rukhin, A. L. "On the theory of estimation for a scale parameter," *Teoriia Veroiatn. Prim.* **XII** (1967), 735–741.

[58] Kagan, A. M. and Shalaevskii, O. V. "Characterization of the normal law by a property of the non-central chi-square distribution," *Litovskii Matem. Sbornik* **VII** (1967), 57–58.

[59] Kagan, A. M. and Shalaeveskii, O. V. "Admissibility of the least squares estimators is the exclusive property of the normal law," *Matem Zametki* **6** (1969), 81–89.

[60] Kagan, A. M. and Zinger, A. A. "Sample mean as an estimator of location parameter in the case of non-quadratic loss functions," *Sankhya, Ser. A* **33** (1971), 351–358.

[61] Kagan, F. M. "An information-theoretic property of the gamma distribution," *Izvestiia AN Uzb. SSR. Ser. Phys.-Mat.* **5** (1967), 67–68.

[62] Kagan, F. M. "Some theorems characterizing gamma distributions and distributions close to them," *Litovskii Matem. Sbornik* **VIII** (1968), 265.

[63] Kagan, F. M. "The condition of optimality of certain estimators for families with scale parameters," *DAN Uzb. SSR* **6** (1968), 3–5.

[64] Kakutani, S. "On the equivalence of infinite product measures," *Ann. Math.* **49** (1948), 214–224.

[65] Kamalov, M. K. *Distribution of Quadratic Forms in Samples from a Normal Population.* (1958). Akad. Nauk Uzbek. SSR Publication, Tashkent.

[66] Kamke, E. *The Lebesgue-Stieltjes Integral.* (1959), Fizmatgiz.

[67] Kovalenko, I. N. "On the reconstruction of an additive type of distribution based on a sequence of independent trials," *Memoirs of the All-Union Conference on Probability Theory and Mathematical Statistics.* (1958), Erevan, Publication of the AN Armenian SSR, Erevan, 1960.

[68] Kovalenko, I. N. "On a statistical characterization of the symmetric stable laws," *DAN SSSR* **170** (1966), 31–33.

[69] Kolmogorov, A. N. "Unbiased estimators," *Izvestiia AN SSSR Ser. Matem.* **14** (1950), 303–326.

[70] Cramer, H. *Random Variables and Probability Distributions*, Cambridge Tracts, Second Edition. (1962), Cambridge University Press, England.

[71] Cramer, H. *Mathematical Methods of Statistics.* (1946), Princeton University Press, Princeton, N.J.

[72] Khatri, C. G. "Classical statistical analysis based on a certain-multi-dimensional complex Gaussian distribution," *Ann. Math. Statist.* **36** (1965), 98–114.

[73] Khatri, C. G. and Rao, C. R. "Some characterizations of the gamma distribution," *Sankhya, Ser. A* **30** (1968), 157–166.

[74] Khatri, C. G. and Rao, C. R. "Solutions to some functional equations and their applications to characterizations of probability distributions," *Sankhya, Ser, A* **30** (1968), 167–180.

[75] Laha, R. G. "On the stochastic independence of a homogeneous quadratic statistic and the sample mean," *Vestnik Leningrad. Univ.* **1** (1956), 25–32.

[76] Levin, B. Ja. *The Distribution of the Zeros of Entire Functions*, GITTL, 1956. (English translation by R. P. Boas *et al.*, Amer. Math. Soc., Providence, R. I. 1964).

[77] Lehmann, E. *Testing of Statistical Hypotheses.* (1959), John Wiley, New York.

[78] Lehmann, E. and Stein, C. "Completeness in the sequential case," *Ann. Math. Statist.* **21** (1951), 376–385.

[79] Linnik, Yu. V. "Remarks on the classical derivation of Maxwell's law," *DAN SSSR* **85** (1952), 1251–1254.

[80] Linnik, Yu. V. "On some identically distributed statistics," *DAN SSSR* **89** (1953), 9–11.

[81] Linnik, Yu. V. "Linear forms and statistical criteria," *Ukrain. Mat. Zhurnal* **5** (1953), 207–243 and **5** (1953) 247–290.

[82] Linnik, Yu. V. "Independently and equally distributed statistics, and analytical problems connected with them." (1954), Mimeographed Notes, Calcutta.

[83] Linnik, Yu. V. "On polynomial statistics in connection with the analytical theory of differential equations," *Vestnik Leningrad. Univ.* **1** (1956), 35–48.

[84] Linnik, Yu. V. "On the problem of finding the parent distribution from that of a statistic," *Teoriia Veroiatn. Prim.* **I** (1956), 466–478.

[85] Linnik, Yu. V. *Decomposition of Probability Laws.* (1964), Oliver and Boyd, Edinburgh.

[86] Linnik, Yu. V. *Statistical Problems with Nuisance Parameters.* (1968), Amer. Math. Soc. Providence, R. I.

[87] Linnik, Yu. V. "On certain connections between algebraic geometry and statistics," *Australian J. Stat.* **9** (1967), 89–92.

[88] Linnik, Yu. V. "On an application of the theory of numbers to mathematical statistics," *Mat. Zametki* **7** (1970), 383–388.

[89] Linnik, Yu. V. and Eidlin, V. L. "On analytical transformations of normal vectors," *Teoriia Veroiatn. Prim.* **XIII** (1968), 751–754.

[90] Linnik, Yu. V., Ruhkin, A. L., and Strelits, Sh. I. "The gamma distribution and partial sufficiency of polynomials," *Trudy Matem. Inst. Steklov. AN SSSR* **111** (1970), 40–51.

[91] Loeve, M. *Probability Theory*, third edition. (1963), Van Nostrand, Princeton, N.J.

[92] Lukacs, E. "A characterization of the normal distribution," *Ann. Math. Statist.* **13** (1942), 91–93.

[93] Lukacs, E. "Recent developments in the theory of characteristic functions," *Proc. Fourth Berkeley Symp. Prob. Statist.* **2** (1956).

[94] Lukacs, E. "Characterization problems for discrete distributions," *Proc. Intern. Symp. Classical and Contagious Dist.* (1963), McGill Univ., Montreal, 65–74.

[95] Lukacs, E. and Laha, R. G. *Applications of Characteristic Functions.* (1964), Hafner Publishing Co., New York.

[96] Lang, S. *Introduction to Algebraic Geometry.* (1955), University of Chicago, Chicago.

[97] Maloshevskii, S. G. "The sharpness of N. A. Sapogov's estimate in the stability problem for Cramer's theorem," *Teoriia Veroiatn. Prim.* **XIII** (1968), 522–525.

[98] Malgrange, B. *Ideals of Differentiable Functions* (1966), Tata Inst. Fund. Res., Bombay and Oxford University Press.

[99] Mamai, L. V. "(Contribution) to the theory of characteristic functions," *Vestnik Leningrad. Univ.* **1** (1960), 85–99.

[100] Mandelbrojt, S. *Séries de Fourier et Classes Quasi-analytiques de Fonctions.* (1935), Gauthier-Villars, Paris.

[101] Marcinkiewicz, J. "Sur une propriete de la loi de Gauss," *Math. Zeitschrift* **44** (1938), 622–638.

[102] Machis, Yu. Yu. "On the stability of decomposition of a distribution law," *Teoriia Veroiatn. Prim.* **XIV** (1969), 715–718.

[103] Meshalkin, L. D. "On the robustness of some characterizations of the normal distribution," *Ann. Math. Statist.* **39** (1968), 1747–1750.

[104] Morozenskii, L. Yu. "Characterization of the normal law through the optimality of criteria based on the sample mean," *Vestnik Leningrad. Univ.* **13** (1971), 61–63.

[105] Netto, E. *Lehrbuch der Combinatorik*, New York, 1927.

[106] Nee Huang Hiu. "On the stability of some characteristic properties of the normal population," *Teoriia Veroiant. Prim.* **XIII** (1968), 308–314.

[107] Parenago, P. P. *A Course of Stellar Astronomy*, GTTI, 1946.

[108] Parthasarathy, K. R., Ranga Rao, R., and Varadhan, S. R. S. "On the category of indecomposable distributions on topological groups," *Trans. Amer. Math. Soc.* **102** (1962), 202–217.

[109] Patil, G. P. and Seshadri, V. "Characterization theorems for some univariate probability distributions," *J. Roy. Stat. Soc., Ser. B* **26** (1964), 286–292.

[110] Pathak, P. K. and Pillai, R. N. "On a characterization of the normal law," *Sankhya, Ser. A* **30** (1968), 141–144.

[111] Petrov, A. A. "Testing of statistical hypotheses on the type of a distribution through small samples," *Teoriia Veroiatn. Prim.* **I** (1956), 248–271.

[112] Pillai, R. N. "On some characterizations of the normal law," *Sankhya, Ser. A* **30** (1968), 145–146.

[113] Pitman, E. J. G. "The estimation of location- and scale- parameters of a continuous population of any given form," *Biometrika* **30** (1938), 390–421.

[114] Polya, G. "Herleitung des Gauss' schen Fehlergesetzes aus einer Funktionalgleichung," *Math. Zeitschrift* **18** (1923), 185–188.

[115] Polya, G. "Remarks on characteristic functions, *Proc. First Berkeley Symp. Prob. Statist.* (1949), 115.

[116] Polya, G. and Szego, G. *Problems and Theorems in Analysis.* (1972), Springer Berlin.

[117] Prokhorov, Yu. V. "Characterization of a class of distributions through the distribution of certain statistics," *Teoriia Veroiatn. Prim.* **X** (1965), 479–487.

[118] Raikov, D. A. "On the decomposition of the Poisson law," *DAN SSSR* **14** (1937), 8–11.

[119] Ramachandran, B. "On characteristic functions and moments", *Sankhya, Ser. A* **31** (1969), 1–12.

[120] Ramachandran, B. Private communication.

[121] Ramachandran, B. *Advanced Theory of Characteristic Functions.* (1967), Statistical Publishing Society, Calcutta,

[122] Ramachandran, B. and Rao, C. R. "Some results on characteristic functions and characterizations of the normal and generalized stable laws," *Sankhya, Ser. A* **30** (1968), 125–140.

[123] Ramachandran, B. and Rao, C. R. "Solutions of functional equations arising in some regression problems and a characterization of the Cauchy law," *Sankhya, Ser. A* **32** (1970), 1–30.

[124] Rao, C. R. "Information and accuracy attainable in the estimation of statistical parameters," *Bull. Calcutta Math. Soc.* **37** (1945), 81–91.

[125] Rao, C. R. "On a problem of Ragnar Frisch," *Econometrika* **15** (1947), 245–249; Correction, *Ibid.* **17** (1949), 212.

[126] Rao, C. R. "Some theorems on minimum variance estimation," *Sankhya, Ser. A* **12** (1952), 27–42.

[127] Rao, C. R. "Sur une caractérisation de la distribution normal établie d'après une propriété optimum des estimations linéaires," *Colloq. Intern. C.N.R.S., France* **87** (1959), 165–171.

[128] Rao, C. R. "Minimum variance and estimation of several parameters, *Proc. Camb. Phil. Soc.* **43** (1947), 280–283.

[129] Rao, C. R. "Characterization of the distribution of random variables in linear structural relations," *Sankhya, Ser. A* **28** (1966), 251–260.

[130] Rao, C. R. "On some characterizations of the normal law," *Sankhya, Ser. A* **29** (1967), 1–14,

[131] Rao, C. R. *Linear Statistical Inference and Its Applications*, John Wiley, New York, Second Edition, 1973.

[132] Rao, C. R. and Rubin, H. "On a characterization of the Poisson distribution," *Sankhya, Ser. A* **26** (1964), 295–298.

[133] Reiersøl, O. "Identifiability of a linear relation between variables which are subject to error," *Econometrika* **18** (1950), 375–389.

[134] Rukhin, A. L. "Strongly symmetric families," *DAN SSSR* **190** (1970), 280–282.

[135] Sakovich, G. H. *Multivariate Stable Distributions*, Candidacy Dissertation, Kiev, 1965.

[136] Sapogov, N. A. "The stability problem for Cramer's theorem," *Izvestiia AN SSSR Ser, Matem.* **15** (1951), 205–218.

[137] Sapogov, N. A. "On the independent components of a sum of random variables with distributions close to normal, *Vestnik Leningrad. Univ.* **19** (1959), 78–105.

[138] Skitovich, V. P. "On a property of the normal distribution," *DAN SSSR* **89** (1953), 217–219.

[139] Skitovich, V. P. "Linear forms in independent random variables and the normal distribution law," *Izvestiia AN SSSR, Ser. Matem.* **18** (1954), 185–200.

[140] Srivastava, R. S. and Srivastava, A. B. L. "On a characterization of the Poisson distribution," *J. Appl. Prob.* **7** (1970), 497–501.

[141] Stein, C. "The admissibility of Pitman's estimator for a single location parameter," *Ann. Math. Statist.* **30** (1959), 970–978.

[142] Strelits, Sh. I. "On a differential equation arising in the theory of sufficient statistics," *Litrovskii Matem. Sbornik* (1972).

[143] Sudakov, V. N. "On measures defined by Markovian moments," *Investigations on the theory of random processes* (Vol. 12 of the Memoirs of the Scientific Seminars of the Leningrad Section of the Steklov Math. Inst., published by Nauka, Moscow and Leningrad, 1969), 157–164.

[144] Tanis, E. A. "Linear forms in the order statistics from the exponential distribution," *Ann. Math. Statist.* **35** (1964), 270–276.

[145] Teicher, H. "Maximum likelihood characterization of distributions," *Ann. Math. Statist.* **32** (1961) 1214–1222.

[146] Titchmarsh, E. C. *Introduction to the Theory of the Fourier Integral*, (1937), Clarendon Press, Oxford.

[147] Titchmarsh, E. C. *Theory of Functions.* (1932), Clarendon Press, Oxford.

[148] Trybula, S. "Sequential estimation in processes with independent increments," *Rozprawy Matematyczny* **LX**, Warszawa, (1968). Dissertation.

[149] Tutuballin, V. N. "On the limiting behaviour of compositions of measures on the plane and in Lobachevski space," *Teoriia Veroiatn. Prim.* **VII** (1962), 197–204.

[150] Widder, D. V. *The Laplace Transform.* (1946), Princeton University Press, Princeton, N.J.

[151] Whitney, H. "Elementary structure of real algebraic varieties," *Ann. Math.* **66** (1957), 545–556.

[152] Ferguson, T. "Location- and scale- parameters in exponential families of distributions," *Ann. Math. Statist.* **33** (1962), 986–1001.

[153] Ferguson, T. "On characterizing distributions by properties of order statistics," *Sankhya, Ser. A* **29** (1967), 265–278.

[154] Fok, V. A. *Theory of Space, Time and Gravity*, Fizmatgiz, 1961.

[155] Fox, M. and Rubin, H. "Admissibility of quantile estimates of a single location parameter," *Ann. Math. Statist.* **35** (1964), 1019–1030.

[156] Fraenkel, Ja. I. *Statistical Physics*, GTTI, 1948.

[157] Farrell, R. "Estimators of a location parameter in the absolutely continuous case," *Ann. Math. Statist.* **35** (1964), 949–998.

[158] Halmos, P. R. and Savage, L. J. "Application of the Radon-Nikodym theorem to the theory of sufficient statistics," *Ann. Math. Statist.* **20** (1949), 225–241.

[159] Hunt, J. "Markov processes and potentials," *Illinois J. Math.* **1** (1957), 44–93 and 316–369; **2** (1958), 151–213.

[160] Heyde, C. C. "Characterizations of the normal law by the symmetry of a certain conditional distribution," *Sankhya, Ser. A* **32** (1970), 115–118.

[161] Hodges, J. L. and Lehmann, E. "Some applications of the Cramer-Rao inequality," *Proc. Second Berkeley Symp. Prob. Stat.* **2** (1952).

[162] Chanda, K. "On some moment properties when two polynomials have independent distributions," *Bull. Calcutta. Statist. Assn.* (1955), 40–48.

[163] Shalaevskii, O. V. "On stability for the theorem of D. A. Raikov," *Vestnik Leningrad. Univ.* **7** (1959), 41–49.

[164] Shimizu, R. "On the decomposition of infinitely divisible characteristic functions with a continuous Poisson spectrum," *Ann. Inst. Statist. Math.* **16** (1964), 387–407.

[165] Shimizu, R. "Characteristic functions satisfying a functional equation—I," *Ann. Inst. Statist. Math.* **20** (1968), 187–209.

[166] Shiryaev, A. N. *Sequential Statistical Analysis* (1969), Nauka, Moscow.

[167] Shohat, J. A. and Tamarkin, J. D. *The Problem of Moments.* (1943), Amer. Math. Soc. Colloquium Publ., New York.

[168] Esseen, C. G. "Fourier analysis of distribution functions," *Acta Math.* **77** (1945), 1–124.

[169] Khatri, C. G. "On characterization of gamma and multivariate normal distributions by solving some functional equations in vector variables," *Multivariate Analysis.* **1** (1971), 70–89.

[170] Khatri, C. G. and Rao, C. R. "Functional equations and Characterization of probability laws through linear functions of random variables," *J. Multivariate Analysis.* **2** (1972), 162–173.

[171] Kotlarski, I. "On characterizing the normal distribution by Student's law," *Biometrika* **53** (1963), 603–606.

[172] Kotlarski, I. "On characterizing the gamma and the normal distributions," *Pacific J. Math.* **20** (1967), 69–76.

[173] Kotlarski, I. "On a characterization of probability distributions by the joint distribution of their linear functions, *Sankhya Ser. A* **33** (1971), 73–80.

[174] Pakshirajan, R. P. and Mohun, N. R. "A characterization of the normal law," *Ann. Inst. Statist. Math.* **21** (1969), 529–532.

[175] Rao, C. R. "A decomposition theorem for vector variables with linear structure," *Ann. Math. Statist.* **40** (1969). 1845–1849.

[176] Rao, C. R. "Characterization of probability laws through linear functions," *Sankhya Ser. A* **33** (1971), 255–259.

[177] Fisher, R. A. "Theory of statistical estimation," *Proc. Camb. Phil. Soc.* **22** (1925), 700–725.

[178] Kagan, A. M. and Karpov, Yu. N. "Bayesian formulation of the estimation problem for the location parameter (in Russian)," *Zapiski Nauchnych Seminarov* Leningrad Section of the Mathematical Institute **29** (1972)

[179] Cacoullos, T. "Characterizations of normality by constant regression of linear statistics on another linear statistic," *Ann. Math. Statist.* **38** (1967), 1894–1898.

[180] Kagan, A. M., Linnik, Yu. V. and Rao, C. R. "Extension of Darmois-Skitovic theorem to functions of random variables satisfying an addition theorem," *Communications in Statistics* **1** (1973).

[181] Ferguson, T. S. "On the existence of linear regression in linear structural relations," *Univ. California Publ. in Statist.* **2** (1955), 143–165.

[182] Laha, R. G. "On a characterization of the normal distribution from properties of suitable linear statistics," *Ann. Math. Statist.* **28** (1957), 126–139.

[183] Rao, C. R. "Sufficient statistics and minimum variance estimates," *Proc. Camb. Phil. Soc.* **45** (1948), 215–218.

Author Index

Subject Index

Applied Probability and Statistics (Continued)

SARD and WEINTRAUB · A Book of Splines
SARHAN and GREENBERG · Contributions to Order Statistics
SEAL · Stochastic Theory of a Risk Business
SEARLE · Linear Models
THOMAS · An Introduction to Applied Probability and Random Processes
WHITTLE · Optimization under Constraints
WILLIAMS · Regression Analysis
WONNACOTT and WONNACOTT · Econometrics
YOUDEN · Statistical Methods for Chemists
ZELLNER · An Introduction to Bayesian Inference in Econometrics

Tracts on Probability and Statistics

BILLINGSLEY · Ergodic Theory and Information
BILLINGSLEY · Convergence of Probability Measures
CRAMÉR and LEADBETTER · Stationary and Related Stochastic Processes
HARDING and KENDALL · Stochastic Geometry
JARDINE and SIBSON · Mathematical Taxonomy
KENDALL and HARDING · Stochastic Analysis
KINGMAN · Regenerative Phenomena
RIORDAN · Combinatorial Identities
TAKACS · Combinatorial Methods in the Theory of Stochastic Processes